QUEENSLAND AIRFIELDS WW2 - 50 YEARS ON

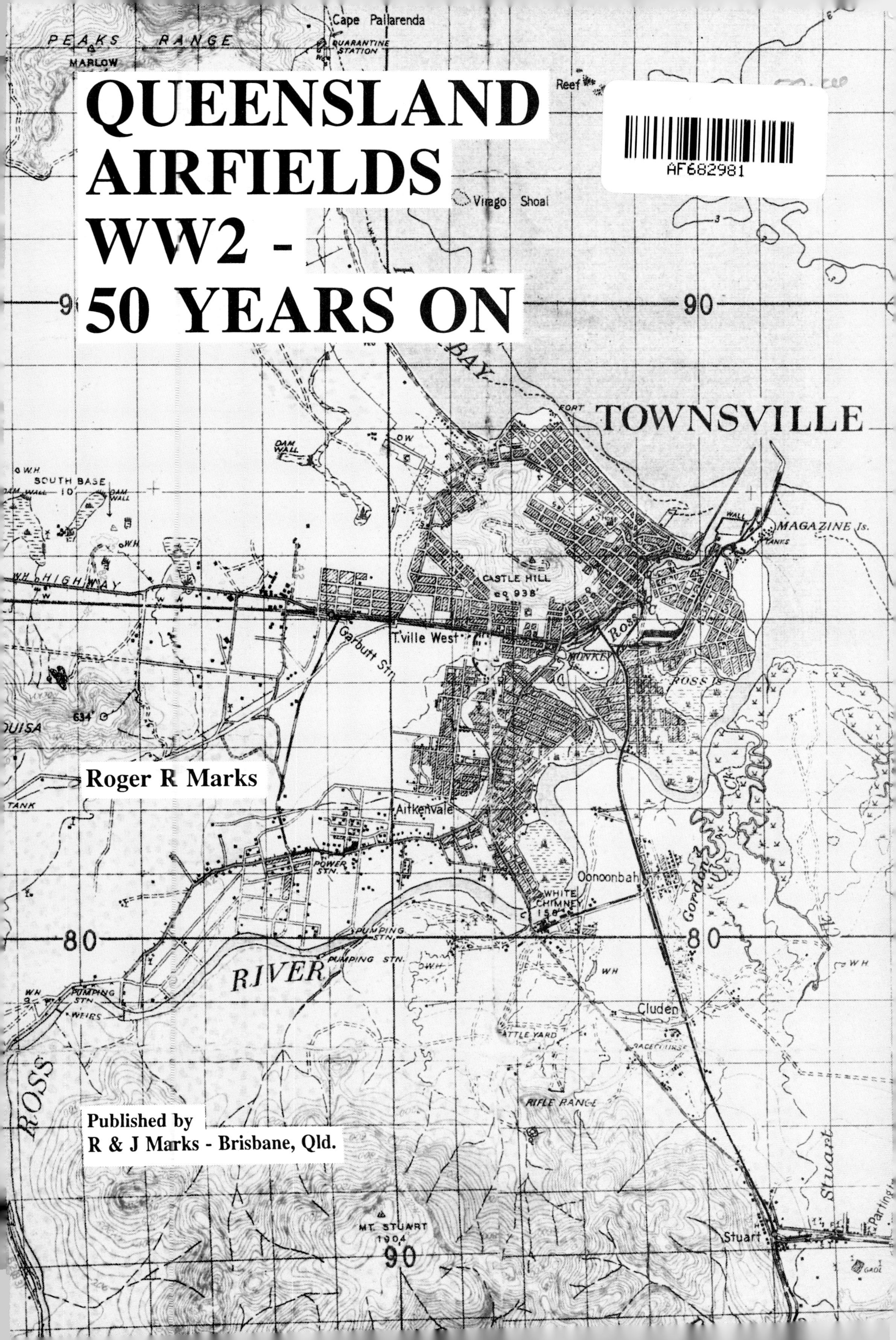

Roger R Marks

Published by
R & J Marks - Brisbane, Qld.

QUEENSLAND AIRFIELDS WW2 - 50 YEARS ON

Author/Compiler Roger Roy Marks
Illustrations
Appendices
Bibliography
Index

Published by R & J Marks
20 Koumala Street
Mansfield Q 4122 Australia
Copies available from the publishers.

First Edition 1994

Design and Layout by:	Roger & Jenny Marks
Typeset by:	Jenny Marks using Word Perfect 5.1 Times Roman 9.5/15
Screened bromide and Lineart preparation by:	Savage Graphics Cass & Co 20 Costin Street Fortitude Valley Q 4006 Australia
Printed & Bound by:	Merino Lithographics 18 Baldock Street Moorooka Q 4105 Australia

Cataloguing-in-Publication Data
National Library of Australia

MARKS, Roger, 1936 -
Queensland Airfields WW2 - 50 Years On
1. Australia. Royal Australian Air Force - History - World War, 1939-1945.
2. United States. Army Air Forces - Foreign service - Queensland.
3. Airports - Queensland - History.
4. Airports - Queensland - Design and construction.
5. World War, 1939-1945 - Aerial operations, Australian.
6. Air defences - Queensland - History.

I. Title. II. Title: ***Queensland Airfields WW2 - 50 Years On***

940.544394

ISBN 0 646 12769 1

SUNDRY ILLUSTRATIONS

NORTH is up the page - By general convention, aerial photographs, maps and drawings are presented as closely as has been found practical, to this orientation.

ON FRONT COVER - clockwise from Title

HIGGINS *(3/6/1943)*
With the US calling this airfield RED ISLAND (POINT) and the RAAF referring to it as JACKY JACKY, the confusion came to a head when some American report showed it as Mutee. This was settled with the announcement by Lt-Gen Kenney, Commander of Allied Air Forces SWPA. *[AWM 54, 13/2/2]*

BUNDABERG *(mid 1943)*
This photograph provided by Harry Edwards shows two Ansons heading for home over Bundaberg city. The two bridges over the Burnett River can be seen each side of the letter 'N'. These aircraft were used extensively by the RAAF's 8 SFTS and Reserve Squadrons carrying out anti-submarine and coastal convoy escort patrols. *[H Edwards ex RAAF]*

LEYBURN *(-/6/1945)*
'Meddlesome Mattie' was the name given by Wing Commander Eric Read to this B-24M A72-183. It was the name his wife earned as a young girl. Eric likened her childhood nature to the clandestine role this and other RAAF 200 SD (Special Duties) Flight aircraft assumed in their sorties over the Netherlands East Indies and nearby areas. Note that aircraft jacks are in place and indeed the starboard main wheel of this 'Lib' has been removed, for a tyre change maybe. Note also the 'duty pilot tower' partly obscured by the tail of the aircraft. *[E Read ex RAAF]*

CHARTERS TOWERS *(ca 1943)*
Looking approximately northeast through this fighter servicing hideout. Note the shadow pattern on the floor which clearly shows the staggered placement of purlin timbers required for both longitudinal stability of the structure and support for camouflage net etc. *[US Photographer Halbert via A Fraley ex USAAF]*

IRON RANGE *(1944)*
B-26 Martin Marauders of the US 22 BG were the first combat aircraft to operate from 'Claudie' Strip at Iron Range. This advanced type underwent significant 'in-service' testing in the SWPA under the most trying conditions. Noted for its high take-off and landing speed, it took the keenest, most adventurous pilots to bring it to maturity. *[Department of Defence]*

BACKGROUND TO FRONTISPIECE -

An extract from 1 : 63 360 Military Survey Map Townsville (special) (7/114) dated June 1942, GARBUTT and several satellite strips are located by grid reference in ***Appendix 1 - Locations****. None were shown on this map!*
[Department of Defence]

ON REAR COVER - clockwise from 3 o'clock

MAREEBA *(5/11/1942)*
A sorry looking B-17 (probably 41-9012) slumped on the ground under one of the many nailed wooden arch bomber hideouts at Mareeba. The loss of this aircraft was due to carelessness on the part of a servicing crew. This hideout was one of the few located west of the N-S runway. Most were located to the east between the runway and the railway line. *[L Moore ex USAAF]*

MAREEBA *(-/7/1942)*
Shortly after members of the 19th Bomb Group arrived at Mareeba this group is pictured by their American style pyramidal tent. The party comprises left to right - Corporal Furman Milo Hickman - Brooking South Dakota, Lawrence J Moore - Riverside California, Master Sergeant Walter B Kolbus - Paradise California, Sergeant John R Simon - Guaydon Louisiana, Staff Sergeant Edward G Vasse - Huntsville Missouri. There is the suggestion of disorder in the organisation here - where else do we find photographs of dashing Americans in Australia dressed in 'bombay bloomers'! Lawrence Moore can't remember why the gas mask kits were being worn - he feels they were more usually ***'mislaid at the bottom of a kit bag'****.* *[L Moore ex USAAF]*

EAGLE FARM *(4/4/1992)*
Inside Hangar 8 this view shows the 'Butler' hangar component of the reworked building. Providing valuable covered storage the whole hangar was, at this time, given over to the storage of bulk fishmeal. *[Author]*

EAGLE FARM *(13/8/1944)*
This Lockheed Lightning P38J-20 aircraft was used in trials with a refurbished Japanese Oscar MkII around August of 1944. Partly obscured by the Lightning's tail is a P47D Thunderbolt also used in the trials. In background is Hangar 8, a 'Butler' pre-fab steel frame structure here fully developed with metal facia cladding but still retaining the canvas doors furled like a giant tent. *[K James ex RAAF]*

EAGLE FARM *(4/3/1943)*
Viewed from the southeast near the main railway line, Hangar 3 is here seen complete with the Duty Pilot or Control Tower erected on the roof at its northern end. Notice the spur railway line feeding along the southern end of Hangar 3 and the neighbouring Hangar 4 (on the right under construction). EAGLE FARM had by this time become both an assembly point for new aircraft shipped to Australia, and a Repair and Salvage Depot for dozens of aircraft as in the foreground of the photograph. Here, row on row of P-39 Airacobra fuselage and other elements await further salvage attention. Hangar 3 survives to this day serving a present role for the storage of new motor vehicles.
[Neg A 30003 USAF Photo Collection (NASM) Smithsonian Institution]

EAGLE FARM *(4/4/1992)*
Viewed from the adjacent railway platform this is the old Hangar 3 now being used as a covered storage by new motor vehicle distributors. The disused vehicle parking area in the immediate foreground was associated with the Brisbane Domestic Airline Terminal, activities of which have since been relocated to a new site northeast of the wartime field. *[Author]*

WOODSTOCK *(26/6/1944)*
The background to rear cover and spine is an extract from RAAF Drawing No 43/44/1925 'work as executed'. The original is at scale of 1" = 600'.
[AA CRS A705/1 Item 171/94/126]

To Adam, Brandon and Cale

... a new generation.

BOY WITH TOY PLANE

With arm outstretched, the toy keeps pace, abreast
        of 'Tango Zulu' climbing clear on 'Runway-one-two'.
When late in bed he hears again that roar, he rests,
and sleep descending, the question teeters on consciousness -
        'don't planes sleep too?'

'Course they do!' Dad says, with mind extended
        chessing out the coming day... *those parts should come*
and 'that' will be done - 'make damn sure' I'll say
...and on the way you can take me near the Airport -
        Dad - are you listening??

Don't worry your Father now -
        finish your breakfast;
        you'll be late for the bus.

What say I take him, Helen? You and Jim are busy.
        I can show him the Hangars - if old George is still out there
we'll drive the perimeter, throw a few shots at the birds -
The visit'll do me good. Ya know, I remember in '45....

[RRM -/6/91]

CONTENTS

Roger Marks.

ABBREVIATIONS

"	inches
'	feet
A-A	Anti-Aircraft; Ack-Ack
AA	Australian Archives
AAC	Army Air Corps
AACS	Army Airways Communications System
AAF	Allied Air Forces
ACH	Area Combined Headquarters
AD	Air Depot
ADG	Air Depot Group
AF	Air Force
AMOE	Air Member for Organisation and Equipment
AMSE	Air Member for Supply and Equipment
ANA	Australian National Airways
ANS	Air Navigation School
AOA	Airlines of Australia
AOB	Advanced Operational Base
AOC	Air Officer Commanding
AOR	Air Operations Room
APU	Aircraft Performance Unit
ARD	Aircraft Repair Depot
ASC	Air Service Command
ASR	Air Sea Rescue
AUF	Australian Ultralight Federation
AUSLIG	Australian Surveying and Land Information Group
AWC	Allied Works Council
AWM	Australian War Memorial
BG	US Bomb Group
BS	US Bomb Squadron
CA	Coastal Artillery
CA/CAA	Civil Aviation / Civil Aviation Authority
CAC	Commonwealth Aircraft Corporation
CAD/DCA	Civil Aviation Department/Dept Civil Aviation
CAS	Chief of Air Staff
CCC	Civil Construction Corps
CE	Corps of Engineers
CIC	possibly... 'Combined Intelligence Centre'
CMF	Citizen Military Forces
CPE	Central Photographic Establishment
CRD	Central Recovery Depot
CSIRO	(Australian) Commonwealth Scientific & Industrial Research Organisation
DAP	Dept of Aircraft Production
DASP	Director of Air Staff Planning
DAT	Directorate of Air Transport
DHS	Department of Home Security
DO	Director of Organisation
DSO	District Security Officer
DWB	Directorate of Works and Buildings
DWO	Divisional Works Office/Officer
EATS	Empire Air Training Scheme
EFTS	Elementary Flying Training School
EGS	Engineer General Service
ELG	Emergency Landing Ground
FB	Flying Boat
FC	Fleet Cooperation
FEAF	Far East Air Force
FEASC	Far East Air Service Command
FG	US Fighter Group
FS	US Fighter Squadron
GHQ	General Head Quarters
GR/FB	General Reconnaissance/Flying Boat
HB	Heavy Bomber
HF	High Frequency
HQ	Headquarters
IFR	Instrument Flight Rules
ITS	Initial Training School
MONAB	Mobile Naval Air Base
MP	Military Police
MPMU	Mobile Parachute Maintenance Unit
MRB	Main Roads Board
MRC	Main Roads Commission (Queensland)
MRS	Medical Receiving Station
MTRS	Motor Transport Repair Section
NAOU	North Australian Observer Unit
NAP	possibly... 'Native Affairs Police'
NEA	North East Area
NEI	Netherlands East Indies
NLA	National Library of Australia
OB/OBU	Operational Base/Unit
OD	Officer of the Day
ORB	Operations Record Book
PEI	Public Estate Improvement (Trust)
PMG	Post Master General - Postal and Telecommunication Department
PSP	Pierced Steel Plank
QEA	Qantas Empire Airways
RAAFA	Royal Australian Air Force Association
RD	Recruit Depot
RDF	Radio Direction Finding
RFDS	Royal Flying Doctor Service
RIMU	Radio Installation and Maintenance Unit
RLG	Relief Landing Ground
RNAS	Royal Naval Air Service
RSU	Repair and Salvage Unit
SDCC	State Defence Camouflage Committee
SFTS	Service Flying Training School
SWPA	South-West Pacific Area
TC	Troop Carrier
TD	Technical Director
TDC	Technical Director of Camouflage
TG	Training Group
U/G	Underground
USAAC	United States Army Air Corps (pre June 1941)
USAAF	United States Army Air Force
USAAS	United States Army Air Services (later ASC)
USAF	United States Air Force (post WW2)
USAFFE	United States Army Forces Far East
USAFIA	United States Army Forces in Australia
USASOS	United States Army Services of Supply
USATC	United States Army Transport Command
USM	possibly... 'United States Military'
VDC	Volunteer Defence Corps
W/T W/R	Wireless transmitting and receiving
WAAAF	Womens' Auxiliary Australian Air Force
WAG	Wireless Air Gunner
WAGS	Wireless Air Gunners School
WMU	Works Maintenance Unit

FOREWORD

How many times have we said 'I wish I could find some record of "such and such" ' - an incident that happened 10, 20, or 50 years ago? At the time of its happening, it is usually considered to be part of everyday events - no one takes much notice of it. Today, more and more responsible people of vision are seeking to put on paper, or tape, history of the past 50 years which will, in 100 years time, be very useful to historians studying our era.

This is a remarkable publication on Queensland's World War II involvement in the Air War against Japanese Air and Naval Forces. In 1942 the invasion threat was real to the service men and women at their posts and to the civilians at work and in their homes. With shipping being sunk within sight of our shores and enemy air raids to our north, we had not today's luxury of hindsight. We are now assured the Japanese plan was never one of total invasion but no such comfort was abroad in 1942.

The author invites us to share his search for information on these airfields - often to read the documents verbatim. He doesn't claim to have found it all. Many of these photographs are published here for the first time though not for their technical excellence. It is in testimony to authenticity that they revel.

Pilots and aircrew are not always those best able to tell of these wartime airstrips, almost certainly not their siting and construction. Much has come from the ground crews and others of the 'great unsung'.

Fewer than half the airfields studied and listed in this book are still in existence but many helped to open up the State to air travel during the past 40 years.

The amounts of time, study and attention to detail put into this publication have been enormous and are a great credit to the author-publishers. I am proud to have been involved, in a very small way, with this huge task which has resulted in such a fine and very useful history of aviation development in this State of Queensland.

Air Commodore (Ret'd) Deryck Kingwell CBE, DSO
Kenmore, Queensland
1994

INTRODUCTION

As some promotional verbiage of our time says - congratulations - if you've read this far, there is hope for you, (and for me?) yet!

Ask yourself, 'What is the value of having written something - a letter, some verse, a report - if no one ever reads it?'

Only philosophers might read on if I examined that statement over several pages. Suffice it to say that in most letter writing I've found if it extends to two pages, you risk losing the reader's interest - certainly the physical loss of one or both pages!

So all sorts of reasons pop up for keeping things brief.

Over recent years since retirement from a Professional Civil Engineering career, friends and others have asked either 'What are you doing?' or 'How's the book going?' It's not always been convenient to answer thoroughly. Some will have heard simply, flippantly, 'another day another chapter!' Others will know that impaired eyesight focussed my twin life-long interests in aviation and hands-on woodwork. I believe I'd love to rebuild a real wooden aeroplane - say, an English 1930's Percival ***Gull*** to flying condition. Recession or no recession, there still seems to be a collector cum enthusiasts' market for such treasures.

Alas, such projects are slow to come by and there are many hurdles to the newcomer.

Ninety plus percent of my working life has been spent in Queensland. Born in Mackay, formally educated in Bundaberg, Gayndah, Maryborough and Brisbane, I then did twenty-seven years 'work experience' over much of my home state.

As a boy in Bundaberg during the early 1940s, our Cub Pack went on bike-hikes out to 'RAAF BUNDABERG', and down past the camouflaged aircraft hideouts to char our saveloys by Branyan Creek. When the famous 'G-George' Lancaster of the RAAF's 460 Squadron visited there on 11 April 1945, my father doubled me on the bar of his bike out to the airfield. It was a stinking hot day, and while Dad was pedaling the bike I remember a fly buzzing around our heads. Dad maintained it was there and then that I asked him 'How fast can a fly fly?'

Despite a continued solid grounding in aviation with scrap wood, bakelite and balsa models in the very cradle town of Queensland's own Bert Hinkler, I recall one of life's early set backs. My registration for late 1950's National Service and my number one stated preference for 'Air Force' saw me at the end of first year University, 'in the Army'!

Engineering, yes, but building roads and bridges; not planes.

Years later, while living in Cairns, I came to realise I was near WW2 period airfields like MAREEBA and I'd heard of BREDDAN, but where was it? What were those airfields like THEN, and what were they like NOW? When I found one, I was impressed with the state of the bitumen seal on pavement remnants, but what was the function of the building which once stood on that concrete slab?

That interest stayed with me, and until I find a Percival *Gull* or similar to channel my energy, I decided there could be others sharing my parallel interest, in the WW2 Airfields of Queensland - 50 years on.

By now you will have flicked through these pages and looked at the 'Contents'. You'll be aware of the copious tabulations near the back. When you want to compare one airfield with another, find and explore them in the tabulations. Some airfields are only mentioned in these tabulations, because of their relative significance, and the economic restraints on this publication.

It is hoped that the reader, either living in or visiting the locality of one or more of these airfields, will be prompted to prepare for such visit(s) and thereby 'relive' some of the atmosphere of 50 years ago.

This statement is not claimed to be a concise history of the various airfields. That is a task well beyond the resources and research patience mustered for the job. This represents the result of several thousand hours of archive and library study - hundreds more in discussion with veterans and others. There comes a time when one has to say - enough is enough - put it on record as rigorously and correctly as you know it to be - note your authority or source and invite the reader to take up loose ends and progress further. It's not fair to oneself nor all one's helpers, to unduly prolong gestation.

Apologies? --- there are none!

--- Not to the many sources upon whose published and unpublished work I've drawn to set the scene. Full credit, yes - but no apology.

--- Not to the adviser who advocated an airfield by airfield approach, writing up the full 'ball by ball' description of events at airfield A, then proceeding to airfield B. The time scale for that is non-commensurate with veteran mortality!

--- Not to the many people to whom I've written, seeking to shake memory cells calcified by age. Thanks yes, for the little gem answers dislodged, but no apology.

If there is an apology due, it's to all who have had to contend with my handwriting.

To the next crop of enquiring minds - may you seek it out and tell it as it was, for that is the nature of history.

My best wishes.

Roger

ROGER MARKS

THE HOW AND THE WHY

'...How did that Aussie get started on the Airfields project? The main thing I remember about the Main Roads outfit when they came to work at Townsville was they asked "when did we boil the billy?".'

This paragraph in a US veteran-to-veteran letter which filtered through to me, poses a legitimate question. It also shows how such an Australian characteristic has stayed firm in the mind of Dr Walter B King Jr, one of the US 46th Engineer Regiment here early in 1942.

The 1978 UK publication ***Airfields of the Eighth - Then and Now*** © *After the Battle Magazine*, with text by Roger A Freeman, has a demoralising one page section on 'How (that) book was produced'. I found it demoralising in that the Editor describes that project's inception, his Magazine's successful approach to Freeman an expert in this field, and the mobilisation of several researchers in the UK and abroad to assemble historic photography both archival and privately held. He tells of the planning of a 600 mile aerial tour route to embrace the chosen seventy odd UK airfields, and the book's production phase. I have read that page a dozen times in the last four years. With the end of this project now in sight, I can see the parallels and the vast differences.

One should never equate the life and death daily grind on those UK fields with the 'home and away from the front' tenor of WW2 Queensland Airfields. Still, a review of the latter is valid. Future generations of Australians should know that HORN ISLAND airfield was bombed and strafed and that Townsville saw a few Japanese planes overhead at night. Sure that's nothing to blanket bombing and the occupation of one's homeland. Yet men lived and worked at, flew, fought and died from, airfields in Queensland during WW2.

In the preparation of **The Airfields** part of this book a number of matters have surfaced - points where I could almost 'hear the reader's cry for guidance'. While it is probably presumptuous of me to think many readers would be curious to know how and why this book was produced, there are other explanations needed. Some background appears in the Introduction. Each Appendix warrants its own lead-in and some opportunities arise there for specifics.

My early airfield visits made without the benefit of information on precise location and layout were generally frustrating. Even impromptu conversation with local residents was often disappointing, with responses ranging from 'I didn't know there was an airstrip here' through to, 'you should'a come last year, before old Billy Smith died - he'd a told you all you wanted to know!'

I resolved to do the necessary 'homework'. First I would obtain a complete list and location details of airfields in Queensland, occupied or used by RAAF and/or Allied Air Forces, during WW2. I bought and read the two relevant AWM published official histories of the RAAF. A thorough check of their Indexes and maps gave me the prominent sites but I wanted all - even the hastily prepared strips which in some cases were little, if ever, used by the Air Forces as the war moved north. After discussion with Staff of the then RAAF Support Unit in Brisbane, it was suggested I write to Canberra, on the basis that 'they would likely have a computer print-out of such information'.

The RAAF Historical Officer replied, in part;

'...No consolidated list exists of the many hundreds of prime, secondary, satellite, dispersal and emergency Allied dromes in your state during the war. Some are still utilised, others have been upgraded, many were abandoned and some have reverted back to agricultural, Crown or Aboriginal title.

Despite persistent rumours and many requests over the years, Air Force Office is quite sure there are no abandoned historic aircraft at or near these sites.

Due to fauna, drug and possible illegal immigrant smuggling in North Queensland you will appreciate our caution in freely releasing comprehensive details on many of the more remote fields. Coupled with this has been the trespass and damage in recent years by people seeking old military sites in the belief that material/aircraft were left behind after the war.'

So much for the easy way, through the RAAF itself!

On numerous occasions, pilots flying throughout Queensland and sharing my interest, have directed my attention to some strip saying *'it was (probably?) a wartime strip'*. Most times, a search of some aerial photo library showed this in error. While many wartime strips were constructed to varying degrees, many more have post war origins, being bull-dozed by miners or pastoralists. In remote North Queensland, one year is usually enough to see such non-maintained strips revert to natural vegetation and nigh on disaster for landing a fixed wing aircraft.

I approached veterans' organisations such as the RAAFA's Sydney office, seeking information on Squadron histories, published or otherwise, assuming these would give valuable insight to Queensland Airfields WW2. I was disappointed with the apparent absence of such histories. Surely all WW2 period Squadrons would have at least an informal written history compiled by one or more of its members astute and caring enough towards posterity. I guess it really depends on effective personal contact being made in the furtherance of such enquiries. It seems greater effort than that which I made, is needed.

For the reader using this book as a reference, I suggest you first consult the **Index**. A possible airstrip site known only to the reader by name should be traceable through the **Index**. Whereas some 190 odd sites are addressed in **Appendix 1 - Locations**, only about sixty are treated in terms of illustrations and specific text.

I hope you'll enjoy the photographs and their lead-in to what I can tell you of each airfield. Photographs are wonderful - they become the windows of memory, both for the veteran who was there and to the new reader. Aerial photos are real magic. Searching old aerials has often proved to be the only way of discovering where a particular airstrip, long since overgrown, actually was!

Most photographs have been incorporated here after re-photographing surviving prints. Few are from original negatives. The drawings were never intended for reduction and inclusion herein as effective educational material. If the reader aims for further research, then personal inspection of originals is recommended. Meanwhile, keep a magnifying glass handy! Where possible, original edge markings of photos have been included. That's great for demonstrating authenticity but makes for smaller subject size - another reason for the magnifying glass!

My review of relevant Military Survey Maps surprised me with the sheer paucity of the coverage when the threat to our continent was greatest. This was undertaken partly at the National Library of Australia (NLA), Canberra, the State Library of New South Wales, Sydney, the Queensland Government Archives, Brisbane, and through kind access to some private collections. The range of curatorial approach to preservation and provision of photocopy extracts also surprised me. While all these sources were sympathetic and accommodating in degrees, my bouquets go to Maura O'Connor's team at NLA and Hugh Pechey, an inveterate private collector.

As an Engineer I was well aware of the power of stereoscopic vertical aerial photography, the heart of modern mapping. My 1985 approach to the Brisbane (SUNMAP) photo library quickly showed me the 50 year old photography was still either with the RAAF or the Commonwealth's then Division of National Mapping - now AUSLIG. I had only little joy with the RAAF and expected little more from National Mapping. My letter to NATMAP, 3 June 1987 came to the attention of Mr Warren Webb. His welcome phone volley meant I'd found an interested soul. I was sorry when later Warren Webb retired. AUSLIG have creditably produced their Flight Diagram (FD) Microfiche. Hopefully its periodic updating will mean extension to include FDs held by AUSLIG of photography for which negatives were, at August 1990, still retained by the RAAF's CPE. The various negative registers referred to in **Appendix 4**, were originally shown to me by Warren Webb. I hope recognition of their historic nature has survived Warren's departure.

Individuals and corporations all have a 'too hard basket'. An early visit to NLA showed me that their extensive aerial photo collection had a component in the 'THB'. On my return home I resolved to write to NLA. I offered my background knowledge of Queensland and my intense interest on some sort of expenses shared basis, to go back and review their 'unsurveyed holdings'. In it for me was a chance of picking up photos of Queensland Airfields WW2 which NLA didn't know they had. NLA liked the proposal, enquired further, made a sensible offer of payment, and accepted my proposal in writing. Before the visit took place NLA wrote cancelling the agreement for any payment but indicating I would be welcome to proceed privately.

Yes I was disappointed - yes, I proceeded and submitted to NLA, December 1990, a comprehensive Report on my two weeks visit. Yes, NLA thanked me and yes I did locate relevant photos otherwise lost in, the 'THB'. In February 1992 I lodged an order for 30 reproduction quality prints via NLA's Reprography Service. The only problem photo was my chosen frame from Run 1 of the 23 July 1943 historic coverage of GARBUTT. I knew NLA had it but they didn't, because it was still in their 'THB'!

GARBUTT *(23/7/1943)*
This is one of some nine frames covering GARBUTT as part of KD 825. Several aircraft can be identified on the airfield, namely a mixture of twin-engine B-25 Mitchell medium bombers, four-engine B-24 Liberator bombers, sundry fighters and a few C-47 Dakota transports. Notice that all evidence of the camouflage effort to disguise part of the airfield as the extension of a suburban development has disappeared. Some aircraft revetments in the characteristically four-sided 'U' shape can still be seen, particularly near the eastern end of the 130° runway. Notice also that the road entry to the Civil Terminal is still the N-S road passing close to this end of that airstrip and in fact crossing the taxiways to the aforementioned revetments.
[NLA (RAAF OFFICIAL)]

In March 1988 I commenced a review of Australian Archives' (AA) sixteen volume holding of the Allied Works Council Minutes, spanning 1942 to 1945. I set out to extract all mentions of airfield works, later dissecting these to build a chronological statement for each field. I completed the extraction but the dissection I've left unfinished at mid 1943. I can certainly vouch for the attentive and helpful assistance of Mr Alex Moutsatsos and the AA Brisbane Reading Room staff while I was a fixture there.

The extension of computerised information retrieval systems to the AA scene nationally is to be applauded. The facility to quickly review the availability and location of material on a chosen subject is great. However, unlike the Interlibrary Loan Scheme, material held in a regional repository is available for study only at that Regional Office.

One should be warned! As powerful as the ANGAM II (Australian National Guide to Archival Material Part II) on-line database is, there are limitations. An item is only included once someone has applied for its clearance, causing a decision on access to be made, one way or the other. There is the basic mass commanded by ANGAM I, where the records and proceedings of various Departments are traceable through each Department's evolutionary stages. It is a formidable maze to the newcomer. Significant help came to me via Mr David Herder, then in the ACT Regional Office. A chance remark by an AWM staff member as to the relevance of camouflage to airfields, led through David Herder to eventual clearance of a Department of Home Security photographic Index and later, substantial minute records of the State Defence Camouflage Committee Queensland, held in AA Sydney. This Photo index gave dates and place names and some idea of the camouflage work being photographed, but no photos. It named the photographers but the names meant nothing. Next day, waiting for the NLA to open its doors, I noticed an exhibition of photography by a Max Dupain. The name rang a bell. In the camouflage register it showed only 'Dupain'.

A couple of phone calls and I was speaking to Dupain! Max kindly put me in touch with two other survivors, Robert Emerson Curtis and Roderick W Voller. Rod has helped through his retention of personal diaries and an energetic interest in my project.

My search for contact with veterans is on-going. I have aimed at an initial written approach giving the recipient every opportunity to quietly consider my requests. The response rate has been around ninety percent, and nearly all have acceded to my wish to 'tape the discussion'. My wife Jenny has toiled to transcribe a chosen fraction of these and is now adept at typing even the (unwanted!) 'trucks passing the front verandah'! In this age when the average person prefers to watch and maybe listen, rather than read, let alone WRITE, audio taping such an interview usually provides a result rather than none, and it's reproducible into the bargain! But it's another call on time, and an addition to the correspondence load if only to the extent of good manners in recording thanks. It's another 'source' to assimilate, to cross refer, and to file. And if one is careful and smart it will be retrievable without a total clean-up of the workplace!

Initially I baulked at the prospect of gathering information from overseas veterans. To America the airmail cycle time and the postage were about three times as long and as much, but then the interest awakened astounded me. I can only feel that to those US veterans, 50 years has indelibly edited what was to most their first adventure overseas. It has been Jenny's and my pleasure to actually meet and share hospitality with several during their recent tourist pilgrimages. About equal in land mass, similar in heritage and ethnicity, both rich in natural resources, our populations differ tenfold. The Pacific War

MAREEBA *(1/8/1991)*
Work on the remanufacture of nailed wooden arch trusses for Syd Beck's new museum. The original trusses, relocated from Townsville were found to have deteriorated irreparably through open air storage. These brand new trusses were manufactured to genuine original design requirements but with the aid of modern tooling. *[Author]*

fused lasting bonds between Australia and the USA but it surprises me nought that each Nation will look after its own people first.

Study trips to Canberra and Sydney, and field trips to a high proportion of these 190 odd 'sacred sites' all had their moments. I'll remember with a wooden posterior the bus travel, the general austerity and the tight schedules. When Jenny could accompany me it was our pleasure to also take my mother. She saw for her first time Cunnamulla and the southwest on her way back to Longreach where she grew up. The Queensland wild flowers bloomed for us. My sons did the cooking for a boys-only trip to Cape York. They wouldn't let me drive. They agreed Fruitbat Falls was heaven to bathe under and sunset at Weipa was better than the postcard, but I thought, wasn't it a shame they didn't like the oysters at Portland Roads!

I'd have liked to review more fully the design background to the major buildings on Queensland Airfields WW2. There is a fascinating story in the evolution of the 'Igloo' with its 3-pin arches of nailed timber construction. This building grew from the brainwave of a French civil engineer who gravitated to Australia with the fall of Singapore. People from the 1942 days in M R Hornibrook's Brisbane office recall Emil Brizay's time there designing a crude structure to hide aeroplanes. Like the American engineers urging him they were to see the idea mature as in the huge US 4th Air Depot's workshops and stores at Townsville. I wonder if there is a Brizay family member living somewhere in Australia now.

I am curious about the four 95′ x 45′ steel workshops resurrected from 'distressed cargoes' in Melbourne, and erected at BREDDAN early in 1943. What did they look like - where did they get to? Are there any photos? Only recently I learned from an interested Townsville resident that Hangar 15 erected at GARBUTT around 1965, was the one originally built near the corner of Duckworth Street and Ingham Road as one of two substantial buildings in the US Project No 1. To trace the history and indeed the present whereabouts of such buildings is a facet I aimed to do. It will probably fall now to someone following me.

I have tended to avoid the field of newspaper archives on the basis that most of the historic development material would have been subject to censorship. I am aware that newspapers may have legitimately covered some of the political debate relating to airfield location and development prior to the Japanese entry to the war.

In an attempt to preserve the authenticity of information, either from a document or veterans' recollections, direct quotation has frequently been employed. Some condensing has been done in the quest to preserve the language and import without descending into inconsequential detail. Much more material remains untapped compared to that which I have tapped, though I believe a reference background has been laid out for each airfield of significance. Surviving veterans I've not reached may well read this and experience some graphic recall. Such a reader is encouraged to commit to record such a contribution, so the fuller history of an individual airfield will eventually be written.

While ***The Airfields of the Eighth...*** team was able to farm sections of their project according to expertise, workload and choice, Jenny and I are reminded of our time in Cairns. There we 'owner-built' our home. When the ceramic tile contractor was 'dragging the chain' I remember him promising that 'next week he would put the whole gang on it!'. That turned out to be he and his offsider!!

Since my first visit to MAREEBA airfield in 1978 I've been back several times - each with the benefit of further study. To indulge one's interest in such study takes money as well as time. It also takes an understanding partner. As the outlay/investment(?) grew, so too the determination to publish the results. My wife Jenny has not only agreed to successive outlays and watched things like house maintenance suffer - she has shouldered the task of typesetting the book!

For years an accomplished typist, Jen's application to this task, in no small way proves Australia can reach 'clever country' status. In a way I was pleased the Australian War Memorial turned down my 1990 project application for a financial 'Grant in Aid'. The process forced some structured planning of the project and estimates of final costs. As usual, out of rejection came a pig-headedness to complete. That has extended to our deciding to 'publish in our own right'. Time will tell if that was smart, but what glorious freedom there has been in concept, design, layout and editing!

One tends to assign brickbats or bouquets according to one's perception of result achieved. Most readers' advisory services at Libraries, score my bouquets but my personal thanks go to Mr Chris Dawkins of the Australian Defence Force Academy Library, Canberra. That library is on two floors and, in Canberra where I've had Certified Mail postings simply vanish, I was staggered to find a Reader Adviser who literally ran up and down the stairs to help me.

GARBUTT *(25/6/1943)*
'Private Otto H Maaser of Tampa Kansas works on a newly installed Allison V-1710 in a Bell P-39 while Pvt John M Pencille of Los Angeles California screws down the wing gas tank plate at the 4th Air Depot Engine Overhaul and Repair Shop Garbutt Field Townsville Australia.' *The hangar in the background is the northern one of the two erected under the US Project No 1. The name board high near the ridge reads - **'Engineering 2nd Service Squadron'** with the original US Army Air Corps red centre on white 5 point star on blue background insignia in place. By contrast the then amended USAAF insignia appears on the fuselage of the Airacobra. The red centre was discarded for obvious reasons in mid-1942. The name on the right side nose of the aircraft looks like 'Clara May'. It is believed this Hangar was relocated to just north of the Civil Airlines Terminal at GARBUTT in 1965 to become Hangar No 15. It remains in service today.*
[Neg A 29952 USAAF Photo Collection NASM, Smithsonian Institution]

ACKNOWLEDGMENTS

My sincere thanks go to the many veterans and others, whether named here or not, who have responded to my written and other approaches. Personally, I've felt better able to gauge the research relevance of a book, if the names of contributors and assistants are given. I have tried to list all who they themselves would feel deserved such mention. Inevitably, I'll have either misjudged or simply forgotten some. For that I do apologise.

I do know some have since passed on. I've decided against trying to indicate that. It would be just my luck to err!

Particular acknowledgments and thanks are due to the following organisations and individuals.

- **Department of Lands**, Queensland - for permission to reproduce latter day aerial photography. 'Today' and some 'earlier' photos of most of Queensland are available by enquiry to - Department of Lands, cnr Main & Vulture Streets, Woolloongabba. Q. 4102 phone 07) 896 3209.
- **AUSLIG** - latter day maps and aerial photography credited to AUSLIG are Crown Copyright and have been reproduced with the permission of the General Manager, Australian Surveying and Land Information Group, Department of Administrative Services, Canberra, ACT.
  Various WW2 period aerial photographs reproduced via NLA's Reprography Services are also embraced by AUSLIG's approval. Sympathetic commercial treatment of Crown copyright fees represents some recognition of likely added sales prompted by this book.
- **Australian War Memorial**, Canberra - for permission to reproduce various photographs and documents, as credited. Reproduction fees have not been waived.
- **Australian Government** through the **Department of Defence**, for permission to reproduce and publish photographs, drawings and extracts from letters, reports, minutes etc. from Australian Archives holdings and extracts from Royal Australian Survey Corp (Military Survey) mapping of WW2 period.
  Also, through the Department of Defence (Air Force Office) for other photographs credited 'RAAF Official'.
- **National Geographic Magazine** - for the use of photos as credited.
- Both the **Smithsonian Institution/NASM** and the **US National Archives** - for permission to use photographs as credited.
- **Universal Press P/L** - for permission to use extracts from 'UBD' mapping.
- **RAAF Museum Point Cook**, Victoria and the **North Queensland Air Force Museum**, Townsville - for the use of photographs and drawings as credited.
- **Northair Surveys** of Mareeba, Q - for various recent aerial photographs as credited.
- **The Arch Fraley Collection** Magnetic Island Qld - reproduction fees apply.

RAAF OR OTHER AUSTRALIAN FORCES (WW2):

Anderson A	Aspley Qld
Atkinson J	Holloways Beach Qld
Bailey A	Culburra Beach NSW
Batts G	North Rockhampton Qld
Bauer G	Fingal NSW
Bell M C	Labrador Qld
Bennett F C	Warrawee NSW
Bentley D B	Wishart Qld
Bigelow J	Coorparoo Qld
Bolton H	Clayfield Qld
Bradley V	Earlville Qld
Brockway G	Bowen Qld
Broome C	Bundaberg Qld
Burley J	Warrimoo NSW
Bussell H	Gympie Qld
Callard O	Frankston Vic
Carr M	Glenbrook NSW
Carroll D	Yarra Glen Vic
Cloake J	Moorooka Qld
Cover W	Toowoomba Qld
Curwen D	North Ipswich Qld
Cutler C	Silkstone Qld
Davies D B	Rockingham WA
Dynes P	Gloucester NSW
Edwards H	Jamboree Heights Qld
Gamble K	Killara NSW
Gillmore G	Culburra Beach NSW
Gow R J	North Gosford NSW
Grantham S	Hendra Qld
Halligan N	Coorparoo Qld
Haughton-James Mrs J	St Lucia Qld
Hecker S	Maryborough Qld
Howsan A	Macgregor Qld
Hutchinson K	Cairns Qld
James K	Claremont WA
Kielly W O J	Brighton Qld
Kingwell D W	Kenmore Qld
Kyle H	Bardon Qld
Lavery M	Tarragindi Qld
Lawrence F	Maryborough Qld
Legge J	Scullin ACT
MacDonald R S	Rockhampton Qld
McBride D	Kiama NSW
McConnel D	Toogoolawah Qld
McDermott R	Bowen Qld
McDonell R	Nerang Qld
McGregor G	Blackburn Vic
McGuffie E	Cooktown Qld
Mason C A	Bunbury WA
Merrotsy R	Bogangar NSW
Millar N F	Coolum Qld
Mills J	Aspley Qld
Munro C	Indooroopilly Qld
Norling H G	Taree NSW
Oliver D	Westleigh NSW
Olorenshaw I	Mt Tamborine N Qld
Parker F	Lowood Qld
Parker H	Coorparoo Qld
Parker W	Redland Bay Qld
Parkinson G	Coorparoo Qld
Payton A	Mt Ommaney Qld
Pope R K	Geebung Qld
Pope R	Hervey Bay Qld
Pratt D	Mackay Qld
Read E V	Sorrento Qld
Redding G	Ballina NSW
Short J	Wagga Wagga NSW
Silver J	Wavell Heights Qld
Steenson J	Mt Kuring-gai NSW
Trench H V	Bilgola Plateau NSW
Waddell V	Geebung South Qld
Whyte T J	East Ipswich Qld
Williamson J	Mt Pleasant Qld
Woodward J	Edge Hill Qld
Yates P K	Sefton NSW
Ziesing R	Magill SA

RNAS (WW2):

Groom P	Maryborough Qld

USAAF OR OTHER US FORCES (WW2):

Allen W B	Crosbyton TX USA
Bentson W	Toowong Qld
Besancon H C	Charleston SC USA
Bruce B B	Ft Belvoir VA USA
Coble G S	Stanley NC USA
Crockett J A	Riverside CA USA
Dietz R	Clarion PA USA
Dudley J	Jacksonville FL USA
Eaton W R	Denton TX USA
Englund D	Racine WI USA
Farrior W	Honolulu HI USA
Fetter T C	Marion OH USA
Fields W	Shamrock TX USA
Fraley A	Magnetic Island Qld
Gaylor W	Mesa AZ USA
Gilpatrick A	Cupertino CA USA
Graf G R	Korora NSW
Haggerty W	San Diego CA USA
Hesse C	Seattle WA USA
King W B Dr	Waco TX USA
Krell W A	Etna CA USA
Kruger D O	Morris IL USA
Lillback J Professor	Marsden Qld
McAteer J	Lake San Marcos CA USA
McFee J	Arlington VA USA
McIntyre J	Utica MS USA
Mahan E	Seven Hills Qld
Maiersperger W P	Carmel CA US
Martin F F	Louisville KY USA
Mills W H	Walla WA USA
Mitchell J	Springfield MO USA
Moore L	Riverside CA USA
Moran W	Macquarie Fields NSW
Peterson B E	Graceville East Qld
Raski H	Royal Oak MI USA
Rhodes E R	Houston TX USA
Somerville A	Orange VA USA
Walker H	Point Piper NSW
Wanielista W	Westchester IL USA
Whistler C F	Dallas TX USA
Woods W O	Chattanooga TN USA

NEI Air Force (WW2):

Beckers H	The Netherlands

NON VETERAN CONTRIBUTORS:

Alford R N	Ludmilla NT
Arnold R P	Mareeba Qld
Baker B	Melbourne Vic
Bassani Mrs D	Mooroobool Qld
Beck S	Mareeba Qld
Birdsall S	Mona Vale NSW
Bjelke-Petersen Sir Joh	Kingaroy Qld
Blackman L	Manunda Qld
Bond Mrs M	Clontarf Qld
Bovelt A	Bondi NSW
Browne Miss W	New Farm Qld
Burry J	Giru Qld
Cardell Dr R	Nambour Qld
Chan Mrs A S	Skokie IL USA
Cooper N	Laverton Vic
Cousins G	Rockhampton Qld
Dawkins C	Campbell ACT
Davidson G	Bowen Qld
Field R	Woodstock Qld
Freier Rev P	Banyo Qld
Frew A	Brisbane Qld
Gill R	Charleville Qld
Gleeson D T J	Kirwan Qld
Gross Mrs S V	Nanango Qld
Head B	Palm Beach Qld
Henderson W	Charters Towers Qld
Holzapfel W	Cooktown Qld
Hoy B	Albany Creek Qld
Jackson T V	Salisbury Qld
Jenkins I D	Weston ACT
Jennings R	Bardon Qld
Johnson D H	Charters Towers Qld
Johnston W	Mareeba Qld
Jones E	Maryborough Qld
Kennard Mrs G	Mittagong NSW
Kenny P	Cranbrook Qld
LeGood D	Blacksoil Qld
Leon T	Giru Qld
Lindorff L	Sunbury Vic
Livingstone R W	Gymea NSW
Lubke W	Lowood Qld
Lynch C	West St Kilda Vic
McIntosh W L	Prairie Qld
Mikesh R C	Temple Hills MD USA
Morris P	Longreach Qld
Mullins I	Condon Qld
Newell Mrs J L	McLean VA USA
O'Connor Ms M	Canberra ACT
Olson J Ms	Mesa AZ USA
Pechey H	Jindalee Qld
Petersen E	Hermit Park Qld
Petersen J	Wooloowin Qld
Ritchie W	Barcaldine Qld
Roberts G	Dover Heights NSW
Rogers H	Pt Cook Vic
Rudd R	Mareeba Qld
Ryan S	Maryborough Qld
Selwood J	Laidley Qld
Shephard K W	Heatley Qld
Sims S J	Jimboomba Qld
Smith F	Burwood Vic
Smith R	Woodstock Qld
Stephenson J	Townsville Qld
Taylor Mrs I	Coen Qld
Todd G	Goondiwindi Qld
Turner L	Wulguru Qld
Vincent D	Highbury SA
Voller R W	Currimundi Qld
Voysey G	Wavell Heights Qld
Warby Rev Canon J	Rockhampton Qld
Watling N	Edge Hill Qld
Weimers Mrs L	Atherton Qld
Wilson C H	Grange Qld
Wilson J	Bundaberg Qld
Wilson R T	Cranbrook Qld

Platitudes too often spring to mind when one attempts to commit written thanks. Apart from some fevered efforts to improve the Index, these few words are 'the last'.

My 'proof readers' have been - Eric Counahan, Bob Livingstone, Keith Pope and Peter Wornham. All have a keen sense of history. As President of the small Canberra based Australian Society of Aero-Historical Preservation, Eric 'worked on' the 1988 restoration of 'G-EAQM', the 1918 built DH9 bi-plane in which Parer and McIntosh struggled from England to Australia in 1920. Bob manages Air Traffic Control at Sydney's Bankstown aerodrome, writes prolifically on Aviation and must read in his sleep to get through what he does. Keith is an RAAF veteran of '112 Mobile Fighter (Control) Sector' at HORN ISLAND who kept a pocket diary and logged hundreds of aircraft 'in and out', unfortunately by type only. Peter, a brother-in-law, is another of those retired professionals who still operates on a 26 hour day. As for 'aeroplanes', he enjoys being just one of the travelling public. Each has brought a valuable balance to the editing job. I do thank them but hasten to stress that the residual errors and the 'commas in the wrong place' are all my own work.

1994 is the International Year of the Family. What better time to be thanking my close and my extended family for their years of patience, encouragement and help! To Jenny, my loving wife coming up thirty-two years, this book wouldn't have happened without you. Love you.

THE AIRFIELDS

... listed North to South, by increasing South Latitude

HORN ISLAND

'...A low earth bank or tar drums filled with earth could be built around these dispersal positions which would afford ample protection to aircraft against fragmentation bombs as has been discovered at the Seven Mile Aerodrome at Port Moresby.'

This is taken from a letter written by Wing Commander Deryck Kingwell, then CO of the RAAF's 32 Squadron, flying Lockheed Hudson twin engine aircraft, based at HORN ISLAND. Deryck Kingwell wrote this request for an urgent review of the dispersal and camouflage arrangements on this airfield on 28 May 1942, some couple of weeks after the fourth Japanese air raid. Coincidentally it was the same day that the RAAF's 100 Squadron first showed its twin engine Beaufort bombers at the Seven Mile strip, Port Moresby.

HORN ISLAND was always a problem airfield. Around January 1937 the Civil Aviation Department set requirements for a landing ground on HORN ISLAND to service the projected Sydney to New Guinea air service. A site was not available on the much smaller centre of establishment of Thursday Island. Air communication with Thursday Island was available through the Flying Boat Services, but with the RAAF's need for an Advanced Operational Base (AOB) the decision was rather reluctantly taken to develop the better of two sites on HORN ISLAND. As events later showed with the twin difficulties of water supply and poor wharf facilities, it's easy to realise that the very retention of an airfield on HORN ISLAND during the Pacific War came under serious review.

Aerial surveys conducted from a RAAF Seagull aircraft were in progress during December of 1938. These included survey of parts of northern Australia for the Military Board and for the Queensland Main Roads Commission (MRC). The Archive record indicates 'photography at Thursday Island and SOMERSET had to be abandoned owing to abnormal weather conditions'. Presumably SOMERSET, a DCA 'numbered' Landing Ground very limited in its potential, was at least under review at that time for development as a useful airfield site. A RAAF Directorate of Works and Buildings (DWB) Minute on file dated 31 January 1941 notes -

'...Advising Civil Aviation expected to complete work on (Horn Island) aerodrome before 8 February 1941 and that hutments will be completed about same date and requesting civilian caretaker be appointed.'

Plans for the proposed siting of bomb dumps at various AOBs including HORN ISLAND appeared during May of 1941. The MRC was involved in the late May construction of a jetty and causeway which was apparently outfitted with some cranage in June. Later, these facilities proved totally inadequate to the needs of handling modern heavy equipment needed to upgrade the HORN ISLAND airstrips.

November of 1941 saw the flurry of activity in construction of the first of the dispersal points, bomb dumps, machine gun posts and petrol storage installations. These works were handled by the MRC at that time under the supervision of resident officer Mr Lord.

With Japan in the war the Works and Services Branch General Index Cards now held in Australian Archives (AA), show 'works in hand' such as -

'...23 December 1941 - aerodrome obstructions, 6 February 1942 - requests for splinter proof traverse walls around wireless receiving and wireless transmitting huts and 2 March 1942 - stockpile 500 cubic yards of gravel or crushed rock and 500 cubic yards good filling and erect notices prohibiting use by unauthorised persons.'

On March 14, HORN ISLAND Airfield received its first Japanese air raid. Eight 'Nell' bombers and nine 'Zero' fighters took part. They were intercepted by P-40 Kittyhawks of the 7 Fighter Squadron, 49 Fighter Group USAAF. Two 'Zeros' were destroyed for the loss of one Kittyhawk and another damaged. One RAAF Hudson of 32 Squadron was destroyed on the ground and another damaged.

On 10 April 1942 a signal from North East Area (NEA) requested provision of -

'...refrigerators for Horn Island... personnel (now on Horn Island) total about 300.'

About this time it was intended to withdraw 32 Squadron from Port Moresby to HORN ISLAND but Archive evidence suggests that preparations at HORN ISLAND were approaching chaos. A response from NEA dated 13 April 1942 stated that -

'...Present living conditions Horn Island disgusting and that AOC will not move 32 Squadron there unless better conditions are provided.'

Barney Davies was a young 19 year old Wirraway pilot with RAAF's 24 Squadron. His was one of three aircraft involved in searching for US P-39 pilots downed on Cape York by bad weather near the end of April 1942. Barney recalls a stopover at HORN ISLAND, prolonged by the Japanese air raids of 30 April and 11 May 1942. The first damaged his and another Wirraway and the second put paid to the efforts of he and his observer to make one of the aircraft flyable. Barney wrote -

'...the biggest ablution and latrine shed I ever saw was at Horn Island - peppered with bullet holes. The Japs must have thought it was a workshop!'

Periodic Reports in documents now in AA on the subject of RAAF AOB 'Hutments' at Horn Island show that progress at 30 May 1942 was -

'...W & T receiving hut additions - 90% complete, splinter proof traverse walls - 60% complete. Splinter proof pens for aircraft - 30% complete. Bitumen surfacing ends of runways and taxiways - 20% complete, (i.e. provision of metal only). RAAF AOB aircraft dispersal and buildings complete. RAAF huts - 60% complete. Machine gun emplacements not commenced.'

This was the position about two weeks after the fourth Japanese air raid on the airfield. Records show that in June of 1942 some moves were made to provide anti-aircraft defence. Batteries A and B of the US 104th Coastal Artillery (A-A) were deployed to HORN ISLAND.

HORN ISLAND (22/5/1940)

Viewed from the northwest the clearing and early work on runway 135 is evident. This photograph is interesting in that it is taken from an RAAF Short Empire class flying boat. The cleared survey line for the second runway can be seen across the middle right of the photograph. Take-off to the north west from this relatively short strip was often hazardous due to changing atmospheric conditions over the mangrove shoreline. [NLA (RAAF Official)]

HORN ISLAND *(6/10/1942)*
Part of a photographic mosaic covering the whole airfield and environs, this photograph was taken after seven Japanese air raids. Clearly the white 'pock' marking in the southern sector of the runway cross is evidence of bomb cratering. Fourteen twin-engine aircraft are evident on the ground, two of these within the protective surround of mounded earth splinter proof pens on the loop taxiway in the centre right of the photograph. *[AUSLIG (RAAF Official)]*

A wide ranging inspection tour by General Brett, then US Commanding General of Allied Air Forces (AAF), in early June prompted the following comments directed to the RAAF -

'...Horn Island, Cooktown, Cairns, Mareeba - in general there is a definite shortage of maintenance crews, also lack of housekeeping equipment such as plates, knives, spoons, stoves, refrigerators etc. Particularly in Horn Island there is lack of water and welfare facilities. In case of water, a bore is being installed 3 miles from the airport. Pipe installations will be required when the bore comes in. Also there is a general need for additional motor transport.'

The Allied Works Council (AWC) was formed in late February 1942 to coordinate and provide for all construction associated with defence works, both those required by the USA Forces and all Australian Forces. The first appearance that HORN ISLAND makes in the minutes of the AWC is in June when requisitions were received for substantial improvements to HORN ISLAND AOB. A note adds,

'...US Engineers to handle the construction of all AOBs in north Queensland'.

On 19 June 1942 Air Vice Marshall Bostock in a memo to Chief of Air Staff RAAF and the Commanding General USAAS advised that a review had been made of operational aerodrome requirements in NEA and the commander of the AAF had directed that various dromes would meet all requirements - 'having regard to Air Forces to be operated in this area'. In respect of HORN ISLAND that requirement was -

'...One General Reconnaissance Squadron. Improve dispersal facilities and provide facilities for 150 men of transient combat crews'.

Camp facilities as such were to take the form of tent camps supplemented by -

'...Semi permanent kitchens, mess halls, latrines and showers. Essential kitchens and mess halls be screened.'

The aim to improve dispersal facilities led to the tip of Cape York Peninsula itself and in turn, creation of JACKY JACKY Airfield.

P-O H V (Jim) Trench, a Works Officer with NEA came into the HORN ISLAND scene on 22 June 1942. Air Commodore Lukis, AOC, NEA advised Air Board Melbourne, that Trench and Lt-Col Mills, US Army Engineers had inspected Horn Island wharf, and found the existing 4′ wide timber structure totally inadequate. Although Mr Redmond of the Department of the Interior had a fresh design proposed, Mills considered it too elaborate for 'purely wartime needs'. Mills instructed Redmond to proceed with a scheme to raise the existing stone causeway above high water mark, extend seaward and then a further 60′ timber structure beyond into sufficiently deep water to enable a ship up about 1 000 tons to tie up. Company 'A' of the US Army's 46 Engineer General Service Regiment arrived at HORN ISLAND by ship on 24 June 1942. Their stay extended until 17 December 1942.

HORN ISLAND was an important staging base. Through here much needed fighter aircraft were transiting to New Guinea. Reconnaissance and offensive bombing missions used the field particularly while Port Moresby was vulnerable to Japanese attack. Ralph Dietz was a ball turret gunner in various 19th Bomb Group B-17 crews operating out of the rear dispersal aerodrome of LONGREACH at this time. Ralph was aboard one of either B-17 41-2655 or 2636, both of which are recorded as having 'ditched' off HORN ISLAND in the early morning of 13 July 1942. It appears all aboard one of these aircraft were lost. Ralph and at least one other did survive their ditching. He recalls being on the first of these aircraft to go in and when he did make it to the shore, still in darkness, there was considerable confusion. Ground staff on the island were not aware that two aircraft had gone in, one only minutes after the other. Both these losses are very likely the result of the treacherous down draught conditions prevailing over the mangrove shoreline of runway 135°.

Strangely, Ralph was also on board B-17 41-2633 as it returned from bombing Lae some few days earlier. On that occasion the aircraft fouled fuel drums on the edge of the runway and finished up with a failed port undercarriage. A photograph of the aircraft in this damaged state on HORN ISLAND has been widely circulated. It was taken by Howell Walker, then on assignment for the National Geographic Magazine and published as at 'an advanced base'. The incident prompted

HORN ISLAND *(early 1946)*
Viewed from the southeast most of the airfield layout is pictured. Thursday Island, top left, and Hammond Island, top right, are evident. The narrow extension at the western end of the east-west runway is visually the only evidence of that extension which was abandoned after serious drainage failures associated with the bridging over Loggy Creek. *[AA CRS A705/1 Item 7/1/1145]*

HORN ISLAND (1943)
US 5AF B-25 Mitchell twin engine medium bomber taking off from 'HORN ISLAND R/W EIGHT-ONE'. Note the dust being blown up by the flight of P-47 Thunderbolt fighters getting reading to take off from the northern end of the cross strip. It was common procedure for such a group of fighters, en route to New Guinea, to be shepherded by an aircraft carrying a dedicated navigator to 'show them the way'. This photograph is taken from the Duty Pilot's Tower (Control Tower) on Horn Island. Note the eastern bluff of Hammond Island on the horizon, just under the Mitchell. [F Smith]

HORN ISLAND (18/8/1990)
Today customs aircraft are frequently in evidence servicing the seaways of Torres Strait. Part of the tarmac is pictured with the eastern bluff of Hammond Island in the far background. [Author]

review of the aircraft's serviceability and whilst it was repaired at a southern depot it did not resume full combat duties. Alternatively it was assigned as the personal transport for General George Kenney, commander of the US 5th Air Force. During this period of its service the aircraft was called 'Sally', named for Sally Chase, wife of Kenney's aide at the time.

Anti Aircraft defences at HORN ISLAND were reviewed during June, according to a report drawn from a history of US Anti Aircraft Artillery in the Pacific War -

> *'...The original dispositions in Northern Queensland were slightly changed on 23 June when Battery H and one platoon of Battery A, 94th CA (A-A), were moved from Woodstock to Horn Island. On 7 July Horn Island was raided by 16 medium bombers. Batteries A and B, 104th CA (A-A) did not go into action as the planes were flying at approx 20 000' and were thus out of range of the .50 calibre machine guns which were the primary weapon of these batteries at that time. Although about 200 anti personnel bombs were dropped in this raid no casualties were sustained by our forces.'*

The RAAF's records show that three RAAF personnel were wounded on HORN ISLAND during that raid.

During early August 1942 USASOS instructed AWC to observe the then nominated priorities for the sealing of runways for operational aerodromes in north Queensland. Both runways 1 and 2 at HORN ISLAND figured in this urgent priority listing.

Rod Voller was the Officer in Charge of Department Home Security Camouflage forces in NEA. Rod took up duty in Townsville early in 1942 and on reviewing his diaries of that period, tells me -

> *'...On 18 November, saw Major Cox (USA) re Horn Island, Cooktown and Mareeba. He stated that no more camouflage was to be done for US or by US in Australia. It was up to RAAF to carry out more camouflage if they wished to - (and this did happen for the next 12 months).'*

Camouflage works had understandably a high priority when the threat to Australia was greatest. Similarly, work was done to deny the enemy use of some airfields if the need arose. AWC Minutes record cancellation on 22 September 1942 of 'aerodrome demolition works' at HORN ISLAND and other Queensland airfield locations. Certainly at least one demolition tunnel had been constructed part way under the intersection of the two runways at HORN ISLAND and in later years this was to prove probably more trouble than its original purpose served. In the post war scene a degree of mystery has attached to such wartime work at various airfields on the Queensland coast; mystery in as much as unexplained depressions were observed in pavement surfaces, leading of course to an urgent need to not just repair but to establish the reason for such depressions.

On 9 December 1942 the No 3 Divisional Works Office (DWO) Townsville is on record advising HQ NEA -

> *'...During a recent inspection of Horn Island, it was noticed that following work had not been completed by the 46th Engineers, i.e. pipe culverts at entrance to hard stands, clearing of rubbish, high stumps above 18" and mounds etc. on flanks of the taxiways and consolidation of the hard standings themselves. In view of the shortage of civilian labour in this locality and as this aerodrome may be required for heavy bombers in the event of trouble occurring with the runways at Iron Range, it is suggested this work be undertaken by the section of Works Maintenance Units (WMU) stationed on this aerodrome.'*

The RAAF's 4 Airfield Construction Squadron had its 4 WMU active on HORN ISLAND at this time. The US Engineer Unit had moved on to Port Moresby. During January of 1943 heavy rain led to the failure of the log timber drainage culvert under the 81° runway rendering the western extension that had been carried out by the US Engineer Unit unusable. This significant failure put paid to plans to operate a heavy bomber unit from HORN ISLAND and underscored the efforts on the mainland to bring JACKY JACKY airstrip into commission.

An Air Board Agenda in early 1943 carries the following reference to the boosting of aircraft dispersal works -

> *'...Air Staff Policy Directive 7 - Aircraft dispersal facilities at OBs, as revised on 18 January 1943, makes provision for dispersal of aircraft in splinter proof pens when required, at aerodromes in certain parts of NQ. It is now considered necessary to provide pens on the scale and at the cost indicated at following RAAF dromes listed in order of proposed construction priority:...'*

HORN ISLAND figures in this listing with sixteen heavy bomber pens and thirty six medium bomber pens to be constructed. The agenda carries on -

> *'...Reconstruction work at Horn Is approved at Agendum 4730, is in progress and pens now proposed are designed to accommodate 36 medium bombers. The 16 heavy bomber pens will accommodate 2 pursuit fighters each. This arrangement is necessary in order that large concentration of aircraft may be adequately dispersed in limited space available. The requirements*

at Jacky Jacky (Red Island) are additional to those approved to Air Board at Agendum 4730 which are under construction, as it is now considered necessary to accommodate 2 bomber squadrons. As these works are important and extremely urgent, approval is sought for provision of funds so that immediate arrangements may be made to transport labour and plant and to commence work as soon as site plans are available.'

Rod Voller's diary notes support this -

'...On 17 March 1943 I discussed with F-O Grounds and Syd Hirst (Com Works) requirements for meeting revised Air Board lists and possible future work for quantity orders to Com Works for 175 to 200 heavy bomber hideouts for Mareeba, Inverleigh, Horn Island, Iron Range and Cooktown. On 18 March F-Lt Farran RAAF stated that splinter proof bays at Jacky Jacky for heavy bombers were being proceeded with and camouflage would be required. Also that 18 medium bomber splinter proof bays were being built at Horn Is and should be camouflaged.'

This work together with reconstruction and upgrading, either caused or delayed by the recent wet, was significant, particularly in the context of all other work in Cape York area. AWC had now to embrace the whole, without any direct contribution by US Engineer units. Hitherto, the AWC had voiced great sensitivity to civilians working in 'active areas'. On 20 March 1943 a signal to RAAF's 62nd Works Wing advised that No 3 WMU was to remain at HORN ISLAND until AWC took over the work.

Construction requirements for air defence of northern Australia were further reviewed by Lt-Gen Kenney's command when a memo under the signature of Brigadier General Donald Wilson, Chief of Staff, dated 22 March 1943 referred to -

'...The growing Japanese strength in northern and north western New Guinea and at the more westerly bases of Ambon and Koepang and therefore the potential for enemy offensive action in the general Darwin, Merauke to York Peninsular area... To meet the immediate tactical requirement the highest priority should be given to dispersal points including hard standing and taxiways at Horn Island to provide at the earliest practicable date, for the security and topping of 32 P-38s, 24 A-20s or Beaufighters and 6 Beauforts.'

Wilson's directive added -

'...The possibility of airborne attack on Horn Island must not be overlooked and necessary defence works should likewise be constructed in the Horn Island area.'

Wing Commander Dale acting for the RAAF's DWB reported to AWC Melbourne on an urgent inspection of early May 1943. Essentially he found the progress of work at HORN ISLAND totally unsatisfactory and indicated that there was a need to send '100 fit men immediately'. Further he indicated that currently, 20% of the men being sent to that area returned medically unfit within one week. This together with already realised difficulties like the lack of shipping and materials prompted the RAAF to instruct AWC to adjust resources.

HORN ISLAND (1943-44)
A twin-engined Beaufort Bomber of 7 Squadron low over the airfield with other fighter aircraft on the ground. Note the American white-tailed Thunderbolt fighter, probably in transit to Port Moresby. [L Pike via D Vincent]

'..Re phone conversation 24 May 1943 between Mr Crawford, Chief Engineer MRC, and W Cdr Towers, it is confirmed work on runways at Toogoolawah may be temporarily suspended in order that construction forces be made available for works of greater priority at Horn Island and Jacky Jacky. This is on understanding that... a recommencement will be made as soon as additional men, already requested, arrive from the south.'

June 16 1943 is the only date on drawing T 1287, one of a series of three preserved in AA holding J153. These drawings deal with at least three projected water supply dams on HORN ISLAND. It's not clear to me how many such dams did actually proceed to completion during WW2, but certainly the MRC History, published in 1949, includes reference to 'a dam' amongst its extensive list of various war effort constructions on HORN ISLAND. On the RAAF's Aerodrome Works File there is an entry dated 28 June 1943 indicating -

'...should it be necessary to decide priority between construction of the dam and the splinter proof pens, then the dam is to receive first priority.'

On splinter proofing of aircraft dispersal positions Rod Voller's notes show that -

'...F-Lt Farran advised on 9 July that splinter proof pens south of Jacky Jacky had been abandoned and Horn Island and Jacky Jacky were the only places where they were to be built.'

The office of the Chief Engineer, SWPA is on record to Gp Capt Knox Director of Works and Buildings RAAF Victoria Barracks Melbourne on 16 July 1943 pointing out that -

'...to camouflage such dispersal areas at these places is unnecessary because enemy bombing target areas are all related and easily found by reference to the coastline and other geographical features.'

Direct experience from Port Moresby is quoted and in view of overall priorities, the letter continues -

'...the advisability of deferring or possibly cancelling this work is suggested.'

This view conflicted with current RAAF policy. Naturally the views of RAAF HQ Forward Echelon were sought and a response under the signature of Gp Capt Charlesworth dated 6 August 1943 advised -

'...Any vital installation such as W/T Station should be camouflaged as thoroughly as possible.'

RAAF Command agreed it was a waste of effort to camouflage runways and taxiways and built up earth pens themselves but believed it was warranted to top cover pens and thus conceal their contents from enemy reconnaissance aircraft.

With its HQ at Townsville, the RAAF's 7 Sqn maintained a detachment at HORN ISLAND over many months of 1943, conducting maritime and anti-submarine patrols. The Squadron recorded a few 'kills' over its time in Cape York. September 1943, a Beaufort crew led by F-O Jim Legge downed a Japanese 'Jake' aircraft, thought at the time to be a float plane version of the Aichi 'Val' dive bomber, remembered well at Pearl Harbor.

It appears 'electric light' may have come to the airfield at HORN ISLAND in late 1943. A Department of Air minute on file dated 23 September refers to -

'...Impending provision of transient facilities for RAAF Station Horn Island, increasing the establishment to more than 500.'

It queried whether 'electric light facilities' were required. The noted recommendation was that

'...The use of only one 25 KVA set should be recommended as that would be consistent with what's been scheduled for Jacky Jacky.'

HORN ISLAND *ca(-/8/1943)*
His Excellency Sir Leslie Orme Wilson GCSI, GCMG, GCIE, DSO, Governor of Queensland, here inspects 7 Squadron Beaufort A9-329 on the ground at HORN ISLAND. State Governors carried out a full and onerous programme of visiting Defence Force and other war-effort locations during WW2. [J Legge ex RAAF]

During September the RAAF was critical of progress being made by AWC on several projects in northern Queensland. The response on file by the Supt. Engineer M K N Johansen dealt extensively with his present commitment of resources. In this wide ranging review the progress of a couple of HORN ISLAND projects are of note, e.g. -

> '...*KS316 Horn Island:dam for water, 5% complete, officer responsible Lowe, 7 men assigned. KS294 Horn Island:runways, dispersals and pens, 65% complete, A1 priority, Lowe, 101 men assigned.*'

A Works Report dated 29 October 1943 includes -

> '...*81° runway sealing 80% complete, 136° runway surface very rough and sealing broken up... taxiways - all gravelling complete, except on those taxiways parallel to runways which have not yet been commenced. Hard standings - 52 exist that are not sealed. Pens - 5 complete, 13 partly complete,... work suspended pending completion more urgent work. Camps for two squadrons and OBU are complete. Water supply is short at the moment but new dam for 13 million gallon is under construction. This is an earth dam with concrete core wall, completion anticipated end November. GENERAL - work on this project has been delayed by an 8-day strike by the CCC.*'

December 1943 saw mention on RAAF Aerodrome Works File of the possibility of providing a building on the airfield to allow for the working of a Plessey belt filling machine for 28 OBU. While the precise workings of this device are not known to me, my contact with veterans certainly shows me that such a machine was not commonly available to OBUs and where they were not, a considerable degree of fatigue and tedium prevailed in the alternative hand filling of metal prideaux clips with machine gun ammunition. The file followed up on 6 January 1944 to clarify that 'all the hut needed was a power point'.

Between Air Board Agendum 5617 dated 24 February 1944 and 5617 supplement dated 6 March 1944, a full review of units in residence at HORN ISLAND and HIGGINS is presented. All of this was apparently set out to justify the chosen siting of a dental section! For example HORN ISLAND was described as hosting 28 OBU, 36 Radar Station, 112 Mobile Fighter Sector HQ, 84 Interceptor Fighter Squadron, 75 Wing HQ, 1 RSU (detachment) and 7 Squadron (detachment). In terms of function, HORN ISLAND is stated as being the -

> '...*Fighter Centre and 84 I/F Squadron and 112 Mobile Fighter Sector HQ are located there for that purpose. 36 Radar Station is located Horn Island for technical reasons associated with siting. 28 OBU provides general base facilities whilst the detachment of No 1 RSU is located on Horn Island so that damaged aircraft can be dealt with on the spot. Horn Island is the point of trans-shipment in Torres Strait and all stores destined for Higgins must be transferred at Horn Island into smaller craft for onward movement to Higgins.*'

The outcome to this was apparently known in May 1944 when the decision was made to construct the dental clinic at RAAF OB HIGGINS FIELD.

A memo of 18 October 1944 to AWC from 3 DWO refers to radio range equipment at Horn Island. It deals with a decision to have RAAF 2 RIMU move the transmitter hut aerial tower marker base etc. and effect completion of the installation at HIGGINS. This was all part of the phasing down of HORN ISLAND in favour of HIGGINS.

During the closing weeks of the war Vic Waddell was on detachment with the RAAF Survey Flight at HORN ISLAND. Vic recalled that the -

> '...*work was the Survey of Cape York. The amount of work we could do each week was governed by the amount of cloud cover. Consequently there was more time spent on the ground than flying. There were no servicing facilities. If assistance was needed one got it by air from Higgins Field (Jacky Jacky). Fuelling facilities were available... For entertainment we had outdoor theatre with films each week. It was supplemented by a visit of someone Lawrence (a boy piano protégé of the day) he and his piano arrived by air! Our leave was limited on non-flying days to visits by ferry to Thursday Is. It was a garrison town so all hotels were closed and the only activity was to "view the sights" or visit the recreation hut.*'

The RAAF did not require HORN ISLAND post war. Certainly the Department of Civil Aviation did. Today the airfield has been taken over by the Local Authority, Torres Shire Council, under the Commonwealth Government's overall plan to replace its airport management role, nationally.

HORN ISLAND *(14/4/1988)*
This recent vertical shows ample evidence of the wartime taxiways still exists.
[Courtesy R Rudd Northair Surveys Mareeba]

HIGGINS

'...Announcement is hereby made that the airdrome now known as Jacky Jacky and/or Red Point, which is located at 10°55'S 142°30'E near the tip of Cape York Peninsula, is officially designated Higgins Field in honour of Flight Lieutenant Brian Hartley Higgins 400620, RAAF, killed on air operations 25 May 1943 - by command of Lieutenant General Kenney.'

A dispersal strip was urgently needed for HORN ISLAND, particularly after it had come under air attack by the Japanese. The Island itself is about 5 miles in diameter and very hilly. As it was not able to provide an effective dispersal strip, attention was turned to the mainland. While mention is made of dispersal strips on the tip of Cape York having been located in a 28 May 1942 report by the AOC North East Area (NEA) to Air Board, Melbourne, it appears in fact that the first effective ground reconnaissance to identify such sites was carried out sometime mid to late June 1942. P-O H V (Jim) Trench RAAF and Lt-Col Mills US Army Engineers did the 'recce', Air Commodore Lukis, AOC NEA, reported on 29 June 1942 -

'...A very suitable area was found approximately 6 miles southeast from Red Island Point, on which two strips could be constructed, one bearing 130° and the other 165°, each 7 000'.'

Back in 1988 after some prompting by myself and members of his own family, Jim Trench committed some recollection of that visit to writing for me:

'...Jacky Jacky was proposed as a substitute for Horn Island airfield, which was limited in size and there was a chronic shortage of water. Col. Mills and myself flew to Horn Island and immediately crossed to Thursday Island where there was an Army base commanded by Col. Meredith. We were able to arrange for the army to supply a motor boat and driver and the advice that the place to start was Red Island Point about 15 miles south of Cape York on the Gulf of Carpentaria side, where the local butcher a Capt Cadenza kept his cattle and we would be able to borrow horses which proved a God send, though neither of us were experienced riders. We set off next morning and on arrival found an Aboriginal settlement and a detachment of the CMF.

What we were after was a ridge at least 7000' long and reasonably straight. We rode the horses along the old overland telegraph line and by counting the posts, which I think were 5 chains apart, discovered that the horses walked at almost exactly 3 mph. This meant if we rode them on a compass guided straight line for half an hour we would have a possible airstrip site.

It was rather frustrating as we found several 'twenty minute ridges' which petered out just when things began to look promising. However on the second afternoon, we located a very likely ridge that we decided to recommend. The catch then was to locate it on the Queensland "4 mile" map, as we were unable to match any of the features shown on the map with those on the ground. These "4 mile" maps were the only ones available and in most areas had obviously been compiled from very scanty information.

The east coast had been charted by Capt Cook and maps were consequently reasonably accurate along the coast for about 1 mile inland. Unfortunately Capt Cook did not sail down the west coast of Cape York. Anyway we located the site as best we could and returned to Horn Island... An American pioneer battalion was despatched to the site a few days later. Col. Mills and I visited the area some weeks later and construction was in full swing - the only trouble was, it was about 5 miles away from the site we had selected!'

Jim recalled that on a later aerial inspection he judged the site under construction was about 5 miles southeast of that originally chosen. It's not clear how it came about that Jim's site was not used. Certainly archive records show that further assessment of Red Island Point as a suitable wharf landing location resulted in such wharf site being shifted further south along the coast to Muttee Head. The same documentation, a report by Herman G Cox, Capt Corps of Engineers to the US authorities dated 8 August 1942, mentions -

'...Investigation of tidal waters at Red Island by naval authorities revealed shipping would be jeopardised. The port of entry was then moved to Mulle Head (I believe he really meant Muttee Head) but the (airfield) site as generally selected was not changed... Two engineer companies are in the process of unloading equipment at the point of this writing.'

Vol VI of ***Engineers of the Southwest Pacific 1941-1945*** prepared in 1951 under the direction of US Maj-Gen H J Casey, covers this aspect -

'...Elements of the 46th Engineers had arrived on Cape York Peninsula in June 1942. Their efforts had been split between initiation of construction at Iron Range and repair of recent bomb damage to the field at Horn Island, as well as clearing for extension of runways at that field. During the following month elements of the 91st Engineer Battalion (Separate) had started construction of the field at Jacky Jacky where far better possibilities of subsequent port development for supply by water had been observed on engineer reconnaissance than existed at the originally proposed site on Red Island Point. Although practically operable by the end of 1942 all three of these fields were not prepared for all weather operation, including necessary facilities, dispersals and housing until late in 1943.'

The journal of the freshly redesignated US 91 EGS regiment, obtained from US National Archives, doesn't clarify the site of that unit's 'timber dock'. October 20, 1942 is there recorded as the date the unit -

'...constructed abutment and two approach trestle bents and drove 15 piles for timber dock at Red Island Point.'

HIGGINS *(15/4/1988)*

Dispersal taxiways are all but overgrown in this recent vertical of HIGGINS now serving as the airport for Bamaga. [Courtesy R Rudd Northair Surveys Mareeba]

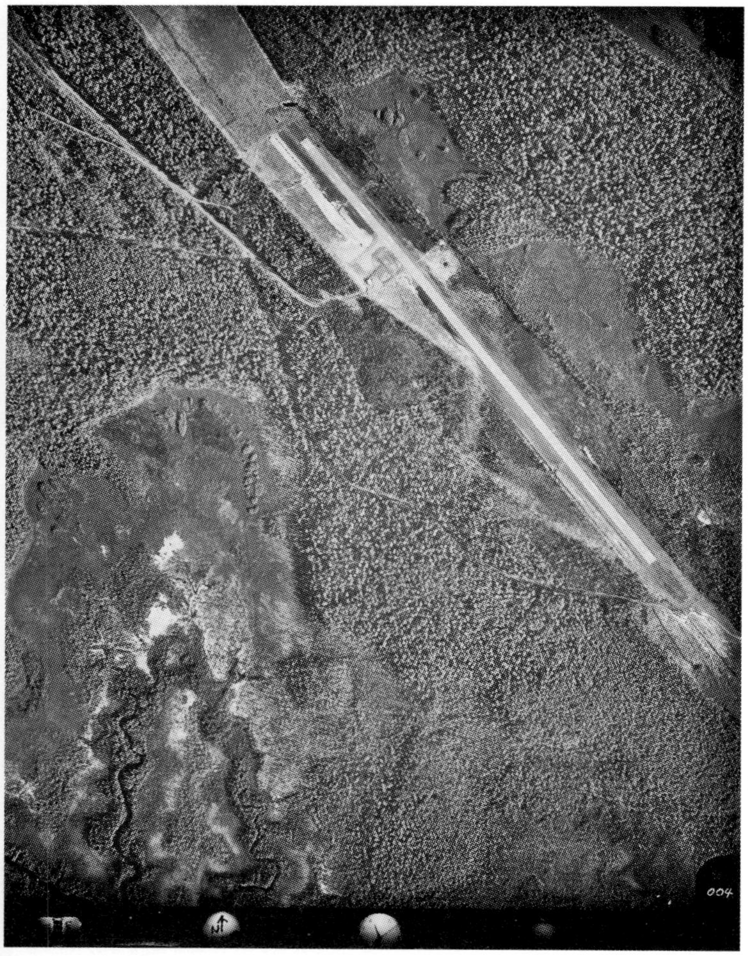

Along with their other works on the connecting road and the airstrip itself, the journal mentions clearing of -

'...19 acres for anti-aircraft gun emplacements... assisted in unloading 1 200 tons of supplies and equipment from ship for 31st Aust. Bn... (and) light anti-aircraft guns, including mounts and parts and transported them from Red Island Point to battery sites.'

So taxed was the water situation on nearby Horn Island that the 91 EGS journal shows during October 1942 -

'...water supply of 8 000 gallons per day for Horn Island developed at Red Island Point.'

October 28 1942 appears in this journal as the day -

'...first B-17 bomber landed on runway. Col. Ramey, AC, and staff made brief inspection of runway and dispersal roads.'

The US Engineer work on the westerly extension runway 81° of HORN ISLAND focussed mainly on the work involved with the diversion of one creek into another and providing for their combined flow through a log culvert across the runway. This culvert proved inadequate in monsoonal rainfall later in the year. This problem with HORN ISLAND strips meant that in January of 1943 the RAAF was urgently reviewing the HORN ISLAND and JACKY JACKY scene. A report on an inspection by F-Lt D C Farran dated 22 January 1943 mentions -

'...In view of the unsatisfactory nature of the foundations of the runway at HORN ISLAND and the difficulty of effecting permanent repairs... the runway at Red Island was inspected. This site is 30 miles SSE from Horn Island and has a 7 000' strip complete with dispersal bays... constructed by the US Army Engineers. The strip would require re-shaping, consolidating, and a small amount of reconditioning the southeastern end with attention to drainage to make it suitable for an AOB. Camp buildings would be required. Water Supply is available at Red Island, whilst at Horn Island it is restricted to 1 gallon per man.'

The US Engineer construction work at JACKY JACKY had ceased during December and the AWC Minutes carries its first reference on 11 November of 1942. US engineer troops were being removed from RED ISLAND POINT and the request was that AWC take over construction work remaining at that aerodrome and its maintenance.

HIGGINS (1945)

This strip is now the Airport for Bamaga on Cape York. Photographed near the end of the war, the complex is viewed from the north western approach. Just off the camera plane's wing tip, Jacky Jacky Creek is seen flowing east to the Coral Sea. With the US calling this airfield RED ISLAND (Point), and the RAAF referring to it as JACKY JACKY, the confusion came to a head when some American report showed it as Muttee. This was settled with the announcement by Lt General Kenney, Commander of Allied Air Forces SWPA on 3 June 1943.

[AA CRS A705/1 Item 171/93/742]

While the RAAF was contemplating its precise needs at JACKY JACKY, an inspection report by F-O Campbell dated 24 February 1943 records a visit made in company with Capt Pierce, O-C Australian Garrison Battalion and Mr L Graham AWC, foreman. Under several headings the Report comments on Red Island wharf in that it was -

'...Recently constructed and about 60% complete, all piles, about 60 of them have been driven, and about 40 have been capped and braced. The wharf is 'L' shaped.'

On the airstrip itself F-O Campbell reported that it had been -

'...cleared 10 000' x 300' wide, due to the rising nature of the terrain and the 60' to 80' height of trees, additional slope clearing, 6 000' x 600' wide has been provided to permit a landing on a 1 in 40 slope at the north westerly end.'

Further on he reports that -

'...the two culverts each 250' long constructed of saplings pass water from the up hill gutter under the runway at two low points in the runway.'

As had happened at HORN ISLAND, these two log culverts under JACKY JACKY runway, proved inadequate and realistic use of the runway was not available until they were totally replaced.

In some general comments, F-O Campbell stated -

'...The site is connected by phone to the overland telegraph line and the Army Hospital 10 miles distant. The possibility of utilising Jacky Jacky River which is a half mile off the strip, to transport petrol and bombs in flat bottom barges was not investigated, but an officer from 28 OBU (Horn Island) has promised to make enquiries and forward a report.'

By 1 March 1943 the RAAF was planning to provide for -

'...Dispersal of aircraft in splinter proof pens at aerodromes in certain parts of North Queensland... Horn Island - 16 heavy bomber pens and 36 medium bomber pens and Jacky Jacky - 18 medium bomber pens.'

This was in line with planning to accommodate two bomber squadrons at JACKY JACKY. Lt-Gen Kenney's advice to the AOC RAAF Command AAF dated 22 March 1943 refers to the growing Japanese strength in northern and north western New Guinea and at the more westerly bases of Ambon and Koepang and the resulting potential for enemy offensive action in the general area of Darwin, Merauke and through to Cape York Peninsula. His advice goes on to indicate -

'...To combat this potential threat we need the completion of Horn Island then completion of an aerodrome in Millimgimbi area of Arnhem Land, the completion of the Jacky Jacky aerodrome and for priorities they to be done in that sequence...'

During May the emphasis was still on review of priorities of work down south so that the maximum effort could be produced to complete works of greater priority such as that at HORN ISLAND and JACKY JACKY. By 5 June 1943 a report on file anticipates 3 000' of runway becoming available for use by 12 June at JACKY JACKY. About this time an assessment of the building requirements for the urgent location of an RSU at JACKY JACKY, to service squadrons nearby and those based on 'Augusta' (Merauke) shows -

'...Current situation where steel transportable workshop hangars as required under the terms of Air Staff Policy Directive 10 are unavailable... necessary to include construction of two transportable timber workshops in the schedule of buildings for the Jacky Jacky project.'

This was also the time Kenney directed this airfield henceforth be named HIGGINS.

With the evolving review of the need for splinter proof aircraft pens, Rod Voller, Chief Camouflage Officer for NEA at the time, recalls that -

'...F-Lt Farran advised on 9 July that splinter proof pens south of Jacky Jacky had been abandoned and

Horn Island and Jacky Jacky were the only places where they were to be built. We continued with camouflage work on other projects, particularly RDF (Radio Direction Finding) and other communication and radio stations - bomb dumps etc.'

Progress on this camouflage question came in a reply by Gp Capt Charlesworth, CO RAAF HQ Forward Echelon, to DASP RAAF HQ Melbourne dated 6 August 1943. His statement included -

'...Any vital installation such as W/T Station should be camouflaged as thoroughly as possible. RAAF Command agrees... waste of effort to... camouflage runways and taxiways... and built up earth pens themselves... (but believe effort is warranted to top cover pens)... thus conceal contents of the pens from enemy reconnaissance aircraft.'

An improvement in serviceability of the strip at HIGGINS was promulgated with the regular notice to RAAF airmen Weekly Edition No 26, dated 17 August 1943 -

'...As at 11 August 1943, 7000' x 100' serviceable for all aircraft. Watch to be kept for marking off pegs.'

An AWC Report of 1 October 1943 lists progress on many projects in North Queensland including -

'...Jacky Jacky: operational base camp for RAAF 25% complete, A2 priority.'

Notice how the old airfield name persists!

Other records show that 33 OBU, arrived at JACKY JACKY on 13 October 1943. It's Operational Record Book (ORB) mentions -

'...Immediately commenced clearing for camp and further work on the erection of huts etc.'

A report dated November 1943 on inspection of RAAF OBs NE and NW Areas by W Cdr G J Towers in conjunction with RAAF command staff is of interest. The party comprised Air Vice Marshal W D Bostock AOC RAAF Command, W Cdr J Thompson Assistant Chief Signals Officer and Sqn Ldr A F E Wilson Admin 2 and the inspection was 25 October to 2 November 1943 -

'...Runway Higgins - approx 5 500' runway serviceable, not yet sealed; approx 1 500' on NW end now being strengthened by placing additional 9" thickness gravel where the culvert crosses the runway.'

Drainage, taxiways and hardstandings were all reviewed. Some 16 were expected to be completed by end December 1943, with an additional 20 to be provided -

'As soon as possible thereafter... CAMP-RSU Camp is complete and unit is in occupation. The OB Unit now lodging on the RSU. OBU and 2 Squadrons will be in 1 camp and work is being concentrated on completion of officers and sergeants mess which will be finished in 2 weeks time and will temporarily cater for all ranks of OBU and 1 squadron. WATERSUPPLY from wells are adequate; piping and retic work still continuing.

GENERAL - telephone advice from DWO Townsville on 3 November indicated 25 more men are expected arrive Higgins this week and attempts are being made to divert some of the men from the USA project at Iron Range, which has recently been considerably pruned regarding extent of work required. It's understood that reconstruction of jetty at Red Island Point is now complete but no information was available on status of repairs of road from there to Higgins.'

Jack Kay was a member of the Australian Army's 56 Composite A-A Regiment, a section sergeant in charge of 'A' Troop's 3 Bofors A-A Guns. Jack has written of his twelve months at HIGGINS arriving there around December 1943 -

'...It was typical of the Army that about 10 days before we pulled out, I had an officer appointed over me but I still had the job of dismantling gun A6, from its 30' tower in the dark and lowering it to the ground. The first 30' tower erected was No 5 (all built by engineers). It had 16 vertical posts 4 groups of 4, but later I was fool enough to tell an engineer officer they could be built with only 4 sloping posts, like we built tank stands on the farm. The remainder of the towers were built with the 4 posts - that officer probably got promotion for that change in construction!. Gun position No 3 was made of 44 gallon drums filled with earth. No 4 was on a man-made hill, built with a dozer. No 1 was a sand-bagged position, which, when the wet season came, (it was late that year too) ended up out in the swamp, with a croc in its vicinity. It needed a raised path out to it for access.

No 2 and No 6 guns were on towers and I think one, if not both, were 40' high - or at least 30' high.

Although we had these high towers for our guns, (higher than those used in Europe which claimed to be the highest!) we were directed to clear large areas of trees to provide a field of fire, much to the annoyance of the RAAF CO, a Gp Capt. At one stage during a heated but one-sided discussion, he placed me under open arrest, because one of my axe men felled a tree over the main track from the strip to their camp as they were going back to dinner.'

Sealing of the airstrip was commenced on 7 January 1944 according to 33 OBU's records.

Air Board Agendum 5617 dated March 1944 Supplement No 1 relates to RAAF HORN ISLAND additional building services and reviews the development of HORN ISLAND and HIGGINS -

'...Under Allied Air HQ Operations Instruction 35 - 16 April 1943 - installations were developed in Cape York area to provide for A and B.

A) accommodation of Fighter and Attack Aircraft at Horn Island

B) accommodation for medium bomber and heavy bomber aircraft at Higgins.

...The Commanding General 5th Air Force was directed to maintain plans for speedy reinforcement of Torres Strait area by Air Forces and plans for transport by air of reinforcing elements of land force.In accordance with these directions RAAF installations have been built up at Horn Island and Higgins. Units at present located Horn Island are shown Agendum 5617. Units located at Higgins are - 75 Wing HQ (recently moved from

HIGGINS *(1945)*
Working inland from Red Island Pt, now known as Seisia, a RAAF (Works) Pilot Officer Jim Trench and Colonel Mills USA Corps of Engineers located a site at the end of June 1942. The US Engineers arrived early August 1942 to commence construction of 'RED ISLAND PT airstrip'. With the onset of the wet season, and progress of the war calling the US Engineers over to New Guinea, construction was halted in December 1942. The AWC took over in early 1943. Work that year was slow, being stifled by lack of manpower, equipment, and shipping to supply the site. A 7 000' runway length became serviceable August 1943. No 1 RSU (Repair & Salvage Unit) occupied the drome from March 1943 until March 1944. No 7 Squadron moved over from HORN ISLAND then and took up residence and sea patrol duty staying until October 1944. [AA CRS A705/1 Item 171/93/742]

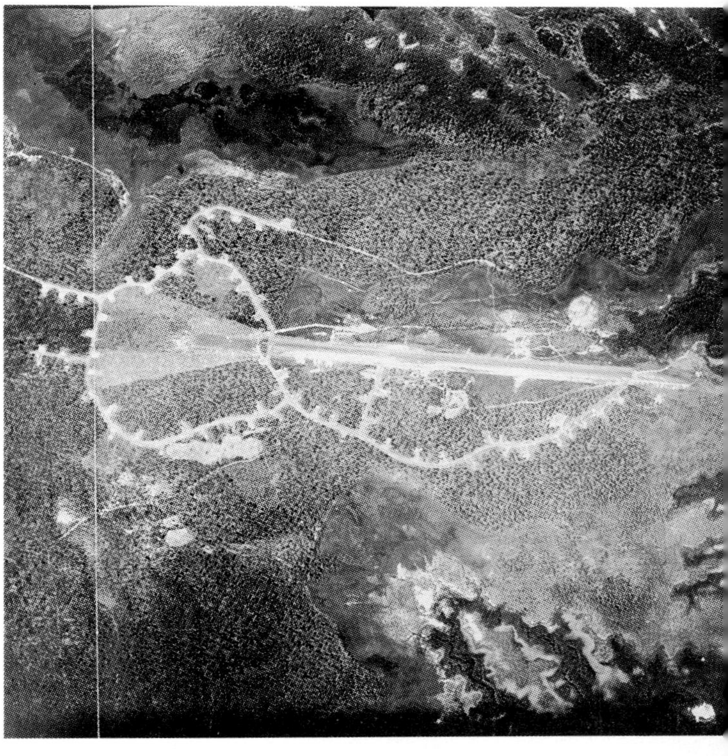

Horn Island); 1 RSU (less Detachments at Merauke and Horn Island); 33 OBU; 7 Sqn, (in transit from Townsville area); 4 Mobile Torpedo Section (MTS), (being withdrawn shortly)...
Horn Island is the point of transhipment in Torres Strait and all stores destined for Higgins must be transferred at Horn island into smaller craft for onward movement to Higgins. 75 Wing has moved to Higgins because -
A) The Army HQ in that area is located near Higgins.
B) Coordination of activities is facilitated. Further, 7 Sqn is moving into Higgins ex Townsville and the Wing HQ is more directly associated with operations of this unit than with 84 Sqn which is operated by 112 Mobile Fighter Sector HQ, on Horn Island.'

These Air Board Agenda papers appeared on a HORN ISLAND OB file as 'background' in relation to proposals for a Dental Building for HORN ISLAND. In May 1944, AWC were instructed it was to go -

'...at Higgins Field, on the northern boundary of the sick quarters, about 50' west of water supply pipeline to building 21A.'

By 22 May 1944, the policy on providing 'top cover', meaning the screening by camouflage netting over dispersal bays, be they bomb splinter proof protected or otherwise, had lapsed. There is on record a report from W Cdr Brogan CO of 3 DWO in Townsville in response to DWB Melbourne's order to cease -

'...HQ NEA memo of 18 May 1944 has directed in respect of Higgins... twenty-four only top covers to dispersal bays are to be provided in lieu of fifty-eight. These top covers to be located in bays most adjacent to administrative and camp buildings, namely on SW side of runway and making use, where possible, of bays which have tongues already constructed.'

Brogan advised that while instructions to cease had been passed on to AWC, he would check 'if any of the well advanced covers should be completed'.

Even the construction of the Dental Clinic at HIGGINS was scrapped by instruction of 26 August 1944. Things were scaling down.

During July of 1944 some relevant comments from the 33 OBU's records show that -

'...a saw bench with a 32" diameter blade was completed during that month, that an armoury 30' x 20' was built of bush timber, that a rest camp was established at the mouth of Larradeenya River and that construction of tennis, badminton, basketball courts and a swimming pool commenced.'

In August of 1944 the same 33 OBU records mention -

'...Twenty mango trees were planted throughout the camp by the CO (F-Lt F J Lucas) and also pawpaw and lime seeds were sown in boxes. A transient camp for airmen was commenced to consist of twenty prefabricated huts with accommodation for one hundred and sixty... A piggery was started, with some escapes and some recaptures... Eight Japanese prisoners passed through; they all seemed quite happy.'

In September of 1944 the same record shows -

'...Aircraft movements through the base as 195 RAAF, 16 American, 16 Dutch, 25 Civil and 1 New Zealand Air Force.'

The RAAF's 7 Squadron moved over from HORN ISLAND in March 1944 and were preparing to leave during September and October, moving on to Tadji in New Guinea. This movement of the squadron, then commanded by Sqn Ldr J O F Barton was notable because it was probably the first occasion a RAAF squadron had been moved entirely by RAAF aircraft. In October ten C-47 transports from 34 Squadron, Parafield, South Australia, 35 Squadron Guilford, Western Australia and 36 Squadron Townsville, were involved transporting the men and equipment in the space of a week. It took about fifty trips to move the 235 000 lbs of equipment and 114 personnel including three 3 ton trucks.

The inevitable aircraft mishaps and various crashes occurred in and around HIGGINS. Notably though, on 5 May 1945 a DC3 'Courier' VH-CXD crashed coming in to land at 5.20 am. The captain was an ANA employee and the crew of four and two US passengers were all killed. The plane was carrying a cargo of frozen meat destined for Merauke, Dutch New Guinea. The wreckage lies to this day on the north western approach to the strip, suitably fenced and available to be viewed by visitors to the area.

Today JACKY JACKY is the airport for the nearby settlement of Bamaga which itself is located near what must be the original site chosen for the airfield by Jim Trench. Red Island Point is more commonly known as Seisia. These are both settlements well known to modern day travellers visiting the very tip of Australia - Cape York.

HIGGINS *(17/8/1990)*
All that remains of this splinter proof aircraft dispersal pen can be seen by the residual encircling earth mound. The height of such mound is indicated by the two figures standing on top. *[Author]*

ELLIOT *(22/7/1950)*
This 87 Squadron Flight Diagram relates to a photograph surviving in the National Library of Australia. It is not yet established when this airstrip was cleared nor for what purpose. This one was overgrown by 1950. Many such strips were constructed to provide basic aircraft access to mineral exploration parties and pastoral development ventures. *[AUSLIG]*

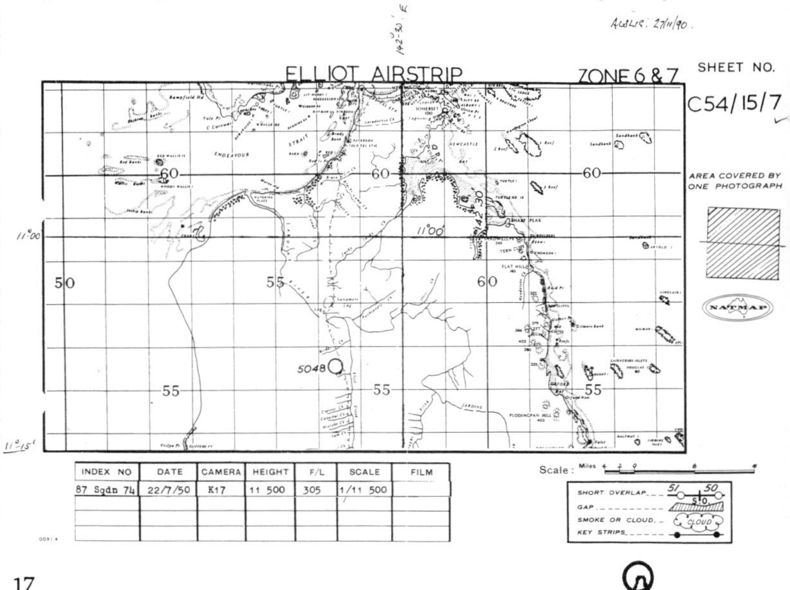

ELLIOT

'...There's the story of the Catholic priest up north who knows where there's a Kittyhawk still standing on an overgrown WW2 Emergency Landing Ground.'

Australia and Queensland in particular have an acknowledged rich aviation heritage. Queenslanders claim prestigious association with QANTAS, the RFDS and 'Bush Pilots'. This State's golden sand beaches are treasured as a birthright. It follows that beach landings and bush landings by aircraft rank high on the glamour scale - none more so than countless such 'set downs' by service aircraft during WW2! However, if there is one aim of this book, it is to promote an accurate and informed view. I remain sceptical of the 'Catholic priest's P-40'.

Many times it's been put to me that 'dozens of Emergency Landing Strips were constructed during WW2 to serve the possible needs of aircraft being flown to and from New Guinea'. Some people, seriously intent on helping my project have directed attention to sites they believed or thought were 'old WW2 strips'. My follow-up research has not led to any additional locations being documented in Appendix 1. Mostly the sites referred have a post war origin in the mineral exploration or pastoral development years. Eric McGuffie has spent most of his life 'north of Cairns'. Eric tells me he served a 'coastwatching' role on Cape York Peninsula during WW2. Post war he was employed for periods by Comalco Pastoral Development, his older brother was employed by Comalco Exploration and his half brother Abbie Seagren owned and operated clearing and earth moving equipment over 'the Cape'. Eric was either involved 'putting in' many of the strips I've referred to him, or he's been able to tell me who did and when.

ELLIOT Airstrip first came to my attention at NLA in September 1990. It was one of two lone vertical prints, edge marked '**Elliot Airstrip S-N (5048) 22 July 1950 305 mm, 11 500'**' and the other, similar except that it was numbered **5046** and titled '**Austin Airstrip**'. NLA didn't have Flight Diagrams (FDs) for these two but had both catalogued for Queensland (Cape York). They could not tell me more. The RAAF ***Aerodrome Guide** (January 1946)* showed Austin as on Melville Island and later I found it to be the WW2 period Snake Bay strip. During a visit to AUSLIG in Canberra, November 1990, I stumbled on both FDs. My efforts to learn more have all drawn a blank. So ELLIOT, with two strips 6 600′ x 600′ and 5 000′ x 600′, both with flared approaches, intersecting at 60°, overgrown to some extent by July 1950, remain in my estimation, at least work done by or for the Defence Forces. Whether during WW2 or not is unanswered.

I have found in RAAF records in Australian Archives (AA) references to ELGs in North Queensland. Capt Herman G Cox, US Engineers in a 13 July 1942 report on a week long inspection of several dispersal strips, wrote -

'...Emergency Landing Strips were located and selected along the coast between Cairns and Portland Roads.'

Cox lists Port Douglas; 'a strip just south of the wharf' and the famed beach itself, Anton; 'a strip approximately 2 miles square', and Princess Charlotte Bay -

'...a salt pan ...3 miles west of intersection N Kennedy and Annie Rivers.'

Anton, obviously a mis-spelling of AYTON. was then a small timber milling settlement near the mouth of the Bloomfield River. On 6 August 1943 the RAAF photographed a clearing in which there is now a 'landing ground' shown on current mapping (15° 53′S; 145° 20′E). Princess Charlotte Bay appears in the Department of Interior's Works and Services Branch (Qld) General (subject/property) Index Cards, in AA, simply as a registered communication of late 1940 'Landing Area - inspection and report'. No precise location is mentioned on the card.

Walter Maiersperger, was the American pilot of the STOCKROUTE based C-40 aircraft which flew Cox on his 27 June 1942 aerial survey of these Emergency Landing Strips. His answer may not be available by 'printing time'. I expect Walter's memory, marvellous as it appears to date, may not bring much joy here.

The RAAF Survey Negative Register, held by AUSLIG (see Appendix 4 for background) lists -

'...P 39 Port Douglas Strips 16°29′S; 145°47′E - 6 August 1943.

P 40-41 Bloomfield Strips 15° 56′S; 145°21′E - 6 August 1943.

P 49-50 Orford Mess Beach Strip 11°15′S; 142°48′E - 7 August 1943.'

The Bloomfield photos are in NLA holdings, but the others I've not seen.

Clearly Orford Mess is Orford Ness and under magnification that's included bottom right of the ELLIOT FD. To show the atmosphere from which tales like the 'Catholic priest's P-40' spring, Orford Ness is the vicinity from which Eric McGuffie tells me of his involvement in the retrieval of four P-39s around 1971. Two of these are undergoing long term restoration for non flying display. Charles Falletta and Walter Harvey both in US 8 FG have visited North Queensland in recent times to see their aircraft. Syd Beck of Mareeba has Falletta's 41-6951 in his collection. In early May 1992 members of the Cairns Aircraft Recovery Team (CART), 'Taffy' Jones, Ian Mullins, Nick Watling and John White, showed Walter Harvey his 41-7215 at John White's cane farm in Mossman, NQ. This was almost 50 years to the day that Charles and Walter and the rest of their flight of six put down in bad weather with fuel gone.

IRON RANGE *(16/8/1990)*
Looking into one of the 3.7 inch A-A gun ammunition stores north of GORDON strip. While the rainforest has certainly reclaimed this area it is hard to realise that the location is the same cleared area at Grid Reference 737697. This A-A location is shown on drawings in Australian Archives as the site of '446 Battery Anti-Aircraft'. Battery 447 is shown at Grid Reference 748647. *[Author]*

IRON RANGE

'...Iron Range was the pits! Right in the middle of the jungle... We lived in six-man pyramidal tents and had one large tent for a mess hall... The weather out of Iron Range to Jap targets in New Guinea and New Britain was horrible. We lost many planes and crews to bad weather.'

These are part of recollections from Ernest Ray Rhodes who was at IRON RANGE from 10 November 1942 through to 1 January 1943. Rhodes was a member of the US 319 BS, 90 BG equipped with B-24D Liberators. The tropical climate, the primitive conditions and the terrible weather, coupled with the long haul direct to target areas are all factors indelibly lodged in the minds of most American IRON RANGE veterans I've contacted. Ernest Rhodes, remarkably, is one of the few who appear not to have had a fear of snakes nor to have remembered the general run of wild life 'critters' of which so many veterans have spoken.

In the Japanese aim to capture Port Moresby and command New Guinea as a base from which to counter the continuing buildup on the Australian mainland, Port Moresby was that key. It at least had some established airfields. These and the Port facility at Port Moresby were the very basis of the Allies' effort to withstand the Japanese sweep into New Guinea. It was clear to the strategic planners at the time that a fall-back airfield was needed on the Queensland coast, a field from which bombing strikes could be mounted direct. While the Japanese could freely bomb Port Moresby, it was not the place to shelter and replenish Allied bombers, even as a staging drome. IRON RANGE lies on the great circle through Port Moresby to Rabaul and it's not coincidental that the airstrips were sited in jungle 'easy of camouflage'.

Early ferrying of much needed aircraft to Port Moresby logically involved staging through HORN ISLAND where a strip already existed. The distance from HORN ISLAND to Port Moresby is virtually equal to that from IRON RANGE to Port Moresby, so clearly when IRON RANGE was available it sometimes became the point of departure from the Australian mainland for aircraft proceeding to Port Moresby.

During the 1930s, gold mining on the then Batavia Goldfield (renamed Wenlock in 1938) and Iron Range area itself expanded justifying construction of a rudimentary road to the coast and calling for a jetty at Portland Roads by which shipping could better service the activity. The ***Queensland Mining Journal*** Volume 37 in 1936 carried an account by S R L Shepherd touching on these transport difficulties -

'...The road over the granite ridges is in fair condition, but due to the prolonged wet season, parts of the road in the scrub and some of the creek crossings require attention. During any wet weather the road for 2 miles from Iron Range is impassible for motor trucks, but loading is either hauled by Caterpillar tractor or packed on horses. At the Port arrangements for landing goods, especially heavy machinery, need improvement. The population of the field is now slightly over 100 including about 12 women. Of this total about 85 are at the Iron Range centre.'

Shepherd reported again around 1938 -

'...Since my previous visit in 1936 the population of the field has been reduced to 50, including 6 or 7 women. At Portland Roads a jetty 810' long to deep water was being constructed and nearing completion. At the end of the jetty the wharf is 60' long by 30' wide with a shed connected by tramline to the beach. This convenience will be of considerable benefit both to Claudie River Goldfield and this section of the peninsula generally.'

AWC drawings dated 16 January 1943, detailing then planned additions to this structure survive in AA.

Also in the 1930s aviation was making its presence felt throughout the country and Portland Roads did not escape. The British aviation magazine ***Flight***, published in 1938 a review of air transport services in Australia as at 8 August 1938. The legendary Tom McDonald's North Queensland Airways is listed with a 'fortnightly DH-84 Dragon service Cairns-Portland Roads'.

In March 1940 when the RAAF was seeking a better site for a drome in the COEN area, sites closer to the better weather of the coast were examined. On the Aerodrome Works file for COEN there is evidence of identification and appraisal of a site at IRON RANGE by Mr V (Val) Augenson, an officer of the Department of Civil Aviation. His 1 April 1940 note states -

'...in my opinion it would be alright to give an assurance to the RAAF that an ELG... suitable for use in all weathers could be prepared on the site.'

His drawing Z-1137 showed this site. It indicated two narrower strips 5 chains (i.e. 330') wide might be prepared, but because winds were consistently ESE-SE, he considered the one strip would be sufficient. Augenson's estimated cost was £1 500 -

'...estimates do not provide for any approach road or telephone to the site from Iron Range where there is a small Radio set... if preparation is carried out in dry season, no difficulty should be experienced in using the old Portland Roads to Wenlock road... some creek crossings might require a bit of temporary corduroy with timber. I think that the mining community at Iron Range would clear a more direct track to the site if it (the airstrip) is prepared.'

A RAAF Survey Negative Register held by AUSLIG in Canberra lists some oblique photography dating back to 27 September 1936. While I have not located any surviving prints, the register shows 'FP877' was a single oblique with title 'Claude River (proposed landing ground)' and later, on the same roll (8085-3) the register lists 'FP879-881', three obliques titled 'Bare Hill landing ground'. The last three clearly relate to the small airstrip cleared at the then Lockhart River Mission Station adjacent Second Red Rocky Point some 25 miles south of Portland Roads. The 'Claude River' listing is clearly a mis-spelling of Claudie and apparently relates to the same site as covered in Augenson's 1940 drawing, Z-1137.

Probably the earliest attention given IRON RANGE by the American Forces was an aerial reconnaissance flown by Walter Maiersperger on or about 18 April 1942. Walter recalls -

'...I had been ordered to make a survey of the coastal airfields (of Queensland) to determine their suitability for B-26 operations. On this trip I landed with the B-26 at Moresby, Horn Island and Coen and surveyed the area of Iron Range from the air only as no construction had yet started there... I never landed at IR and yes, all I saw from the air (and still see it) was the rainforest and the river, not a big one, winding down to the coast.'

A tattered print of the Queensland 4-mile Series map dated 1941 attached to a report from HQ NEA to Air Board Melbourne dated 28 May 1942, survives in AA. It carries markings to indicate a 'selected area' just east of Claudie River. The report, submitted under the signature of Air Commodore Lukis then AOC NEA includes -

'...P-O Trench made an inspection of the country in the vicinity of Portland Roads. This inspection was made in conjunction with Col. Mills and Capt Cox of the USA Forces. Portland Roads... (has a) jetty... constructed in '39 and in good condition. This extends over the reef into... water at least 12' at low tide and 20' at high tide... no suitable site for an aerodrome was located (near the jetty). There is however, a fairly extensive

area just to the west of Iron Range and east of the Claudie River where several strips could be constructed on sound country... At present, access to area selected is rather bad. The road from the jetty to Iron Range is followed for about 9 miles and then there is an old road leading to the area itself. It will be necessary however to relocate and form the road for the whole distance in order to make it easily trafficable under all conditions... The selected site is situated on slightly undulating country providing good drainage... it will be necessary to gravel the runways. It is suggested that each runway be cleared, formed and gravelled, and if possible sealed... gravel (is) obtainable in the vicinity... after sealing, underground side drains will be necessary. Owing to the extremely heavy rainfall, a certain amount of local flooding is bound to occur during the wet season... Plentiful supply of water... in Claudie River,... suitable camp site..., available ½ mile from this river.... Practically the entire white population of the district has been evacuated... Army have also removed all useable motor vehicles and mining machinery. The only means of communication is by the Wireless installation at Iron Range, but at present there is no Operator available. Restoration Beach situated immediately south of Cape Weymouth is suitable for landing any type of light aircraft at low tide. Map showing the proposed sites is enclosed.'

Capt Herman G Cox, then Works officer of HQ AAF in Townsville, submitted an independent report on the reconnaissance. His recommendations were -

'a) that site No 1 be developed into major drome and sites 4 and 5 used for dispersal strips if this location be used for AOB

b) that immediate steps be made to prepare jetty for demolition in case of emergency, as this is the only Port now known to exist between Cooktown and Horn Island

c) that radio set now located at Iron Range and owned by State of Queensland be moved to a look-out post in vicinity of Jetty for defence as well as communication purposes. This set was of great convenience recently when several US aircraft were forced down some 40 miles to north (of) Portland Roads.'

Site No 1 became CLAUDIE strip and GORDON strip appears to have been site No 4.

Jim Trench's account to me in 1989 expanded on this:

'...The first night out was spent at Coen, where there was a RAAF Operational Base. The plan then was for the aircraft to endeavour to set down on a beach a few miles south of Portland Roads... There was thought to be a trafficable road from Coen to Wenlock where there was a small civil airstrip and hopefully a track from there to Portland Roads, whose other name was Weymouth Bay...

A detachment from Coen would set out by motor truck to meet us at Portland Roads if both parties were able to make the trip. We started out by air next morning but after about two hours one of the Rapide's two engines failed and it was necessary to return to Coen. It was quite worrying looking out at the propeller of one engine which was stationary and then at the ground which was very rough... a sort of peneplain with small hills with steep sides and flat tops a few acres in extent. I think we all wondered whether, if the other engine stopped, the flat tops of the hills would be big enough to land on. Luckily we never did find out. The return trip

IRON RANGE *(1944)*
This combined extract from the Lloyd Bay and Cape Weymouth sheets in WW2 period (1 inch = 1 mile) Military Survey mapping shows three of the four strips constructed at IRON RANGE. The first one was at Grid Reference 711728 in the area shown as 'Rifle Range'. It was the first strip cleared so that communications aircraft at least could land in the early construction period. The 120° 'Gordon Strip' on this map is the only one in use today. *[Department of Defence]*

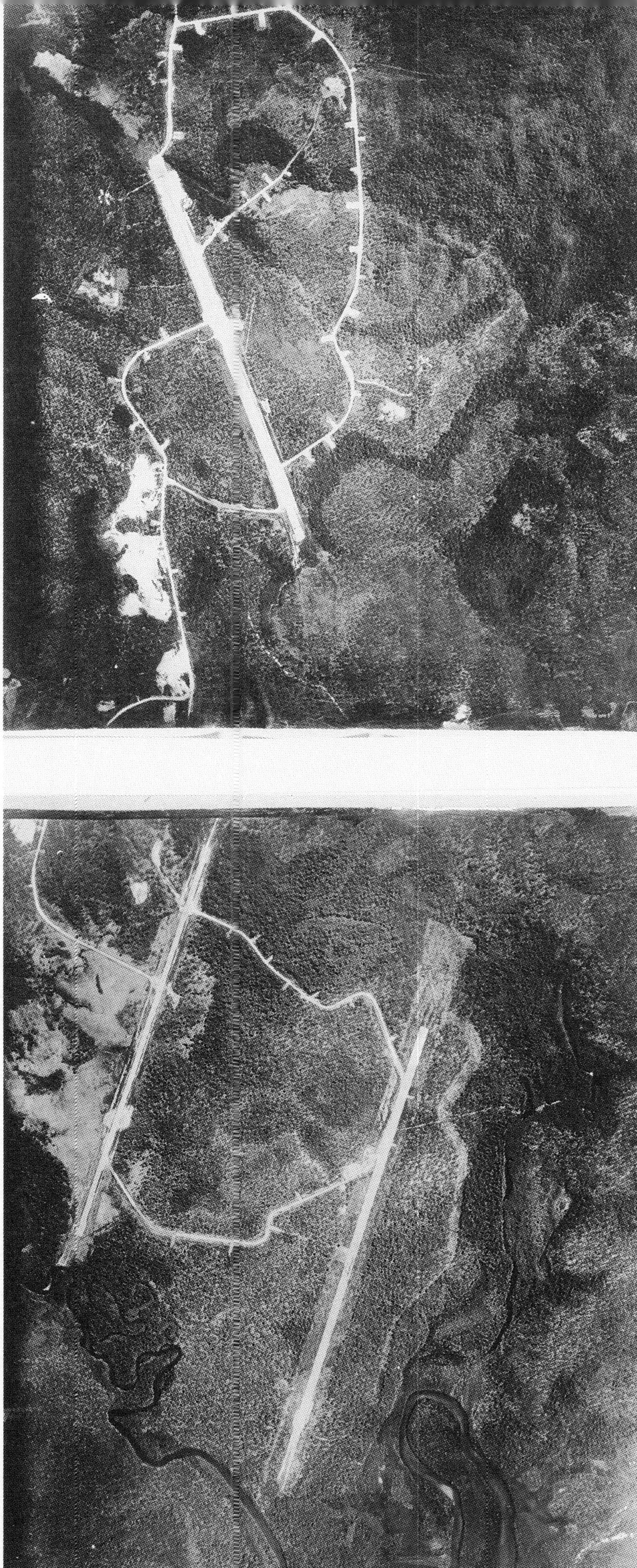

IRON RANGE (1944-6?)
'Pin point' vertical of GORDON (top) shows layout of taxiways and dispersal bays. At bottom is 'pin point' of the parallel strips, OLD CLAUDIE connected via taxiway to NEW CLAUDIE. OLD CLAUDIE proved flood-prone and work ceased on NEW CLAUDIE before it was completed. [AA CRS A705/1 Item 171/94/167]

took about an hour longer than the forward journey. A mechanic was flown up from Cairns, the Rapide being a civil aircraft. It was declared fit about 10 a.m. next morning although the mechanic's statement that the aircraft "seemed to be all right" was not completely reassuring.

It was necessary to arrive at the beach where it was proposed to land, at low tide, which we did. There was a black line running along the centre of the beach parallel to the sea and the pilot could not make up his mind whether it was seaweed or rocks. He decided to land between the black line and the sea. Everything went well until the plane started to slow down, when the tail swung inland and the aircraft slowly made its way into the sea. This was exciting enough until the tips of the propellers hit the top of the water and the whole aeroplane was deluged. However the pilot, with I think considerable skill, managed to manoeuvre the plane back onto the beach. It happened 47 years ago but I can still remember the look in the faces of the other passengers when we started out to sea and the look of relief when we pulled up on the beach. No doubt my face showed similar emotions.

We then set off for Portland Roads thought to be about 5 miles away across what showed on the Queensland 4 mile map as blank. Actually our path was intersected by a number of creeks bounded by mangrove swamps and no doubt infested by thousands of snakes although I don't remember seeing any. The walk which we thought would take one to two hours actually took four or five.

We were met at Portland Roads by the Harbour Master and his assistant although there was no sign of shipping or of any other inhabitants of the area. We were pleased to see the jetty still intact. Before leaving Townsville we were warned that the Army intended to blow up the jetty and if they arrived while we were there we were to prevent them... The Coen detachment arrived that evening with everything apparently in order. The Harbour Master advised that the only person available who knew anything about the Iron Range area was Mr Jack Gordon who had a small Iron mine in the area. The mine had been closed down by the Army and the machinery removed apparently with the object of preventing the Japs setting up an Iron industry there. Jack Gordon had set out from Portland Roads to walk the 18 miles to the mine where he still lived the day before. We drove to Iron Range first thing next morning and found Jack in residence. He advised that he knew an area a few miles away that he thought would be very suitable. The only trouble was there was no road and he had sore feet due to his walk from Portland Roads. Anyway he agreed to give it a go. After three or four hours fast walking we arrived at the area which, on investigation, seemed to be a very suitable site for the proposed airfield with two ridges that could be developed into runways. After walking briskly back to the Iron mine I think every member of the party wondered how far we would have walked if Jack's feet had not been sore. We returned to Portland Roads that evening.

With the help of an Aborigine we made our way back to the Rapide by a far more direct way than we arrived. Without a word the Aborigine led us through a sheet of water about half a mile wide and about 5' deep in the middle. We took off without trouble and we managed to land, also without trouble, at the proposed airfield site. After a few hours further investigation the plane dropped us off at the small civil strip at Wenlock. We had a wait of about an hour when we were picked up by the regular Saturday afternoon service from Horn Island to Cairns. The strip was covered with small anthills about 18" high. We spent the hour knocking some of them over. However there was no trouble there and the plane arrived I think on time.

IRON RANGE *(15/8/1990)*
Chili Beach looking south to Albatross Cove. This is where Bill Hill landed Jim Trench's party on 22 May 1942. The idyllic palm tree studded beaches to the Pacific Ocean hide the fact that airmen based at IRON RANGE found it such an inhospitable environment. That's not hard to understand as their camps and the airstrip, their place of work, were some '5 miles from the nearest beach'. [H Wirth]

IRON RANGE *ca(-/10/1942)*
DeHavilland DH-89 Dragon Rapide VH-UU0 on charter from Airlines of Australia (AOA) to US Engineers on the CLAUDIE STRIP. Brice Allen recalled he once did the round trip to Brisbane to get a 'payroll for the 46 Engineers'. This is the same type aircraft in which Bill Hill landed Jim Trench's party on Chili Beach three miles south of Portland Roads. [W (Brice) Allen ex US Engineers]

IRON RANGE *(-/9/1942)*
Disembarking from the Liberty? ship Jesse Applegate at Portland Roads Jetty. David Pratt was the only Australian member of the 33 BS advance party which sailed from Townsville to set up camp at IRON RANGE. He recalls being ***'at anchor two or three days before some Dutch ships vacated the wharf allowing our ship to tie up'****. [D Pratt ex RAAF]*

I visited Iron Range with Col. Mills a few weeks later and there were two American battalions of Pioneer Engineers hard at work with all the equipment in the world. In about three months two 7000′ sealed runways and 13 miles of sealed taxiways had been completed. I next visited Iron Range in January 1943 in the middle of the wet season and at that stage only 3000′ of runway was serviceable for light aircraft and none of the 13 miles as sealed taxiway. The rest had completely failed due to lack of drainage. American Pioneer battalions could run up a serviceable runway in the matter of days. The idea was, having done that, someone else would take over if the runways were required for long term use.'

The day after Jim's party returned to Townsville, the AOC presided over a meeting attended by Col. Matthews US Army Engineers, RAAF officers W Cdr Dale DWB Air Board, F-Lt Dawkins Area Works Officer, P-O Trench NEA, P-O Campbell Air Board, and Mr E Doig Dept of Interior. The current Priority list was discussed and the AOC's directions included:

'...Construction of Defence Bases in NQ: clarification is (sought) as to whether supervision is to be RAAF or by US Engineers. Sites have been selected at Mareeba, where No 100 Squadron is now camped, Wrotham Park, a new strip 3 miles west of Cooktown, an aerodrome in the vicinity of Portland Road and Dispersal strips on the top of Cape York. It is desired that all... have very high priority and it has been directed that a Battalion of US Engineers proceed as soon as possible to Portland Road in order that work may commence there at an early date. It should further be pointed out that, at direction of Col. Matthews work has commenced at Wrotham Park and Cooktown.'

The AOC continued with an urgent plea to have,

'...further plant for aerodrome construction forwarded to this area... At the moment there is insufficient equipment... and at Giru and Upper Ross coloured US Engineer Corp troops are working with practically no equipment at all and it is necessary for them to do a tremendous amount of tree felling by hand... this Area is making strenuous efforts to send further equipment for aerodrome construction to Port Moresby where the need also is very acute. There is no spare equipment at all in this area and your advice... is requested.'

As always, communication between IRON RANGE construction forces and NEA was vital. The RAAF's 26 Operational Base Unit (OBU) was formed and attached to the US 46 Engineer Regiment's party headed for Portland Roads. This advance party boarded Burns Philp's SS *Wandana* in Townsville on 7 June 1942. By the evening of 10 June, 26 OBU was in wireless communication with Townsville.

Several veterans of the 46 Engineers have provided historical detail they've credited as 'from the file(s) of John McFee', also of their Regiment. This shows 'movements' as:

'...June 14 - Companys A, B and C embarked from Townsville (aboard the MS ***Dona Nati*** *a medium cargo vessel manned by a Philippino crew) en route to Portland Roads Qld (Iron Range Operational Base) arriving there June 16, 1942.*
June 23 - Company A sailed from Portland Roads (on the MS ***Dona Nati****) and arrived at Horn Island on June 24.*
July 10 - HQs & Service (H & S) Company and Company D left Townsville and arrived at Portland Roads July 12.
October 27 - HQs Detachment left Townsville after operating there since June 1 1942 and arrived at Iron Range Operational Base October 29.'

Archive evidence is clear that on 19 June 1942 the strategic requirement was, 'Portland Roads - new strips to provide for one light bombardment group'. This starting point coupled with the general appreciation of the heavily

IRON RANGE *ca(-/10/1942)*
Two US 22 BG men photographed at the extreme southern end of CLAUDIE strip. Looking north along the unsealed strip, 'Nest' hill is seen on the western side. David Pratt, who snapped the photo, was an Australian (RAAF) co-pilot assigned to the 33 BS flying B-26 'Marauders'. *[D Pratt ex RAAF]*

jungled area as easily meeting camouflage needs, was to spell disaster some months later when the utilisation of IRON RANGE turned to heavy bombers. The B-24 Liberator's wing span at 110′ was much more than the 65′ of the B-26 Martin Marauder for which the strips were initially cleared - 250′ wide.

The 46 Engineers' archive material includes a drawing titled - ***IRON RANGE (progress) as of November 2, 1942.*** It shows both the 120° and 160° runways, by legend, gravelled and the southern half of 160° runway 'prime coated only'. All the taxiways are shown cleared with most gravelled and about 80% prime coated.

Three camp sites, each shown for 400 men, appear between the two strips and a little to the west. An earth fill dam is indicated on the river, level with the northern end of Claudie Strip. The progress notes say -

'...Estimated completion dates - 160° strip November 4th, 120° strip November 22nd, taxiways and bays December 5. This... assumes ideal conditions. Controlling factor is screenings. Delay... very likely, particularly if wet weather...'

Several photographs are included taken by the Regimental photographer. One dated 14 July, shows road clearing and log culvert construction. Two photographs showing bridge construction are dated 28 August 1942. The text continues -

'...Company B, in addition to improvements to road and bridges, strengthened the jetty by placing additional stringers between the pilings and reinforced the floor boards. After supplies and equipment finally reached the campsite, Company C commenced clearing the dense jungle area for (extension of the) "plain strip".'

This was where W H (Bill) Hill landed the chartered ANA Rapide on 24 May 1942 with Jim Trench's party. This extension facilitated communications air traffic. Interestingly, much later in 1963 the Australian Defence Forces mounted what was termed 'Operation Blowdown' only yards northeast of this site. This low level air burst of a large conventional explosive charge was sited near to a range of military equipment to study the blast protection afforded by the jungle.

The 46 Engineers' history continues -

'...Work was (soon) abandoned (there) and efforts diverted to the clearing for the Gordon strip, named after the famous gold prospector, Jack Gordon, who operated a gold mine nearby. A lack of equipment (meant) clearing this site by hand with small tractors and trucks.

On July 12 1942, H & S Company and Company D which had been making final improvements on the runways at Reid River, arrived at Portland Roads aboard the ***Dona Nati****. The ship not having sufficient crew to man the winches to unload the equipment, H & S Company furnished winch operators and in conjunction with Company D, unloaded the ship. The equipment was loaded onto a lighter and then unloaded from the lighter onto the jetty.'*

With this tedious procedure, it is easy to see the subsequent move to double the wharf deck area as in the January 1943 AWC drawings -

'...Along with the normal equipment of the two companies, several Caterpillar D8 bulldozers, five Caterpillar D12 road graders, and five heavy Sconfier rooters were also brought along. A few weeks later four 12 yard "carry-alls" arrived. This was the first time any units of the regiment had sufficient heavy equipment to do the jobs assigned to them... On their way from the jetty to Iron Range, the bulldozers and graders, widened and improved the road to a width of 25′.

H & S Company commenced operating as a Line Company and in conjunction with Companys C and D, completed the clearing for the Gordon Strip and started construction on dispersal roads and parking bays. Clearing of the Claudie strip was started shortly thereafter.

Word had reached HQ 1st Battalion that the entire project at Iron Range Operational Base may be abandoned, but due to the foresightedness and initiative of Major Herman W Blackwood, Battalion Commander, work was continued on the runways and dispersal strips. This action saved valuable time and enabled planes to begin operations from this base much sooner than would have been possible otherwise.

Work was rushed to complete the paving and to make the Gordon Strip ready for use as soon as possible. Dispersals were also rushed to completion.'

A cypher message dated 14 August 1942 from NEA to AAF Brisbane announced -

'...FOLLOWING FROM IRON RANGE PROGRESS REPORT. BUILDING MESS HALL OPERATIONS FIVE ROOMS LATRINE 12 CAPACITY SHOWER AND WASH ALL COMPLETED WATER INSTALLATIONS MADE FIELD BAKER SERVICE AVAILABLE BUTCHER PERSONNEL PRESENT, HERD EN ROUTE. STRIP 6 800′ COMPLETE AND READY FOR LANDING ALTHOUGH FINAL SURFACINGS NOT FINISHED. GASOLINE IN DRUMS AVAILABLE. ROAD 2 TRACKS GRADED FROM PORTLAND ROADS TO CAMP AND STRIP NOT COMPLETE BUT READILY PASSABLE. PIER OUTSIDE AT PORTLAND ROADS SINGLE TRACK PILE TRESTLES 16 TON CAPACITY 16′ WATER AT LOW TIDE.'

With the practice of words all typed in capitals and full stops sparingly used, such messages were cryptic but welcome at HQ.

The Engineers' history carries on -

'...On September 9, 1942, the 19th BS of the 22nd BG flying Martin Marauders (B-26), arrived at Iron Range and began immediate operations from the Gordon Strip. The 33rd BS of the same group arrived two weeks later. During the last week of September 1942, a squadron of Flying Fortresses (B-17s) of the 43rd BG began operations from this strip. On November 7, 1942, one squadron of Liberators (B-24s) of the 90th BG had arrived.'

Dave Pratt, an Australian, was one of several RAAF pilots seconded to the 22 BG as co-pilots in May 1942. The move served two purposes. It allowed US co-pilots to leave established units and use their combat experience to seed fresh crews from new arrivals. It was also tangible evidence of the Australian - US cooperation needed in the Pacific War. History shows that Lt-Gen George Kenney, in forming all USAAF under his command into the new

IRON RANGE (1944-6?)
Viewed from the north the GORDON strip (runway 120°) is pictured with loop taxiway 'E' in the foreground. In the background can be seen the OLD CLAUDIE strip in the middle and parallel and away to the east (left) is the NEW CLAUDIE strip. [AA CRS A705/1 Item 171/94/167]

US 5 AF in August 1942, was about to separate these front seat pairs, though not to damage the friendships, some of which last to this day.

Dave was with the 33 BS. He recalled his arrival around 18 September 1942 -

'...I was sent up - the only Aussie with the USA ground crew, in charge of the Aussie gear, per the Liberty ship ***Jesse Applegate****. We landed at Portland Roads, boarded trucks and travelled over half finished road to Iron Range. The two strips were unfinished by far.*
Some aircraft started to use the northern strip - I came out as a passenger on a B-24 on October 11, 1942...'

RAAF Historical notes, US veteran sources and the 46 Engineers' history combine to record a landing mishap at the northwest end of GORDON on 13 September 1942. Returning from a mission in New Guinea, one of the B-26s, serial 40-1433 'Kansas Comet' piloted by Walt Krell, landed short shearing off the right landing gear strut against an isolated termite mound. Out of control, the aircraft crashed into a compressor truck parked on the western side of the strip. The RAAF co-pilot, P-O G Robertson and a 46 Engineers C Company man on the ground private R G Rodrigues, were killed. The aircraft burned. Krell was hospitalised for several months after his valiant attempts to extricate the trapped co-pilot.

On bitumen sealing of the strips, the 46 Engineers' history continues -

'...It was a problem to find a suitable source of material from which gravel or crushed rock could be obtained for use in the sealing of the landing strips and dispersal roads. Gravel from a creek bed containing 50 to 60% clay and waste was the only material available. An improvised screening plant was constructed and operated until a portable (one) was brought in... At the stock piles, grizzlies were constructed and after sufficient quantity of material had accumulated the aggregate was loaded into fan-tailed trucks which spread the material uniformly on hot bitumen laid down by a bitumen distributor. The tar and bitumen was heated by seven 400 gallon and one 1100 gallon bitumen heaters. This proved to be hot and disagreeable, but Company D did a creditable job of heating tar and bitumen.'

Various photographs record these sealing operations on CLAUDIE strip and serve to date its completion around mid November 1942. Photos also show -

'...the huge drainage project in connection with the construction of Claudie Runway. Drainage tile (8" diameter earthenware) was placed in a ditch on each side of the runway the full length of the strip. The ditch was then backfilled with (broken metal and creek) gravel, compacted and levelled off.'

Rough bush timber was used for drainage culvert construction and for some of the 'building work'. The men improvised further, with -

'...A small sawmill plant (which) cut lumber for most of the buildings constructed at Iron Range. The plant was constructed by Company C shortly after (it) arrived at Iron Range. The (log) carriage... was formerly an ore car used in the Gordon Mine located about 6 miles from the campsite. The wheels were removed from the ore car, the axle extended and placed on the frame. The track (rails)... was also procured at the Gordon Gold Mine.'

Tales also exist of improvisation in producing 'moonshine' spirits. Dried fruits as part of the mess rations are said to have rarely seen the mess table. While not necessarily connected, the following appears in the 46 Engineers' history -

'...November 16 1942, a Board of Officers was appointed at Iron Range... to investigate the deaths and illness of fourteen members of this command, seven of whom died. It was... discovered that the cause was from ptomaine poisoning, medical term botulism. Six were from Company B and one from Company C.'

IRON RANGE (-/10/1942)
'One of our dispersal bays on the new strip at Iron Range'*, is the handwritten caption on this photograph by Franklin Allen (19 BS, 22 BG). The meagre clearance of the dispersals in the rain forest is evident in this classic shot of a B-26 'warming up'. Brice Allen (46 Engineers) recalled* ***'riding in one of these on a test flight in November 1942. We landed real fast on Claudie Strip and just managed to stop before hitting the rock crusher in the river!'****.*
[F Allen ex USAAF via Jeanne Newell]

IRON RANGE (9/11/1942)
'Bitumen kettles' - oil fired heaters, let into the ground so that drums of solid bitumen can be rolled for immersion. Caption/identification on the photo indicates it's one of a record series taken by the 46 Engineers' Regimental photographer. Location is believed to be between GORDON and CLAUDIE. [W (Brice) Allen ex US Engineers]

IRON RANGE (24/3/1943)
The Liberty ship Payne Wingate loading at the Portland Roads Jetty. This was a period of embarkation for various Army and Aircorps Ground Echelon elements. Visible in this photograph on the extreme right are three shallow draught landing barges moored to a floating pontoon. They were likely the most favoured form of quick supply transfer from ship to shore. Although not very clear, just beyond the right hand six-wheeler truck, there is what appears to be the top of a US staff car. It must have been brought ashore from shipping and doubtless taken out by ship as it would have had difficulty making it by road to Portland Roads in those days. That's so even today! After the war, the jetty was little used. It was demolished in the 1970s. *[Neg SC 223516 US National Archives]*

The AWC was a civilian organisation charged with the coordination of all war effort construction. Where this involved work in or near a potential war zone, it was reluctant to schedule its Civil Construction Corp (CCC) workers. The AWC feared for their safety both as unarmed civilians and 'armed' as some advocates urged. In this context, AWC in a 13 October 1942 query to RAAF DWO Brisbane, raised the matter of maintenance after the airfield construction, under direct control of USAFIA, was complete -

'...PEI Branch of MRC has a civil construction force assisting USA Engineers in building Portland Roads airfield. This (PEI) organisation built Coen airfield, completion of which has been abandoned. In accordance with an arrangement made by AWC with USAFIA and RAAF, a gang of civil workers is to be left on each airfield in the combat zone for... maintenance purposes.'

AWC wanted clarification on responsibility for ration supply to such civilians. RAAF Brisbane reacted in surprise as it understood such maintenance work would be carried out by RAAF Works and Maintenance Units (WMU).

Departure of the 46 Engineers from IRON RANGE is noted in the history, as -

'...November 19, 1942, Company's B and C, the medical detachment and a small detachment of H & S left Iron Range aboard the US liberty ship ***Willie van de Vanter*** *and arrived at Port Moresby on November 20... H & S and Company D embarked aboard the Dutch ship* ***Vandersun*** *on 2 January 1943 and arrived at Port Moresby on January 6.'*

Some US veterans recall journeying to Australia in 1942, on the *Queen Mary* and the *Queen Elizabeth*, both a far cry from the likes of the *Vandersun*. Some provincial Queensland historical circles carry an anecdotal account of the birth of IRON RANGE airfield. It is claimed therein that US troops were ferried ashore from the *Queen Mary* standing out of sight off the Great Barrier Reef. Recognised published accounts put the two famous troop ships half a world away at that time, doing anything but standing still in open sea.

George Coble was Engineer in Lt Paul Johnson's crew aboard B-24D 41-23719 'Hellzapoppin' on its way back to IRON RANGE around mid-day on 15 November 1942. George counts this as Mission No 1 in the diary he kept of his early days with the 319 BS in SWPA. This was the 90 BG's first combat mission, aircraft staging through Port Moresby. Returning direct from Bougainville, they made landfall -

'...Went up the coast 'til we got to our base about 3 or 4 miles in - base is called Iron Range - we lost one ship off the southern coast of New Guinea... The 400 BS lost a ship right on our coast here. Landed on the beach here, all crew OK - gas trouble. We landed at 1300 hours.'

Tom Fetter was bombardier on the 400 BS aircraft which landed wheels up on the beach just east of the airfield. In his letter of 29 July 1992, Tom referred to his flight log and explained -

'...The plane's number was 41-23824... The flight engineer was Rudy Vetter... the assistant flight engineer was Oscar "Slow John" Sjolin, who is still with us and my warm friend. The co-pilot was Lewis Jones... I talked with Sjolin twice on the phone and also with Col. Harry Bullis who was the CO of the 400 BS at the time and also took part in the mission. Both Sjolin and I remember that we were aware we would probably ditch for quite some time and that we threw out everything we could to lighten the ship. We had a bomb bay tank and Rudy could not get the gasoline to transfer. The starboard engines were out and the landing was made with the wheels up. The picture you have, was probably made the next day and with the tide out. Dale (Thornhill, the pilot) had lost so much altitude that we had no choice but to put down. There was gasoline in the bomb bay tank. There was a controversy about this crash with some of the ground crew taking the position that Rudy should have been able to transfer the gas. Oscar Sjolin cannot recall the details of this flap, nor can Col. Bullis. Col. Bullis did recall that he was unhappy about the loss. I cannot resolve this controversy other than to point out that Rudy was experienced and had been through the drill before...'

Jim Dudley was 'in the maintenance' side of the US 22 BG's 33 BS at IRON RANGE. That Group's 19 BS shared GORDON strip with the B-24s of the newly arrived 90 BG while its 33 Squadron appears to have used CLAUDIE. The ground echelon (maintenance) of the 90 BG had not yet arrived at IRON RANGE and their absence placed extra strain on the flight crews. Jim and several other surviving veterans remember the calamitous night of 16-17 November 1942.

Consider IRON RANGE at this time. CLAUDIE was being bitumen sealed. GORDON strip was still gravel probably requiring considerable watering effort to control dust, and water was in short supply. A November 2 status plan shows 48 dispersal bays, most gravel surfaced. The 22 BG had nominally two Squadrons in residence with around 20 Marauders. The 90 BG on paper, appears to have had three of its four Squadrons with an estimated 20 Liberators on the airfield.

At least one B-17 of the 43 BG was there too, which meant full utilisation of all gravelled bays. With limited use of CLAUDIE during sealing, it is easy to believe veterans' recollections of aircraft being parked on both sides of the GORDON strip. Remember, this runway was about 100′ wide in the centre of a 250-300′ cleared strip. With a B-24D parked nose facing the runway, the clearance to the wing tip of another taking off would be about three feet! Allowing that the planes may have been parked 'tail into the trees' and that the pilot would try to use the offside of the runway, it's still precision flying.

IRON RANGE (16/11/1942)
*Frank Allen's handwritten caption to this photograph reads - **'This joker forgot to transfer gas when 5 miles from the field. Result: forced landing 3 miles from field with 800 gals. of gas.'** The site is Quintel Beach opposite the present day Lockhart River Aboriginal Community, with Lloyd Island faintly showing on extreme right. The B-24 was flown by Capt Dale Thornhill of the 400 BS, 90 BG. The US 28 Service Squadron personnel laboured for some 36 hours straight in an effort to recover the aircraft. In the end most of it was lost. Thornhill's crew, as with most of the 90 BG were 'green'. This was the Group's first bombing mission, whereas Frank Allen's 22 BG were relative veterans. However, Thornhill and his crew had flown a B-24 across the Pacific, and as Wiley Woods, 90 BG Historian says, **'you can't do that without "transferring gas"'**. [F Allen ex USAAF via Jeanne Newell]*

IRON RANGE (-/12/1942?)
90 BG airmen in their tropical camp area at IRON RANGE. Their smiles certainly belie the recollections they now have of this Operational Base. From left to right: Cliff Davis, Johnny Johnson and Jim McAteer. W E (Johnny) Johnson was one of three crew members of B-24D '50 Cal Gal' who persevered and successfully completed pilot training after their tour of combat from IRON RANGE and New Guinea. [J McAteer ex USAAF]

Enter 5 AF Bomber Command with orders for the 90 BG to strike Rabaul **direct** from IRON RANGE in the dawn of 17 November 1942. This meant take-off around midnight, or put another way it called for precision flying of fully loaded aircraft from a narrow congested strip by relatively 'green' crews, at night. Surviving records available to me are few and veterans' recollections, even those of eyewitnesses, suffer the usual variations. The formal 'Attack Report' written two days after by Capt R S Holt, the Intelligence Officer of 320 BS, concentrated on that Squadron's involvement and says in part -

'...The mission was planned as a joint operation of three squadrons under direct command of the (Group) Commander. The operation was based on total tactical strength of the 320th, consisting of 7 airplanes with combat crews plus 4 each airplanes and crews of the 319th and 400th BSs. Plan of operation was to take-off at intervals commencing 2300 hours EAST 16/11/42, airplanes to cruise at various indicated air speeds (from 150 to 190) so as to arrive over the target at 0400 EAST 17 November 1942.

Briefing commenced at 1800 EAST 16 November 1942 and continued for various elements of the combat crews until shortly before designated take-off time.

IRON RANGE ca(-/12/1942)
Probably photographed looking south on CLAUDIE strip, this visiting LB-30 of the US 7th AF (Hawaii) is parked nose-on to the edge of the bitumen seal. A magnifying glass on the original print shows serial AL626 and the name on nose as 'Southern Cross Airways No 3'. Bob Livingstone's several years research on the B-24 type has explained its background as an early USAAF transport pioneer following in Kingsford-Smith's path over the Pacific. [W (Brice) Allen ex US Engineers]

Navigators, bombardiers, radio operators, etc. were briefed severally at different times and places. This resulted in inevitable delays in preparing the airplanes of the 320th for take-off. It had been necessary to install bomb-bay tanks in the airplanes, load bombs, service the planes, and taxi them to suitable positions for take-off on the taxi strip along the runway; with only two maintenance men per airplane available, the non availability of combat crews on the line during the major part of the evening militated against an orderly and thorough preparation of equipment for the mission. The first airplane took off as scheduled at 2300 hours, EAST 16/11 the next airplane did not take off until 2314, substantial confusion existed thereafter, due to the fact that certain airplanes with assigned number in order of take-off were not ready, no method of communication on the line had been worked out, no method of control in dispatching was arranged; hence pilots did not know whether to taxi out to the runway or not. The first airplanes took off with landing lights; those that followed did not, and accordingly came dangerously close to the airplanes parked along the taxi strip on east side of runway. Runway lights were placed too far apart, leaving inadequate space between the east string (which the pilots had inevitably to follow due to their position in the cockpit) and the airplanes parked on the taxi strip. The affair culminated in the disastrous crash of the 11th airplane (piloted by Lt Larsen (sic) of the 400th) as already reported by Group. Remaining airplanes, including two of the 320th failed to take off. One of these (41-23751) was slightly, the other (41-23765) materially (damaged), in the crash of Lt Larsen's airplane.'

Two things were clear by the time Holt wrote his report. The first plane to take-off, 41-11902 'Punjab', had vanished without trace taking the CO of 320 BS Major Raymond S Morse, his crew, and the Group Commander Col. Arthur W Meehan, who was along as co-pilot for the mission. Secondly, the 90 BG, new to combat, had a serious morale problem.

Coincidentally, Capt Holt had set out to go on the mission too. Leroy Iverson, at the controls of 41-23751 'Big Emma' had Holt on board as an observer. Iverson explained that because of a problem in fuel transfer, he had to abort. Holt quickly exited to climb aboard Larson's plane. When it crashed Holt apparently 'felt cool air, made for it and hid behind a B-17 before the B-24's gas tanks exploded'. He was the only survivor.

Capt Holt understandably addressed the mission from his own 320 BS view point. The 90 BG (Group) 'report' to which Holt refers is not available to me, if in fact, it still exists. Such information would possibly disclose the serial(s) of any other aircraft damaged in the crash, along with total loss of life at the GORDON strip that night. Photographic evidence of damage to aircraft apparently waiting to take-off makes it incredible that all the crew would escape!

William R (Bill) Eaton had been at Pearl Harbor December 1941, spent time in Australia in first the 19 BG then the 403 BS of the 43 BG. Bill's diary mention of 21 November 1942 on return to IRON RANGE, included -

'...A B-24 blew up (here) on take-off the other day and killed several men. The crash totally ruined 4 other airplanes on the runway and killed or injured several bystanders...'

Bill Moran was a Photo Interpreter attached to S-2 in the 90 BG while at IRON RANGE -

'...I arrived at Townsville about 11/42 and flew up to Iron Range as a member of the Advanced Echelon. We arrived before most of the ground crews did, only to find our aircraft had already been involved in combat with the enemy... as to the conditions at Iron Range. It seems it was always raining,... Mildew seeped into everything... The food was "bully beef" in every which-way possible to cook. The coffee (?) was something else, and what dehydrated foods we had was impossible to eat... Not far from us, the Aussie outfits had erected a bakery, in which they made bread... but not for us... I can assure you the aroma from the bakery was mouth watering...'

Bill Moran also mentioned going down onto the beach and on occasions having the use of an outrigger canoe and blasting fish with grenades. He added -

'...The aboriginals on the beach with their fires generally got to retrieve and eat most of the fish...
My most notable memory of Iron Range was the night Lt Larson's aircraft blew up on contact with another B-24. A mate and I had gone down to the airstrip to watch take-off but the dust and darkness made it almost impossible. When the accident happened we were far enough away not to be hurt. Early next day we went back and were involved in the filling of body bags.'

On further questioning Bill made two points. Dust generated on take-off with full throttle and flaps took 'about 15 minutes to clear'. He recalled a 'lineman' sitting on the wing of a stationary aircraft had been struck and killed, and that the total loss of life was likely fifteen. Several veterans recalled being warned away from the immediate disaster scene on the basis that they 'weren't needed there'. Group morale had plunged and some comment that they never were told this total loss of life.

Jack Atkinson, in his unpublished autobiography titled ***Aircraft Assistant to Chief Engineer and Airport Manager 1931-1956***, mentions -

'...It was in 1942 that I received an order from Brisbane to report to Colonel Bertrandias, the Commanding Officer of all American Forces in the North. I had no idea for what reason, but was told to take my tools. It turned out that he wanted a Liberator bomber salvaged at Iron Range by building 8' of nose section of wood and fabric covered so as to fly it south for permanent repairs. They gave me open order on their stores, they loaded the material I required into a B-25 Bomber together with myself and two off-siders, an apprentice named Robert Merrotsy and a Tradesman's Assistant named Harold Blake, both from ANA and off we went to Iron Range and despite repeated requests from my wife as to my whereabouts, no one would disclose the

IRON RANGE *(11/1942)*
Severe nose damage to a B-24 very likely caused by the wing of another aircraft impacting this aircraft while taking off. The caption on the rear face of this original photograph, in Frank Allen's handwriting reads - ***'One night a B-24 got a little off the strip on take-off at Iron Range - result 3 B-24s and one B-17, one gas dump and 12 men.'*** *[F Allen ex USAAF via Jeanne Newell]*

information. Anyhow after being billeted with the Yanks for 4 weeks, the job was completed and the plane flew south and later returned into action. This job was at a time when bombers were doing nightly bombing runs to New Guinea from Iron Range and it was a common sight to see dead and wounded being unloaded on return, i.e. if they did so.'

I later located Robert Merrotsy. He certainly knew Jack but explained he'd joined the RAAF mid-1942.

Jack Atkinson does not know the serial of the 'Lib' he helped mend. A 'nose only' photo appears in a 1984 Reunion Booklet produced by the 47/479 Service Squadrons veterans' organisation. In the context of its 1942 designation, as 28 Service Sqn, this event is described as -

'...Another job at Iron Range requiring tenacity of purpose in the overcoming of many and difficult obstacles was B-24 airplane, 41-23812. This airplane was damaged severely from the cockpit forward. It was first decided to salvage it. Consequently, the bomb squadrons immediately and indiscriminately started stripping it of all instruments and almost all hydraulic units. Instrument connections were cut off as were most radio parts. Inspection by ASC Rep APO 922, requested repair at Iron Range. Three Australians from AHA (ANA actually!) were sent to build false nose section. All instruments and missing parts had to be rounded up from various squadrons. The inspector, S-Sgt Pizzuto and Meaders, did an exceptionally fine job in this respect. Tail section, damaged in the explosion of the airplane that damaged the nose, was replaced from an airplane that landed on a beach. Great hydraulic trouble was encountered. With the experienced help of Mr Zebora, Consolidated representative, [the] airplane was put in a flyable condition. Capt Scott, 90th BG, showed cool courage and faith in our workmanship, by ferrying the airplane some thousand miles south to the depot.'

The Service Squadron's booklet touches on Thornhill's B-24 on the beach -

'...At the same time, (as the incident in which 41-23812 was damaged) a B-24 made a crash landing on the beach, five miles from the strip. Many men worked 36 hours straight in an attempt to save the airplane. However only the engines and some parts not damaged by salt water were saved because of insufficient foundation for the jacks. There is no doubt that this airplane could have been saved had we a C-2 wrecker.

All serviceable parts, were very well used...'

Paperwork took a backseat with much of the early 5 AF frontline effort. Records now are such that the serial identity of Jack Atkinson's 'Lib' may never be known. The 47/479 Service Squadron booklet shows 41-23812 with Capt Scott flying it South. Bill Haggerty, Line Chief for the 319 BS tells me it was 41-23714 '8-Ball' with John Arant flying it south. From Capt R S Holt's report. there is a good chance it was 41-23765 'materially damaged' that night. Apparently this aircraft was later adopted by Col. Arthur H Rogers, who was in the forefront of efforts to improve the frontal defensive fire power of the B-24D. The fact is that '1 000 miles south of Iron Range', Brig-Gen Connell's ASC bought Art Rogers' idea. Their 81 ADG through the 'Repair Section' at ARCHERFIELD, fitted a tail turret from a wrecked B-24 to the nose of Rogers' 41-23765. It was promptly named 'Connell's Special' and went back to New Guinea early March 1943 to prove the idea in combat. Rogers' leadership of the 90 BG gave it the name, 'The Jolly Rogers'. The veterans, to this day claim the title 'Best Damn Heavy Bomb Group in the World'.

One of the 320 BS pilots, Charles Hesse arrived at IRON RANGE around 22 November 1942 after a hospital spell in Brisbane -

'...The enlisted personnel on my flight crew had not yet arrived; they were coming by boat from Hawaii. In the meantime, I was appointed Mess Officer... my biggest problem... was finding a way to dispose of the garbage. It seemed that... the men who were in charge of this detail were simply taking it by the truck load to some place 5 or 10 miles from camp and dumping it into a huge hole. When the Aussies discovered what we were doing, they let us know that if we didn't stop that immediately, we'd be in real trouble. Next, we tried burning it, with the help of gasoline. That took forever. Thanksgiving was 26 November, and we had a dinner to be remembered: turkey, cranberries, and all the rest. On 29 November our ground troops arrived along with the rest of my flight crew and on 1 December I was relieved of my duty as temporary Mess Officer.

On 3 December I was assigned my first flight in more than a month. Our mission was to fly a search for Lt Norman Crosson and his crew who had failed to return from an attack mission against four Japanese destroyers on 1 December. Our search was to concentrate on the coast of New Guinea from Hood Point to Milne Bay. We saw no signs of his plane, "Little Eva". On 4 December we flew from Iron Range to Garbutt airdrome just west of Townsville. Lt Foster was in charge of the huge repair depot there. The following day, we flew back to Iron Range, carrying 32 sacks of mail and an ample supply of liquor for the officer's club... The rainy season was expected to hit our part of Australia at any time soon, and sure enough, we experienced a huge downpour on 10 Dec. At 1300 hours on that day I completed my 24 hour job as airdrome officer; what a pain in the neck that job was... On the evening of 12 December, various people in the 90th BG put on a talent show that was quite good. On 14 December we flew down to Townsville, stopping first at Mareeba to pick up Col. Roger Ramey, CO of the 43 BG, and other passengers. At Townsville, I ate ice creams and drank milk shakes as if they were going out of style. The food at Garbutt field was excellent. I visited Lt John Bily at the hospital in Townsville. The next day I met Lt Foster at the warehouse at Garbutt field and made arrangements to pick up officer's mess chests, footlockers, etc. We flew back to Iron Range through much rain, carrying 16 bags of mail, ten cases of liquor and Chaplain Bebe's portable organ. This day I received three letters from home; one took only 18 days to get to Iron Range from Chicago'.

IRON RANGE ca(-/12/1942)
'The home of the 19th Bomb Squadron, 22nd Bomb Group for 8 months at Iron Range. This picture shows the whole squadron area.' With these few words Frank Allen said it all. There would clearly have been little need for camouflage in this camp area! [F Allen ex USAAF via Jeanne Newell]

A couple of days later, the 90 BG was to lose a B-24 to this bad weather on the same leg. By noon on 18 December, 41-23825 'Texas Terror' and twelve men were overdue at IRON RANGE. The aircraft was brand new, on delivery to the Group. Lt James Gumaer his four crewmen and the seven passengers they picked up at GARBUTT remained 'missing' until the end of 1943.

Forty years later, armed with my National Parks camping permit, a 'mud map' drawn by my good friend (the late) Donald Hector Johnson, and the eager company of my youthful sons, I set out to climb to the wreck on Hinchinbrook Island's Mt Stralock. Several rolls of bright coloured plastic survey tape and almost the same length of 'Wait-a-While' scratches later, we conceded defeat. Looking out over the passage to Lucinda Point with Ingham in the distance, I realised we'd strayed out of 'Johnson's ravine'. I resolved that next time I'd make sure Don came too.

Six months before his untimely death Don published ***Torres Strait to Coral Sea.*** He described Gumaer's passengers on the 'last day of their war' -

'...most senior of them was Col. Carroll G Riggs, a West Pointer commanding the 197th Coastal Artillery. He had held the appointment since 26 June, when the Regiment had been deployed for anti-aircraft defence around Perth, Western Australia. The Regiment was now fulfilling the same function in Townsville, but Batteries "B" and "E" had been deployed to Iron Range. Col. Riggs was paying his first visit to his detachments. Accompanying him was Lt Raymond F Dakin, also of the 197th, carrying money for the gunners, who had not received any pay since their detachment in August.

Capts Peter E Kiple and Carl H Silber were both members of the 8th FG, which was then stationed in New Guinea. Lt John E Cooper jr was on attachment to the 19th BS, 22 BG, one of the medium bomber squadrons at the Claudie strip near Iron Range. The last member of the services to board the aircraft was Technician 4th Grade Michael M Goldstrop of the 1156th QM Co. One civilian completed the passenger list. He was Robert C Trevithick, a representative of the Pratt & Whitney Division of the United Aircraft Corp, whose motors powered the B-24.'

Don went on to report -

'...late 1943... two aborigines searching the gullies on the flank of Mt Stralock for stream tin, reported (finding) some burned US currency. Two experienced rock climbers were engaged to search for the wreck.'

At GARBUTT, in mid December Charles Hesse learned -

'...Lt Crosson who we had searched for on 3 December had been found and was in the hospital at Burkeville, wherever that may be. Also, two of his crew were known to be dead and six were still missing.'

Hesse's 'Burkeville' was Burketown near the Gulf of Carpentaria coast. Remnants of the crashed and part burned 'Little Eva' remain near there to this day. The epic 110 day survival of crew member S-Sgt Grady Gaston has been addressed in Terry Gwynn-Jones' ***Heroic Australian Air Stories.***

Heimo Raski was in H & S Company of the 46 EGS Regiment. He recalls involvement of some Australians -

'...We had quite a number of Aussies in the advance party to Iron Range. An amusing thing, though not at the time, was an Aussie who tried to be helpful when I

IRON RANGE *(-/9/1942)*
'Morningstar and Crell at mess Iron Range September 1942' *is Frank Allen's handwritten caption to this photograph. The uncleared area in which this unit was camped is evident. The open air mess table is clearly two planks laid on some support.* *[F Allen ex USAAF via Jeanne Newell]*

was washing my clothes. Since the clothes were extremely dirty, I soaked them in gasoline, rinsed them off the best I could and put them in a half barrel... I went to a creek to get water when my Aussie friend came by and he thought I went to find some matches, so he was going to do me a favour and light the fire. My clothes caught on fire and I wound up wearing half Aussie half US clothes until new ones arrived... I was in charge of heavy equipment and I remember several incidents where we were called upon to help with planes that were stuck in the mud or ran off the runway.'

Dale Kruger was in one of the US Service Squadrons maintaining aircraft at IRON RANGE. He wrote -

'...We had two closed trailers for a machine shop and we mostly devised or made our own shop buildings - at Iron Range it rained and rained so we had it pretty miserable... One time out of the blue here came 2 Aussies and 3 of the natives driving several ranch cattle through the jungle. We couldn't believe our eyes. Here we were in a war area and these guys didn't seem to care.'

They were probably headed for the 'slaughter yard', grid reference 786688 on the Military Survey map.

Jack Atkinson had difficulty dating his four weeks at IRON RANGE until his wife reminded him that he was home just in time for Christmas (1942). Jack did recall two other matters -

'...I bet you did not find in the Archives in Canberra any record of the "battle of Iron Range" when one night wild pigs ran amuk among the audience at an open air picture show. Not only Yanks but there were also a lot of RAAF personnel there too. Some bright spark yelled "Japs" and the panic button was hit and a stampede was started. The COs had a lot to say the next morning.

I also was there when 2 refuelling crew men were sitting on top of a huge mobile tanker and one lit a cigarette and naturally they well and truly cooked up and died instantly.'

As part of his fourth mission George Coble's diary describes his introduction to the static electricity phenomenon of St Elmo's fire. They had departed IRON RANGE around midnight -

'...We hit stormy weather just on take-off. We had to go almost half the way on oxygen, whenever we dropped below 1500′ we would get into stormy weather... then I encountered a new experience - St Elmo's fire - the whole ship was bathed with it. My top turret was completely covered with streaks of lightning, my gun barrels were glowing and sparks jumping off my guns. The props had a complete circle around them, the pitot tubes were shooting out a blue flame almost a foot long, the wing tips and aerials had fireballs on them.'

Claudie River had ceased running in November, and December brought expected storms. With CLAUDIE strip sealed but flood prone and GORDON unsealed, it's not surprising to find the following comment dated 9 December 1942, on a RAAF file for HORN ISLAND -

'...In view of the shortage of civilian labour in this locality and as this aerodrome (Horn Island) may be required for heavy bombers in the event of trouble occurring with the runways at Iron Range, it is suggested this work be undertaken by the section of Works Maintenance Units stationed on this aerodrome...'

With the 46 Engineers set to depart in December 1942, a formal request was made to Allied Works Council to complete the works at IRON RANGE. Naturally AWC caused a review to be made of the work to that date. A detailed report dated 23 December 1942 appears on RAAF files in AA, saying in part -

'...The reported state of the airfield... has come from notes by MRC, Northern Engineer's Rep and particularly from information obtained from the PEI Branch Foreman Mr Moulden, who has been on the job with the civil force assisting the USA 46th Engineers since inception of the job and who left the job on approved leave on 11 December 1942.'

The technical description of the two strips is invaluable. Drainage culverts are recorded as built with mostly tea tree round bush timbers and some open ended 44 gallon drums welded together. Importantly, GORDON strip width is recorded as 250′ with CLAUDIE described as 250′-300′. It confirms the surfaces as, CLAUDIE - 'sealed' and GORDON - 'gravel, very thin - about 2" ' (with) US Engineers about to bitumen surface it. AWC was sensibly pressing for instruction as to whether a minimum cost, short life runway surface was required or something more permanent.

Of CLAUDIE, a warning is sounded -

'...This strip is in wet low country and it's considered that Claudie River overflow will submerge one end for some considerable distance.'

George Coble's diary sets the Christmas scene -

'...On the alert, the day before Christmas - nothing happened. Had a Christmas, what a Christmas for this neck of the woods. Had church services December 24 at 7.30. Went off the alert 1600 hours. December 25 had liquor for the boys. What a surprise! Nearly everyone got tight. On December 26 we were told we were to take off at 2230, (direct, to attack Rabaul - again) had supper at 2100 hours, went to ship (719) taxied out to runway ready for take off. Wet rainy night, the second ship that took off, was right wing heavy and no more than got off the runway than it crashed into the trees. The ship burned, three big explosions from bombs going off. It was terrible, all the crew was killed. One boy lived 1½ hours, these things really get on your nerves. Hope if I get mine it will be like Frandsen, on combat. Command, after the fire asked the pilots if they still wanted to go. Several, including ours, Lt P C Johnson, volunteered. He, our pilot, told us anyone who did not want to go, did not have to. Of course no one backed out, none of this crew will ever back out on Johnson. We took off at 00 hours December 27. Had bad weather all the way over. Flew at 15-17 000′ most of the way over. Had to chew oxygen most of the time. No 2 engine acting up...'

And so it continues. This is the occasion when Roy Kendrick and his crew were lost. Again, take-off was from the northwest end of GORDON, at night. It's likely the B-24s to follow Kendrick were again waiting on the left side. Kendrick probably set out to keep to the right!

Charles Hesse in the 320 BS also recorded Christmas -

'...That evening I attended a Protestant communion service conducted by Chaplain Bebe... received a gift package from the American Red Cross... and best of all, eleven letters from family and friends.'

The wet rainy nights like the 26-27 December were settling in. Hesse added -

'...By (January) 6, the rains had become so severe and so heavy that we were compelled to move our B-24s to Claudie strip, a weather proof strip. We had to taxi the planes there on a road which connected the two runways. I was assigned to taxi airplane 094 of the 400 squadron to the new strip. We stationed a man sitting on the wing near each tip to holler if it appeared we were to hit something. With less than 100 yards to go to the point where I was to park the plane, my left wing tip struck a wire that was supporting a post. It all but sheared off the wing tip; apparently, the man sitting near the left wing tip had been shouting his head off for me to stop, but I didn't hear him with the engine noise and all. I wasn't the only casualty. Lt Gottke had to swerve suddenly to dodge a GI truck that drove onto the taxiway from a side road. His right landing gear got onto the soft shoulder and the wheel began settling rapidly. It took about an hour to get him towed out. That was a close call; for a moment I thought we might lose the airplane. We flew to Port Moresby on 7 January.'

Camouflage was easily solved in the earliest days of IRON RANGE in that camp areas for example, were so primitively organised in quite dense tree cover. When the Americans moved on to New Guinea the RAAF's 26 OBU to some extent had the pick of the various camp sites. On file in AA is an interesting report dated 10 March 1943 prepared by W Bryan, a camouflage adviser, after inspection of a new location for the OBU (in fact, the old US 28 Service Squadron's camp) between the two strips. In a clear reference to the broken outline of roofing for buildings, the report includes -

'..RAAF personnel are building the camp themselves and wavey eaves are incorporated in the standard drawings being used. It is suggested that, out of possible surplus iron, wavey eaves be added to the existing American buildings in accordance with a design left on the job.'

By 20 March 1943 there was some confusion in RAAF circles as to what the future for IRON RANGE really was. If it was to be used for heavy bombers, should 26 OBU act as a servicing party for the 5 AF units concerned, or would the US provide their own? The RAAF saw 26 OBU possibly being better deployed elsewhere, say at JACKY JACKY.

In wartime, with its inevitable censorship and martial authority, truth often suffers while rumours grow to legend. Most Japanese 'aircraft sightings' defy logic in respect of their operating radius. Certainly expeditionary landings were most feared. 'RAAF Historical' notes for IRON RANGE, mention that -

'...on 13 April 1943, a report was received (by 26 OBU) that a party of Japanese had landed on the Gulf coast. On 17 April a party of 12 Guards and a detachment from 13 Garrison Battalion were sent to patrol Tozer's Gap, only to return 2 days later without sighting anything suspicious.'

IRON RANGE (1944?)
This is Dom McBride's description of RAAF Signals Station at IRON RANGE. **'Connected to the Overland Telegraph Line, it had about five wireless receiving sets (including AR7's). The transmitter was located a short distance away. The building had four rooms including "a bit of a kitchen", a Cypher Decoding Room and the Wireless Room. Located in dense rainforest some miles from 26 OBU Base Camp and 200 yards off the Portland Roads road this was home to the RAAF Signals Unit.'** *Location may have been Grid Reference 722775, and the name of the architect is unknown.* *[D McBride ex RAAF]*

Commonwealth Security Service files in AA, contain a 22 April 1943 advice by a Northern Command Intelligence Officer (IO) to his HQ Brisbane. It reads, in part -

'...The following is a copy of a signal message brought to IO Townsville by Capt R Crosby, Sig Officer, Yorkforce, and it is forwarded for your information:-

"SECRET. FOLLOWING SIGNAL FROM 13 GRN BN TIMED 2115 20 APRIL READS 197 CA A/A BTY REPORTS OFFR OPERATING SCR 194 WALKIE TALKIE RANGE APPROX 25 MILES INTERCEPTED TWO WAY TALK IN JAPANESE 1930 HOURS 20 APR ON 39 MEGACYCLES. COULD HEAR CLICK OF SET AS SWITCHED FROM TRANSMIT TO LISTEN. ENDEAVOURING TO LOCATE WITH RAAF. IN CASE OPERATING ON LAND REQUEST SUITABLE MONITOR BE FLOWN IRON RANGE IMMEDIATELY. ACK. ENDS."

Capt Crosby requested direction as to the Section in Townsville most likely to be interested in the report.

Capt Crosby was taken to the RAAF Intel and thence to that Section of the RAAF which keeps watch on this type of communications. This Section is most hush-hush and does not disclose its designation.'

About a month after the first report of 'Japanese radio intercepts', the Deputy Director of Security, Qld, Lt-Col Wake advised his superiors in Canberra that -

'...Corporal-Investigator A M McDonald and Mr Richards of the US CIC Cairns are proceeding to Iron Range to conduct an investigation...'

He was later to regret 'the futility of sending non technical men on investigations of a semi-technical nature'. However, McDonald's report dated 24 May 1943, gives interesting background.

Flares had been reported by personnel of the 197 Coast Artillery US Army and efforts had been made to check the reliability of those reports. The 13 Garrison Battalion 1 Aust Army CMF sent out patrols and generally reported no confirmation of the alleged sightings. To test this the CO of the Battalion, Major Cunningham, arranged for flares to be let off unannounced at different positions and times. These arranged flares were seen. Generally the reports of flares coincided with periods of brightest moonlight. However, after discussion with 'responsible persons' the investigators concluded 'that the sighting of flares were without foundation'.

On the subject of two-way radio conversation in language thought to be Japanese again, because of conflicting opinions encountered and the belief 'that would be feasible under favourable atmospheric conditions', it was concluded that 'enquiries did not disclose any reason for the belief that any transmission was made from the area'.

The investigators commented briefly on 'OBJECTS WASHED ASHORE'. Reports at the time related to two mines 'presumably our own' which had been suitably attended to and an unidentified object which turned out to be the belly tank of a fighter aircraft.

The CCC presence came under scrutiny, on two counts. The first related to industrial unrest resulting apparently from some men recently transferred to IRON RANGE from Tolga on the Atherton Tableland. It seemed their views on the 'travelling time' condition of the O'Mara Award had influenced most of the men to rebel. Mr C H (Clem) Wilson the AWC Resident Engineer 'signalled a request for an Industrial magistrate to visit... 18 May'. Apparently there was no further development by 21 May when the investigators departed IRON RANGE.

A J (Jack) Bond, more widely known in the MRC/AWC as 'Bondie' had a reputation for improvisation and 'getting the job done'. His area embraced the Atherton Tableland and his secretary during those WW2 days Mrs Laurie Wiemers recalled

'...Townsville MRC asked Bondie to send someone to get the work going again. He sent Jim O'Reilly of Ravenshoe whose vocabulary was most colourful - colourful enough to get the men back to work in double quick time.'

The lack of censorship on outgoing mail from CCC personnel caused the investigators some concern for 'security of information'. Their report included -

'...Seven Aboriginals and half-castes, some of whom had been previously employed by CCC recently returned from Mapoon Mission on the Gulf side of the Peninsular and were re-employed. It's understood that they freely discussed Australian and US troop concentrations in the Wyper River area and there does not appear to be any form of restriction on the periodic movements of this type of itinerants.'

With strict censorship of all Services' mail, the investigators suggestion for at least a trial period of censorship of CCC mail from IRON RANGE was adopted.

Civilians left in the area also attracted scrutiny. They numbered seven whites and a few others. Some of these Aboriginals were among the natives who dispersed from Lockhart River Mission at the time of the first Japanese raid on Horn Island -

'...Owing to the failure of Government rations to arrive, a short time ago a large number of natives have again "gone bush" for sustenance in the south of the area.'

McDonald concluded his investigation report with some comments on 'general conditions' apparent at IRON RANGE -

'..Relations between US and Australian troops, all ranks, were on a good footing and that they cooperated to their mutual advantage. The American 18th Station Hospital serves all services and the CCC... Adverse reports regarding the morale of troops of some of the US Army units remaining in the area, at the time of the supposed Japanese landing near Galbraith Station on the Gulf side, were heard from several sources. The impression was gained that, in one unit, poor morale was, in some measure, due to the tendency of other formations transferring their culls as replacements to units in "out of touch" areas. As the Iron Range airstrip has been vacated by service aircraft at present, the control of air transport is in the hands of two enlisted men of the USATC Air Transport Command and there is no check of passengers alighting. Those disembarking from ships at Portland Roads are checked by the guards from US Port Detachment A on duty on the jetty.'

The Deputy Director of Security for Queensland Lt-Col R F B Wake, in a letter dated 29 June 1943 to DG Security Canberra, reviewed the matter -

'...It now appears that conversations were heard over an extended period of approximately two or three weeks, at about the same time, and for an hour on end. At times when challenged the keys would click, this of course, may have been coincidence, but as it apparently occurred on several occasions, it is unlikely that it was coincidence.

Whilst it is difficult to assess with any degree of reliability the distance of Iron Range from the transmitting sets, it would be necessary for them to be within 50 miles in order to hear the challenge which was made on an ordinary "walkey talkey".

The foregoing I consider would discount the theory of Japanese aircraft, but not necessarily discount the possibility of two Japanese submarines conversing with each other. It will be recalled that at about this time one Japanese submarine was known to be south of the area and therefore in my estimation adds to the possibility of a two-way conversation between an agent on land and a submarine on reconnaissance, for the purpose of speaking to the agent...

Attention is invited to the fact that these conversations took place about the time of the report of Japanese landing in the vicinity of Staaten River on the east coast of the Gulf of Carpentaria. Although the rumour was out of all proportion to alleged events, it seems to have had as a basis a young Australian aboriginal seeing something strange in the area and which was probably a submarine. Of course there may be no connection between the conversations and the alleged Japanese submarine in the Gulf, but it is a link worthwhile connecting at the present stage'

Apparently a submarine was sighted 47 miles NNE of Great Palm Island on April 21 1943, the day after the first conversation was heard at IRON RANGE.

During the 1930s, Japanese fishing and pearling boats are known to have landed parties on the mainland for reasons of sustenance. Certainly such contact with isolated residents and Aboriginals occurred, in some cases to engage them as crew. Allied authorities were understandably cautious about the loyalty of such remote residents and Aboriginals where they were known to have been disenchanted.

A Security Service report into alleged espionage activity in COOKTOWN, dated 7 April 1944, in AA, makes now most humorous reading in its oblique reference to a Lockhart River Mission official transiting south through COOKTOWN -

'..Rev Father James Edwards. A permit appears in this man's name to "visit Cairns en route to Sydney on instructions from his Bishop". Permit dated 18 April 1943. Identity card number N7-M91-24529.

He is alleged to be from the Lockhart River Mission (a Protestant organisation).

He is quite unknown in Cooktown and I suggest his history is checked. Why a Roman Catholic priest should be at the Protestant Lockhart River Mission and proceed from there to Sydney on instructions from his Bishop is not clear.'

What is clear is the investigating officer came most likely from Sydney or Melbourne where 'Father' was synonymous with Roman Catholicism only.

In July of 1943 the Intelligence Officer at Cairns advised his HQ in Brisbane of the following extract of a letter which came to his notice on 12 July, written by a CCC worker at IRON RANGE -

'...I don't know if I mentioned in previous letter about (blank) finding about 12 ton of rice and 1 ton of dried fish and a lot of other gear all done up in drums, on the coast just north of here - first they thought it was wreckage from the Bismark Sea battle, but now they think the Japs landed it quietly at night, perhaps for future use. The drums all branded "Burma".'

The report carried on with related information -

'...a civilian miner named "Densley" had reported on 15 June to 13 Grn Bn Iron Range, that on 31 May 1943 he had seen strewn on the beach about 40 to 50 drums containing rice and fish. There was also a quantity of rifle ammunition, semi-rimless, approx. .303 in sealed cases.'

Lt-Col Wake Deputy Director of Security for Queensland wrote on 20 July 1943 to his Canberra HQ on this subject of IRON RANGE He added -

'..Alas I thought this was news, but find to my horror that the DSO Cairns, possessed similar knowledge about "Flotsam and Jetsam" of this type being washed up along the shores of Portland Roads, but had not reported same.'

Attached was this excerpt from a censored letter apparently written by a CCC man at IRON RANGE -

'Two old timers on their way down from the Cape to the hospital found a Jap boat hidden in the Olive River just up the coast a bit from here. It had about 40 tons of rice on it, some tea, candles, powdered milk and salted meat. No doubt it has been used as a supply depot for one or more subs and here's hoping that they come back for some more in the near future, because I am sure they will get a pretty warm reception. The last

week we have had a hell of a lot of new planes and squadrons passing through here and Tojo is going to get some sting shortly. The Kittyhawks are better and faster than the ones we used to see down there and the Australian made Boomerang is a wow! It makes the old Cobras look like they are in reverse or standing still and talk about guns, gee they sure have their share. I could say a lot about military matters that are big news up here, but you understand how that is.'

Col. H R Cole, for Rear Echelon USAFFE wrote to RAAF DWB, 24 June 1943 announcing a submission to expand IRON RANGE. The letter sought formal Commonwealth Government Ministerial approval and provision of funds in the context of an urgent requirement of USASOS -

'...The project includes the construction of a new 160° strip and dispersal area for four heavy bombardment squadrons...'

Approval came on 12 July 1943, care being taken to emphasise this was an 'American project.'

In essence this build-up aimed at 'forward' yet safe mainland facilities for long range US bombers, a requirement which was to recede with the successful blockading of Rabaul and the gains of MacArthur's island hopping campaign.

Around 14 July 1943 AWC files carry mention of a Lt Parker 'an engineering officer maintained at IRON RANGE by USASOS as their inspecting officer'. Lt Parker had been asked to cooperate with MRC resident engineer Mr C H Wilson on various matters pertaining to reconciliation and dissection of costs. While I've not been able to verify, I believe this Lt Parker is likely to be Lt John C Parker Jr who, back about August 1942 was involved in the supervision of US work at CHARLEVILLE and later in the Rockhampton district.

Owen Callard was a member of 26 OBU at IRON RANGE from August 1943 to January 1944. He is somewhat typical of RAAF veterans of the period who recall it as an inactive airfield. Owen's duties 'in stores and equipment' did not place him at the strip proper. During an infrequent visit to the strip, he did at least recall one aircraft there, 'it was called "Bucket of Bolts" '.

Dominic (Dom) McBride trained as a Wireless Operator at Point Cook Victoria. After qualifying he was sent to Townsville where there was a 'pool', a base from which personnel were posted for service overseas, mainly New Guinea. He recalled -

'...I duly received advice of my posting to some unit in Hollandia, but as I was about to board a DC3 at Garbutt, some clown discovered that I was still only 18 years old and thus was barred from overseas service. As a "punishment" I found myself several days later being off-loaded at a place I had never heard of - Iron Range. This would have been about September 1943.

The aircraft which took me there immediately took off and I walked with my kit to a small hut situated on the southern side of the Gordon Strip. There I met the Duty Corporal - a pleasant young man by the name of Ted Smiley, I think. He offered his commiserations but seemed convinced that I must have done something very wrong to be sent there. There were no aircraft to be seen and no sign of any habitation whatsoever.

A telephone call was made and subsequently a jeep appeared out of thick bush and I was driven to my home of the next 13 months - a small camp of timber huts and American style army tents, situated in a dense rainforest off the SW end of the airstrip. My new unit as best I remember was 26 OBU. It numbered less than 40 men. Our CO was F-Lt Brian Chirlian, a journalist..., 2 i/c was F-O "Swampy" Reed a solicitor from Eastwood, Sydney. The cook, a genius with bully beef, was "Derby" Munro from Newcastle. We had a total of 3 wireless operators to maintain a 24 hour "watch" 7 days a week. The other 2 were I think, Sam Last from Sydney and Jack Matthews from Tamworth. A number of cypher decoders, transport drivers, armourers, guards, general hands, medical orderly, refuellers and mechanics (including wireless mechanics) made up the numbers. I was the youngest of this "family"...

Our role was to provide weather information, wireless relays and refuelling to aircraft flying from Townsville to New Guinea. Weather information was given to us by a lonely little group of Americans who had a small camp by the airstrip. They even played cricket with us. I can only recall the surname of one of them - Manning. We had one visit by a team of cricketers - most of whom had played for Australia, O'Reilly, McCabe etc. The Captain I think was a Jack Chegwyn. They didn't like Iron Range. We told rather lurid tales of snakes etc.

A small Army bakery about 3 men - provided fresh bread for a time and at Portland Roads there was a small detachment of Army - I think they spent a lot of time fishing. Somewhere in the bush near the strip there was a camp of CCC (Civil Construction Corps). They maintained the roads and airstrip and also built a beautiful big camp for a long since disappeared American Army. It was never used and we were not allowed to move in. It was burnt to the ground by a bush fire about the middle of 1944.

Our Wireless hut was a tin shed situated about 3 or 4 miles from our main camp. It too was in a dense rainforest... I still recall with real dread, sitting in that shed in a pitch black night entirely by myself with the rain pelting down during the wet season. One was literally armed to the teeth - a tommy gun and a machete close by, God help any poor soul who could have wandered in unannounced. We were trigger happy.

There was a linesman, Bill Fernley, who maintained the section of the Overland Telegraph which ran through Iron Range. He drove a jeep and was a real bushman. We regarded him with some awe. I thought him to be quite old but he was probably aged about 40. He was also a crack shot and a hair-raising driver. Our supplies were brought in by a small ship the ***Santa Maria****. It was lost with all crew in a cyclone in early 1944. We lost all our tents and most of our clothing. We just moved into the mess hut and lived there. The rain continued for weeks and weeks. After that I think everything came in by DC3. The mail plane was an ANA Dragon Rapide VH UVT.*

Wireless reception was always bad due we were told, to the Iron in Iron Range. Static was severe and we were fortunate to have the Overland Telegraph. We often used the Overland Telegraph to speak to RAAF in Townsville. This was on a much more personal level than morse code. The Sergeant in charge at Townsville (located in the Townsville convent) was Syd Kennedy who after the war worked with me at the Registrar Generals Office. I often spoke to a Gloria Robins who came from Texas Qld, I never met her but still

IRON RANGE (15/8/1990)
This is the OLD CLAUDIE strip viewed from its northern end with residual bitumen seal still evident. The Lockhart River Aboriginal Community have in recent times taken advantage of this cleared line to locate one of their runs of barbed wire fencing for a cattle rearing venture. It is still possible to visualise the narrow clearing that must have represented a hazardous operational airstrip. [Author]

remember her voice received in a dark little place in North Queensland. Many aircraft used the strip in transit... There were a few terrible accidents - one in particular involving a Kittyhawk and a Thunderbolt. The pilot of the P-40 was killed. (25 February 1944 as related in ***The Secret Airfields : RAAF News December 1990****; by R K Piper)...*
There were two other airstrips, fighter strips I believe. They were then well overgrown and rather inaccessible. Because of wild pigs and snakes we did not venture into kunai grassed areas.
We often went down to Lloyd Bay just near the mouth of the Claudie River. There were huge boulders in the sea there. We swam there completely unaware of the existence of stingers but very wary of sharks and crocs. We always had a sentry sitting on a rock who looked out for such predators. He would be armed with a .303 with armour piercing bullets... We blew up fish with hand grenades and cooked them on the beach. Crayfish at certain times of the year were in abundance near the long jetty at Portland Roads...
For some reason our "picture theatre" was situated about half-way between our camp and Portland Roads. It was only a sloping hillside by the road. We went there on the hottest nights and the wettest nights - travelling in open trucks with some very crazy drivers. The films of that vintage contained no sex and very little violence but certainly an abundance of American patriotism and malarky. As we returned to our camp we drove up the airstrip and it was quite amazing to see the number of snakes wriggling across the airstrip and seemingly heading for our camp. The drivers had considerable fun trying to skid their wheels on the reptiles. Iron Range fostered in me a real fear of snakes - they even got into our tents as did scorpions and other creatures. Three people were killed by snakes while I was at Iron Range - a stockman and two aboriginal children...
Last year Greg my son flew me up to Iron Range and left me for some hours in the care of the gentleman who runs the canteen at the Lockhart River Aboriginal Community. He kindly drove me around but I could find no trace of my former home. The place has changed - much bush has been removed.
I have quite a memory of our camp area, even to a stage and a boxing ring we built. We probably thought a concert party might visit us - it never did. George Huxtable who was about the same size as Mike Tyson was the "brains' behind the boxing ring. He didn't get any contenders for his unofficial crown of heavyweight champion of Iron Range. Life was hard enough without wearing broken noses, black eyes and cauliflower ears.'

The RAAF took over all construction and maintenance responsibility connected with military air installations at IRON RANGE at the same time as CHARTERS TOWERS and MAREEBA and were effectively totally in command there 30 June 1944 onwards.

The Operations Record Book (ORB) for 26 OBU, shows that during June 1944, a record number of 154 aircraft were handled on GORDON strip - RAAF, US and Civil. Also during June, the 3.7" A-A defences were withdrawn, and work was commenced on the installation of Aeradio, which became operational mid July.

MINUTE 52 - dated 11 August 1944 on the RAAF's IRON RANGE aerodrome works file reads, in part -

'..Iron Range is not now required as an operational base but, nevertheless, may be used from time to time by aircraft proceeding up and down the coastal route... Gordon Strip (only) is to be maintained in serviceable condition.'

On 11 August 1945, a few days before War's end a brief report of an inspection dated 12 July 1945 filtered through to the then Department of Works and Housing in Brisbane. Mr L G Redmond, Aerodrome engineer for that Department wrote -

IRON RANGE *(-/4/1991)*
This recent view replicates the wartime shot of GORDON strip seen from the north. With Lloyd Bay clear in background it's now virtually impossible to see the parallel strips of OLD and NEW CLAUDIE. [Courtesy R Rudd Northair Surveys Mareeba]

'..At Iron Range there has been constructed 4 landing strips. One 3 300' long, the Gordon Strip 6 000'x100', the Old Claudie Strip 6 800'x100' and the new Claudie Strip 7 200'x150'. The 3 300' strip was abandoned at an early stage of the Japanese war. The old Claudie Strip is flooded during heavy wet seasons and now is abandoned. Owing to the reduced activity in this area, the Gordon Strip is the only one being used at present. It is in excellent condition...'

Redmond quite rightly addressed the road between the aerodrome and Portland Roads. While he was not able to inspect it personally, he reported it to be -

'...in very poor condition. Many culverts washed out and the road badly broken up and corrugated. RAAF personnel are experiencing great difficulty in hauling fuel to the airstrip...'

He added that -

'...no funds were available to repair this road. RAAF had not assumed responsibility for its maintenance as a great portion of it was a mining access road and State allocations of funds were insufficient to carry out repairs.'

By way of explanation of cost overruns in the early postwar efforts to settle accounts, an Air Board Agendum dated 14 June 1946, states, in part -

'...Conditions in North Queensland at the time this work was carried out, were generally speaking subject to many delays and set backs due to the remote locality and infrequent calls by shipping... This was particularly noticed on this job where so many works were authorised and then cancelled and at the final cancellation a large number of men was idle awaiting evacuation... The cancellation of all works was dated 15 May 1944, but all plant and personnel were not evacuated until 28 August 1944.'

COEN

'...When the Fortress was ready to leave, the drums were all shifted and the drome cleared of small trees and antbeds. All the Coen folk gathered to bid them farewell and it was a beautiful sight watching that big plane lift gracefully off the strip. Our best wishes went with them and the drums returned as our protection.'

These drums were likely 44 gallon fuel drums part filled with concrete or earth and placed to 'obstacle' the old DCA landing ground in the town area of COEN. The scarcity of accurate mapping and charts available to US Air Corps crews on 1 June 1942 meant that Fred Eaton's crew in B-17E 41-2434 scored a chart which showed this old DCA landing ground in preference to the AOB constructed during 1941 some 13 miles north of COEN. Doris Bassani, now living in Cairns, was living in COEN during the war years. She was then newly married to the local pack horse mail man Alec Nichol. She has written her recollection of that occasion of the 'big aircraft' landing in COEN. I believe her 'farewell' as the aircraft departed after some days of repair, epitomises the way in which the outback people look upon those wartime years. The townsfolk were not to know that this aircraft, some weeks later, would be lost with all its crew off Cairns. Dean Hoevet, its pilot, would be honoured by the USAAF's naming of the new airfield then being developed at MAREEBA.

Kevin Shephard was nearly five when he witnessed the same landing. His unsolicited and unprompted account reads -

'I do recall it coming in from the western side of the township, missing the PMG powerlines which run from the Police Station south and running the full length of the flat in the Shephard's cow yard and my Grandfather's house and ending up on the hill to the east of the township. It was there for several days and I became fairly friendly with some of the people who were either guarding it or stationed with it during that period and I do recall actually receiving a sheepskin lined coat because I was quite infatuated with these sheepskin lined coats and one of the men gave me one so that I could have it for myself. They also actually supplied me with a, what I called in those days "a parachute" which I suspect today was a flare chute and had replaced it with the lead from one of the armour piercing bullets and I do recall this quite vividly because it was silk and I used to have great fun as a little fellow firing it from my shanghai into the air and letting it float down.'

COEN *(7/8/1943)*
One of a set of verticals flown by RAAF Survey Flight DH Dragon A34-82, this photo shows two of the developed aircraft dispersal revetments. Another partly developed revetment lies to the north of the runway intersection. That area was used for ammunition and fuel dumps. In the bottom right the camouflage painted roof of a narrow building can be seen. At '2 o'clock' on the right hand side of this photograph, 'tracks' radiate from the area where the airfield's Construction Camp was sited.
[CPE RAAF OFFICIAL]

COEN *(13/8/1990)*
As you enter Coen from the south and look across towards the east, this part of the township was originally DCA numbered landing ground No 598. It is on this old ground that the USAAF B-17E Flying Fortress No 41-2434 landed on 1 June 1942. Their story was more fortunate than some other crews finding the need for emergency landings in Queensland during WW2. *[Author]*

Armed with these two accounts of this event in COEN 'sometime in 1942', I discussed the matter with Bruce Hoy who has a keen interest in the WW2 US Air Force involvement, in SWPA. Bruce tabled information on a few occurrences. The one which most fits these independent recollections from COEN residents at the time is that of a single B-17 bombing mission to Lae on the northeast coast of New Guinea. It was mounted from Townsville staging through HORN ISLAND. The following excerpt from an archived Narrative Report of the mission of 1 June 1942, signed by Fred C Eaton, Captain, Air Corps, Pilot, reads -

'...Para 7 - the return trip to Horn Island was made and on the final landing approach a red light was flashed from the ground. This was followed by light code saying that "there was an air raid alert" - the wheels were retracted and the ship was pulled up close to the clouds and headed south towards Coen. A code radio message was sent to Horn Island asking if it were clear but the answer received said the air raid was still in progress. A formation of 3 unidentified ships were seen about 10 000' above, but they did not attack. Another single unidentified ship was observed some distance off flying through the clouds.

Para 8 - The gasoline supply was getting very low so the pilot elected to fly down the coast where a crash landing could be made on a beach or in the water. There was in his possession the airport map of Coen which was followed. This map showed the Army aerodrome to be between the mountain ranges and the town of Coen. This is wrong, however, as the aerodrome is 13 miles away, over the next mountain range. When the ship reached the town of Coen, there was practically no gasoline left so it was decided to make a crash landing at Coen, but with wheels down. This was accomplished in a space approximately 800'. The tail wheel assembly was smashed due to the wheel not remaining locked or possibly not having been locked. It was not checked previous to landing. However, there was no further damage to the airplane or crew.

Para 9 - The airplane was temporarily repaired and after 2 days spent in clearing a runway the ship was flown back to Garbutt field by Capt Spieth.

Para 10 - The condition of the secret RAAF Intelligence Map COE/1A showing COEN aerodrome has been brought to the attention of ACH and NEA.'

I spoke later with a US veteran of 435 BS (19 BG) John Lillback, now living in Brisbane. John told me that certainly he flew as a member of Fred Eaton's crew many times. He felt this occasion of landing in COEN was connected with the ventral (Sperry Ball) turret gunner, George Ryan from Texas, having somehow damaged the tail wheel with his guns.

Memory is a frail thing and in John's case that occasion may not have been the particular landing in COEN itself. Certainly B-17 Flying Fortress aircraft landed at the regular airfield north of COEN both by intention and under conditions of emergency. Who knows - John's occasion may have related to one of those or, on the other hand it may even have been John who presented the very small Kevin Shephard with a flare parachute that day in June 1942.

Doris Bassani's story goes back to early 1941, when she was cooking for a group of miners at a 'ranch on Wenlock Gold Fields'. She remembers the little grass strip at WENLOCK used in those days by bi-planes and the like, probably operated by the legendary Tom McDonald. Doris remembers the cessation of that work at WENLOCK being connected somehow to a threat of Japanese invasion. With the loss of mining income in the district, she says many of the men obtained work on construction of the new airfield.

I have found archival evidence of the RAAF's searching for an AOB in the COEN vicinity as early as June 1939. The file on COEN aerodrome documents the situation in March of 1940 where decisions needed to be made on various sites being considered. Aerial navigation to and from COEN then was by visual observation of the ground - contact navigation. Radio aids were years away. The main criticism of the site 13 miles north of COEN was its tendency to be difficult to find in bad weather, which spanned the whole spectrum of smoke from bush fires through to rain and storm conditions. There were alternative sites being considered, one in fact a few miles south of the Iron Range Goldfield.

March in the COEN area was still part of the wet season and the file shows an apparent conclusion -

'...that the selected site 13 miles north of Coen be proceeded with after the wet season'.

On 2 April 1940, the Director General of Civil Aviation wrote to Department of Air Melbourne supporting construction of the NW-SE runway shown on Drawing Z-1137 (Iron Range). He requested early advice on Department of Air's wishes, because the wet season was then ending and MRC were making arrangements to commence work on the COEN site.

The RAAF's Wing Commander F M Bladin is on record in a Memo dated 4 April 1940 to the Director of Works and Buildings, pointing out that he understood the aerodrome site north of Coen was not so liable to fog as the old site nearer the town. But he continued that -

'...Coen district is hard to get at by contact flying in all weathers owing to the mountainous nature of country between Coen and the coast.'

He summarised saying that -

'...We require to be able to operate from a base or bases at or near Coen. This has been laid down on Policy File for Advance Landing Grounds. If Coen is not accessible in all weathers, due to cloud on the mountains, then an ELG which is accessible should be made. If the ELG is accessible in all weathers it would be preferable to have that ground only, provided petrol, oil and telephonic communication could be got at that ground. If (these)...cannot be provided except at excessive cost, then the best arrangement is to have two grounds, Coen for normal use and the ELG so that aircraft can at least get down after the end of their patrols.'

The History of the Queensland Main Roads Commission during World War II 1939-45, published in 1949, mentions COEN as an urgent RAAF requirement in

COEN *(13/8/1990)*

Used now as an equipment shed by the airfield caretaker, this building remains from the wartime period. Seen in the 1943 vertical with camouflage painted roof, the building has not changed much since then. *[Author]*

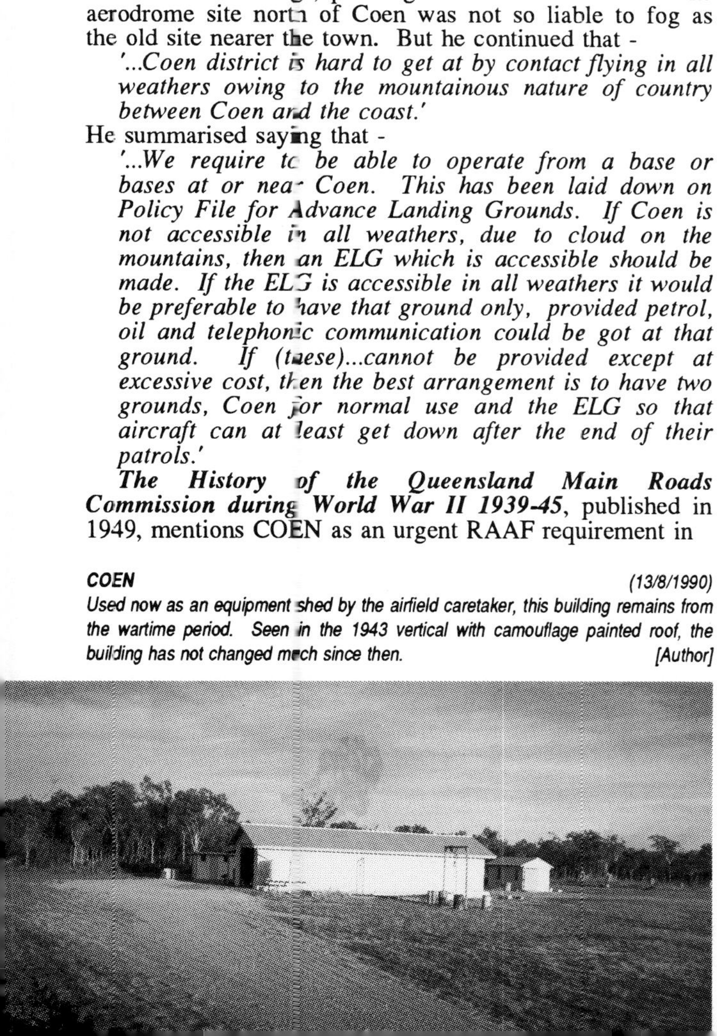

1941. It refers to the great difficulties associated with transport of plant, materials and supplies to the site from Port Stewart on the coast, and overland from Cooktown. The ***MRC History*** states -

'...3 runways were laid down, 2 large buildings erected, and a deviation of the overland telegraph line effected. Taxiways, dispersal points, workshop, bomb dump, and pyrotechnic stores were provided and the drome fenced to keep out the wild pigs which infest the district. This work was undertaken in the first instance by a unit of the PEI (Public Estates Improvement) Board of Lands Department, which later merged with the MRC.'

A RAAF file I found in archives deals with construction of various 'hutments' at RAAF AOBs. There is an entry dated 6 July 1941 which reads -

'..AOB COEN - work being carried out by departmental day labour. Buildings complete and base occupied 6 July 1941. Erection of masts, stays etc almost completed. All materials for erection have been forwarded.'

Around October of 1941 there is mention of the delivery of 'concrete picket blocks' for aircraft at COEN. I was puzzled for some time as to quite what these were! I later learned their role was to provide tie-down of aircraft against strong wind.

November 1941, the circumstances attaching to an Ambulance mercy flight to COEN brought home some of the earlier mentioned difficulty of visual navigation to COEN, albeit in darkness. A 17 November report by the PEI Engineer in charge to the Queensland Land Administration Board reads in part -

'...Serious accident occurred... Coen aerodrome 12 November 1941... resulted in death of mechanic named Strang (who) was working on a benzine pump in a well when it exploded and he was badly burnt. Within half an hour of receiving an urgent telegram from Coen works, PEI Northern District Engineer, Mr Pheasant, had arranged for the large Stinson plane to leave Cairns for Coen with Dr Mansfield and an Ambulance officer aboard. The Stinson plane (AOA) which was the only one capable of making a night landing, left Cairns 7 p.m. and returned next morning 3 a.m. with the injured man and his wife. The unfortunate man died mid-day the same day. The pilot of the plane stated that owing to presence of hundreds of small bush fires, he had the greatest difficulty in picking up the landing flares and in fact was flying around for some time and was actually on his way back to Cooktown to wait for daylight, when he sighted the Coen drome. This illustrates the danger of night landing on Coen drome during the bush fire season and Engineer Pheasant has suggested that either or both following precautions be adopted immediately.

- *Supply at least 1 dozen powerful red signal rockets*
- *Supply a small portable search light about 1200 watts for hooking up to 30 volt portable lighting sets and directing a beam into the sky.'*

This aerodrome Works file does not show whether the suggestions were taken up. The 17 November report did reveal that hire of the Stinson plane was around £75 and that -

'...By telephone today, Engineer Pheasant advised that an amount of £50 was collected by employees of Coen drome (project) for the late Mechanic Strang's widow and there will be an amount due to her under Workers' Compensation.'

Three weeks later the Japanese attacked Pearl Harbor.

Dated 18 February 1942 there is an interesting mention in the NEA HQ Unit History Sheets (Form A50) held at the Australian War Memorial, Canberra. It reads -

'...Port of origin of flight - Port Moresby, 1 Hudson - search for missing (Genairco) Monospar, Horn Island, Wenlock, Coen, Lockhart Mission'.

My efforts to learn more of the circumstances surrounding this aerial search for a missing Monospar aircraft have come to nothing. I know not whether it was ever found.

On 12 March 1942, I find the earliest mention of camouflage painting work proceeding on buildings at COEN Airfield. Similarly a cypher message on RAAF Coen file dated 22 April 1942 indicates the earliest record of United States B-26 Marauder aircraft operating through COEN -

'...COEN AERODROME. AIR STAFF NEA ADVISE URGENT OPERATIONAL NECESSITY THAT RUNWAYS BE IMMEDIATELY EXTENDED TO 6 000' FOR B-26 AIRCRAFT. DR BRADFIELD ARRIVING COEN 25 APRIL LEAVING 26. REQUEST URGENT APPROVAL BE GRANTED FOR HIM WHILE ON SITE TO INSTRUCT CONSTRUCTION TO PROCEED WITHOUT DELAY AND SUPERVISE COMMENCEMENT. PEI STILL AVAILABLE. APPROACHES SHOULD ALSO BE CLEARED AS DANGEROUS FOR B-26.'

Walter Maiersperger's flight log shows he took B-26 Marauder 40-1437 into and out of COEN strip on 19 April 1942. While he can't recall much else of the visit, he says he and his Flight Engineer have never forgotten the hair-raising take-off. This approval for extension of runways at COEN was granted and recorded in a Minute dated 24 April 1942.

On 26 April 1942 a further mention in the Form A50 for NEA HQ notes -

'...APRIL 26 - 3 B-25's from Moresby attacked aircraft and installations at Lae - aircraft to return to Coen or Horn Is. Dropped bombs on drome, several aircraft destroyed on ground, 2 bombers and 1 Zero taking off destroyed. Attacked by 5 Zeros. Shot 1 Zero down. All 3 aircraft landed at Coen.'

The unit involved would have been either the 13 or 90 BS of the US 3 BG operating from CHARTERS TOWERS. They were hastily equipped with B-25C aircraft side tracked from the Netherland East Indies AF's 18 Squadron allocation.

Doris Bassani/Nichol recalled that a few days before Pearl Harbor, she and her husband Alec had flown to Coen to live. Alec was to resume his activity as a pack- horse mail man working northward to Moreton Telegraph Station on the then named Batavia River. Doris wrote initially to me of her recollection of a trip 'down the telegraph line' with her husband when -

'...we saw Jap planes flying overhead going south. They made no attempt to harm us, probably thinking it was just a mob of horses'.

My closer discussion with Doris in a later taped interview, revealed that the time was 'early' in 1942. I tried to tease from Doris' memory her best description of the aircraft and why she felt they may have been Japanese. She related -

'...We got to Moreton and stayed overnight and then coming back the next day, of course with the hobble chains and that swinging on the horses necks (making) a bit of a noise, you couldn't hear. I was towards the back of the pack horses. The black boy was in front, my brother on one side and my husband on the other and... then all of a sudden my husband came tearing around and he's pointing and cracking the whip and all. He pushed us all over into the trees and I said - what's wrong, what's wrong and there were Jap planes flying just ahead of us...'

Doris felt there were four planes flying towards Coen. On my pressing for why the party believed they were Japanese -

'...It was the sound of the planes that was different to our own,... it was a different noise altogether... they weren't as big as the American planes...
American planes always seemed to be so much bigger...'

It is probable that for Doris Bassani, the close contact with the B-17 in COEN township confirmed in her mind this feeling of relative size. It is also probable that the four Jap planes were a flight of B-25 Mitchells of the 3 BG or B-26 Marauders of the 22 BG. Both would have been a new visual and aural experience to four Australians on horseback making their way south along the Overland Telegraph Line early in 1942.

A further progress statement dated 1 May 1942 on the erection of RAAF AOB hutments mentions -

'...CONSTRUCTION OF AERODROME - practically complete, in full use. Extensions now required. RAAF AOB AIRCRAFT DISPERSAL SCHEME - progress 50%, dispersals well advanced. BUILDINGS PYROTECHNIC STORE, BOMB STORE & WORKSHOP progress - temporary buildings erected. SPLINTER PROOF PROTECTION TO DISPERSAL POINTS - progress 80%, approaching completion. ROAD TO AERODROME FROM COEN TOWNSHIP - progress, just commenced.'

This was the position at the time of the Coral Sea Battle.

The Civil Aviation Department's Dr Bradfield stressed the need to bitumen seal the gravel runway after noting that B-25 Bombers operating from COEN -

'...blew out a very considerable amount of dust from the runway...'

On 9 May 1942 P-O John Keays carried out a site inspection with P-O Campbell. His report, apart from general details relating to both the 105° and 60° runways includes the following statement -

*'...**PLANT** - existing plant consists of 1 tractor, (broken down) with 3 cubic yard scoop, together with 2 tractors, several ¾ cubic yard scoops. Four trucks are available with a tractor drawn grader. It is obvious that any program requires that this plant be considerably augmented and it must be brought in over very bad roads after transportation by sea. A major requirement to be considered in any gravelling program is supply of water. The Constructing Engineer estimates existing sources will supply the camp for 3-4 months. The RAAF Officer in Charge located a boring plant at Wenlock and has transported it to site. Trial bores are to be commenced immediately,... pumps and piping will have to be freighted in by Service Aircraft.*

***CAMP** - existing accommodation and Signals Huts are in large cleared area immediately adjacent centre of main runway. Good timber cover exists from ¼ - ½ mile behind and buildings should be moved there immediately. Personnel in the RAAF establishment now number approx 100, so it is suggested that a new kitchen and mess hut be erected and that existing buildings be re-erected as Operations Room, Signals Hut and Officers Mess. Tents have been dispersed but should be located at a greater distance from runway*

COEN (-/6/1985)
Around July of 1992 the airport ownership and responsibility for management passed from Department of Transport and Communication (DOTAC) to Cook Shire Council.
[Courtesy R Rudd Northair Surveys Mareeba]

and dispersed to greater extent.
STOCKPILES *- Instructions were given to Construction Engineer to proceed immediately on transportation of gravel... for emergency repairs to runways.*
SURVEYS *- The Constructing Engineer is carrying out surveys of proposed extensions and will forward information as soon as available.'*

Airfield development plans in Australian Archives, show machine gun pits, several in number, around COEN Airfield. Not too surprisingly I have so far failed to find anyone who can recall actually seeing these. An official US Army publication of 1946, extract copies of which I obtained from the Cairns Historical Society, shows -

'...HQ's and Battery D, 104th CA Battalion (A-A) furnished protection for the airstrip at Cooktown north of Cairns while Battery C of the same Battalion covered an airstrip at Coen still further up the Cape York Peninsula. Batteries A and B were used to defend the airstrips on Horn Island at the tip of the Peninsular...'

Around 19 June 1942, the 104th Coastal Artillery (A-A) Battalion had approximately 170 men at COEN, 300 at COOKTOWN and about 350 at HORN ISLAND.

NEA was pressing for a decision on the upgrading of COEN, a decision which was taking considerable time to be made. The COEN Aerodrome Works File contains a Minute dated 4 June 1942 recording copy of a signal sent to NEA stating that -

'...Plans for Coen are being prepared by this HQ and US will execute.'

At this time Civil Aviation were trying to reach agreement for the RAAF to take control of COEN drome. Civil Aviation wanted such a policy clarification before RAAF simply assumed control. However, significant expansion at COEN was discarded with the move to construct IRON RANGE.

A Cypher message dated 13 September 1942 from NEA to Air Board reads -

'...OWING EXTREME SUPPLY DIFFICULTIES, AIR STAFF CLOSING DOWN COEN AS OPERATIONAL BASE. WILL BE RETAINED FOR EMERGENCY LANDING ONLY. NO ACTION BEING TAKEN TO PROVIDE FURTHER BUILDINGS.'

While COEN was closed down as an Operational Base in October of 1942, it remained with a detachment through to February 1945. A file reference dated 17 February 1945 refers to this decision in that -

'...HQ NEA advised 3 DWO, in respect of RAAF DROMES QUEENSLAND - RAAF has no further use for aerodromes at Inverleigh, Winton and Coen,... understood Inverleigh has already been disposed of. Winton has never been occupied by RAAF... Coen is still occupied but as soon as removal of equipment and fuel is complete in a few weeks, all RAAF personnel are to be withdrawn.'

Ted Petersen in Townsville told me of his early days from 1939 onwards during the war as a young 16 or 17 year old employee of Airlines of Australia (AOA) at Cairns. His duties became those of 'Aircraft Dispatcher', being concerned with the detailed loading of aircraft to ensure correct trim. Ted recalled hearing from AOA pilots an unexplained cache of 100 octane fuel at the old WENLOCK strip. It appears it arrived there sometime before the Japanese entered the war. Most likely intended

COEN *(1944)*
RAAF 36 Squadron aircraft being refuelled. Crew members, from left F-Sgt Noel McClelland, F-O Alf Aydon, F-Lt Bill McCallum, Sgt Dudley Wood, Sgt Fred(?) Walker and F-O Don Cox. George Batts remembers the squadron being involved in relocating ordnance (bombs etc.) from a dump 'just south of Coen' to Kiriwina Island. *[G Batts ex RAAF]*

for Batavia NEI it probably arrived at WENLOCK because that was previously called BATAVIA! Ted recalled a Vacuum oil 'testing fellow' named Bob Hare being involved in checking the fuel which was prone to deterioration and accumulation of water. He suspected that as most Australian aircraft then in use could not use 100 octane fuel, and as those that could, were not realistically able to use WENLOCK strip, this fuel dump surely had to relate to some provision for modern US aircraft.

This excerpt from ***The (US) Army Air Forces in World War II*** by W F Craven and J L Cate Vol 1, p.230 states -

'...Fortunately, preliminary steps had been taken prior to the war for coordination of defensive efforts between American and Australian authorities as occasion might require. General Brereton had visited Australia in November 1941 and plans for the South Pacific air route had required arrangements for use of airfields and other facilities within Australian territory. It was against this background, then, that the first of several Allied conferences met on 28 December at Amberley Field to consider common problems and opportunities for mutual assistance. It was agreed that American officers would assume actual responsibility for the erection of their planes, but that to assure proper coordination of the effort with plans for movement of the assembled planes an Australian officer would be put in general charge. Since the aircraft would have to be ferried overland for a distance of over 2 000 miles before they reached a jumping-off place at Darwin, it was decided to establish refuelling depots at Charleville, Cloncurry, Daly Waters and Darwin. Though 100 octane gasoline could be procured from the Netherlands East Indies, this fuel was of so high an aromatic content that it destroyed the leakproof lining of fuel tanks and consequently forced consideration of the problem of importing American fuel. In addition to limited stocks built up before the war, the steamship ***Mauna Loa*** *was on the way to Brisbane with a load of 400 000 gallons, but even so the supply would fall far short of the prewar goal of 10 000 000 gallons.'*

The quantity of 100 octane fuel found at Batavia/WENLOCK is not known. George Redding, a RAAF 36 Squadron veteran certainly recalled their aircraft being directed to call at COEN both for the task of transferring aerial bombs from a COEN dump, north to the islands and also to simply use fuel already stored there. George had no knowledge of the Wenlock fuel let alone whether it was ever trucked to COEN for use there.

Such details become more intriguing with each failure to find a satisfactory explanation. For a while there I wondered if the 'Wenlock fuel' may have been an 'error' in preparation for Amelia Earhart's Lockheed Electra transit eastward through Java in 1937. However, she landed elsewhere in Java and probably didn't use 100 octane! Certainly Maj-Gen Lewis Brereton was instructed by Gen MacArthur to survey aircraft ferry routes linking the Philippines with the NEI. Surely it only needed two places with the same name to cause this minor misplacement! (Jenny liked this better when the fuel might have been for Amelia!).

COEN *(13/8/1990)*
While the structure remains from the wartime days it has been upgraded for current use as the caretaker's residence. Fibreglass rainwater tanks and solar hot water roof-mounted panels are improvements that WW2 veterans with experience at Coen would well appreciate. *[Author]*

COOKTOWN

'...At that time it was proposed to spend a considerable sum on improving the existing civil aerodrome at Cooktown. On the way back either the (Rail-motor) driver or the guard, I never did find out which was which, suggested that we look at a site used as a Mission Station which was located adjacent to the railway about 5 miles from Cooktown. The Mission was in the process of being closed down. An inspection of the area confirmed that it offered an excellent site for a possible airstrip. In my adverse report on Laura I suggested that the Mission area be investigated, as a possible substitute for the existing Cooktown Aerodrome. The airstrip was constructed and known as Mission Strip and is, I think, still used as (the) Civil Aerodrome for Cooktown.'

This is Jim Trench's brief recollection of his part in the identification of the COOKTOWN MISSION strip site around mid 1942. RAAF files now in AA put this inspection date around 19 May 1942. The same RAAF files reach back to January of 1941 in their coverage of works proposals to provide an AOB facility at the then COOKTOWN CIVIL drome some 3 or 4 miles southwest of the township.

RAAF's 24 Squadron based at Townsville was instrumental in attempting to expedite construction of things such as bomb dumps, fuel facilities, accommodation and importantly the fresh water supplies at COOKTOWN CIVIL drome site. However, the file reads as a story of

COOKTOWN *(16/5/1942)*
This vertical is part of a Trimet test conducted about one week after the Battle of the Coral Sea. The test photographic mission originated at Cairns and included this brief run over the only airfield then existing at Cooktown. On the ground in the loop dispersal taxiway by the bottom left corner of this photograph, it is possible (with a magnifying glass on the print!) to discern the outline of a twin-engined B-26. The largest of the buildings survives today, currently the simple residence of the private land holder. The extreme right of the photograph shows a prominent white criss-cross pattern over the early aircraft landing field at Cooktown, i.e. the racecourse of the time. The 'white lines' are trenches hurriedly cut across the open flat land by earth moving equipment brought from as far afield as Coen to deny the possibility of successfully landing enemy aircraft. [NLA (RAAF Official)]

COOKTOWN *(12/8/1990)*
This is virtually all that remains of the wartime buildings at the old COOKTOWN CIVIL landing ground. While it served pre-war as a Radio and Meteorological Building, it is now a private residence. [Author]

some lethargy in the prosecution of these works. Fl Lt John Yeaman inspected the drome for DWB on 30 November 1941 and his report refers, amongst other works, to that of planned aircraft dispersal points -

'...Mr Jennings is preparing an estimate for this job, which should be available by 4 December 1941... Seems... insufficient money available in original estimate for this work, but I have instructed Mr Jennings to proceed with and complete the work at the earliest possible date.'

Reg Jennings was at that time the engineer responsible for Main Roads Commission works at Cooktown. While he was in fact living in Cairns, he recalls the urgency attaching to the outbreak of the Pacific War. That was Reg's first ever flight in a plane. He thinks it was 'either a single or twin engine Dragon Rapide type of plane'. They flew low over Bloomfield River timber milling settlement to drop a message, requesting the 'Bulldozer be taken to Cooktown post haste'.

The very resourceful Rollo D'Arcy Gallop ('RDG' in MRC circles) worked on this crash upgrading of COOKTOWN with Reg Jennings. In her biography of RDG, Alexa Gallop refers to this COOKTOWN job -

'...Within two or three days work started with a bulldozer which had been shipped from Bloomfield, two tractors freighted by land and sea from Coen and labour from Cooktown.

As the work neared completion, a few Wirraways arrived on their way to Port Moresby and stayed for a day or two awaiting last minute instructions. The pilots were warmly entertained and farewelled when they left for Port Moresby. It is understood that one survived their initial contact with the Zeros a few days later.

The four weeks work was carried out under fear of having to build a new airstrip to a much higher standard and this no doubt was a spur to the rapid completion.

As the Dragon carrying RDG back to Cairns touched down at the airport, an officer from Control rushed across with a message addressed to Cooktown to await urgent instructions. Tom McDonald's three engined "bitser" plane was about to leave for Townsville, so RDG returned there and was able to report at District Office (MRC) the following morning.'

Of course as soon as these dispersal points were constructed, they were, as an urgent cypher message from NEA to Air Board dated 28 January 1942, declared -

'...PLAINLY VISIBLE FROM AIR OWING WHITENESS OF TAXIWAYS. REQUEST URGENT ACTION BE TAKEN.'

Rod Voller, in charge of the Camouflage Section of Department Home Security in Townsville at the time, is on record in a report for the period to 25 March 1942 -

'...AOB's - paint has now been sent from Townsville to Horn Island, Coen, Cooktown, Cairns, Bowen and Mackay for the painting of aerodrome buildings, and as far as we know the work is proceeding.'

It was a little prophetic to comment on the progress of such painting, for as Rod's later report to 20 May reveals -

'We discovered that the Cooktown Aeradio building had not been painted with the others, the OC being of the

opinion that drinking water would thus be poisoned. A wire was sent instructing that the painting of this building be put in hand at once.'

The mode of preparation for demolition of the runways at COEN was by drilling a pattern of 4" diameter holes several feet vertically into the runway and placing a cap over the hole. A Bangalore torpedo style sausage charge was prepared and stored ready for speedy placement in the hole if needed. This method was recommended for use in a 23 April 1942 report by E Markham, CO of 17 Battalion VDC Cairns. Mr Markham's task was to urgently provide 'guards to the drome' and other vital points. His three page report in AWM archives, reads as though 'demolition' was a foregone conclusion. Whether or not such preparations were actually carried out at COOKTOWN CIVIL drome is not clear to me. Certainly the AWC minutes for 22 September 1942 record 'cancellation of aerodrome demolition at Cooktown'.

This was around the time that Jim Trench was detailed to investigate sites inland from Cooktown. Interestingly, the planners had seen the Cooktown to Laura Railway as some thing of strategic value. Jim described it as -

'...LAURA - this was a good example of a suggested site that was impossible from almost all points of view. Laura was then situated 67 miles by rail inland from Cooktown. The railway was only capable of taking a Rail Motor. There was no road at all. I was accompanied on this inspection by Mr (Staff) Bettridge, who was the MRC engineer stationed at Cooktown. The RAAF hired a so-called special train, but it consisted simply of a fettler's van which was not more than 6' long, but nevertheless was supplied with a driver and a guard. Although the 67 miles consisted almost entirely of curves there wasn't a curved rail on the line. On the curves the normal 25' rails were cut in half and the curves consisted of 12.5' straights. The little contraption we were on nearly rattled itself to pieces as well as the passengers. On one quite long wooden bridge over the Kennedy River, the rail stepped to one side about a foot in the middle of the bridge for about 25' and then stepped back again. The van stopped for water and I looked over the side of the bridge and the two middle piers had completely rotted off near the water and so far as I could see the 25' of rail was held up by the fish plates.'

It is as well to put to rest now the matter of what appears in some published histories of the wartime activities in the SWPA as the AOB LAURA. Certainly, sites were identified, but as the MRC History, published in 1949 records, quite minimal work was done before the site at LAURA was abandoned.

In his NEA report of 26 May 1942 to Air Board, the AOC advised that among various new airfield sites, one had been selected '3 miles west of (the present) Cooktown (drome)...' The report sought urgent clarification of responsibility for supervision of fresh construction; i.e. by RAAF or the US Engineer Units moving to start actual work. The AOC added that -

'...at direction of Col. Matthews work has commenced at Wrotham Park and Cooktown.'

Halfway through July, a report by P-O Campbell to Air Board in Melbourne, addressing the subject of an RSU site being sought in the vicinity of Lappa Junction, shows that a site about 4 miles south of Mt Garnet had then been chosen. His report outlines possible moves with respect to MRC plant and workforce and finishes with the statement -

'...In event of approval not being given (for Mt Garnet site) within a fortnight, Mr Bond intends to convey the tractors direct to the Laura drome works.'

This indicates the little work which was done at LAURA, may have been done after mid July.

George Gates, now living in retirement in Mareeba, was in the group carrying out that construction work at WROTHAM PARK. George in fact, described to me where that same party carried out the minimal work on the chosen site at LAURA before moving on to the job at COOKTOWN.

The CCC organisation of about 24 in number at Wrotham Park, grew to about 250 at Cooktown, with an influx of southern draftees. George recalled the CCC people were at Cooktown aerodrome for about 7 months during which he was 'mates' with the US 104 (or 94?) Coastal Artillery (A-A) of whom there were 'several hundred, (about 300) camped over under those mango trees and the CCC people were under the other mango trees'. The CCC were banned from the town as a result of serious brawls with servicemen. George recalled US guards posted on the Endeavour River bridge to police the ban.

The A-A troops, in his words, 'fired at balloons', high above the airfield site, presumably for practice.

Progress Reports on the construction of 'hutments' and various other improvements to RAAF airfields show the following against Cooktown for 30 June 1942 -

'...Machine gun posts and emplacements - under consideration. Widening and lengthening runways, extra thickness main runway, splinter proof pens and obstructions - (all bracketed and stated as) temporarily abandoned for dispersal strip at Endeavour River, which is 60% complete at 30 June.'

Progress at MISSION STRIP as at 8 August 1942 is recorded briefly by Herman G Cox, Captain, Corps of Engineers, Air Engineer (Works), in a report addressed to Lt-Col W T Abbott, Air Engineer HQ USAAS -

'...Cooktown Mission - the strip is cleared for 7 700', graded for 7 000', first course gravelling 3" entire length, second course gravelling for approx 3 000'. Camp facilities (shower and latrines only) for two squadrons are under construction and about 20% complete. Camoufleur is on job and is assisting in selection of dispersal bays and preparing plans for approval of this HQ for dispersal. Several items of equipment needed to expedite completion of work were arranged for on my return. September 1 (is date set for completion of) water sealed runway and bays. The camp should be completed except for water pipe (for) which transportation is now being arranged.'

COOKTOWN *(8/7/1943)*
This vertical is one of the systematic runs over the Cooktown area used for topographic mapping during WW2. It shows most of the Cooktown Mission landing ground constructed by the Allied Works Council in 1942. South of the airfield, over the Endeavour River and the Cooktown to Laura railway line can be seen the winding access road into what became an extensive RAAF camp area. The prominent trees in the cleared area close to the airfield are large mango trees which were part of the Lutheran Church Mission development taken over by the authorities before airfield construction. *[AUSLIG]*

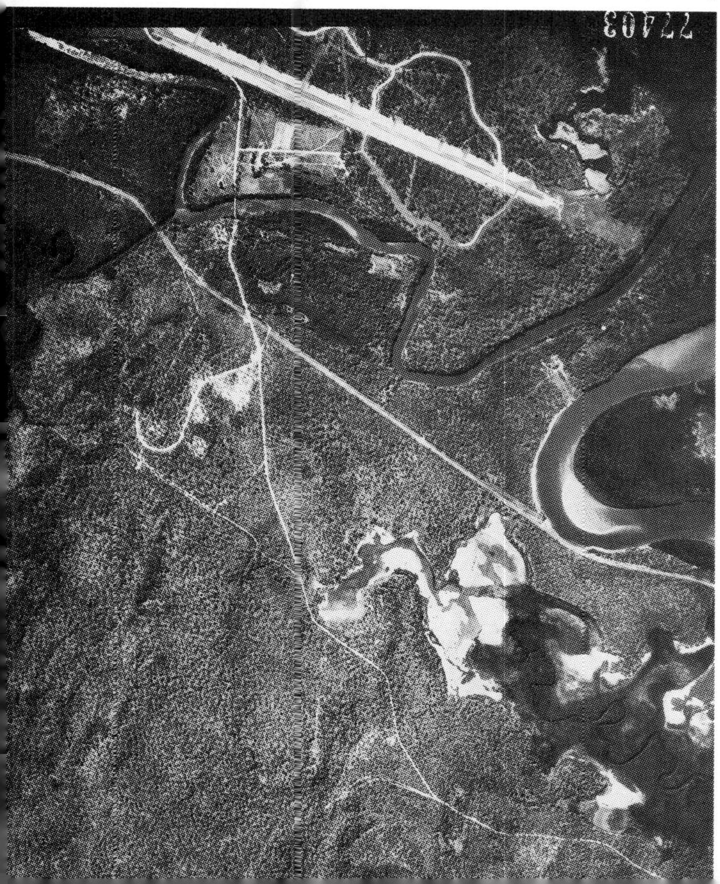

Cox's 'water sealed' runway meant bitumen sealed.

The young George Gates recalled resisting transfer northward to Horn Island when the construction crew was due to move on after about 9 months at Cooktown. He left on a ship to Cairns and then a train towards Wrotham Park (WP). When he got off at Chillagoe he found a policeman waiting for him to take him back to Cooktown. He promised he'd return to Cooktown after he'd been home (WP) for his 21st Birthday. While he did report back he absconded again. A ship was in from Cairns - 'a bitumen boat' - with a cargo of bitumen in drums. He was involved in the off-loading, and talked the ship's captain into taking him back to Cairns. He said he 'took off', swam the Walsh River and others to return to WP to resume work. The authorities never caught up with him.

It seems from its inception COOKTOWN MISSION Strip was used predominantly by USAAF aircraft as a refuelling stop. RAAF and USAAF planes continued to use the inferior strip(s) of COOKTOWN CIVIL, where practical and necessary through congestion at the better strip.

By October 1942 operating conditions for all aircraft, including those civil aircraft using the COOKTOWN CIVIL drome, had deteriorated badly. Jim Trench is on record as concluding that the RAAF's AOB should be relocated to the MISSION strip -

'...The Cooktown Mission strip has been completely formed, gravelled and sealed for a length of 7 000' x 100' wide. The flight strips have been cleared to 400' wide. The bearing of the strip is 104° into prevailing wind. Area is high, well drained, excellent cover with dispersal and camp sites. Pipeline has been laid from large rock pool higher up Endeavour River giving ample water supply all times. It would appear most desirable to relocate the OB at the Mission Strip. The total personnel attached is 65, who are at present accommodated in tents. As most of the equipment is mobile, little difficulty should be met in effecting the move. The distance by road from Civil Aerodrome to Mission Strip is 4.5 miles. Provided the Mission strip is not over-crowded with operational aircraft it is strongly recommended that Cooktown Operational Base be relocated there.'

P-O Trench pre-empted possible US objection to this amalgamation of Service needs, leaving the old drome to civilian use only. He emphasised the disproportionate cost of upgrading the saltpan site, pointing to an alternative site-

'...10 miles by road from Cooktown close to the railway and to the pipeline supplying Mission Strip... could be fully formed and gravelled at about half the cost of extension works at Civil Aerodrome with much more satisfactory result.'

By 11 November Jim Trench's recommendation had been adopted and the work of shifting the OB's equipment and organisation began.

Rod Voller's mention on camouflage matters show US current thinking -

COOKTOWN *(-/11/1943)*

This is the 1943 scene at the old Cooktown Mission buildings adjacent the strip, 'occupied by OB (Operational Base Unit Staff) and US Mets Unit.' The residence in the middle was that of the Mission Superintendent. Note the Observation Tower behind the building. It is uncertain whether this tower was used for weather and other meteorological observations or to support an elevated water tank. No tank appears in the photo. Neither this Mission residence nor any of the WW2 buildings remain on COOKTOWN strip today. [AA CRS A705/1 Item 171/93/743]

'...On 18 November, 1942 saw Major Cox (USA) re Horn Island, Cooktown and Mareeba. He stated that no more camouflage was to be done for them or by them in Australia. It was up to RAAF to carry out more camouflage if they wished to and this did happen for the next 12 months.'

In March 1943, with the resurgence of Japanese air activity and apparent strength building in the Dutch New Guinea (Irian Jaya) area, the RAAF persevered with planning and construction of aircraft dispersal facilities including splinter proof pens at several of the North Queensland airfield sites. COOKTOWN MISSION was included in this programme, 18 heavy bomber pens being proposed. Formal approval for these was issued on 28 May 1943. The need subsided and like many similar projects in North Queensland, this was cancelled on 7 July 1943.

Jim Legge of Canberra was a member of the RAAF's 7 Squadron. He described landing at COOKTOWN MISSION on 11 June 1943 -

'...we were carrying out an anti-submarine patrol ahead of a convoy travelling up the Queensland coast between Cairns and Cooktown and were to leave the convoy at dusk to make a night landing at Cooktown Mission strip... The weather completely closed in around the convoy and also along the coast by the time we got back there. As there was a radio beacon at the Civil aerodrome some distance from the Mission Strip, we were able to determine our position accurately but could not find the flare path at the Mission strip because of cloud and rain. With the hilly nature of the country, Mt Cook itself located near the Civil drome being over 1000' high, we had to keep on flying above 1000' over the area where we knew the Mission strip to be until we could get a glimpse of the flare path. After about 2 hours it showed up sufficiently for us to get a line on it. We flew along it for several minutes and then did a couple of circuits losing height and timing each leg of the circuit to bring us in along the line of the flare path each time to get another glimpse of it. Third time round, in fairly heavy rain we were low enough to attempt a landing and we got down safely. As flying "blind" requires the pilot to fly by his instruments all the time, and as he cannot see the horizon, it is a great strain. When the instruments show he is flying straight and level, his body often feels that he's tilted sideways in a turn. You have to keep telling yourself to believe the instruments and not give in to a desire to move the controls the wrong way. This was the most difficult bit of flying I ever had to do.'

Towards the end of 1943 the MISSION strip served primarily as a refuelling stop for aircraft traffic north and south. A report on file by the Divisional Property Inspector dated 23 November 1943 records -

'...Interviewed the CO F-O Birch... No general repairs are recommended, but provision of grease traps and fly proofing of kitchen (are) both items of very urgent nature... There are 116 RAAF personnel of No 27 OBU and 33 personnel of an American Weather and Signals Units of which 19 are Sergeants. The whole camp area and buildings are clean and tidy and well looked after.'

During the wet season early 1944, some serious pavement failures occurred. There was then an urgent need for repairs to be completed before elements of the RAAF's 13 Squadron took up residence at COOKTOWN.

In the sparsely populated Cape York Area where (inevitably) 'word got around' and the inhabitants tended to know the business of neighbours, in some cases 50 to 100 miles away, it's not surprising that rumours abounded. In April of 1944 the Commonwealth Security Service saw fit to investigate some reports of alleged espionage activity in or around Cooktown. A twenty page report by Major K L Murray on file in AA, would certainly be of interest to surviving elderly residents of Cooktown. His findings read in part -

'...After 2 weeks 2 days in Cooktown I'm satisfied there's no tangible evidence that this town was being used for any espionage activities. Minor happenings in the past detailed subsequently in this report would give an air that espionage may have taken place. But there is no trace of any evidence in support. The matter is mostly rumours and gossip aimed at persons who are unlike the general type of local residents in that town. While there are some people here whose activities in the past may be questionable, it is possible that there is the potential for espionage, but I must repeat no evidence other than wildest gossip is available. However I have made a thorough inspection trying to examine possible habits, activities and reputations of persons now there.

Cooktown is a ghost town. Buildings both in the town shopping centre and residential area are dilapidated and vacant and valueless. A good home based on Cooktown's financial standing can be bought for less than £50. Many buildings appear as though occupants simply walked out and left them and this in fact is what happened.

Cooktown's population is less than 300 with some 60 of this 300 attending the local school. There's a RAAF Unit at the strip 8 miles out. A Field Survey Unit of the Army occupies a Hotel and the US Navy have a small unit in the convent. There are also the staff of the civil aerodrome, the civil aeradio station and the AWA Coastal Station. Also the Cooktown Hospital and staff, and an NAP launch and crew.

Intermarriage amongst the residents is large and nearly everybody is related to each other. Mixed religious marriages are common with their attendant gossip and ill feeling. The inhabitants are divided into cliques each clique engaged in the detrimental slander of the others.

At present the war is far away and any discussions which take place are based on the daily paper received every 2 or 3 days from Cairns or the events surrounding the evacuation during the Japanese advances.

In the gossip surrounding each person, the usual opening statement is that he or she is disloyal. How or why such a person is disloyal is never explained.

A main feature of my survey was to investigate possibility of Cooktown being linked with stories of espionage activities at Iron Range, Portland Roads, Coen or elsewhere on York Peninsular. Cooktown area covers some 50 000 square miles, extending west to Palmer River area and on towards Gulf of Carpentaria and northwards to Princess Charlotte Bay.'

Dozens of residents of Cooktown and several 'PIR' (permanent itinerant residents), came under scrutiny. I wonder if the highly regarded modernist painter Ian Fairweather, who painted and lived the life of a recluse for years on Bribie Island before his death in 1974 would have known of his mention -

'...Ian Fairweather- This man left the district by boat on 10 December 1943. He stated he was going to Melbourne. His Identity Card Number is Q6/M91/38025. This man did not fit in Cooktown society and as he was alleged to be an artist he suffered the usual allegations that he was a "spy". Nobody could produce any evidence or incident which would support it.'

The statutory 30 year closed period on this report probably meant Ian Fairweather died blissfully unaware that Big Brother had 'the eye' on him in April of 1944.

Ron Ziesing was a Medical Orderly at COOKTOWN with the RAAF 13 Sqn. Ron recalled June of 1944 -

'...When the Squadron was due to arrive in force, I had the task of setting up the medical section. I had to argue about accommodation for that section... we had to have... some kind of shelter and, after a lot of discussion we were given three tents. They were pitched each with a dirt filled log surround (raised earth floor) about a foot high and were ready for occupation when the doctor arrived. One was for the MO's examination room, one the treatment tent and one for the sick bay. When Verner W Threlkeld arrived and we were about to do a sick parade the tents were whisked away and we were without accommodation. But what were we to do for sick parade? On an occasion like this, the doctor said, we just "muddle through" and do the job on hand. We had six panniers: 2 medical, 2 surgical, 1 fracture and 1 hygiene. We opened what we needed and did what we had to do then and there, in the open air. The "doctrine of necessity" dictated that we "muddle through" and it worked well enough for the immediate needs.

Within days one of the buildings that had been used as sleeping quarters was emptied of personnel and the medical section moved in...

We had use of the hospital in the town for the more serious cases, but none of the orderlies were expected to work in the hospital; the hospital had its own civilian staff. When we had to take someone to hospital for admission we were always treated to cups of tea.'

Ron Ziesing went on to tell of the actual arrival of 13 Squadron -

'...The Squadron eventually occupied the camp site. I was in a tent up the slope and to the left of the cook house. Further to the left from us was an American "Radiosonde" unit. They would send up a balloon each night to track the upper level air currents, temperature and so on, or something like that. As I recall it, their names were Glazer on the bottom rung (of rank), Hanson on the next rung up the ladder of rank and someone else, a sergeant of some kind, who was in change and responsible to an officer who visited them from time to time.'

Ron could not remember the Sergeant's name, but did recall the dissatisfaction this 'grounded pilot felt, stuck out on a hill at the end of the earth'. Ron said -

COOKTOWN ca(-/8/1944)
Inside the Medical Hut, Dr Threlkeld is pictured. Note the 'common or garden' beer bottles arrayed on the medicine shelf, no doubt serving an approved medical purpose. [R Ziesing ex RAAF]

COOKTOWN ca(-/8/1944)
The Medical Hut and some of the RAAF Staff at the COOKTOWN MISSION strip. In the middle is Dr Threlkeld with his dog. On his right is F-Sgt G G Peachey and on his left is Sgt Ron Ziesing. Ron can't recall the name of the man on the extreme left but he 'had a good singing voice'. The 'architect' is unknown, but clearly someone's spare time has gone into the balustrading. [R Ziesing ex RAAF]

'...I remember this Sergeant telling me some fantastic story claiming it to be the proven truth and I responded with "you're not fair dinkum are you?" His face lit up. He said he had read that expression in a book of facts about Australia but could not see how it could be used. It seemed simple enough to me but then they used to say things I didn't understand.'

With its Lockheed Ventura aircraft deploying over enemy territory for the first time, 13 Squadron detached seven aircraft to the AOB at Merauke early in August.

Its interesting how veterans have struggled to 'date' various events. Ron Ziesing for example had keyed certain recollections of COOKTOWN to the following -

'...I may be wrong but I seem to recall that there was an election, or voting of some kind held while we were at Cooktown and there was something of a campaign. Politics was beyond me in those days but someone may recall something about it. Gough Whitlam was with the Squadron at Gove I'm sure and he may have been at Cooktown too. He may know something about it, if there was voting (in it).'

This was the 19 August 1944 '14 Powers Referendum' whereby the Australian Commonwealth Government sought wide-ranging Post-War powers. Like most referenda it was lost.

Apparently in June of 1945, brief consideration was given to accommodating a WAAAF detachment at COOKTOWN. While some of the accommodation was deemed suitable, the proposal, for whatever reasons, did not proceed.

COOKTOWN MISSION strip after the cessation of hostilities fell into the category which the RAAF termed as 'airfields planned to be "acquired only" and the runways in existence will not be maintained'. On this basis the Department of Civil Aviation did assume responsibility for COOKTOWN, up until mid 1992 when it was taken over by Cook Shire Council. It is the current airport for Cooktown.

MITCHELL RIVER MISSION *(5/7/1942)*
Happy to be down safely, the crew of B-17E 41-2417 smile for the camera. Believed piloted by Jack Thompson, the aircraft belly landed some 15 miles northeast of MITCHELL RIVER MISSION, heading the Navigator believed, for Normanton. Bruce Hoy's advice is that the rest of the crew were: ? Schmid (CP), ? Dawson (Nav), Feaster Norwood (Bomb), ? Sexton (Eng), ? Norton (Asst Eng), ? Prince (Rad), R D Copley (Asst Rad-W Gun), ? Secor (Gun). It seems likely this photo was given to Fred Cooke by Feaster Norwood. *[F Cooke via B Hoy]*

MITCHELL RIVER MISSION *(-/7/1942)*
With 12417 just discernible on the tail, this 19 BG aircraft is inspected by US recovery team. By design, the main wheels of a B-17 while fully retracted, protruded below the engine nacelle to provide weight carrying contact on belly landing. In this case the left wheel passed directly over a 'melon hole' or sink hole roughly enough to call for extensive repair. *[W Maiersperger ex USAAF]*

MITCHELL RIVER MISSION

'Blacks save Big Plane.'

So stated the Brisbane ***Sunday Mail*** newspaper, January 3, 1943. In a 5 'column inch' report hampered by the censorship of the day, the ***Sunday Mail*** quoted W F Mackenzie, in charge of a Presbyterian Mission in the 'wild bush of the north' -

'...not long ago a Flying Fortress with an American crew made a forced landing.
She pancaked on to a natural saltpan clearing, but she was in pretty bad shape,... She was located by a Catalina, and a message was sent through to the superintendent of a river station where there are about 700 blacks. The head stockman set out with a team of horses and brought in the crew.
The pilot had a rest and something to eat, then asked could he borrow some men to take him back, because he would have to destroy the plane. And that meant £80,000 worth of Flying Fortress.'

The report continued with how the Mission Superintendent (Alec) McLeod saw the potential for the aircraft's recovery given the help of 700 'strong healthy natives'.

The Australian Board of Mission's ***ABM Review***, published monthly even during the war, carried the following item in its March 1 1943 edition under the heading of MITCHELL RIVER MISSION -

'...Even with war conditions so noticeable in the north the work of the Mission goes on, but is more varied than usual. Its variation consists in rescuing airmen, building by native voluntary labour a 17 mile road to where these planes came down, and constructing a runaway (sic) to take off again. The airmen were amazed at the capacity of the natives to find their way about in the dark, and also refusing payment for the work. The 17 miles of road was put through in a week, and the landing strip of 1200 yards in five days. To get to this plane, four rivers had to be crossed and many creeks (crossings) had to be made so that they would carry 8 ton trucks. The natives had only a few shovels and picks; the rest worked with their hands. It was a case of the natives showing the Yanks how to work.'

MITCHELL RIVER MISSION airstrip was in existence at the outbreak of the war. It catered to aircraft visiting and servicing the then Church of England (now Anglican Church of Australia) Mission, itself part of the activities of the Diocese of Carpentaria.

My inclusion of MITCHELL RIVER MISSION in the 'Airfields' section of this book was sealed with my successful contact with two US veterans in 1992. One forwarded extracts from a 1984 Reunion Booklet of the 47/479 Service Squadron which told of this B-17 recovery from the American viewpoint -

'...On or about July 22 1942 Lt Schmick with twelve enlisted men of the 47th Service Squadron and two enlisted men of Company B, 33rd QM Regiment (T) were ordered to proceed overland to Mungana (west of Cairns) and then to Mitchell River Mission on the east coast of the Gulf of Carpentaria where a B-17 of the 28th Bomb Squadron (19 BG) from Cloncurry had crash landed. The pilot was returning from a mission and lost his bearings. After a good deal of flying he ran out of gas and was forced to land his ship. He sent a telegram to his unit giving the location of the plane and then left on a bi-weekly commercial plane from the strip near the mission. This information was communicated to the group and immediately road maps were consulted and plans were made to send a detachment out to repair or salvage the aeroplane.
It was known there were aborigine natives in the vicinity of the mission and a supply of rations, flour,

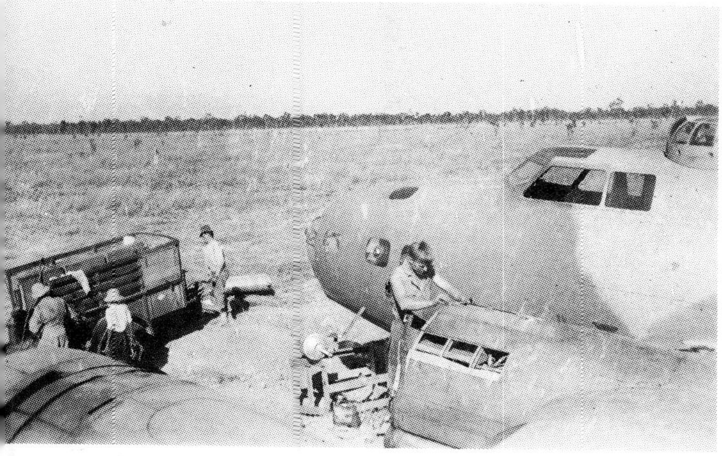

MITCHELL RIVER MISSION (-/8/1942)
*Rebuilding the port inner nacelle, crown piece and landing gear damaged by wheels-up landing. The trailer was built up by the US Service Squadron specially to handle repairs on recovery expeditions. It is known the team had the assistance of a '**14 year old boy from the Mission named Peter**'. This was probably Peter de Rizey and it may be Peter in white shirt and US pith helmet by the trailer. Peter - where are you now?* [W Maiersperger ex USAAF]

tobacco and trinkets were assembled beforehand. Sgt Joseph Pizzuto flew to the scene and pickled the engines, while the rest of the detachment travelled overland by truck. Work had to be done on the propellers and landing gear. At the same time a runway had to be constructed. The plane was jacked up and work on the landing gear and props proceeded. Natives cut a runway out of the thick undergrowth and jungle and were paid with flour, tobacco and trinkets. Several weeks afterwards Capt Walter Maiersperger, the new Assistant Base Engineering Officer, who had recently checked out in a B-17, flew the B-17 to Charters Towers where it was worked on some more and then flown to the depot in Brisbane.'

My contact with the surviving Captain, now Lt-Col (Ret'd) Walter P Maiersperger, struck gold. Walter was able to give more detail and loan a treasured set of fifty-six private photos of the recovery expedition. Too often, the only 'notable' event at or near an airfield is a crash resulting in death. This opportunity to tell even a little of the story of an aircraft recovery tied to Mitchell River Mission, has been most inviting.

Th Rev'd Philip Freier, Rector of the Anglican Parish of St Oswald, Brisbane has studied the long term history of the Mission. Philip seeks contact with any Service personnel involved in the recovery to gain their impressions of the Mission, its policies and the impact of the servicemen's brief presence. His own research had uncovered the following extract from RAAF's 11 Squadron ORB -

'...5 July 1943: Search by Catalina aircraft A24-11, Captain:Squadron Leader R A Atkinson DFC (RAF 70030), Second Pilot: Pilot Officer P L B Marsh (RAAF 408191) of 11 Squadron, for missing B-17 aircraft, and to drop food and water. A24-11 was airborne, ex Cairns, at 2030 Z/6. The missing aircraft was located in position 1522 S, 14156 E (sic). Visibility was good, broken cumulus throughout and light winds. A24-11 returned to Cairns and was waterborne at 0716 Z/7.'

Where the Service Squadron Reunion Booklet mentions Sgt Pizzuto flying to the scene, the RAAF's 24 Squadron ORB states -

'...8 July 1942: RAAF Wirraway A20-167 departed Garbutt to convey maintenance engineer to survey damaged B-17 aircraft at "Rutland Plains".'

Walter's description of the expedition supported by the photos, showed the party's route: from Charters Towers through Cairns to the Mungana rail terminus where the vehicles were off-loaded. A native policeman was engaged to guide the expedition northwest through 'Wrotham Park' and 350 km down the Mitchell River. Walter related -

'...they set up camp alongside the river near the plane and kept me informed of progress through the Mission's land-line telegraph... in this way I learned of the need for a new battery and differential bearing for the truck and delivered both with the A-24 (utility aircraft). I remained overnight in their camp and had anything but Australian Army rations for dinner that night. They had learned to supplement that diet by hunting and fishing and shooting crocodiles in the river for steaks! After dinner in that peaceful camp, which also included a native boy from the Mission, called Peter, the B-17 life raft was used to go crocodile hunting...'

The aircraft was restored to support on its undercarriage by hand excavating a sloping trench under each main wheel. The wheels were then lowered and eventually the aircraft was winched forward to emerge on its wheels. The natives gathered to watch. Walter says -

'...when the B-17 rose out of its ditch... the natives scattered, afraid of the big white bird! Little Peter, 14 years old, had picked up some engineering and GI language from our men. He announced to all, "don't be afraid, the goddamn thing has no battery in it!".'

Stripped of unnecessary weight, with minimal fuel and short of oil for all superchargers, the take-off began. The plan was -

'...to use full power supercharged on the inboard engines and full power on outboard engines without superchargers (on), minimising directional control problems should any engine fail...'

Normally following a belly landing, both props and engines would be changed for fear of sudden prop stoppage causing engine and reduction gear damage. Here the calculated risk of prop change only paid off. The flight back to CHARTERS TOWERS was uneventful up to the halfway point, when -

'...there was a loud bang in the rear and the control column started a violent fore and aft oscillation... I ordered Sgt Pizzuto back to investigate... there was another loud bang and the control column shaking stopped... part of the rubber de-icer boot on the right stabiliser had torn loose and flapped in the slipstream, excited the elevator and caused the stick shaking.'

MITCHELL RIVER MISSION (10/9/1942)
B-17 41-2417 takes off again after its nine week unscheduled stay on the ground. The task of preparing a take-off strip was not so much clearing of timber as the need to fill, compact and grade over 'melon hole' depressions. Dressed down tree trunks were dragged behind a truck to 'grade' the surface. None of that stopped the dust! [W Maiersperger ex USAAF]

As Walter found on landing the rubber boot had deteriorated due to the sun's rays, indicating the boots in all leading edges were due for renewal. He believes the aircraft was fully restored 'but never again flew operationally'.

With the outbreak of the Pacific War, there was concern for how the Aboriginals in remote parts of Australia would react to the threat of invasion by the Japanese. It became very important to the authorities to properly appreciate these Mission communities of Aboriginals.

R A Hall in his book ***The Black Diggers*** covers well the differing points of view. On the one hand it was feared the indigenous people had developed an affinity for the Japanese through the activities of pearling and fishing luggers along the northern Australian coast. This translated to a conviction that, were the Japanese to land, they would find the Aboriginals helpful to their cause.

The other side of the argument was put by observers with a longer history of close contact with the Aboriginal people. They claimed the Aboriginals could be relied on and were in fact loyal. It's probably true that such loyalty was directed to the Mission staff and that it could not be automatically transferred to all newcomers such as members of the Armed Forces.

Certainly as the threat of invasion receded, the Government found it unnecessary to further consider various recommendations for wholesale evacuation of such Mission populations. In fact, because of the logistics of building and supplying any Armed Forces outpost, one thing became obvious to the RAAF. All things considered, if a Radar station was needed in an area generally, then best it be located near to a Mission centre. Such was the case for Radar Station No 320, which was set up just east of 115° runway at MITCHELL RIVER MISSION.

Nowadays, the Mitchell River Mission settlement is known by its traditional Aboriginal name of Kowanyama. The airstrip continues to provide aerial service facilities, particularly for the Royal Flying Doctor Service in north Queensland.

WROTHAM PARK *(5/8/1943)*
This vertical shows the western end of the E-W strip hurriedly cleared and prepared in June of 1942. WROTHAM PARK was little, if ever, used by combat aircraft.
[AUSLIG]

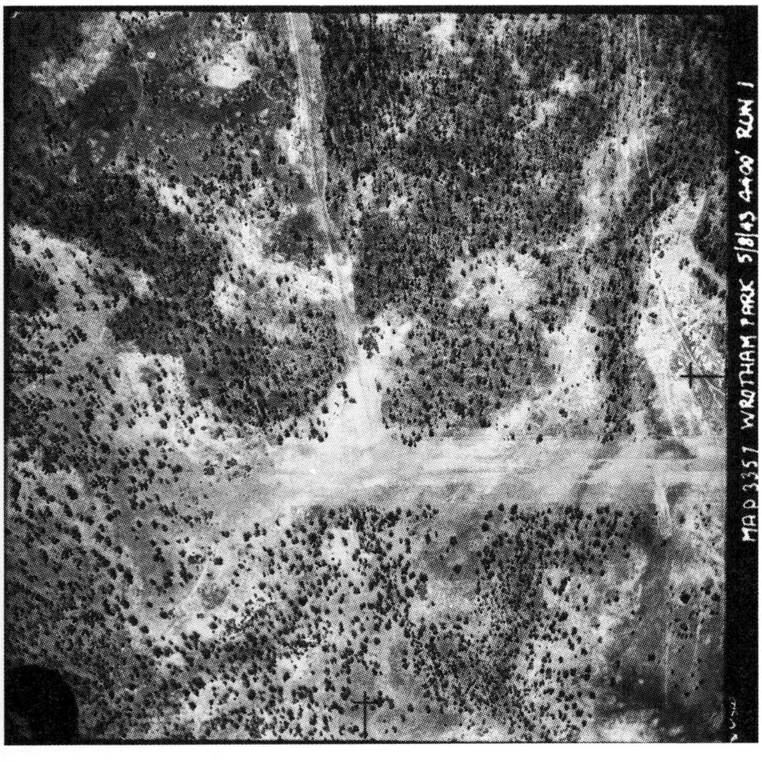

WROTHAM PARK

'...Both runways could be made useable in a few days but a skin of gravel about 1" thick will be required if strips are to be used continuously...'

It seems naive at the best to have thought in terms of a 1" thickness of gravel sufficing on these airstrips. Still that is part of the report on the WROTHAM PARK site filed back to Air Board on 14 May 1942. P-O Jim Trench, accompanied by MRC District Engineer Mr A J Bond and Mr Andrew an Architect from the Department of the Interior, had been searching for a strategically required site in the MUNGANA area. The site chosen was approximately 3 miles west of WROTHAM PARK homestead and 45 miles by very poor road from Mungana.

Under dry conditions the natural claypan and red sandy loam were quick to turn to dust. The ingestion of dust into aero engine cylinders is fundamentally taboo. So, laying, watering and maintaining a compacted gravel layer over the natural soils was the only option. Equipment, manpower and time constraints limited the thickness during this period of critical shortage of bitumen.

WROTHAM PARK *(23/8/1990)*
Discarded and forgotten at the western end of the E-W strip this improvised roller bears a lone testament to the wartime origin of this now disused airstrip. [B Marks]

Jim Trench figured in site reconnaissance for several new airstrips and on his eventual return to Townsville, some decisions were finalised. The AOC NEA, by letter dated 28 May 1942 addressed the urgent matter of 'Construction of Defence Bases in North Queensland'. Apart from moves forward by US Engineer units to commence crucial work, mention was made that -

'...at (the) direction of Col. Matthews work has commenced at Wrotham Park and Cooktown.'

The AWC Minutes of 30 July 1942 record that 'Wrotham Park is the aerodrome to be built in the Mungana area'. This tends to dispel the element of mystery attaching to 'MUNGANA' as a WW2 airstrip. It shows on a map in the 1944 issue (US) 'Air Pilot Manual of Pacific Islands' by legend, as 'Civil Ground - RAAF used'. There was confusion over the actual name assigned to some strips and difficulty of keeping all parties appropriately informed on development progress. It often took months to bring such 'maps' up to date.

The road itself northward from Mungana towards Laura and 'the Cape' was largely non existent. So when the initial construction work was finished at WROTHAM PARK, the crew numbering some twenty four men, were

WROTHAM PARK *(23/8/1990)*
The inscription clearly visible in the infill concrete of the roller, reads ***'Joe Hooper Wrotham Park 6/6/1942'****. George Gates, a young ringer in the district at the time was one of the construction crew on this airfield. He recalls that Joe Hooper was* ***'the roller driver'****.* *[B Marks]*

applied with their equipment to the task of improving that section of the road northward to Laura.

Post war, WROTHAM PARK strip was used to some extent by aircraft up to DC3 size. Sited across Elizabeth Creek from the homestead, it was found inconvenient on those occasions when Elizabeth Creek was running water to an impassable depth. Today, air service needs to the property are satisfied by a smaller strip much closer to the homestead.

'Wrotham Park' during the war years ran cattle on a holding through which the mighty Mitchell River flowed for 'a hundred miles'. The homestead was a welcome 'stop' to convoys of soldiers then. It still shares an outback welcome with today's travellers so long as they prepare responsibly for their journey.

CAIRNS *(1991)*
Today, the triangle of wartime runways can still be seen in the southern quarter of the airfield. The new International Terminal is on extreme left and the busy Domestic Terminal is located between there and the main runway '15'.
[Courtesy R Rudd Northair Surveys Mareeba]

CAIRNS

'...In preliminary discussions between your engineer, Mr O F Anderson, and the Officers of the Department, Mr Anderson said that provided the work was started before the end of August he would complete it before the wet season in December at a cost of £20 000. ...The total cost is 0.5% over the estimate and had there been no hold up in the delivery of flood gates it would have been under the estimate... More important still is the question of time. If this stage had not been completed before the wet season, Cairns Aerodrome would have been out of action to the RAAF and all but the lighter planes for weeks if not months... By your efforts the aerodrome has been open to all RAAF and civil aircraft throughout the wet season... at no time during construction was normal air traffic impeded in any way. In this case I would like to refer specifically to... your engineer at Cairns, Mr O F Anderson,... his organisation has been masterly and all officers of this Department who have come in contact with him speak of his cooperation in all matters. Furthermore, he did what he said he would do.'

This is the gist of a letter of appreciation addressed by Mr A B Corbett, then Director of Civil Aviation, to the Secretary of the MRC. The occasion was the completion of the first section of extensive work on CAIRNS aerodrome during the last four months of 1940 at a cost of £20 081.

Near Tom McDonald's coal-ash surfaced strip of saltpan, the Cairns City Council had constructed its first aerodrome. It needed improvement at the outbreak of WW2. The site was restricted by tidal influence to its low lying ground and alternative sites were considered. Called San Remo, a site about 2 km northwest of the present aerodrome was seriously considered. However with the urgency of defence needs and the fact that an embryo airfield with support infrastructure did exist, the present site was confirmed in February 1940. Extensive filling within levee banks was committed.

In mid 1941, 24 Squadron stationed at Townsville was naturally concerned to monitor progress on various requisitioned works at its several AOBs on the Queensland coast. CAIRNS was one of these and it was due for such mundane requirements as water tanks along with more substantial works such as bomb storehouses. RAAF records held in AA refer to these bomb dump requirements in a letter dated 12 July 1941, with comment such as -

*'...**Cairns** - this site is excellent and is situated on a small sand ridge, close to the aircraft assembly area. Tropical foliage covers the spot at present and this should be left untouched to conceal the hut and to help keep it cool. It is suggested that the access road to the store house be camouflaged and a cheap but obvious "decoy" hut, with an equally conspicuous access road, should be built on the eastern or southern side of the aerodrome... The MRB are at present at work on the aerodrome runways and it is recommended that, to save time and money, every effort should be made to start work on the aircraft dispersal positions and bomb storehouse, before these men are dispersed.'*

With Japan's entry into the war, the pace of life, work and general activity at the aerodrome stepped up remarkably. Jack Atkinson was at the time an Engineer with AOA a civil airline operator at CAIRNS. He recalls in his unpublished autobiography that -

'...when the Japanese attacked Pearl Harbor and started... south towards Singapore and New Guinea... things got hotter in many ways. The first order from

CAIRNS *(16/4/1991)*

This overlapping pair is taken from approximately the same vantage point as occupied by the Coastal Observation Post 41 during WW2. Since then Cairns airfield has grown enormously to reach international gateway status. In the area where Mitchells once stood, Boeing 767s now take to the air. The original AOA/ANA hangar can still be seen in centre left. What has not changed over the last fifty years is the presence of the once ubiquitous DC3 (lower right). *[Author]*

the Defence Dept was to camouflage the hangar. None of us had any idea on how to do it so we bought a lot of paint of all colours and painted all sorts of designs all over the roof and walls. It was a work of art.'

Rod Voller was in charge of the Townsville office of the Department of Home Security's Camouflage section. Reports in Australian Archives (AA), include one for period 12-25 March 1942 mentioning -

'...AOB's - paint has now been sent from Townsville to Horn Island, Coen, Cooktown, Cairns, Bowen and Mackay for the painting of aerodrome buildings, and as far as we know the work is proceeding...'

Jack Atkinson refers to a stepping up of runway works early in 1942 as allowing -

'...bigger aircraft to use the airport which they did by the hundreds. Civil aircraft such as tri-motored Junkers, tri-motor Fords, single engined Junkers, even (Charles) Ulm's old Fokker "Faith in Australia". All these had been used in New Guinea in transporting mining equipment up to the highlands, but were now being used evacuating civilians. I don't know how it came about but one "Ford" arrived in Cairns with only the pilot aboard. The plane was commandeered, we camouflaged it as best we could and he was sent back to help bring out more civilians. At one stage I witnessed bars of gold unloaded from a plane. There must have been 40 or 50 bars each about 16" long and about 3"x 2" thick and the funny thing was, they were left stacked on the tarmac for quite a long time with no guard because they were too heavy to carry away manually.'

Archive records include reports dated 15 April 1942 by P-O H V (Jim) Trench on inspection of CAIRNS aerodrome and dated 15 May 1942 by P-O J F Keays on aerodromes in Townsville area including CAIRNS. The US 46 Engineer Regiment had arrived in the Townsville area during March and its Commanding Officer Col. A G Matthews had by April of 1942, been appointed as the Senior US Officer in NEA. Both Jim Trench and John Keays worked closely with Matthews in various reviews and planning for upgrading of airfields in NEA. Jim Trench's report of inspection dated 12 April 1942 effectively said that there was no easy solution to the upgrading to all-weather use of the main 150° runway at CAIRNS. In this context Col. Matthews was to make the momentous decision on work at MAREEBA.

In its planned role as an AOB, CAIRNS required a number of aircraft dispersals. It was a site where the opportunity for construction of taxiways leading away from the airfield to hideout areas was just not available. For this reason, as at BOWEN and MACKAY, attempts were made to provide that protection via splinter proof bays in full view from the air, and close to the established building area. Progress reports on the construction of such 'RAAF hutments' show that for 30 June 1942 the -

'...Provision of additional buildings and services - work 92% complete, nearly finished. 12 splinter proof pens - work 40% complete. Obstruction barriers - work 80% complete.'

CAIRNS quickly became a major refuelling point on the coastal air service route with a huge influx in military traffic, particularly US Air Force. Some insight to the daily activity is evident in an RAAF report in AA dated 29 May 1942 to the AOC NE Area, attributed to Group Captain F R W Scherger, Allied HQ. Scherger advised -

'...In my capacity as Director of Defence, I visited the Operational Base at Cairns on 25/5/42. Little work had been done on aerodrome defence because of following:

a) there were very few available to do any work and no officer to direct them

b) apparently no quarters were available and the men who arrived, first had to erect their own quarters, shower rooms and latrines.'

In short, Scherger found, or concluded, that the only officer at the base was 'the Cypher Officer, said to be a P-O Doyle'. Scherger said Doyle had obviously no interest in the base or the men posted there, and that only the energetic leadership of Cpl Baggott had 'saved the place from complete chaos'. Numbers of men had increased from about 14 to 40 without warning. Many American aircraft used the base from time to time without prior warning. Despite this Cpl Baggott had been able to -

'...rapidly refuel visiting aircraft, feed and arrange accommodation for crews and accommodate and usefully employ the relatively large numbers of RAAF personnel who have, without prior warning, been posted there.'

CAIRNS *(12/1/1944)*

Australian War Memorial's caption to this photograph reads - ***'Q116289 Private V T McKinnon of the 17th Battalion Volunteer Defence Corps manning coastal observation post 41, checking an incoming aircraft'.*** *On the extreme left of the photograph the single high gabled hangar of ANA is seen amongst the building group. Aircraft parked near the far end of the runway are North American B-25 Mitchells probably transiting north. Nearer the southern end are C-47 Transport aircraft, the Military version of the DC3 airliner of the day.* *[AWM Neg 62844]*

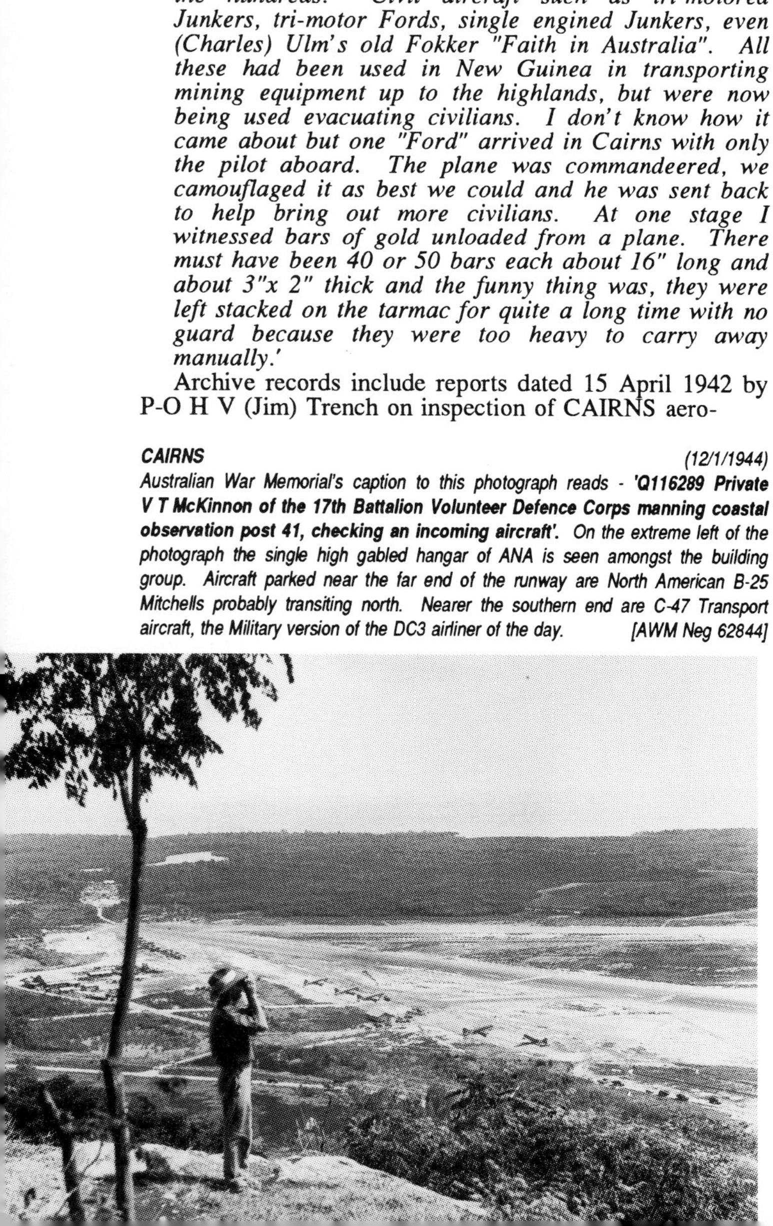

Clearly P-O Doyle had been invited to respond to this 'adverse report'. In his 3 page reply of 22 June 1942, he spelled out in greater detail the unannounced arrivals and general strain on resources, even down to food. He added -

'...since Signals Office was removed from the drome (its) work became so heavy that I have not personally had much time to devote to (the airfield) side...'

P-O Doyle summarised -

'...Generally, I was attached here for duty as Cypher Officer. I could not obtain any indication of... the aim of the Area in regard to this base while I was in HQ, nor what developmental work was desired. I arrived here towards the end of February, to find a detachment from 24 Squadron in occupation, (this being one of their AOB's) with F-Lt Wright in charge. Until that detachment was recalled in March, I was under the command and the base under the control of that Officer, except the Signal Office which was my responsibility.'

On the matter of Aerodrome Defence, Doyle said, he'd forwarded a report to HQ with no response -

'...With a personnel of 10 men and 3 rifles nothing could be done except to plan a rapid withdrawal.'

Incidentally, the RAAF's new 'bare base' being constructed east of Weipa on Cape York Peninsula will be named **Scherger.**

In June of 1942 the immediate role for CAIRNS was dictated in an NEA review of operational aerodrome requirements. Air Vice Marshal Bostock, in a memo to Chief of Air Staff RAAF, and the Commanding General USAAS, indicated that CAIRNS together with COOKTOWN were both to be retained as refuelling points. A further result of this strategic review was confirmation of instruction to cease work on splinter-proof revetments at CAIRNS. Indeed construction of these 15' high timber faced earth mounds had proven a real pain in their siting on what was, effectively mangrove mud.

In a move by the Director General of Civil Aviation to positively address the problem of an all-weather runway at CAIRNS, a proposal was made to reconstruct or treat the surface by soil-cement methods. This was the recommendation by Dr K N E Bradfield in his 20 July 1942 letter to Department of Air. Extremes of wet and dry seasons made unsealed gravel surfaces impracticable at CAIRNS. Probably in recognition of underlying mangrove mud, he noted that -

'...Bitumen even if it were available is not considered very satisfactory due to peculiar sub grade conditions at Cairns aerodrome. There is a considerable quantity of cement at Townsville which was part of a cargo destined for Netherlands East Indies and it is understood sufficient of this can be made available for treatment of the runway.'

Dr Bradfield briefly reviewed the complementary merits of this aerodrome and MAREEBA 'being developed on the Tableland at the back of Cairns'. He referred to the last wet season -

'...when on more than one occasion heavy bombers returning from attacks on enemy territory with crews tired and petrol low, made use of Cairns aerodrome... returning aircraft flew on a course until they reached the Australian coast and then flew by contact down the coastline until they reached a coastal aerodrome such as Cooktown and Cairns...'

CAIRNS (1942-43?)
*This group of RAAF men is pictured in front of a fire tender with the Duty Pilot's Tower in the background. The prominent sign on the tower reads - **'all pilots are to report on landing and before taking off'.*** *[K Hutchinson ex RAAF]*

CAIRNS (early 1942)
'Six of the first RAAF men to be stationed at CAIRNS airfield'. *Likely a group photo not to have been promoted by the Australian Government's Department of Information during the war.* *[K Hutchinson ex RAAF]*

On 6 October 1942 a report to Air Board appears under the signature of W Cdr Gordon Grant for Gp Capt Commanding NEA -

'...Following instructions P-O Trench visited Cairns and inspected the aerodrome on 28 September 1942 in conjunction with Mr Rudge, Main Roads Engineer and Mr Redmond of Dept of Interior. Further inspection was made on 30 September in conjunction with Mr Bond and Mr Pollock, MRC Engineers. It is estimated an average of 25 transient aircraft use the aerodrome daily and as it is proposed to locate fighter squadrons at Cairns, it is considered essential that one runway be put into all-weather condition for the wet season... mid December.'

Jim Trench's deliberations covered some ways to upgrade the surface but come back to the need for some degree of bitumen sealing. The AWC minutes of 24 September 1942 indicated the soil-cement work at CAIRNS was 'unlikely to proceed until equipment is released from (other) road and drome works in North Queensland'.

Jim's reviews addressed the stalled work on aircraft revetments -

'...As all the (75 Sqn) aircraft will be located within a few hundred feet of the runway, proper splinter proofing is most desirable. The pens could probably be completed by constructing the back with A-frame sections instead of the open (dumped fill) section specified and it is suggested this procedure be adopted.'

By October of 1942 an interesting insight into the tenor of activity at CAIRNS aerodrome is gained by reference to a Security file in AA. Headed, 'Report on aerodrome defences Cairns and Mareeba', only that section applying to CAIRNS appears. In part it reads -

'...Cairns - aerodrome defence officer P-O Bromley was absent on leave and expected back Wednesday 14 October 1942. Guards are primarily used for labouring work and are not performing their duties as guards. The Local Garrison Battalion provides the sentry on the main entrance. The local VDC provide the guards for the Bomb Dumps. Two only could be contacted and this number is inadequate for the area to be patrolled owing to the dense nature of the scrub. There is a serious risk of fire as the grass is high and dry. The northern edge of the dump is bounded by a river, navigable by small craft at high tide and there is nothing to prevent a party from moving upstream from the sea for purposes of sabotage. The whole of the eastern and southern perimeter of the area, including drome and dumps is bounded by mangrove swamps which afford a concealed approach with little fear of detection. It is suggested the mangroves be cleared for a distance of at least 50 yards beyond the aerodrome perimeter, in order to give a field of fire for the defending troops. Gun positions for both anti-aircraft

and ground defence have only recently been commenced and only one is approaching completion. There was no person available who could give me any information as to the siting of the defences. In the past the VDC have constructed Section posts on the southern end of drome perimeter. These are only death traps as they are not sited for all round defence and the scrub closes in to within a few feet of them. These must be destroyed. Arrangements were made with the Military authorities to provide Medium Machine Guns and Light Machine Guns for defence against low flying aircraft while the aerodrome was being used by the Bostons (A-20s) which arrived on 10 October. Plans are in course of preparation by the Military Authorities for the tactical defence of the Cairns area including the aerodrome.'

By October 1942 the runway surface at CAIRNS had deteriorated. The AOC NEA advised Air Board that -

'...Civil dromes at Bowen and particularly Cairns and Cooktown have been used extensively during recent months by service aircraft on operational and transport duties with result that runways are in need of attention. Gravelled surfaces are loose and corrugated and require grading, watering and rolling with addition of paving material in places of maximum wear...'

Positive action on all-weather sealing continued to be delayed. Clearly the Department of Air and the Civil Aviation Department (CAD) didn't want to share responsibility for CAIRNS. It was a CAD drome, with an expensive runway problem, which was just going to have to muddle along 'til its priority came up.

Dated 6 November 1942, the DWO, F-Lt B A Brogan forwarded a letter to DWB Melbourne with a copy of drawing AT42/43/17. He described the drawing as -

'...showing the location of 26 hard standings being constructed... for use by a US Transport Squadron. This work has been put in hand by the US Base Section Engineers at the request of GHQ. Instructions from GHQ state that bomb protection and wide dispersals are unnecessary. The scheme has been prepared in this office in conjunction with Dr Bradfield of Civil Aviation Department.'

The drawing shows, on one of the old hardstandings off the 80° taxiway - 'Bofors gun position - to be removed'. The hardstandings were required for the C-47s of US 33 Troop Carrier (TC) Squadron, of the 374 TC Group. So started the association with CAIRNS of various US TC Squadrons, often involved training paratroopers in the Gordonvale area.

During December of 1942 the US Forces moved to have a major arch type (Igloo) hangar erected at CAIRNS aerodrome. The project aimed to provide under-cover maintenance facilities generally for transient US aircraft. The archive records show that some indecision attached to the progress of the project because the site chosen was, of necessity, to be close to, and within the established 'building area'. The Civil Aviation authorities viewed the proposal in a 'post war' setting and therefore were anxious to ensure that anything that was done was not a wasted effort.

With the concentration on maintenance facilities for the Pacific Air Terminal and later the huge expansion of the 4 AD at Townsville, the US Authorities advised on 18 January 1943 that this CAIRNS Hangar was no longer required. It was to be another two years before a major hangar structure did proceed for ANA.

The RAAF's 75 Squadron had pulled out of New Guinea to HORN ISLAND in October 1942. Keith Gamble remembers it as a short stay -

'...there was a water shortage there, thank goodness, so they sent us to Cairns; it was better - there was a brewery there!'

Amongst some old snapshots, Keith produced one of Lex 'Pixie' Winten a 75 Squadron Flight Commander, 'beating up' the 'drome at CAIRNS in a Kittyhawk. Keith dated it as 'about December 1942'.

CAIRNS *(25/12/1942)*
A RAAF 75 Squadron P-40 Kittyhawk piloted by Lex Winten screams over the AOA gable roofed hangar, with Lumley Hill unmistakable in the background. The Squadron was reorganising after front line duty at Milne Bay. (K Gamble ex RAAF)

Joe Mills was in RAAF's 7 Squadron. He recalled a trip up to Port Moresby -

'...with some cheer for someone's Christmas. The weather and other things made it a difficult run back... it was about 3 o'clock in the afternoon we finally got into Cairns. They'd eaten all the Christmas Dinner - all we got was a bit of ice cream... an' Lex Winten was there. I hadn't seen him since I knew him at Evans Head. He must've had a good Christmas dinner - someone got him out of the kite to sign the authorisation book which they reckoned sobered him up a bit, you know - gave him a bit of time. Well we'd done a torpedo (bombing) course - in that you fly through the masts and under the bow... you know, evasive action... but the shoot up that Lex Winten did was nobody's business. He'd roll off the top, come down the mountain there at Cairns, across the drome and out towards Green Island... do a "split-arse turn" as we used to call it, back and up again - boy, did he shoot the place up!...'

A three page report dated 17 January 1943 by Gp Capt Thomas, Senior Air Staff Officer, provides an insight to 75

CAIRNS *(5/8/1943)*
One of a set of overlapping verticals this photo shows the dispersal taxiways and revetments near the building complex at that time. The track leading northwards off the end of the main strip leads to the bomb dumps. The largest aircraft visible is a B-17 Flying Fortress of the US Army Air Force. The largest of the buildings was the high gabled roof of the ANA (ex AOA) hangar. [NLA (RAAF Official)]

CAIRNS *(-/8/1943?)*
USAAF B-17E 41-2408 is photographed on the ground with the undoubted background of the southeast spur running up to Mt Whitfield. The date is not positively known but by coincidence a similar aircraft is seen in this position, in the 5 August 1943 verticals. [B Penglase via D Vincent]

Squadron's preparations for its return to Milne Bay. Addressed to the AOC NEA it spelled out some personnel matters deserving attention, but generally found the unit 'completely ready to move to its new destination'.

Thomas' report extended to review the local 25 OBU -

'...Matter of moving OB personnel from the aerodrome where they are very cramped to the camp site to be vacated by 75 Squadron was discussed with Sqn Ldr Parker and it was decided this be put into force soon as 75 moves out. This will allow approx 7 houses now being occupied in vicinity of aerodrome by OB personnel to be handed back to their owners. Two houses will be retained, one which is now the OB Store and another which is the OB Operations Room. These will be held until the reinforced concrete operations room is completed.'

I don't know where this new Ops Room was located!

Gp Capt Thomas reviewed the RAAF's 9 Mobile Fighter Sector unit too. He praised its state of equipment and readiness to move on 'soon'. Australian-American rivalry shows through in his mention of -

'...The camp site of this unit at Cairns is very nearly a perfect one and so it will not be occupied by the Americans, its intended to place guards from the OB temporarily there until next Fighter Sector, believed to be possibly, No 11, comes forward to Cairns...'

CAIRNS *(1945)*
This vertical shows the development reached by war's end. The taxiways and revetments had long outlived their requirements and logically became the foundations for service roads in the airfield's post war development.
[AA CRS A705/1 Item 7/1/963]

CAIRNS *(1945)*
The steel frame structure of a large modern Maintenance Hangar for ANA rises with the then familiar sight at Armed Forces locations, of the mobile cinema vehicle here towing its generator trailer. The southeast skyline is False Cape. [E Petersen]

Some reason for such rivalry shows later in Thomas' report -

'...the set up of US 565th Air Warning Battalion was inspected at their location in Abbott Street, contact being made with Lt Blackstock, Technical Officer in Charge. The equipment of this unit is nothing short of amazing. Every conceivable point for air warning and also air/ground control for what we know as a Fighter Sector has been covered with the most modern Western Electric equipment and it is indeed a model set up and well worth inspection of any officer of this area passing through Cairns. It is of the most elaborate design and capable of handling up to 8 squadrons for Fighter Control as well as the many watching points, gun control, army liaison etc, which goes to make up an efficient Fighter Sector. The unit is absolutely complete in itself as far as equipment is concerned, but lacks sufficient personnel to fully man it for operational use. It will be ready to go into operation by the time No 9 Fighter Sector moves out and at the moment has sufficient personnel to control it for the very limited use it will receive in the Cairns area.'

Earl Mahan was an American Air Force air engineer with an acting Co-pilot role on DC3 VH-CFU. From December 1942 to November of 1943 he served with the 41 Troop Carrier Squadron, mainly at GARBUTT. Earl recalls a significant time in CAIRNS, in his words -

'...hauling paratroopers for jump training, picking them up at Cairns airport and dropping them over Gordonvale or whatever.'

Most of his loads were food and troops to New Guinea and the return to the mainland of wounded, bound for

CAIRNS *(29/8/1990)*
By war's end these two buildings were the largest at the airfield. Nearly fifty years on they remain, now serving the general aviation industry. [Author]

hospitals at Cooktown, Cairns and Townsville.

Eventually, CAIRNS runway was described in a 2 November 1943 inspection report as -

'...recently completed blade-mix seal coat on main runway was inspected; generally satisfactory... some imperfections were noted... This is a CAD job, but was inspected, owing to probable use of similar methods on RAAF works, namely Runway Strengthening Amberley.'

By March of 1944, construction of a new 'bomb handling' jetty at CAIRNS was in hand to serve the flying boat operations from the waters of Trinity Inlet. Interestingly the site of the initial timber slipway at CAIRNS to serve the Catalinas, was very close to the long established public ramp near the Radisson Plaza Hotel.

Alan Payton was involved in the 'forming up' of RAAF's 10 Local Air Supply unit. He recalls working on an engine change to a Beaufreighter which had been 'road hauled down to the Flying Boat Base at Smith's Creek.' He remembers the RAAF's 41 (T/Sea) Squadron's Martin Mariner Flying Boats there in late 1944.

A January 1946 RAAF review of airfield requirements post war, indicated the Flying Boat Base was to continue as an RAAF responsibility. No such mention was made of CAIRNS airfield itself. Flying Boats have long since disappeared. However the civil role of CAIRNS has multiplied to its present day International Gateway status in the forefront of catering to booming tourist numbers.

MAREEBA *(1943)*
This extract from the 1:63 360 Military Survey Map shows the layout and location of the airfield with respect to the town. Two runways, connecting taxiways and dispersals appear with several 'camp areas'. Dispersal taxiways were made 'winding' to make it more difficult for attacking aircraft to strafe an orderly line of aircraft.
[Department of Defence]

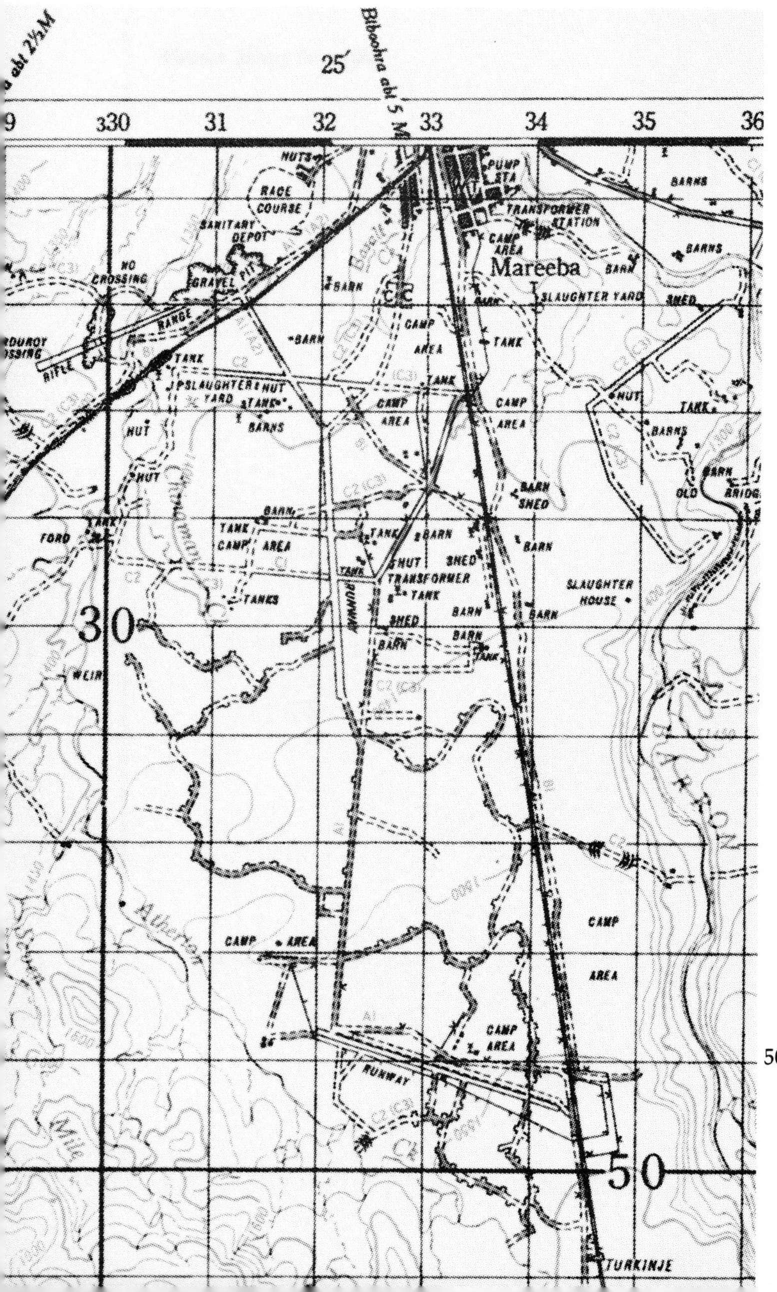

MAREEBA

'...I was driven from Cairns to Mareeba by Mr Partridge,... from Civil Aviation. On the way he showed me a site just outside Mareeba on which Civil had pegged the centrelines of two suggested runways. The idea was to use the runways as an alternative to Cairns which was subject to fogs and the aerodrome there was below high tide level and not very satisfactory...'

P-O H V (Jim) Trench recalled further his part in the initiation of this important airfield -

'...I arrived back in Cairns on the Sunday of the weekend that the Coral Sea battle was fought and I stayed the night there. During the evening I had a drink with John Keays, another RAAF engineer whom I knew very well and Colonel Matthews who was Base Engineer and Commanding Officer of the Americans in the Townsville area. Col. Matthews asked me to have a look at Cairns aerodrome with the idea of suggesting any improvements that could be carried out quickly as it was proposed to move (RAAF's) 100 Squadron there in two weeks time. I then had one of my infrequent brainwaves and suggested that the Colonel consider developing the Mareeba site. At that time the Colonel had the right to spend up to £5 000 on any one "work". He seized on the idea and issued six separate chits each for £5 000 and work at Mareeba started almost immediately. 100 Squadron moved there... Within a few weeks Mareeba was taken over by the Americans and at one stage there was a headquarters and five operational squadrons stationed there with thousands of men and hundreds of aircraft. As a matter of interest Air Commodore Lukis AOC Townsville RAAF and the senior Allied officer in the area was able to spend up to £50 without reference to anyone, compared with Col. Matthews' £5 000.'

During the 1970s, with some awakening to the WW2 historical significance of MAREEBA, there were claims of it being involved in the Battle of the Coral Sea. While historians are comfortable with that action extending over the week ending 11 May 1942, North Queenslanders remember a much longer period of anxiety. To many this stretched into early 1943, long enough for their memories of RAAF then USAAF activity in the sky around MAREEBA to fuse with that historic naval-air engagement.

Mr F J 'Snow' Baker told me a little before he died in November 1989, that work had started at MAREEBA on Monday 11 May 1942. If both Jim Trench and 'Snow' Baker were correct then those were very busy hours since Col. Matthews 'seized on the idea and wrote out his six chits' some time on the previous Sunday evening! 'Snow' Baker had spent the better part of four months with MRC on the drome's construction, under the direction of Mr Geof A Griffiths an MRC Engineer.

RAAF files in Australian Archives (AA) include a report to Air Board, Melbourne, dated 14 May 1942 under signature of Sqn Ldr Garnsey Potts, for the AOC NEA -

'...on instructions... inspection was made by P-O Trench of the area adjacent to the railway system running inland from Cairns, with a view to selecting suitable aerodrome sites in strategic positions in that area. This inspection was made in conjunction with Mr Bond, MRC District Engineer and Mr Andrew, Architect from Dept of Interior.'

MAREEBA was one of four sites described in this five page report. The trip must have taken all the week of the Coral Sea Battle. It covered about 700 kms over 'rough tracks'. As Jim Trench told me -

'...it was the policy of the Queensland Government not

to bother with a road where there was a railway!...'
The report concluded -

'...Mr Bond advises that there is sufficient plant available to construct the aerodromes in a reasonable time. In the event of the full scheme being approved, it is suggested that two construction gangs be formed to construct Mareeba and then Wrotham Park and one to construct Einasleigh and then Prestwood.'

Mrs Laurie A Wiemers is one of the few participants who can today draw on personal memories in this matter. As secretary to the MRC Herberton District Engineer, A J Bond, Laurie is vigorous in her praise for the important part 'Bondie' played. As Laurie tells it -

'...Mr Bond was given a description of the land and following the 5 am phone call from (Sir) John Kemp, he inspected the site which he found to be a potato farm...'

Ralph B Arnold continues from there. He was then Shire Clerk at Mareeba. He tells of Alf Bond arriving on his doorstep at 6 am with news of the urgent development. Arnold's part was to summon the Shire's own plant and additional equipment where possible from other Atherton Tableland Shires. By midnight that day forces were gathered and clearing work commenced the following day.

The earliest 'plan' I've located of MAREEBA was apparently drawn to record 'Hideout positions chosen by W Cdr Balmer RAAF, and Douglas Annand, Senior Camoufleur, 23 May 1942.' This was the day after Sam Balmer's 100 Squadron arrived on the field.

The main road from Mareeba to Atherton at that time departed the present highway opposite the northern end of N-S strip and carried along what in fact was developed and used as the interconnecting taxiway down to the western end of the E-W strip. It then zigzagged eastward to rejoin the present highway near the current railway level crossing at the eastern end. A note on the drawing says -

'...as main road will be used as taxiway and possible dispersal area it may be advisable to divert public traffic to track CD.'

That wound amongst the trees keeping east of the railway line and southward to the existing level crossing. Other notings show the Camoufleur envisaged the gravel runways

MAREEBA *(3/11/1943)*
This photograph shows the northern end of the 7 550' Runway 171°. Right of that northern end can be seen one of the 3.7" A-A Batteries, some reinforced concrete structures of which remain today. The road leading from the northern end of the runway was a haul road for gravel from near the rifle range. *[AUSLIG]*

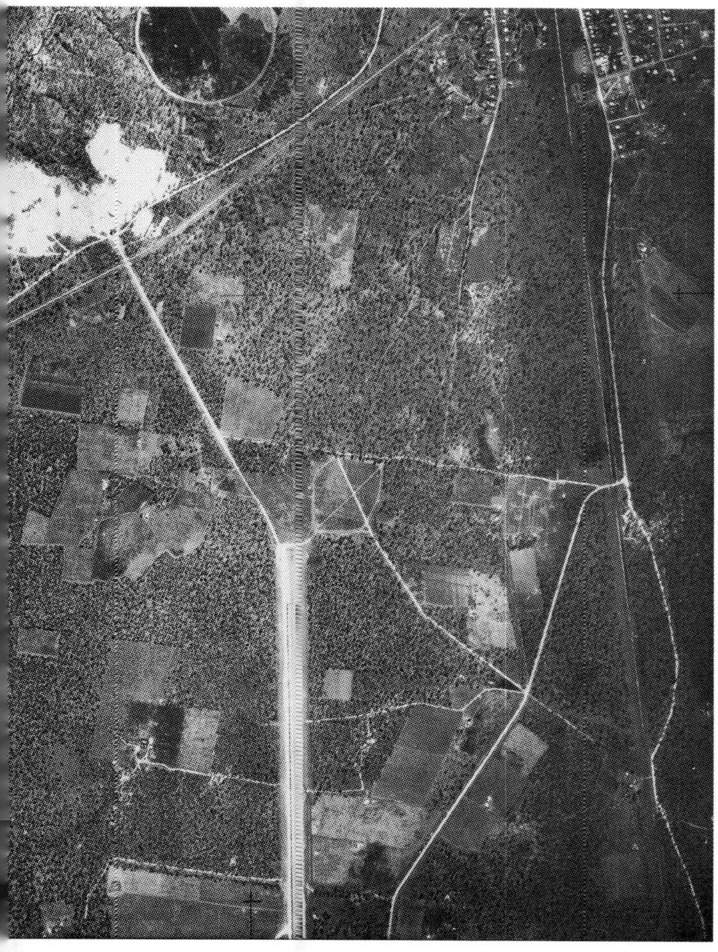

MAREEBA *(5/8/1943)*
One of approximately eighty photographs comprising a systematic coverage of the airfield, this shot homes in on the site of the northernmost 3.7" A-A Battery. The emplacements can be seen centre right of the photo grouped around a Control Bunker in what may be otherwise taken for agricultural land. For the reader who is particularly interested in MAREEBA Airfield, this large group of photos is well worth careful study. At 5400' altitude, the photos show areas of interest, such as 3.7" A-A batteries at grid references 330570 (the one shown in this photograph) and 342494. Light A-A emplacements appear to be located in this series of photos at grid references 317551, 317536, 336548, 335514 and 316535. At least one aircraft salvage area is discernible at grid reference 338534. Those frames show what appears to be a B-24 being salvaged, or scrapped? *[NLA (RAAF Official)]*

blending with surrounding broken outlined clearings by 'sowing grass or crop (such as)... Red Natal grass' -

'...The gravel of initial runway is decomposed granite of a pinkish-white colour and by addition of a dusting of (local) red earth, will tone in with Red Natal grass.'

The words 'RAAF Camp' appear on the drawing (grid reference 316521) suggesting this was the only camp at that time. While I can't be certain, this drawing by Annand points to the E-W strip being the first one brought into use. This is likely where 100 Squadron landed and this is today's Mareeba Airport.

100 Squadron aircraft made a first appearance in the sky over Port Moresby some days later. Burton Graham and Frank Smyth described this in their 1946 publication, ***A Nation Grew Wings*** -

'...these were the first two Australian built Beauforts of No 100 Squadron RAAF which had recently been employed as an Operational Training Unit on loan from the RAF for conversion to twin engine bombers. The squadron, the first to be equipped with Australian built Beauforts, had moved to Mareeba, NQ, for a period of training and convoy work. Its CO, W Cdr Sam Balmer was anxious to "tee up" at least one operational strike for his squadron before it returned south for reforming prior to going into solid action,and for this reason, on May 28, he had come to Moresby.'

The far reaching conference Jim Trench attended in Townsville which led to Lukis' report to Air Board dated 28 May 1942 has been touched on in several other relevant sections in this book. In respect of construction of MAREEBA and 'other Defence Bases in NQ' it was unclear to the AOC whether the RAAF or US Engineers were to be responsible for supervision. The US 46 Engineer Regiment was being pushed north to repair bomb damage and upgrade HORN ISLAND and to build IRON

MAREEBA *(28/8/1990)*
One of the reinforced concrete A-A gun pits in residential land fronting Cater Street, Mareeba. Presumably, the owner recognises some potential for its development as a sunken garden. Note the heavy steel fixing bolts for the gun base in the floor of the pit along with entrances to covered ammunition storage. *[Author]*

RANGE. The AWC with its Civil Construction Corps (CCC) either had its hands full with other work or its hands tied wrestling with the morality of working civilians in a 'war zone'. Australian-American cooperation was being strained.

A 1946 US Army publication titled ***Anti Aircraft Artillery Activities in the Pacific War***, gives some indication of initial A-A defence of MAREEBA airfield -

'...In June (1942)... HQ 1st Battalion of the US 94 Coastal Artillery (A-A) Regiment, with Batteries A (less 1 platoon), B, C, I and K... and the Australian 33rd Light A-A Battery defended the airstrips at Mareeba...'

Apparently five US CA (A-A) Battalions remained at MAREEBA for about 12 months. They were lightly equipped with Cal .50 machine guns, totally inadequate against high flying bombers. Events at Darwin, Port Moresby and HORN ISLAND showed this. AWC Minutes first mention the contract assignment of (reinforced concrete) work on 'A-A Batteries' at MAREEBA, to T J Watkins on 25 February 1943. This goes part way to explaining why several US and RAAF veterans can't recall being aware of any such defences there.

Requirements for Aircraft Maintenance Units in NEA were seen, mid June 1942, as major Depots at both CHARTERS TOWERS and GARBUTT, a Repair and Salvage Unit (RSU) near Port Moresby and another in the vicinity of CAIRNS or MAREEBA. Such RSUs needed runways either dedicated or shared.

Similarly, operational airfield needs were reviewed. Dated 19 June 1942, MAREEBA was to cater for -

'...a three squadron Heavy Bombardment Group. Present long taxi strip to be changed to a take-off strip. Powerline to be moved...'

Minimal sealing of 600′ lengths at each end of the two strips was prescribed, various building requirements spelled out and even detail of rudimentary refrigeration offered -

'...constructed simply of 2 board walls 8" apart packed tight with sawdust, properly ceiled and with sawdust on top of the boarding. For this, petrol motors with compressors will be required...'

Rod Voller, the State Defence Camouflage Committee's Officer in Charge, NEA, notes his earliest involvement at MAREEBA, as -

'...(Townsville inspections were) followed on 20 June with a visit to Mareeba with Hughes, Doug Annand (camoufleur), Doig (Com Works) and others re dispersal areas and possible use of the arches.'

In a Report to Air Board dated 21 June 1942 P-O J W Campbell addressed progress then current -

'...Inspected 20 June 1942, those attending included Colonels Mills and Matthews, Messrs Doig, Pollock and Bond (MRC) P-Os Trench, and Campbell and Camoufleurs...'

Both runways are mentioned, with 104° reported 'in use' and 172° 'ready for use'. The taxiway connecting the two -

'...follows existing roadway which has been widened on western side to approximately 50′. A powerline exists along the eastern side of this road. Dispersal hideouts off this taxiway are being cleared...'

Priority areas for 'sealing with whatever primer is available' are detailed. Four camp sites have been selected between the Cairns railway and the Barron River and meagre buildings and basic facilities are outlined. Regarding camouflaged dispersal -

'...A number of camoufleurs also made inspection and quite a number of suggestions were discussed. Mr Hughes represented US Army and Messrs Voller and Annand for RAAF, but no satisfactory scheme was definitely approved...'

The arrival of American squadrons at MAREEBA was heralded by a document issued under the signature of Millard L Haskin Lt-Col Air Corps Chief of Staff HQ AAF NEA to the Air Engineer Officer HQ USAAS, APO501. Dated 2 June 1942, on Aerodromes in NEA, it read -

'...Entire 19th Group (H) will be moved to Mareeba (this includes 28th and 30th, 93rd Tactical Squadrons and the HQ and HQ Squadron). Additional buildings have been requested for this site to (meet) housing needs of this group. The two runways, taxiway and dispersal points will be sealed and all work is to be completed by July 1 if possible...'

Simultaneously, HQ USAAS moved to improve coordination of this construction. After detailing NEA

MAREEBA *(August 1943)*
This key diagram (KD), is typical of prepared records of systematic parallel flight runs over airfield complexes. *[AUSLIG]*

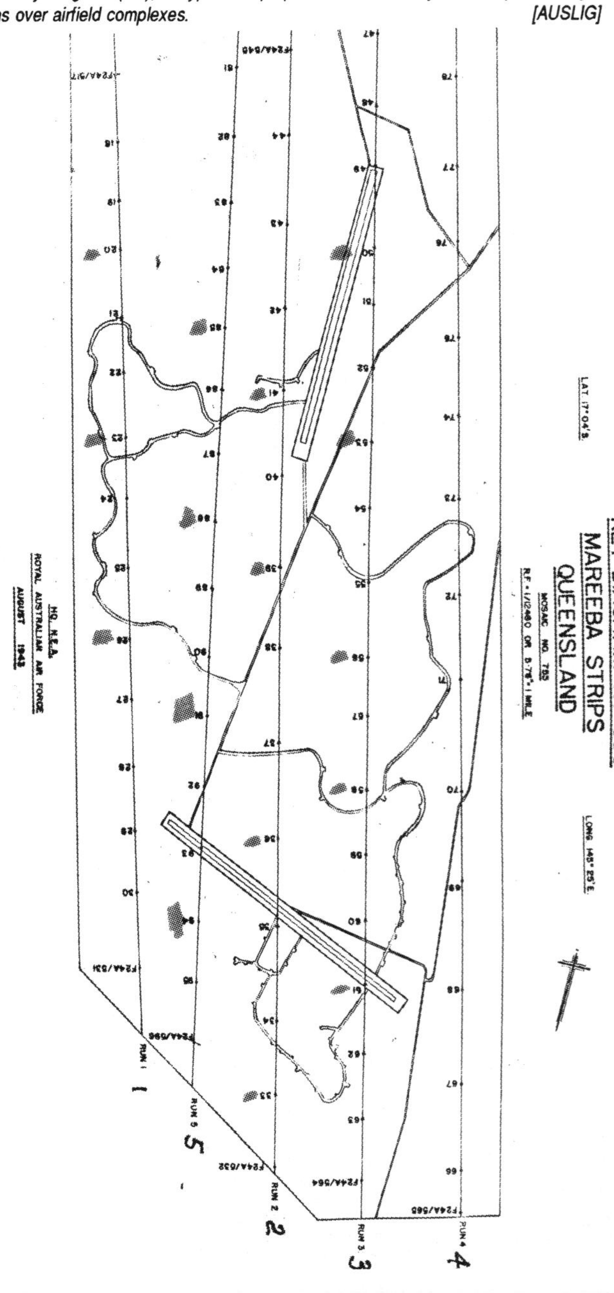

works in New Guinea and Queensland, it was stated -

'...All construction necessary to accomplish this programme has been designated a US responsibility and control thereof will be exercised by HQ USAFIA. Location of all new sites required by this programme is the responsibility of AOC NEA...'

Basic camp facilities were emphasised as -

'...tent camps supplemented by semi-permanent kitchens, mess halls, latrines and showers... Essential (that) kitchens and mess halls be screened... Requested that all new airdromes and camp facilities desired be completed concurrently with the aerodrome being made useable. On fields already constructed and lacking the necessary camp facilities, it is desired... (they) be provided as soon as possible...'

An interesting report on RAAF files dated around 5 July 1942, by (US) Capt Herman G Cox, CE describes strips visited in the preceding days -

*'...**Mareeba** - inspected 28 June and again 2 July -*

a) On 16 June after a conference with the AOC, Engineer HQ was instructed to provide housing facilities for the 19th (H) Bomb group together with other improvements, all of which was to be completed between 1 July and 8 July. Main Roads was given the Works Order and was to proceed at once on the buildings and to continue the work on runways.

b) On 28 June I met Mr Bond, Herberton District MRC manager in Mareeba, when he advised that the housing units would be ready for occupancy, together with a complete water layout from the Barron River on 2 July. Runways also were to be sealed 200 yards at each end by this date.

c) On 2 July in company with Col. Elsmore (...?Chief of Staff) a trip was made to Mareeba to ascertain the degree of completion and if the group could move at the planned date. The following conditions were found to exist:

Only 2 of 3 mess huts were complete. No shower buildings, no latrines and no Ops buildings were complete. The water system had not even begun... pumps were on the ground but no tanks were at hand and only part of pipe requirement was also available. Runways were not sealed and not enough equipment was available. This apparently was not receiving any supervision and no effort had been, or was being made, to secure more equipment. The job generally showed signs of lack of supervision and interest on the part of the foreman. Men had never worked on Sundays on this job to my knowledge, and I have yet to find a foreman on the drome when I arrive at the field. They have always been in Mareeba. These conditions... were reported to Col. Matthews on my return. Gen Brett when here around 29-30 was not satisfied with the work being done by the Engineer in this area as well as Mareeba in particular. The writer was instructed to pass this information on to Col. Matthews.'

MAREEBA (-/8/1942)

Mareeba and the B-17 are somewhat synonymous in the early Southwest Pacific period of WW2. Here an aircraft (believed to be 41-2621 'The Daylight Limited' piloted by Lt Ken Casper) is in difficulty after undercarriage failure while being towed to dispersal. Lawrence Moore recalls the aircraft he photographed had been damaged in action and its right landing gear had either collapsed on landing or could not be lowered for landing. The photograph was taken somewhere on the taxiway east of the runway between the airfield and the railway. [L Moore ex USAAF]

Air Board reacted promptly. W Cdr G J Towers visited around 14-16 July 1942. Towers put on record the RAAF NEA staff appreciation of the US Base Engineer's problem - not enough men to properly fulfil the assigned responsibility for airfield works supervision -

'...The (US) Base Section Engineer definitely does not interest himself in any of those works recently accepted by US Engineer HQ as their responsibility, and what is happening at the present time is that MRC is proceeding with these works to their own designs, and generally without direction, except such guidance as can be given from time to time by (RAAF) Area Works Staff, which, however is not sufficiently strong in personnel to give these works the attention they warrant. Dept of Interior Supt Engineer (Mr E T Doig) advised me that US Engineers dealt directly with MRC and they very often executed major works without his knowledge and then came to him for provision of funds. This lack of supervision and direction on jobs accepted as their responsibility by the USA Engineer HQ, is regarded as a serious matter, as it will probably lead to wasted effort (at a time when the shortage of manpower and plant is acute) and extravagance in construction. In a recent report, Capt Cox of the US Engineers complained of the lack of supervision at Mareeba, apparently not knowing that supervision was the responsibility of his own HQ. In addition, the danger exists of improper design on many of these works, and it is feared that for this reason there is some possibility of newly constructed aerodromes becoming unserviceable in the wet season, and it is submitted that this should be seriously regarded from the operational aspect...'

Progress on the RSU shows in a July 6, 1942 Cypher message from P-O Campbell to Air Board -

'...LAPPA JUNCTION UNSUITABLE RSU. RECOMMENDING SITE WITHIN 4 MILES RAILWAY, ELECTRIC POWER, ALSO WORKSHOPS TABLELAND DREDGING COMPANY, MT GARNET...'

On 17 July 1942, AOC NEA forwarded to Air Board Melbourne, plans and schedules of Mareeba Dispersal Scheme. A covering statement was made that -

'...Dispersal roads were located by P-O Trench in conjunction with Mr Annand (Camoufleur) of Dept of Interior. It is proposed to construct a total of 30 hideouts all suitable for heavy bombers. It was decided to locate the hideouts approx 400 yards apart and not less than 350 yards from the runways. The hideouts were located so that the maximum distance from hideout to runway was approx 1 mile.'

This rationale was a balance between adequate dispersal and the need to avoid aircraft taxiing long distances -

'...Since the schedule was written, construction has begun on one experimental arch for heavy bombers. It appears very doubtful if arches of this size can be satisfactorily hidden and whether all or any of the arches at Mareeba are constructed depends on the result of this experiment.'

Lawrence J Moore was an electronics specialist with the US 19 BG. Questioned about any 'defences' he recalled -

'...At Mareeba, we dug (foxholes) near the tents, which was also near our workshop. Some of the ground was so hard we had to use black power and dynamite... Other times we poured in water and let that soak in

overnight and dug a few inches and put in more water and more digging etc. The only A-A defences I ever saw were at Mareeba. The guns sometimes practised... shrapnel came down in the tent and work areas and tore holes in the tents - no one was hit that I know of...'

The NEA Works Officer produced a lengthy review of Aerodrome construction in the Townsville and MAREEBA areas in July 1942. It pointed to the urgent need to augment both men and equipment if the outstanding 18 runways were to be useable during the next wet season. Some of the answer seemed to lie 'in the south' -

'...Four heavy bomber fields are being constructed west of Brisbane, each one with a complete and separate construction organisation. Construction of the parent fields should be almost complete and the organisations now constructing satellite dispersal strips. It is contended that the construction of parent field in this (NEA) area should take priority over construction of these satellite strips. Further, at a conference with Mr J R Kemp in Brisbane several months ago, Mr Borrie of the Dept of the Interior advised that he could arrange for the transfer of at least one complete organisation from the NSW Main Roads Department. It should still be possible to do this...'

Of these 18 runways, about two thirds did proceed. MAREEBA figured high in priority. Capt Cox reported on a subsequent visit to MAREEBA, 8 August 1942 -

'...This job is progressing slowly but surely. Work on strengthening 170° runway has just started. It is planned to have this runway sealed (water tight) by September 1. The 280° strip will then be strengthened and sealed. Five arched hideouts are completed except for garlands and work is in progress on the sixth one. On my arrival, it was found that the MRC office here had not forwarded our priority list to the foreman in charge. Priorities for both aerodrome and buildings were outlined on the ground and were verified by the local MRC office on my return...'

By 12 August, RAAF files in AA show a site for '26 RSU' was still undecided. In support of the Mt Garnet site, a comment is made that the Lappa Junction site was initially chosen purely on 'area' consideration. A handwritten endorsement on the bottom of the Minute Paper, carries the signature of W Cdr Poole, (RAAF) Director of Aircraft Maintenance -

MAREEBA *(-/6/1989)*
Mareeba Shire Council owns and operates the present airport. The original E-W runway has been shortened at its western end. The winding taxiways are not difficult to pick. While some are back to furrowed ground others continue to serve as farm equipment access routes. East of the railway line in the wartime US camp area, Syd Beck has a property on which he is developing an Aviation and Military Equipment Museum. *[Courtesy R Rudd Northair Surveys Mareeba]*

'...It is recommended that the RSU be sited in the Atherton area in view of the mountainous nature of this (Mt Garnet) country.'

William R (Bill) Eaton was 'in and out' of MAREEBA over August and September of 1942. The time was spent partly with the 93 BS and partly the 435 BS, 19 BG (H). Bill returned from a ferry run to the Depot at CHARLEVILLE on 27 August. His diary notes read -

'...Take off from Charleville, 1000, landed Mareeba, 1425. Really glad to get out of Charleville. We are (now) assigned to the 93rd Bomb Squadron and maybe this time we'll get some action. I learned that T-Sgt Harman T Randal and Sgt Coy Adams of the old 2nd Provisional Squadron had been killed on a hop over Cairns, Q. Both good men... they were killed in an experimental flight out of Mareeba, testing bomb rack adaptors for big flares for night bombing...'

Maj Dean 'Pinky' Hoevet, CO of the 30 BS at the time, was piloting the test aircraft, B-17 41-2434, on 17 August 1942. A flare exploded inside the plane. It crashed into the sea killing all on board. Whether officially named so or not MAREEBA airfield has been referred to by some historians as HOEVET field.

Some publications relate the severe shortage of bitumen, mid 1942, to the sinking of a shipload in the Pacific. True or not, dust was so troublesome on unsealed pavements that some improvisation was needed. ***Engineers of the Southwest Pacific 1941-1945*, Volume VI *Airfields and Base Development***, mentions -

'...Prior to World War II an experimental stretch of road was constructed near Sydney, Australia, by using a 90% water-insoluble molasses which was mixed with bitumen - about 2 parts molasses to 1 part bitumen. Using this pre-war Australian experience as background, molasses was used successfully at Mareeba in NQ in connection with soil stabilisation on airfield runways and taxiways, and for dust control...'

A report by Col. Leif J Sverdrup, CE, in his US armed forces appointment as Chief of Construction OCE, GHQ, SWPA, has been quoted in part by the Wild River Times, in its issue of July 1976. The extract is undated, but by comment made, would appear to relate to the third quarter of 1942. In respect of MAREEBA, Sverdrup said -

'...Mr Griffiths of MRC is in charge of this project. Two runways, one E-W and one N-S have been constructed and the two runways have been connected with a strip which can be used by Fighters for landing or take off purposes. Both runways are 6 000' long. The E-W runway which is the most frequently used this time of the year has night lighting equipment, while the N-S runway which is out of operation at present time due to construction work, does not have such facilities. Both runways are fairly dusty, and outside of the immediate runway area dust is very heavy. Three applications of molasses mixed 1:8 have been used on the E-W runway, and has in Mr Griffith's opinion been 40% effective... Mr Griffith estimates that he can finish the bitumen surface on the N-S runways in 3 weeks and that he can place all the gravel on both runways from 4-5 weeks. He has as yet not received his bitumen but has all the necessary plant. A small supply of bitumen is on hand... but not sufficient to start operations... 230 men started work on the job: 300 are working there at present time so the supply of labour seems to be plentiful. In a dispersal area plane pens with an arch type camouflage are being built. 55 carpenters are working on this arch project. Most of the carpenters have been sent in from NSW, 17 carpenters and 3 labourers are working each arch and they estimate that they can complete one a week. The camouflage men are falling behind the carpenters and attention should be paid to this.'

Col. Sverdrup finished with -

'...Recommendation: the same old story - "get rid of the dust". Bitumen must be expedited to the job in order to get the work done.'

Allied Works Council Minutes of 27 August 1942 record the 'spreading of molasses to allay dust at Mareeba'.

Early August of 1942, Lt A T Longland an Australian Army Liaison Officer attached to the SDCC reported on an inspection of MAREEBA Camouflage work. His descriptions provide a clue to the Australian A-A unit there at the time and the task of hiding the **hideouts**!

'...New A-A Site - made inspection with OC 224 Light A-A, Camoufleur O'Harte and Works Supervisor, Department of Interior. Decided on a suitable working road, track plan and preliminary arrangements to prevent despoilation of the area. The site is most suitable for effective camouflage treatment. It is not adjacent any landmarks and should not present any difficulties as regard concealment.

Mareeba Aerodrome: Work proceeding on construction of arched hideouts, placed adjacent to runways. Camoufleur O'Harte about to proceed with erection of pre-scrimmed netting. Concerned with difficulty of disguising the great arch ribs and proposed to break the curves into a series of planes with extra timbering as a basis for dummy trees of a scale to match the growing timber in the vicinity.'

Jim Trench came back into the scene with this extract from a 6 October 1942 NEA advice to Air Board, Melbourne -

'...P-O Trench visited Mareeba (to examine) possibilities of locating the detachment of No 10 and 12 RSU, either in the vicinity of the existing aerodrome or of locating a site suitable for constructing a new strip (for) this detachment...An average of 80 heavy bomber aircraft are at present occupying the existing aerodrome and between 2 000 and 2 500 personnel are camped in the vicinity. Under the circumstances it appears ***inadvisable to concentrate additional personnel and*** *aircraft in the vicinity of this aerodrome while present squadrons are in occupation. A site suitable for constructing a new runway was located about four miles south of the existing aerodrome adjacent to the Mareeba-Atherton road. Bearing...95°...'*

This Maude Creek site is described further. In respect of road access, the report says -

'...The Mareeba-Atherton road adjacent to the suggested site is useable in all-weather. The Cairns-Mareeba road has now been given a high priority and it is anticipated it will be completed before the wet season...'

Instead of providing for detachments of 10 and 12 RSU, by 20 October 1942 this was suggested as the site for 14 RSU to be formed in October 1943.

In March 1943, increased Japanese activity in the western New Guinea area renewed RAAF fears of bombing of North Queensland airfields. An extensive programme of splinter proof pens was initiated. MAREEBA's assignment

MAREEBA (17/11/1942)
The crew of a Boeing B-17F Flying Fortress of the 64 BS, 43 BG, photographed by their plane at Mareeba after returning from a mission. Like most such photos, this is one of a series of at least three 'posed' on 17 November 1942. This is Lt Raymond B Holsey's crew. Recent contact with Navigator, Lt John A 'Davy' Crockett and cross reference on caption provided in Negative A29956, show personnel as, left to right: ,—, Sgt J B Young, Sgt John W Rosenberger Radio Operator, —, Lt Jack Ryan Co-pilot, Crockett, — (behind), 2nd Lt Jacob H Franz Bombardier, and Holsey.
[Neg A29570 USAF Photo Collection NASM, Smithsonian Institution]

MAREEBA (17/11/1942)
Members of the 64 BS, 43 BG, shown gathered around the Operations Board to see who is on the 'alert call' at Mareeba. Ray Holsey is in the middle, with face turned 'left profile' probably talking to 'Davy' Crockett second on his left. Clearly the 64th Squadron have taken over the 'board' vacated by 30 BS, 19 BG.
[Neg A29957 USAF Photo Collection NASM, Smithsonian Institution]

was eighteen medium bomber pens. This work was later cancelled quite dramatically after official correspondence between General MacArthur and the Australian Prime Minister, Mr Curtin.

Rod Voller's notes of the same period record his involvement and his concern for the sheer quantity of associated camouflage work -

'...On 17 March 1943 I discussed with F-O Grounds and Syd Hirst (Com Works) requirements for meeting revised Air Board lists and possible future work for quantity orders to Com Works for 175 to 200 heavy bomber hideouts for Mareeba, Inverleigh, Horn Island, Iron Range and Cooktown...'

This development should be viewed in the context of Rod Voller's November 1942 notes which clearly made him aware that the US Forces had, by then, discarded all thought of doing further camouflage work on the mainland. While this may simply reflect a more informed US view of the strategic situation, no doubt the RAAF was simply looking to ensure security of its own aircraft.

In March of 1943 the US 5th Air Force was approached on the possibility of the RAAF's 14 RSU sharing MAREEBA runway facilities. Most of the US 43 BG had left by then. As a result the area between the N-S runway and the railway/powerline was made available. After all the effort put into finding a site for the RSU at MAREEBA, a 'Minute' dated 5 November 1943 simply notes, '...Mareeba is not (now) required as an R & SU site'.

Constructional control of many NQ aerodromes was passed to the RAAF late in 1942. The 'serviceability date' of 31 January 1943 on RAAF 1000′ = 1" drawing 42/43/1061 shows 178° Runway - Sealed, and the 110° as 'gravelled, priming 50%'. MAREEBA stayed under the control of USAFFE for purposes of responsibility for construction and maintenance, well into 1944. In the first half of 1943, the US 8 FG withdrew to MAREEBA from New Guinea. The well earned rest from the malarial conditions in New Guinea was spent in re-equipping and training with P-38 Lightning and P-40 Kittyhawk fighters.

Colin Munro was CO of the RAAF's 5 Squadron when it was due to move from KINGAROY to MAREEBA. He recalls this move as the first for which he was personally responsible. Great care was taken in its planning, to ensure both men and equipment were transported efficiently and expeditiously. Some 73 vehicles and 23 aircraft were involved. Most vehicles were transported by train. Colin made the trip in a Chevrolet staff car with his Adjutant and a driver. They carried spare cans of petrol

MAREEBA (4/1/93)
Doesn't every man (and his dog!) have a P-39 in their farm shed? Unfortunately, not so. Here Syd Beck is showing the author's eldest son Tim Marks, the P-39, 41-6591 (...Charles Falletta 35 PS) that Syd recovered from Cape York. Meticulous restoration has marked time in recent years while Syd has embarked on his Museum relocation from Townsville to Mareeba. Here the aircraft is secure under cover of Syd's re-manufactured 'Igloo' hangar awaiting completion of concrete floor and all else involved in what should make a first rate Museum. [Author]

and drove virtually non-stop, sharing the driving, one hour on and two hours off. Travelling north through Clermont, Charters Towers and Townsville, they reached Mareeba around 2 a.m. The Squadron completed the shift on 8 June 1943. It took up residence in an area which Colin described as a camp 'hurriedly vacated by the Americans'.

Equipped with CAC Boomerang and Wirraway fighters, 5 Squadron stayed at MAREEBA for approximately seventeen months. It served a training role in ground support for the Australian Army formations on the Atherton Tableland.

A total of 15 arch-type hideouts, 160′ span x 84′ long, appear on Drawing 43/44/1955 (AT 43/44/519). All were very likely finished by September 1942 when most of the NSW carpenters were released to more urgent work at GARBUTT. In October of 1943 the RAAF sought US permission to dismantle and relocate six of these structures. They intended to add roof sheeting and floors and so provide hangar/workshop space at STOCKROUTE, GARBUTT and BOWEN. As versatile as Emil Brizay's 'igloo' shaped hideouts had proven, RAAF NEA staff were to learn some limitations. Unlike pre-fabricated steel 'Butler' hangar frames, the nailed wooden arches could be freshly built for equal cost. Syd Beck has reconstructed such an 'igloo' hangar at MAREEBA during 1991. He believed his careful road transport of the original WW2 arches from Townsville was successful but their age and unscheduled open-air storage placed a question mark over their re-erection.

I was fortunate to interview Mr W J (Bill) Johnston who was involved as a carpenter on projects on MAREEBA drome. Bill told of construction of the hideout arch frames in Lawson's sawmill yards in MAREEBA. He recalled personally building formwork for the A-A gun emplacements. I was also pleased to later contact Mr Jack Kay who personally 'manned' one of these batteries for a short while, in October 1943. As a member of the Australian Army's 56 Composite A-A Regiment, Jack wrote -

'...We were at Mareeba airstrip before moving up to Cape York but only for a little while, and as I was then in charge of one gun I didn't get a chance to move away from my gun position.'

Winter on the Atherton Tableland can be cold. While most veterans will remember the cold water rigours of such camp life they may not realise that 'someone' was trying for improvement. Minute 34 of 28 July 1944 on a RAAF Mareeba Buildings and Services file reads -

'...Before considering proposals (for hot water showers) I would be glad to know if there is any likelihood of No 5 Squadron being moved from Mareeba in the near future... (signed)...AMSE'

A week later, Minute 36 indicates -

'...it is anticipated that an Army Co-op Squadron will remain at Mareeba for co-operation with Army units in the Atherton district for the duration of the Japanese war. It is therefore recommended that the works proceed.'

I have found great value in periodic aerodrome inspection reports. F-Lt Gilbert filed one on MAREEBA dated 4 October 1944. To matters such as condition of bitumen seal, 'obstructions' and fencing, he added -

'...The Atherton-Mareeba road is virtually completed and through traffic should not now need to use the taxiway or aerodrome roads...'

F-Lt Gilbert's office followed up to NEA HQ on 18 October, particularly on the runway obstructions -

'...noted during recent inspection, that certain obstructions (as below) were existing close to the runways at this station.
1) Concrete machine gun nest (believed constructed by US Authority) at a few feet from the end of the 351° runway (bitumen) on the N extension of the centre line and about 18" above ground level.
2) A slit trench in a similar position at the E end of the 103° runway.
3) A gun pit (believed construction by Australian Army Unit) at the SE corner of this bitumen runway.
4) A similar gun pit at the edge of the bitumen about 650′ W of that in (3) above.
5) A timber obstruction just off the bitumen runway about 700′ W of gunpit in (4) above.
6) Timber notice boards on the S overrun of 351° and on the taxiway at the S end of this runway.
The presence of these obstructions was pointed out to F-Lt Sinclair, CO of 24 OBU who proceeded to arrange for their removal...'

The hazard of straying stock on runways has always been real. The AOC NEA Air Cmdre J H Summers pressed Townsville DWO on 28 November 1944 -

'...Mareeba is classed as an L/A Base, and as such must be able to handle one Squadron during max operations. Indications are that the Atherton Tableland will be the Training ground for Army forces for a long time to come, resulting in a fairly extensive flying programme more or less continuously - e.g. at the present time six (6) Vengeance (Aircraft) of No 5 Communication Unit and six (6) Ansons of No 8 SFTS are now operating from Mareeba. It is also proposed to carry out searchlight training which will necessitate additional night flying to that already required by normal Army co-op work. Above operations could be greatly hindered by wandering stock.'

Here the AOC was referring to 5 (Comm) Unit. The RAAF's 5 Squadron, by then termed Tactical Reconnaissance (TAC/R) had moved on to Bougainville.

Since May 1944 No1 MPMU had been quartered at MAREEBA, basing its work area and camp on one of the disused bomber hideouts. The Australian Army's No1 Parachute Battalion was camped east of the railway line in the original USAAF camp site.

In the RAAF's post-war review of airfield needs, some such as MAREEBA were 'acquired only, with runways in existence not to be maintained'.

KARUMBA (1/12/1943) →
This is an oblique taken from Catalina A24-44. It shows the buildings associated with Anderson's meatworks and further towards the bend in the Norman River, closer to its mouth, the QANTAS Empire Airways flying boat base. The clay pan beyond was the designated RAAF Emergency Landing Ground. One of the aircraft of 43 Squadron RAAF is seen at its river mooring. [NLA (RAAF Official)]

KARUMBA

'...Currumbin - yeh, I landed there in a flying boat.'

With place names vague in the memory and sounding like others more commonly heard, it's not surprising this comment came from a RAAF veteran. Currumbin is part of Queensland's Gold Coast, but this WW2 flying boat base is now a well known base for the Gulf of Carpentaria prawning fleets.

At the start of WW2, KARUMBA, along with Groote Eylandt in the Gulf, were already established refuelling stops for the QANTAS (QEA) Short Empire flying boats. In August 1939, just prior to the outbreak of war, the Defence Committee meeting in Canberra, considered submissions on the defence of KARUMBA and Groote Eylandt. They recommended that a 'Rifle Club Vulnerable Point Guard' be established at both centres comprising flying boat base personnel and that the 'Military Board' should issue necessary rifles, ammunition and equipment. Members who complied with the provisions for efficiency would thereby be granted an 'efficiency allowance of £1 per annum under the Australian Rifle Club regulations'. To be fair, these measures were seen as inadequate against a raiding party, but reasonable protection from 'marauders or sabotage'.

This Club formed the only protection until February 1942 when the weapons were handed over to a detachment of VDC newly formed in the area. The AWM publication ***On Guard with the Volunteer Defence Corps*** tells of this VDC group challenging and firing on two boats in March 1942. The party proved to be Australian and Dutch military personnel escaping from the Japanese at Ambon. Fortunately the VDC fire produced no casualties, and the occupants were flown to Brisbane in Dutch Dornier Do24K flying boats.

Token camouflage painting was applied to buildings at the base in May 1942. After the April 1943 'Gulf scare' in which it was feared a Japanese landing had occurred on the coast north of KARUMBA, Air Staff policy was revised. The RAAF's 43 Squadron General Reconnaissance Flying Boat (GR/FB) Catalinas were based here from August 1943, when a full scheme of camouflage was instituted.

AWC Minutes first mention KARUMBA around the end of March 1943 with 'provision of buildings and services for the Operational Base'. A significant concrete slipway, plus hangar and workshop followed, though a RAAF report of March 1944 mentions these as 'incomplete' to some degree.

DUNK ISLAND

'...As for the DUNK ISLAND strip, it didn't come into use until after the war.'

Such was the reply I received from one of the Historical Societies in respect of the wartime relevance of the strip on DUNK ISLAND. By contrast, DCA licensed a strip there at least by March 1941. The John Oxley Library's photographic holdings include an aerial oblique of the island dated June of 1939. The caption reads -

'...showing new aerodrome; Bramo Bay in foreground with sugar boat and launches. New landing strip to be seen in middle foreground. The white dot on extreme right end of runway is the covering on John Collins' Vega Gull while he was on the Island with Lady Gowrie and party.'

Camouflage of radar installations remained important after its discontinuance on airfields. A 17 January 1943 report by Gp Capt Thomas reads -

'...At Dunk Island the Station was very difficult to locate and it was not until the aircraft was directly over it, within 200' that it was picked up...'

The 27 Radar Station's ORB prophetically tells of several new orchid varieties being noted in the natural camouflage of the island. It quotes the island's famous beachcomber/author E J Banfield -

'...This is veritably an orchid grower's paradise.'

Today the island still boasts this natural beauty and comparative tourist privacy.

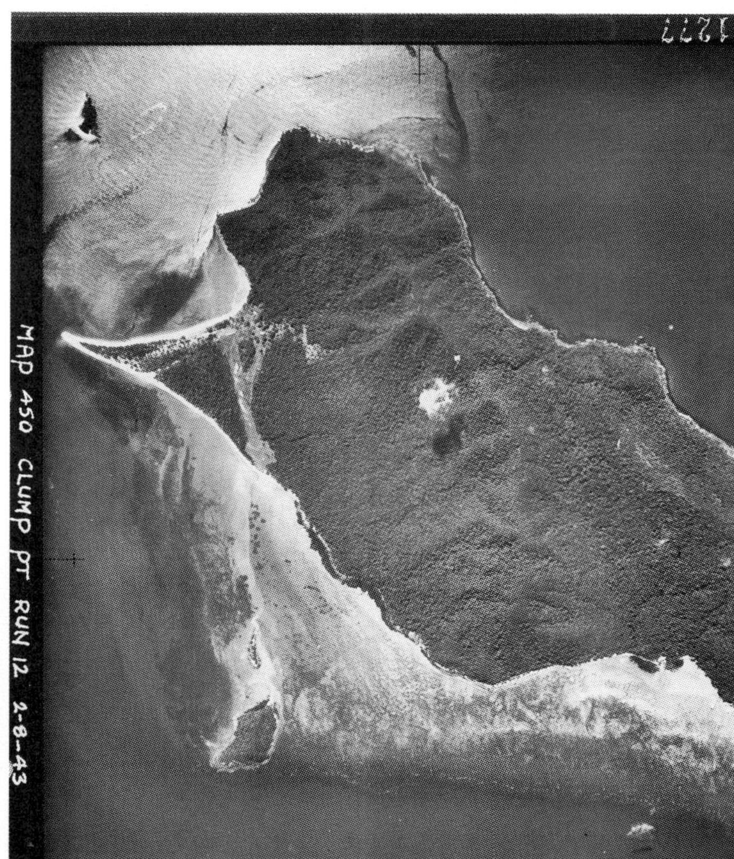

DUNK ISLAND *(2/8/1943)*
One of KD 842, the systematic coverage for 'Clump Point' map, this photograph shows the landing ground on the island. RAAF radar and weather observation facilities were serviced by light aircraft using this grass strip. *[AUSLIG]*

INVERLEIGH

'...Just a strip in the wilderness. We went out shooting bush turkeys, and I remember the most extraordinary huts were there - they'd been painted by Jolliffe.'

It has not been easy to find veterans who have 'spent time' at INVERLEIGH. Ron McDonell, ex RAAF 83 Squadron, recalled stopping over there because of bad weather on a flight with several other Boomerang pilots, en route to their station at Gove. It was there early in 1944, that Ron saw Eric Jolliffe's work on the walls of the huts. Jolliffe was then Australia's favourite cartoonist who specialised in portraying the Aboriginal people in their outback element. Retired now in Bateau Bay on the NSW central coast, he had been seconded to the Department of Home Security's band of camoufleurs. They were men with artistic skills and the ability to sketch and commit ideas and designs quickly to paper.

INVERLEIGH is an airstrip which stemmed from pre-war days. A civil strip with three 'intersecting runs' existed just west of the 'L' Creek. It served Inverleigh Station then owned and managed by the Chandler family. Jack Atkinson, a gentleman who could be rightly aligned with the likes of Tom McDonald in the pioneering of aviation in North Queensland, told me of his experience at INVERLEIGH repairing damage to an AOA DH-84 Dragon aircraft VH-URG during the war. More recently I located and spoke with Robert Merrotsy who was Jack's apprentice/helper on that excursion. Robert's photos of the time showed the aircraft had been damaged on the old civil strip. Jack's best estimate of date was 'when (he) heard on the radio of Italy's entry into the war!'. That was June 1940.

When the RAAF came to grips with the prospect of AUGUSTUS DOWNS not serving well during the wet season, attention shifted initially to the INVERLEIGH civil strip. Ex RAAF Staff Officer Works, Jim Trench, when questioned about INVERLEIGH wrote -

'...I flew there with two Senior Air Staff Officers who were most anxious to have an ELG constructed, mainly I think, because the owner of the nearby homestead and his wife were particularly hospitable people. The area proposed was flat, well drained and consisted of excellent gravel with no overburden. All that appeared necessary was to grade the strip to produce a serviceable emergency landing runway. Air Cmdre Knox who was in charge of DWB issued a chit for I think, £5,000, which should have been sufficient to cover the cost of the work that I thought would be required. Anyway, MRC set off with quite an elaborate convoy of equipment and was caught by the wet season... I think far more work was done there than originally proposed; hence the cost of £85,800 shown on your list. I don't know how the accounting was adjusted to cover the extra cost amounting to about seventeen times the original estimate!'

INVERLEIGH *(10/8/1943)*
'V815', one of 19 frames in two parallel runs over the airfield. It is a classic case where great costs were incurred by the AWC in such areas isolated by the lack of effective transport over country prone to wet weather. Notice the earth mounding associated with dispersal bays. INVERLEIGH replaced AUGUSTUS DOWNS as the AOB ***'in the Gulf'.*** *[NLA (RAAF Official)]*

Jim's inspection visit was actually dated 25 November 1942. A cypher message dated 4 December 1942 from NEA to RAAF HQ RAAF Command reads -

'OWING TO UNSERVICEABILITY OB AUGUSTUS DOWNS AFTER RAIN AND NECESSITY PROVIDE ALL-WEATHER BASE GULF AREA, DESIRE TO MOVE OB AUGUSTUS DOWNS TO INVERLEIGH WHERE NATURAL ALL-WEATHER STRIP OF UP TO 3 000 YARDS CAN BE CONSTRUCTED BY MINIMUM EFFORT WITHIN SHORT PERIOD. PROPOSE SET UP INVERLEIGH AS MAIN OB GULF AREA REDUCING AUGUSTUS DOWNS TO AOB STATUS DRY WEATHER USE AND AS SATELLITE TO OB INVERLEIGH. WILL RETAIN FUEL, BOMBS, RATIONS, AT AUGUSTUS DOWNS WITH SMALL CARETAKING STAFF... ANTICIPATE STRIP COMPLETED AND OB TRANSFERRED BY END DECEMBER.'

The wet season intervened. A small party transferred from AUGUSTUS DOWNS permanently in early March 1943, but it was not until July 1943 that 29 OBU commenced full operation at INVERLEIGH.

Archive records show that Eric Jolliffe was dispatched to INVERLEIGH around 15 March 1943. NEA had of course been advised and requested to provide 'Mr Jolliffe with every assistance'.

MRC Engineer Mr Alec M Black, was responsible for both INVERLEIGH and AUGUSTUS DOWNS as well as a vast road building programme. The Rail service from Cloncurry north to Dobbyn was so unsatisfactory that in May 1943, Alec Black took the initiative of buying a house in Cloncurry. This served as a depot for the trucks picking up off-loaded rail freight for the long haul to the airstrips.

INVERLEIGH was one of several NQ AOBs at which splinter proof aircraft dispersal pens were proposed. Dozer pushing-up of the curved earth walls was well advanced by the time the 18 June 1943 'approval' arrived. No doubt the Overseer didn't like his alternative of booking 'idle hire' on equipment already committed to this isolated job. Like many others, these pens were cancelled after General MacArthur's personal letter to Prime Minister John Curtin.

From an NEA Senior Officer's conference on 7 July 1943 the following directions came -

'...• construction of pens not to proceed
• sealing of strip to be deferred
• taxiways and hardstandings to be completed
• all construction forces now at Inverleigh to be dispatched as rapidly as possible to Higgins and Horn Island to expedite those dromes
• cancel camouflage of dispersal areas
Augustus Downs - aerodrome to be abandoned.'

For INVERLEIGH the rest of the war was down hill. The white ants are terribly hungry in that part of the country and the record shows that few of the buildings escaped. Some were sold for relocation to Doomadgee Mission away to the north west, nearer the Gulf.

The airfield itself has been licensed since 1945 to the Inverleigh Pastoral Company. The Royal Flying Doctor Service and the civil Air Mail service sought and were granted permission to use it in August 1943.

AUGUSTUS DOWNS

'...24 December 1942 AUGUSTUS DOWNS? - oh yes, all I remember of that is the CO sent me there to pick up the pigs for Christmas.'

Only his log book entry prompted Jim Legge, ex RAAF 7 Squadron, to remember anything about AUGUSTUS DOWNS.

Early in 1942 the Allied Air Forces sought an AOB 'in the Gulf of Carpentaria region'. P-O Jim Trench, then a Works Officer attached to NEA Townsville, made an inspection in company with a Lt MacKenzie of the US Forces on 25 April 1942. Jim knew proximity to a railway line was important in this site selection. He also found that the 106 miles of particularly bad road to the nearest railhead, Dobbyn, was virtually impassable during the wet season. Still such sites were required and work went ahead. By the end of May Jim Trench had reported -

'...One strip 6 500' has been cleared and graded by Burke Shire Council... propose to consolidate runway with skin of gravel... buildings of temporary nature only required... improvements required to access road... there is no plan of the area but a plan traced from County maps is attached herewith... propose to construct latrines, ablutions, mess huts, kitchen, combined operational cypher and signals room and an underground building for W/T.'

During June the decision was taken in discussions between General Brett of the US Forces and the RAAF's AOC NEA, to reduce to a minimum the building policy for such new aerodromes. As with other remote area strips, it would fall to the MRC to complete the work as few (US) Engineer troops were in the Townsville area at that time.

The RAAF's P-O J W Campbell inspected work on 27 and 28 June 1942 in conjunction with Camoufleur Knight of the Department of Home Security. The MRC history published in 1949, records AUGUSTUS DOWNS aerodrome as having involved two landing strips. In this early July report by P-O Campbell there is mention of an Emergency Strip having been cleared full length of the strip proper and parallel and adjacent to it -

'...This strip will be useable in dry weather while gravelling is in progress on the main strip.'

P-O Campbell experienced the tedious ten hour plus (under ideal conditions) truck journey from Cloncurry to the site. This difficult supply line contributed to the huge cost of work in such isolated areas. Such costs were to be argued for months to come in the financial reconciliation of airfield work by the AWC. Poor communications were also to blame. Campbell's report of 2 July included comment -

'...Communications: The mail truck passes the camp once per week but messages may be sent and received by portable radio set from Augustus Downs Homestead.'

Communications and hence supervision difficulties worsened with the US assuming responsibility for 'all AOBs in North Queensland'.

The surrounding area held only scattered timber. This presented a challenge to Camoufleur William C Knight, and his drawing NEA-W-104 dated 3 July 1942, shows careful use of the timbered creek fringes to locate '14 hideout pens'. Such flair and imagination came to nought as the 1 April 1944 'Serviceability date' of drawing 42/43/582C shows 10 hardstandings loosely hugging the strip proper.

On 24 November 1942, one month to the day before Jim Legge was told to - 'fly to AUGUSTUS DOWNS to pick up the pigs for Christmas', the CO of 7 Squadron, W Cdr J M Lerew visited and inspected AUGUSTUS DOWNS. As one of this party of five RAAF officers, F-O Jim Trench, then Staff Officer Works, found less than half

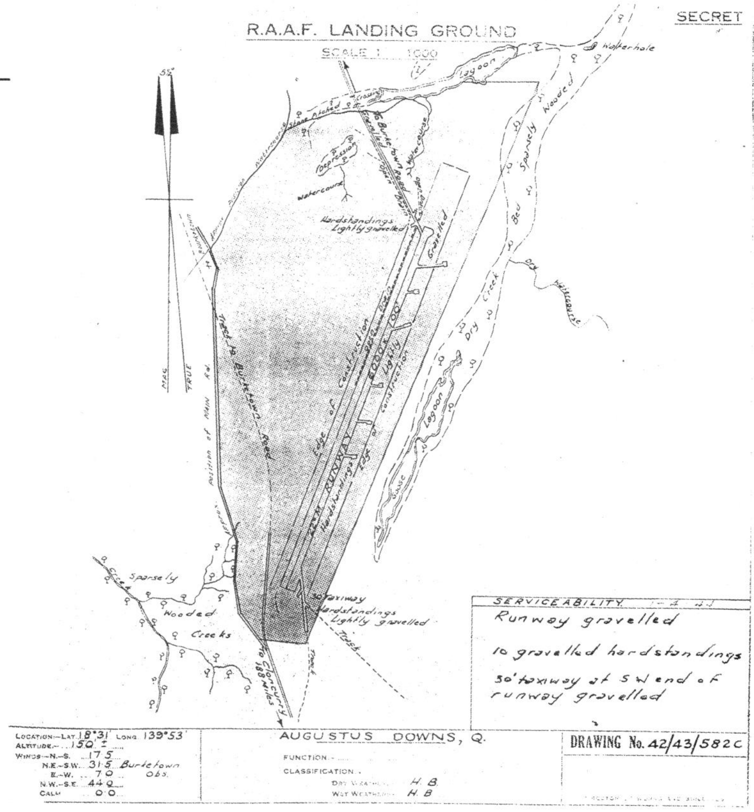

AUGUSTUS DOWNS ***(1/4/1944)***
This is the serviceability date shown on above drawing 42/43/582C. It is typical of RAAF 1000' : 1" airfield reference drawings. Notes on serviceability were made as a result of periodic inspections. *[AA CRS A705 Item 171/93/1095]*

the specified thickness of gravel paving in place. While the runway appeared satisfactory, the Base could not be used operationally in the wet because the lack of hardstandings meant aircraft could only be parked on the runway! Jim found the camp 'complete and well camouflaged', but the 4 mile connecting road impassable when wet. Water was 'cloudy', being pumped from a 'large pool in the Leichhardt River'. With no means of adding chlorine (for purification), he concluded 'the water supply is not altogether satisfactory'.

P-O Campbell was also a member of that inspection party. His report dated 26 November 1942, makes clear that the RAAF was having second thoughts about this Gulf AOB. Campbell concludes with recommendations -

'...To make Augustus Downs an all-weather strip will cost many thousands of pounds more than Inverleigh... recommended ...consideration be given;
a) abandon Augustus Downs and shift buildings immediately to Inverleigh...'

For W Cdr Lerew, the visit of 24 November, was as a reconnaissance. His Squadron was to operate from there in a few days time. Some real deficiencies and difficulties of operation from this AOB, were then revealed -

'...On 29 and 30 November 1942, operations ordered by W/T from AOR NEA were carried out by a detached flight of 3 Beauforts of No 7 Squadron based on OB 29 (Augustus Downs). The aircraft arrived at Augustus Downs 0930 L on 28 November 1942 and at approx 1600 L on same day, decyphering was completed of signal form green ordering 2 searches on 29 November 1942. Owing to lack of maps, plotting instruments and tables and the fact that the only operations officer at the OB had no experience in operations room work, most of the work preparatory to briefing had to be done by personnel of the detached flight,... and crews could not be briefed until 2130 L, the same night. This was undesirable since crews had to be called at 0430 next morning in order to take off in time, and at 0415 the following morning.
Bombing up was carried out by aircraft crews themselves, with the assistance of 1 fitter-armourer

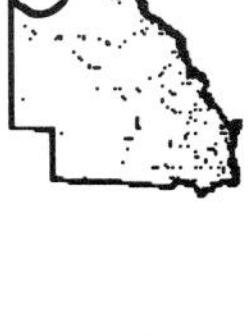

stationed at OB 29, and went smoothly except that the lack of bomb trolleys necessitated the use of 1 truck... which was needed elsewhere at the time....

Request for a weather forecast for route Karumba-Groote Eylandt for the next day was made in the evening, and sent to Townsville by W/T. No reply to this signal was received, and the 3 aircraft took off the next day without any idea of the weather conditions they were likely to encounter on their searches. On 29 November, a further signal requesting a forecast for 30 November was sent, but again no reply received, and 2 aircraft took off on searches the next morning without any weather information.'

Only two of the aircraft operated on the second day's search/patrol through the illness of the remaining Captain. He suffered a 'mild form of dysentery and stayed in bed during the day'. Practically all personnel of the flight were affected by this complaint. It was thought to have been due to the poorly treated water or the 'superabundance of flies at OB 29', and their effect on food preparation.

W Cdr Lerew drew some obvious conclusions -

'...1) W/T facilities at OB 29 are insufficient for satisfactory operations. More powerful transmitters and steel aerial masts are necessary.

2) (OB) Staff is totally inadequate for operations of more than 1 day. One (untrained) Ops Officer, one cypher officer and two W/T operators worked practically continuously while No 7 Squadron detachment was at OB 29.

3) Bomb trolleys or alternatively, another truck are necessary since transports must frequently proceed to Cloncurry, 200 miles distant, for supplies.

4) Better facilities for purification of water.

5) Some organisation for placing a transmitter on loop frequency for homing in bad weather should be established. Without that, approach (to Augustus Downs) in bad weather should be to go to the mouth of the Leichhardt River and fly up the river.'

With the decision to move the operational base to INVERLEIGH further to the northeast, AUGUSTUS DOWNS was to be vacated. But it was to see the flurry of further activity when on 11 April 1943 six Boomerangs were dispatched urgently by Townsville to the AUGUSTUS DOWNS strip. Their temporary stationing there was an immediate response to what has been referred to by several writers as 'the Gulf scare'. The incident stemmed from the report by an Aboriginal stockman of his having seen a Japanese barge discharging men and equipment near Inkerman Station on the Gulf coast north of Karumba. No doubt these Boomerang aircraft were used from AUGUSTUS DOWNS in search efforts to locate such a landing party. After several days of searching, VDC Units reported no lasting evidence of such a landing.

The tremors of the incident were vigorous and widespread. It is interesting to read individual depositions made by Signals Office Staff in the NEA Headquarters Sydney Street, Townsville. Such depositions were sought by Intelligence Investigators keen to establish how rumours of a possible Japanese landing had been leaked to the public. Most of the Signals Staff were well aware of the large 'situation map' on the wall, but when asked to commit to writing their recollections of events, they generally recorded 'seeing nobody look unduly at the map on the wall.' While no doubt it was deadly serious stuff at the time, reading such depositions puts me in mind of the portly German stalag guard named Schultz in the popular television series of 'Hogan's Heroes'. His ready vocal defence was that he - 'saw nuthink'!

By August of 1943, the airstrip was abandoned leaving two guards on site for safe keeping of RAAF installations. A year later the RAAF had determined no further need for AUGUSTUS DOWNS. Since early in 1946, the Civil Aviation Department, has licensed its use to the Augustus Downs Pastoral Company.

CARPENTARIA DOWNS

'But there's nothing here, Dad.'

This was my son Ben's comment when we visited in 1981. Only the 41° strip appears on today's mapping and while I had no knowledge then of its wartime configuration, we explored some of the nearby 'tracks'.

Early May 1942, P-O H V (Jim) Trench, working out of Townsville, accompanied the MRC's Herberton District Engineer, Mr A J 'Bondie' Bond and an Architect from the Department of Interior, Mr Andrew, on inspection of this and other possible sites for strategic airstrips. CARPENTARIA DOWNS was referred to initially as EINASLEIGH because the planners wanted a strip near that small township on the railway west of Cairns. During this same trip the party endeavoured to check a suggested site 16 miles east of Abingdon Downs. The track north west from the railhead at Forsayth was found to be 'practically out of the question', even in May. In the wet that crossing of the Etheridge River was reputed to be 'a swim for a horse for six weeks'! An alternative site on Prestwood Station west of Georgetown was chosen.

Bondie's crew cleared the 108° strip at PRESTWOOD but probably only pegged the 168° strip. That was enough to put PRESTWOOD 'into the system', and in turn, into the US '***Air Pilot Manual of the Pacific Islands***' - (see Appendix 1). However, while CARPENTARIA DOWNS was developed more than PRESTWOOD, they both had their 108° strips incorporated into post-war road construction.

Both sites figured in a letter dated 3 June 1942 from the General Manager of the Australian Stock Breeders Co Ltd, Brisbane, to Air Board, Melbourne -

'...We have been informed by our Pastoral Inspector that a site is being cleared under the supervision of the Air Force for the erection of an aerodrome on our "Lyndhurst" Station via Einasleigh, North Queensland. We understand that the work has been going on for some time and is nearing completion... also that sites have been selected on our "Forest Home" Station, near Georgetown, with a possibility that Goose Lagoons will be used as a Seaplane Base. If this is the Board's intention we should be much obliged to hear from you in the matter. If (this) is confirmed, certain alterations in the arrangements for working our cattle... will have to be made. Thanking you for an immediate reply....'

One month after the Battle of the Coral Sea was still a time of great urgency. Airfield site selection was an overall confused picture, made so by poor mapping availability and some overlapping efforts by the US Army, its civilian agents, the RAAF and the MRC. There is no

CARPENTARIA DOWNS *(17/6/1952)*

It is not difficult to see how unneeded airstrips such as this became part of the post-war road system. The Gregory Developmental Road has since utilised all of the 108° strip and the connecting taxiway in its leg between Greenvale and Kidston, both mining centres of some note. Einasleigh River is nearby. *[AUSLIG]*

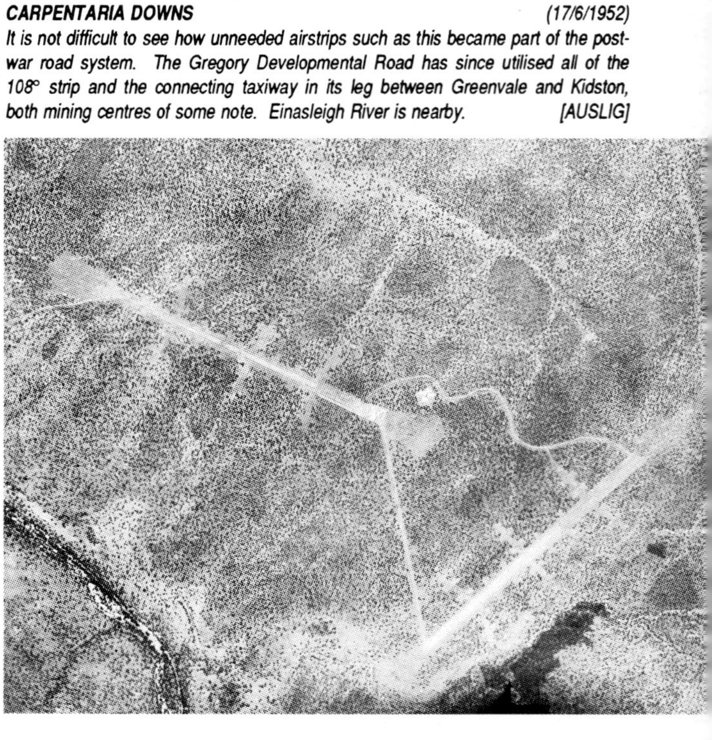

prize for the reader guessing the answer to this letter. Dated 19 June 1942 it stated -

'...no information is available at this Head Quarters...'

When I finally contacted Mr F J 'Snow' Baker who had worked on this strip, he was only too pleased to try to recall that time. Alas, my next letter was from Mrs Julia Baker to sadly inform me that 'Snow' had died in the interim.

Mr R W (Rod) Voller, Chief Camouflage Officer for NEA found in his diaries, that he had visited CARPENTARIA DOWNS on at least one occasion, for the purpose of planning camouflage treatment for aircraft concealment sites. He noted, 8 July 1942 -

'...One strip was ready for use and dispersal bays 150' wide were completed, all surfaces gravel...'

The original aim was that CARPENTARIA DOWNS should serve as an 'Operational Base - Bombardment Field - Parent'. While it is quite possible that aircraft used the strip during WW2, either through intention or emergency, I've not found any evidence.

A review of airfield needs in December 1944 refers to the 'Serviceability 16 December 1942' notes on drawing 42/43/2150A -

'..Data from US Intel. Plan... 42° R/W Gravel Skin. 12 Hardstandings and taxiways thereto, gravelled. 110° Strip and rest of taxiways thereto, cleared only. Abandoned by US and work stopped...'

After the war, DCA saw some value in the 42° strip as an ELG, for aircraft flying between Townsville and Normanton. Known locally as Mywyn strip, Etheridge Shire upgraded it in the 1960s, to cater to possible mining exploration traffic. However, it received very little use and today, healthy suckers grow through the gravel. The reserve has since passed to private pastoral ownership.

GARBUTT - STOCKROUTE (-/7/1941)

Frame V 33904 (extract) is part of the systematic aerial photography conducted for map production. The large triangular block bounded by Duckworth Street (unformed) on the west, Bayswater Street on the north and the wide relatively cleared strip which was physically a ***'route along which stock were driven on foot'****, was to become in 1942, a vast aircraft Repair and Modification Depot. Known first as the 4th Air Depot, it was later termed Depot No 2. GARBUTT RAAF Station appears at the top of the photo. The major buildings are in place. The civil airlines apron is unsealed on the east and the white fingering trackwork off to the west of the airfield leads to the early bomb dumps.* [AUSLIG]

GARBUTT

'...(Townsville City) Council is prepared to incur this expenditure (on sub stations etc.) provided your (Defence) Department will guarantee an annual consumption of electricity equal to £56/-/- and enter into an agreement to use electricity for a period of five years...'

In a letter dated 2 May 1939 the Town Clerk confirmed the various arrangements for services under which the Defence Department would reimburse Townsville City Council for the taking over of its airfield near Melrose Park, on what was known as the Town Common. It certainly proved a 'sure bet' for the Council! While the wartime needs of the GARBUTT complex were not satisfied wholly by the Council's electricity supply, those five years have extended to the present, with a Defence Force commitment well into the future.

Townsville City Council was the constructing authority during 1940 and some of 1941 in the early development of this permanent RAAF station. The general building design and layout was clustered and geometric. Drawing from its UK advice, the RAAF gave little consideration to the Base having to serve under a combat situation.

By June of 1941 the RAAF's 24 Squadron, equipped with Wirraways, was based at GARBUTT. An interesting report by its commander, then Sqn Ldr J M Lerew, is on RAAF files held in AA. Addressed to HQ NEA -

'...On 15 June 1941 a very realistic exercise was carried out with this Squadron, in conjunction with the Army. The Army made a night landing on the coast, and was both supported, and attacked by aircraft. Home defence units were also employed... outstanding lesson to be learned was the rapidity with which even non-mechanised troops reached the aerodrome. In addition... it made one clearly realise the urgent necessity for a high state of preparedness against ground and air attack... important to realise that the following recommendations are based on the assumption that a raiding force, including ship-borne aircraft, could attack this aerodrome by surprise or at least at short notice. This is a very real possibility and may happen now...'

Sqn Ldr Lerew listed various works necessary to 'meet the emergency'. He stressed the urgency, pointing out he'd already discussed a 'scheme of protection' with the 'Main Roads Board'. Lerew praised the MRC's District Engineer J C 'Jumbo Jack' Mathison's grasp of the situation and tried his best to facilitate action. He even suggested -

'...with my previous experience as Staff Officer Works & Buildings, I could handle this work and have it carried out quickly and economically, thus relieving DWB of a considerable amount of work. Rough sketches of latest ideas would be appreciated...'

This report highlights the very slow progress that had been made on many of the 'requisitions' that concerned 24 Squadron's ability to operate from its AOBs and even its home base at Townsville. Detail such as locating underground 'petrol tanks in dispersal areas rather than at the sides of runways' greatly concerned Lerew.

At least by October of 1941 some further urgency had crept into the scene. The Director of Works (Air Services) wrote to the Director General of Works, Department of Interior, Defence Liaison, Melbourne, requesting that Department to approach the Local Authority in Townsville to establish their willingness or otherwise to undertake 'repair work' in an emergency. Two weeks before Pearl Harbor, the Town Clerk replied, generally accepting this commitment, saying in part -

'..Further reference your letters (of various dates) and my reply... I have by direction to advise, that the

GARBUTT (27/7/1942)
Eight months into the Pacific War, some idea of the expansion of GARBUTT can be gleaned from this photograph. It is part of Frame 723 V, a vertical of a Trimet Run. The southwest taxiway leading off from the southern tip of runway 21 has crossed Ingham Road and extended into the first major building of the US Air Project No 1 which was for the erection of crated fighter aircraft. The hangar went on to be used as an early engine overhaul centre. It became part of the huge 4th Air Depot to be erected south along Duckworth Street and around STOCKROUTE Airstrip. In this photo STOCKROUTE is the white strip lying at an angle SW to NE, some two and a half large blocks south of GARBUTT. Of particular interest but barely visible in this photograph, is a strange rectangular grid of what appears as white lines near the southernmost intersection of runways at GARBUTT. This puzzling pattern was eventually explained by finding reference to - ***'moves to camouflage this part of GARBUTT airfield to induce an enemy reconnaissance sortie to believe that such was just a suburban development.'*** *Away on the left or western edge of this photograph can be seen the approximately N-S emergency strip known as MT ST JOHN. Several Splinter Proof Pens for aircraft are studded around GARBUTT Airfield.* [AUSLIG]

Council decided at its meeting on 20th instant, to adopt the suggestion that your Department hold a certain quantity of material in stock for repairs to electrical services rendered necessary by enemy action, and is prepared to undertake the repair work, but reserves the right to use its own discretion as to order of priority. As regards repairs to water and sewerage services the City Engineer reports that Council's stocks are adequate to carry out the emergency repairs to these services, but the Council reserves the right to use its discretion as regards priority.'

The Australian Government welcomed the October 1941 US approach for upgrading of the runways at Townsville and other works vital for supply to the Philippines. Mr H Barker, Works Director, Department of Interior in his letter to the MRC dated 30 October 1941, indicated that the desired work should commence on 3 November and be completed by 15 December 1941. The ***Main Roads Commission History*** records this construction period of six weeks was achieved, but only by the use of floodlighting and 'working around the clock'.

24 Sqn was never backward in offering its suggestions for more effective dispersal works. A 14 November 1941 report by its CO W Cdr Lerew touches on the -

'...Type of dispersal hangars decided upon makes camouflage of these practically impossible. Lessons learnt from Middle East, France and Germany show that unless these are fully camouflaged aircraft will be lost. It is considered that the policy should be strongly questioned and an alternative scheme as follows be adopted.

a) arched-roof single aircraft hangars with earth against walls blending with contours of roof

b) roof sprayed with tar and sprinkled with earth

c) heaps of selected material for runway repair made in the same shape and size, sprayed with tar and sprinkled with earth

d) patches of same size, bladed with grader, treated with weed killer and tar and sprinkled with earth giving in all, say 100 similar shaped mounds or patches of earth of which only 24 are aircraft positions. Prepared lead-in to each one necessary in case of heaps of repair material, and dummy for the other patches.'

Two weeks into December 1941, just after Pearl Harbor, the RAAF was wrestling with its plans to locate a second unit, 32 Squadron, at GARBUTT, while still accommodating the recent US commitment. To avoid looming congestion the RAAF turned attention to CHARTERS TOWERS as a base for its second Squadron.

The following initial mention of SDCC activities at Townsville appears in a report for the week ending 20 December 1941 -

'...The TD (Technical Director) phoned telling of allocation of money for two aerodromes in Queensland to be camouflaged - Townsville and Brisbane. Supervision by a camoufleur was advised in each case... Mr R W Voller was chosen as the camoufleur to proceed to Townsville, had accepted his commission and was present at the meeting. The Chairman (is) to contact F-Lt Dawkins and enquire whether the drome scheme had arrived and when Mr Voller was wanted...'

Rod Voller spent all of 1942 and 1943 in Townsville. In a personal letter to the TD, Professor W J Dakin, dated 23 March 1944, he addressed some of the more sensitive aspects not spelled out in his formal reports.

'...When I attended a meeting of State Defence Camouflage Committee (SDCC) in Brisbane on 26 December 1941, I was told that the committee would advise who I was to be employed by, whether by Interior, Home Security, or the Committee. My salary and expenses were left for Messrs McInnis, McColl and Dawkins to settle with the Works Director. I left Brisbane that evening, so you can see that on arrival at Townsville nobody knew just what Department I came under. Dawkins was the representative of the SDCC in Townsville and I was sent to assist him and work more or less under him. When I met Mr Beattie at the Commonwealth Works Dept he had been informed of my coming and had been asked to assist me and make available all facilities for my work. He did not know what my job was or how I fitted into his Department and neither did anybody else....'

The efforts of these civilian camoufleurs, have to some extent, been denigrated by Armed Services representatives

GARBUTT (1942-43?)
An unidentified B-17E, thought to be parked on the north side of 'Hangar 76'. Likely date is 1942 as the white star (underwing) insignia still has the red centre, which was phased out from July 1942. [W J Bennett ex RAAF]

over the War years and since. If nothing else, they certainly demonstrated initiative. In contrast to some of the established departmental systems in which officers more or less did things 'by numbers and procedures', these camouflage advisers were largely people with artistic and creative flair. The minutes of 5 January 1942 in respect of Townsville, record -

'...Mr Voller has discussed schemes with F-Lt Dudley on the spot. The main buildings were being toned down with "paint" made of ant bed, cement and oil; generally the colour is not dark enough.'

As Rod Voller carried on in his March 1944 review to Professor Dakin, general supplies and availability of such basics as paint, were quite a problem.

'...After some frustration and backed by F-Lt Dawkins it became evident that I must take action to get work moving at the aerodrome. MRC was contacted direct and asked to proceed with the formation of dummy roads on the aerodrome. I became dissatisfied with the efforts of Mr Beattie's office to locate paint and other materials and make continuous visits to the Railway Goods Yards in search of supplies which, according to Brisbane office, had been sent to Townsville...'

This creativity in extending the nearby road pattern over the drome, actually included a proposal to incorporate a dummy cemetery! Numerous photographs both oblique and vertical survive to show this proved one of the less effective moves. Voller's letter to Prof Dakin carries on -

'...As the aerodrome was the most important project and required most... time I established an office there. Transport became difficult as more demand was made by the whole station on its small transport section and some relief, not counting the sweat, was effected by acquiring loan of a bicycle from the CO...'

With those two years of hindsight, Rod explained -

'...The old peace time machinery of Commonwealth Works was there and was practically useless. There was no prospect of any special arrangements being made to have work carried out at once as required by RAAF and Army, even though the threat of invasion became greater each day. F-Lt Dawkins, the Area Works Officer of RAAF demanded that red tape be cut, that work of an essential nature should proceed as soon as DWB approved allotment of money... and that the usual delay and waiting for orders to filter through from Com Works, Brisbane or Melbourne, should not be tolerated... It was possible by working in with the Painter's Union at Townsville to have painters working on private jobs drawn off for work at the aerodrome. Labour disputes and upsets threatened but were avoided. When labour gangs for ground treatment at the drome were required I had to mobilise them and arrange for their pay. The pace soon became too hot and failing necessary action from Com Works Office I engaged a public accountant to make up time sheets and pay envelopes and informed the District Works Inspector accordingly. With the one clerk from the Com Works and this accountant, wages for all the painters and labourers working on all jobs in Townsville area were paid. This continued for many weeks until the employment of men was transferred to MRC...'

Voller goes on with a succinct description of those daily activities in his mention of 'hectic campaigning for materials, tools, brushes and labour as he searched Townsville each day for everything available'. He refers also to his first contact with the Americans -

'...Contact was first made with the Americans on February 11 and thereafter a certain amount of work was carried out for them. On April 2nd I took delivery of a new Bedford utility which was a great help in getting about and in transporting materials and men from town to various jobs or from job to job...'

John Wallace Fields is an American, who arrived in Townsville 20 February 1942 flying as co-pilot with 1st Lt Harry Spieth -

'...our group consisted of 12 combat crews from the 7 BG originally stationed at Fort Douglas Utah but were destined for Mindanao of the Philippine Is as a permanent change of station prior to the beginning of WW2. Some of these 12 crews were among those landing during the attack on Hawaii, including Frank Bostrom who later flew General MacArthur out of Mindanao in a B-17...'

Wallace recalls their 'quarters' for a while were those of the RAAF's enlisted men at Townsville and that -

'...we had access to the Oz Officer's Club on the base and frequently stopped in for a can of ale or two en route to our mess...'

In a brief reference to their 'work' he says -

'...Our Squadron (after 22 April 1942, redesignated 435 BS, 19 BG(H)) *operated as an armed recon Squadron whose first objective was recon work...on some of the missions (we carried a) complete load of bombs as for the five missions in six days during the Coral Sea Battle - you will no doubt realise that our Squadron insignia was created in Townsville giving it the flavour of being born in Oz, hence the Kangaroo with a bomb in its tail looking through a spy glass - thus the recon capability and the cloud in the background which many times was our greatest defence against Jap fighters...'*

Rod Voller's Townsville section camouflage report for the period to 25 March 1942, included the following -

'...We are proceeding with ground work as fast as possible on runways and other areas using a lime wash, coal tar etc. In an interview with the CO it was decided to cover with netting one of the splinter proof pens,

GARBUTT (4/9/1987)

This oblique shows relatively up-date layout and development of the current day RAAF Base Townsville. The new runway 01/19 has been developed roughly parallel to the original runway 21 which is now a parallel taxiway. Only the southwestern part of runway 07/25 remains in use and all of the old runway 135 has now been discarded? The tarmac and building development in the right foreground is associated with the Australian Army's Aviation Regiment's Black Hawk helicopter force. RAAF 35 Sqn using Caribou transports is based at the airfield. The prominent long white building in the rear centre is the combined International and Domestic terminal for Townsville. [RAAF Official]

near trees over by the Ingham Road and for the present to place decoy planes in the remainder. We have painted three of these enclosures with sump oil and khaki green which makes them less conspicuous.'

Rod also refers in that report to progress on -

'..AMERICAN BUILDINGS - Project No 1 - painters are continuing to follow on the heels of the builders of the 2 hangars and huts and the whole job will be completed within a week after the period of this report.'

Apparently 'Building 25', was being built as an HQ for the US Army. Rod described it as a concrete building with concrete splinter proof wall all round and flat concrete roof. His report added -

'...arrangements have already been made to make provision for sockets let into concrete to take tent fittings (for camouflage!). The building will be completed in about 1 months time...'

So sturdy was 'Building 25' that it survives to this day. McIntyre and Associates, a well known firm of Consulting Engineers, have redeveloped it as their Townsville office, at 3 Ramsay Street, Garbutt, just a few blocks on the city side of the airfield.

Merv Bell had completed his RAAF 'service flying training' at AMBERLEY, early December 1941. Unlike previous intakes which marked the occasion with a ceremonial passing-out parade, the salute being taken by an appropriate dignitary, Merv remembers his 'lot were quietly told to go to the Q store and draw their wings'! He spent the next few months at Evans Head, towing target drogues behind Fairey 'Battles'. Merv explained -

'...about mid April, out of the blue I was called to the Orderly Office...I was on a posting to NEA Townsville...another lad George Andrews and I were the only two from Evans Head...we'd heard the Navy wanted some target drogue towing done up there, but we had no real idea what it was about...we had to report to Brisbane. Once there we joined four other fellows headed north too... when we got out on to the platform at Townsville, the PA system called our names...the Railway Transport Officer showed us to a back room and there was an American Colonel and a RAAF Wing Commander...we found we were seconded to the USAAC...they said there was a staff car for us and on the way we asked the driver...he said "435 Heavy Bombardment Squadron"..I said what type of aircraft...he said "Flying Fortresses"!'

Merv Bell's 'multi engine' experience at this point was on twin-engined Avro Ansons. He described the ever so brief few days 'conversion' he received to start him in his role as co-pilot on the B-17D. Looking over his log book reveals he flew co-pilot to Lt (John Wallace) Fields on occasions, and from late June 1942, he teamed up with Fred Eaton. They flew together until November 1942. Merv Bell and Fred Eaton were reunited, by phone, in 1989 when a newspaper 're-ran' the saga of Fred's B-17E 'still there' in the kunai grass swamp in New Guinea.

When P-O H V (Jim) Trench arrived in Townsville early in 1942, his posting was as a Works Officer at NEA HQ. Jim recalls -

GARBUTT (approx 8/1943)
General Douglas MacArthur greeting Wilbur Beasley, pilot of the B-17E 'Sally', General George Kenney's personal transport aircraft. General MacArthur frequently used this aircraft on trips to and from New Guinea. In the background is a Beech 18 twin-engine transport and an American Kittyhawk with at least one RAAF Beaufort visible under 'Sally's' outer engine. Note also that not all US staff cars were the modern American vehicles often pictured. [R Graf ex USAAF]

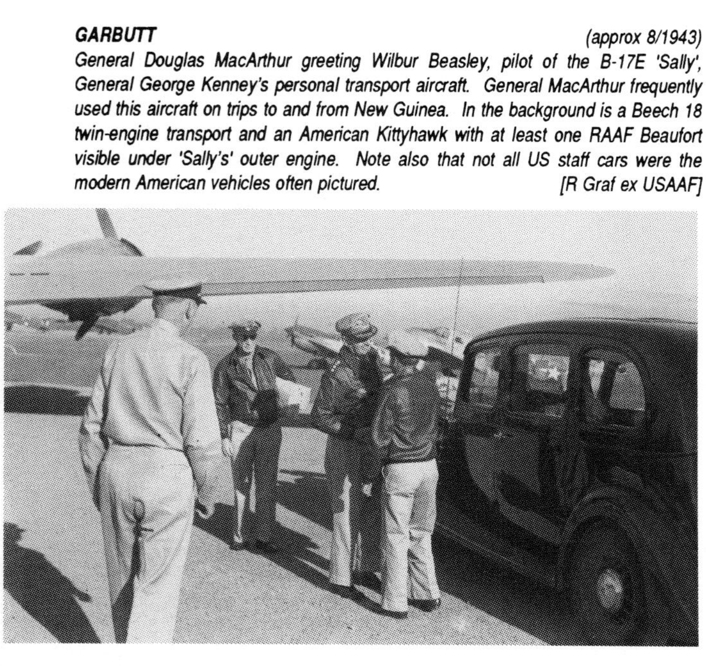

GARBUTT (1945?)
With handrail appearing across the frame, this photo has been taken from the observation platform atop the Headquarters building a little northeast of the main GARBUTT 'Hangar 76'. In the background are the maintenance structures occupied by RAAF 36 Squadron. It is of interest to compare the two igloo structures. The one in the centre above the taxiing US C-47, has its top arch hinge made substantially rigid with extra framing. The igloo on the extreme left or to the north, is typical of the majority with only nominal ridgecap framing erected above the top hinge joint. The centre structure collapsed during a cyclone in 1971. It was found simply to have been insecurely anchored at its foundations. [C A Mason ex RAAF]

'...Townsville was then the headquarters of the North Queensland Division of the Department of the Interior with Mr Beattie as Works Director. Until war broke out the Works Branch was mainly responsible for maintenance and construction of Commonwealth buildings, mainly Post Offices and Commonwealth Banks. Sometime after, but prior to Japan's entry, a Works Section of the Department was set up in Townsville under Mr Eric Doig who was able to start new construction without some of the red tape normally associated with such matters. There is no doubt that Mr Doig was a first class expediter and I am sure that Mr Beattie was horrified by some of the things that happened...'

While I'm not able to expand and Jim didn't specify, one thing is obvious. Certainly drafting staff were few and no doubt hard working. My review of those tattered surviving tracings show that not only did Eric Doig 'approve' most of those early drawings, there are some I've found which were actually drawn by him!

Rod Voller's camouflage report for the week ending 20 May 1942, mentions work on the western fringes of GARBUTT -

*'...**A-A Battery Mt St John** - number of labourers employed - 4, week's work consisted garnishing netting and large quantity is now ready for erection. Practically the whole area has been covered with dark soil to tone down the establishment. And American machine gunners are making good progress with the netting of their positions round the main battery. **St John Zoo** - 3 men are employed at this strip erecting nets for covering 2 aircraft.'*

The following excerpt from ***Anti Aircraft Artillery Activities in the Pacific War*** by General HQ US Army Pacific 1946, notes -

'...Townsville - numerous batteries from the 208th (Coastal Artillery - Anti Aircraft) the 94th, the 197th, the 743rd and the 104th and the 40th Brigade were in and around Townsville generally from mid-42 'til as late as September 1943...'

From 17 April 1942, when Gen MacArthur appointed Lt-Gen Brett as Commander of the Allied Air Forces in SWPA, serious efforts began in the search for a close-working alliance, RAAF with USAAF. All the sourcing of material, its distribution and transport, the furnishing of airfields, accommodation and maintenance, in most cases the physical building of such installations - these services and others were all necessary to create a fighting force out of disarray and deficiency. Brett's answer to this was to set up on 27 April 1942, the US Army Air Services (USAAS) with Maj-Gen Rush B Lincoln in command. Lincoln's priority list of airfield works was the prime

GARBUTT *ca(-/3/1944)*

Dating and placing this photo has been difficult. Received from Norm Parker, ex RAAF 36 Squadron, I am advised the scene is STOCKROUTE. However, certainly such a cantilever hangar was in place on GARBUTT's eastern boundary about then and it is possible that's where this maintenance on VH-CTM and CDK is being carried out. George Redding, who passed on the photo adds - ***'CTM, an RAAF C47, was received by 36 Squadron June 1943 and was retained beyond 1944. CDK was on loan from the USAAF to 36 Squadron from early 1943 to April 1944. Hence the photo was likely dated between June 1943 and April 1944.'*** *Inspection of the original suggests there is the eastward down-sloping side of a hill showing between the aircraft and the hangar. This could be Many Peaks Range near Cape Pallarenda and if so, then the scene is GARBUTT. [N Parker ex RAAF]*

subject of discussion at the AOC NEA's meeting in Townsville on 26 May 1942. Attended by Col. Matthews US Army Engineers, W Cdr Dale DWB Air Board, F-Lt Dawkins Area Works Officer, Mr Doig Dept Interior, P-O Trench NEA and P-O Campbell Air Board, the resulting direction for GARBUTT was -

'...all work (of runway upgrading) has been stopped on this field, the remaining strengthening of the runway to be completed on low priority. Camouflage work will be required to be completed at the dispersal strips.'

P-O John Keays had been active in NEA from late March 1942, primarily in the design of the USA Pursuit Squadron fields just south of Townsville. Keays had encountered confusion in priority lists issued by US and NEA RAAF authorities. NEA itself was concerned with overlapping efforts in further airfield site selection. In its 29 April 1942 plea to Air Board, RAAF HQ NEA stated -

'...understand that the finalisation of sites in this area was to be left to W Cdr Lerew who has been working with Col. Matthews of the 46th Engineers, Mr Doig of the Dept of the Interior or his representative, and the MRC. P-O Keays was also in this area... working with the above named officers and officials. Prior to the departure of W Cdr Lerew for the Brisbane Area, Messrs Connellin (sic) and Moffat, representing Sverdrup and Parcel, made an aerial reconnaissance of the Hughenden-Cloncurry-Winton area... following their departure, Capt Wetstone and Mr Fredman of the staff of General Johns (then Chief of the Engineer section, USAFIA) also arrived to do aerial reconnaissance in the same area. At the same time Messrs Coleman and Fraser were carrying out investigation in the Hughenden and Cloncurry areas respectively. So far no reports have come from the above mentioned sources and there seems to be some duplication of effort, as it is understood from Col. Matthews that he has also had officers of his Regiment moving in the same locality...'

GARBUTT *(1945?)*

This photo is typical of several which have been accumulated from uncertain sources. The original photographer is not known, nor is the date, though its estimated as early 1945. The identification serials on the C-47 transports, cannot be read, but they are likely 36 Squadron aircraft. It is clear that the Liberator in the foreground is an RAAF machine. The gable entrance to the Butler hangar on the left and the skyline together prove the location as GARBUTT. [K Hopper]

The AOC's letter, under the signature of W Cdr Gordon Grant, sought coordination and an effective 'clearing house', saying it was unimportant whether the 'correlation of information is directed to the US Base HQ in Townsville or to this (RAAF NEA) HQ'!

John Keays and Jim Trench were professionally associated before WW2. Early in July 1942, they were to resume this connection, when P-O Keays was posted to NEA to replace Sqn Ldr Dawkins, on transfer to Eastern Area. Keays brought with him a proven track record of cooperation with the Dept of Home Security's Camouflage team, most recently in work on the preparation of heavy bomber airfields on the Darling Downs.

Keays, in his new role as Area Works Officer NEA, came to realise that camouflage and dispersal had gotten out of hand. Planning for and constructing the dispersal taxiways was lagging the completion of the arched hideouts. W Cdr G J Towers reported on his two day visit, mid July 1942 -

'...In one case, one of these hideouts had been constructed in a position from which it was quite impossible to reach it from the nearest landing strip.'

This certainly puts the meaning into 'hideout', when in fact you can't even reach it with an aircraft.

An appreciation of the US camouflage approach is given by the following extract from the SDCC report for week ending 2 August 1942 -

'...On 30 July, Capt Murphy (US) came and discussed their present policy regarding camouflage. He said their work would be confined to a strip 100 miles wide along the coast of Queensland northwards from 50 miles south of Brisbane and including the whole of Cape York Peninsula and shores of the Gulf of Carpentaria... He also said that an order would be issued from Allied HQ regarding dispersal of planes, which should overcome the present difficulty. It stated that the maximum dispersal justifiable should be obtained in all instances. He stated that Mr Fredman would be given a commission and attached to NEA in Townsville. He would then have control of all (US) Air Force camouflage in that area....'

It is perhaps unfortunate that some information can only come via written correspondence. I have found a more satisfactory relationship when I've been able to meet veterans and discuss face to face their various recollections. However, when we are separated by the whole Pacific Ocean, some snippets can only come by letter. Jack McIntyre for example has given some lasting recollections of his time with the 435 BS -

'...Great days in Queensland in 1942 and 1943 - the lassies were right on! Bought a wing of Queen's Hotel. We rented a 4 bedroom home from an Aussie wholesaler - lived there and not the quarters at Garbutt - had 4 lassies to keep house for us when on missions; kept store room full of Abbots Lager and Johnnie Red Scotch. The lassies furnished the food, etc. Those were great days for me and I still think of them. As for food at Garbutt mess, we bought our mess sergeant a 53 gallon keg of Dewars 12 year old Scotch - placed in his tent. Our mutton turned to heavy beef. Our food super and we got water instead of "tea" with our meal(?)...'

Problems in coordination extended to the supervision of aerodrome works in NQ. Col. Matthews, the US Base Section Commander had suggested all works south of the 19th parallel be delegated to the RAAF and those works north be directed by the US Engineers. The RAAF DWB wrote on 20 August 1942 suggesting a modification - that all Allied Engineer Units in NQ be directed by US Engineer HQ with works being executed by civil engineers to be directed by RAAF. That letter concluded -

'...it is felt... essential... whatever scheme to define responsibility be adopted, it should be put on a definite official basis, in order to eliminate duplication and overlapping by the responsible officers in the field.'

Jim Trench was posted north to Port Moresby on

3 September 1942. His replacement was F-Lt Farran. Rod Voller's diary for 18 January 1943 mentions -

'...with F-O Grounds we saw F-Lt Farran at RAAF HQ Townsville re the splinter proofing of "ready flights". He showed us an Airforce directive stating that all aircraft dispersal positions on airfields on Cape York Peninsula as far south as Charters Towers and up to 300 miles inland from the coast were to get splinter proofed. This means reconsideration of camouflage policy. F-O Grounds advised that all high RAAF Officers were of the opinion that overhead cover of some sort for aircraft was essential...'

The RAAF's concern for Japanese air activity in the West New Guinea area apparently triggered such a directive. It came at a time when the Dept of Home Security's Camouflage organisation was realising its own need for rationalisation. It abolished its State Directorates in Tasmania and South Australia, reduced the activity of others and boosted that of northern areas. A new Directorate, to be known as NEA Directorate was established at Townsville. As the next few months were to reveal, any realistic need for **full** camouflage concealment at airfields had vanished. The Technical Director of Camouflage, Prof W J Dakin in his report to a meeting of Deputy Directors, at Canberra 6 February 1943, explained -

'...The Department of Air has recently informed us that concealment of aircraft in hideouts, even though the position of the latter is indicated, is a great advantage for the following reasons:

1. Types of aircraft, numbers, etc. etc. are hidden.

2. Occupation of the hideout is not given away.

And so we have followed instructions and policy, whilst continually giving our own opinions.

In the meantime we were discovering from England that hideouts were not being used for such large aircraft as B-17's and B-24's. And gradually the cost of construction in Australia was recognised to be unduly high. The advantage of simpler methods became clearer. The possibility of using nets was then communicated to us and with the sparse data sent to us we made up our own example, drew detailed plans of its construction with minute descriptions, took photographs from the air at low levels and forwarded all the data to the Air Board. A special discussion evidently took place. The matter was looked into and the most gratifying acknowledgment of our work ever received resulted from it. The use of nets in forward areas has been formally approved and efforts are being made to construct them - chiefly in Sydney...'

On the scaling down of camouflage effort, Rod Voller recalls from his diaries that -

'...8 June, F-Lt Farran stated that splinter proofing of aircraft positions in Townsville area was to be wiped and nothing south of Mareeba and Cairns was to be done... F-Lt Farran then advised on 9 July that splinter proof pens south of Jacky Jacky had been abandoned and Horn Island and Jacky Jacky were the only places where they were to be built. We continued with camouflage work on other projects, particularly RDF (Radio Direction Finding) and other communication and radio stations - bomb dumps etc.'

There were several notable developments during the latter half of 1942. MacArthur engineered the establishment of his 5 AF and appointed Lt-Gen George C Kenney as its Commander. Aircraft commissioning and maintenance capacity was significantly advanced, with EAGLE FARM and the Townsville 4AD complexes nearing fruition. At year's end the US partner in the AAF was savouring some confidence from this supply improvement and operational gains in New Guinea and the Solomon Islands. The time had come for the 5AF to move forward. Through the crucial period RAAF squadrons had played no small part in the New Guinea scene but a suitable RAAF Command structure upon which the junior partner could proceed overseas from Australia was absent. The long established RAAF HQ and the Air Board system denied full and complete control to any expeditionary RAAF force. MacArthur's forces were growing stronger and he could afford to leave behind him some of the irritant of cooperational alliance on the Australian mainland. GARBUTT was one of some five airfields in NQ which Kenney addressed in his 5 June 1943 direction -

'...responsibility for construction and maintenance of following military airfields in NQ has been delegated to commanding general USAFFE. Garbutt, Charters Towers, Iron Range, Woodstock, Reid River, Mareeba. You are advised of this so you can arrange for construction and maintenance of such other military airfields in NQ as maybe required for operation by RAAF.'

Air Cmdre A H Cobby, then AOC NEA queried Air Board on the full import of this assignment of responsibility. He cited 23 OBU's role at GARBUTT in an informative letter dated 28 July 1943 -

'..A large detachment of American personnel are based on 23 OBU, while five civilian airline companies operate from this base (ANA and AOA, QANTAS, Ansett, Guinea and Carpenter). The constant amount of construction work being carried out on this aerodrome necessitates the employment of a large number of CCC workers, while the guarding of all American and RAAF aircraft on the drome overnight (averaging 42 per night) is the responsibility of this unit as well as the guarding of Bohle River, Stockroute and Weir Strips when not occupied. The formation of Wings and Units takes place here, while a Met section, now operating Radio Sonde, also operates from this base. Other RAAF units now lodging at 23 OBU are - No 5 Communication Flight (occupying a hangar) and No 42 RDF Wing (occupying portion of the sick quarters). A detachment of No 1 Air Ambulance Unit, and a section of 3 DWO is attached. No 42 RDF Wing will shortly be moved to a new building of its own, but personnel will require to retain their sleeping quarters at Garbutt, there being no other quarters available. Its not practicable to remove 5 Communication Flight from its present situation. Due to proposed move of No 5 MRS to "Ginger" at an early date, it will be necessary to reopen the sick quarters at Garbutt. ..In addition to the lodger units noted, No 12 RSU has detachment at Garbutt. This detachment... services all transient aircraft under direction CO 23 OBU... The detachment occupies 1 complete hangar. It is also proposed, as a temporary measure to install a section of the MTRS of No 13 ARD at Garbutt pending erection of a new building for it elsewhere in the Townsville area. This

GARBUTT (27/7/1991)

'Hangar 76' as it is today. Notice the significant extension on the long southern side and the disappearance of the upper level control tower on the near north western corner. Standing in foreground is F-O Pat Kenny of RAAF's 27 Reserve Squadron Townsville. Pat was Curator of the North Queensland Air Force Museum established on the base and regularly open to public access. [Author]

GARBUTT (14/8/1944)

'Control Tower and Hangar, Garbutt Field, APO 922', showing the western end of the basically RAF type 'C' saw-tooth roofed 'Hangar 76'. [A Fraley ex USAAF]

GARBUTT (24/4/1944)
From about twenty frames covering GARBUTT, this is part of KD 1026. The frame reaches south along Duckworth Street to show the southernmost one of the two hangars of US Project No 1 with a single Butler hangar erected on its hard-stand entrance. One other Butler hangar appears on the eastern boundary where maintenance facilities are in course of erection for the RAAF's 36 Transport Squadron. Just north of that Butler hangar, work is underway with on-site construction of the nailed wooden arch frames for an igloo hangar. A small cantilever maintenance hangar is located between this Butler and the igloo under construction. Dozens of aircraft can be identified in this frame, C-47s comprising the majority. The white disk appearing to left of the intersection of runways 21 and 130 is a compass swinging platform. On the western side of Duckworth Street an impressive maze of 'tracks' can be seen providing shortest path routes in and around the cluster of accommodation and other hutments. The camouflage authorities were forever busy impressing a need for 'track discipline' on personnel in areas as vital as GARBUTT and its surrounds. Logically, enemy aerial reconnaissance would note the development of such tracks and interpret the value of such areas as targets. However by the date of this photograph, all needs and concern for camouflage had vanished from this area. [AUSLIG]

detachment will take over the whole of the building now occupied by the transport section of 23 OBU and portion of which is occupied by the 4th Air Depot of USAAC and another portion by the Army for Mail Room purposes... functions at present performed by 23 OBU are...two-fold:... accommodation for large numbers transient RAAF personnel and secondly ...routine inspection of RAAF aircraft which call or are staged (at Garbutt) pending a further journey. (This sometimes involves) Aircraft of complete squadrons, e.g. 79, 76, 12 and 77. Refuelling of all RAAF aircraft is also done by 23 OBU from an American tanker... (All these services) are essential ones that must be performed in Townsville for the RAAF. Other strips exist at which aircraft could be serviced... (but there would be difficulties involved due to) no accommodation, ops room, or suitable hangar space for servicing parties, (etc.)... '

Cobby requested further information in respect of -

'...i) Are the buildings occupied by RAAF now to be occupied subject to the pleasure of the local representative of USAAC?
ii) May the CO. 23 OBU make alterations or additions to buildings presently occupied by RAAF which he considers necessary?
iii) May the local rep of USAAC erect new buildings on Garbutt without permission CO 23 OBU?...
iv) Is the responsibility for maintenance of all the buildings, whether occupied by USAAC or RAAF, the responsibility of the USAAC?
The present practice existing at Garbutt is that the CO, 23 OBU is responsible for internal security of the aerodrome and buildings but not for operational control. In practice, members of USAAC do not erect a building without the permission of the CO, 23 OBU, who has on several occasions refused that permission, and such refusal has been acquiesced in by the Americans. The maintenance of the existing buildings has been the responsibility of the Service which has occupied it. It is emphasised that present practice is not covered by any existing orders, but is one that has grown up based on practical experience, and its continuance has been entirely due to the tact and personality of the CO's of the respective Services.'

A memo dated 26 August 1943 from the Office of the Chief Engineer GHQ SWPA, addressed to Gp Capt Knox, DWB RAAF HQ Melbourne, dispelled most of the concern. So clear was this explanation that the original abrupt direction could even be construed as mischievous -

'...The reason for establishment of responsibility between USAAF and RAAF was to prevent conflict in the improvement, maintenance and construction of airfields between the services involved. As you realise, there are many fields throughout Australia on which work was started for the US and for which need no longer exists due to the changing tactical picture. Similarly, a conflict arose at Amberley due to the requirements for an additional strip. Both the (US) Air Service Command and RAAF were working on the plans for... an additional strip. In order to clarify the situation, it was decided to assign a definite responsibility for the work on the fields themselves throughout this theatre. It was not desired to convey the impression that such responsibility would extend to barracks or other installations which naturally are a responsibility of the occupying units. This is... brought informally to your attention... to obtain your views... prior to... issuance of clarifying instructions.'

Typical of the many supply and transport functions emanating from the giant GARBUTT facility is this recollection by Earl Mahan a member of the US 317 Troop Carrier (TC) Group -

'...The most important load was 6 000 lbs of blood plasma that we hauled from Brisbane to Port Moresby and Hollandia in New Guinea. At times we would leave our base at Garbutt and go to Port Moresby, from there to all over New Guinea where troops or supplies were needed. We would be gone for 5 or 6 days at a time, sometimes up and back in one day...One of the most interesting loads carried by the 41st Squadron was during American Thanksgiving and Christmas of 1943. It was fresh turkey with all of the

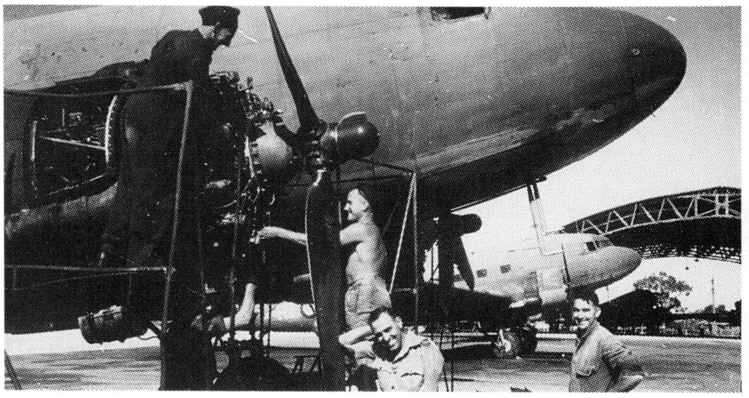

GARBUTT (1944)
Left to right. Ern Cobb, Alvin Colley, Vyv Plumb and (?) McFadden, RAAF 36 Squadron members at work on their aircraft's maintenance in front of one of the Igloo hangars erected on GARBUTT's eastern boundary. Note the igloo's design modification at the top. The top and bottom chord members have been continued through the hinge, to strengthen and improve the load hoisting capability. [F Vonarx ex RAAF]

GARBUTT *1945?*
Here the photographer and date are both unknown but clearly this B-25 Mitchell N5-233 is one of the Netherlands East Indies (NEI) No 18 Squadron photographed during some visit to GARBUTT. *[D Johnson]*

trimmings, cranberry sauce, celery and long awaited mail. This was brought to the troops by parachute drops along the front lines.'

Sufficient and efficient air transport was always a pressing need in the SWPA. The RAAF's 33 Squadron, with a motley array of aircraft, some civil types impressed into service, was the first 'Land/Transport' unit here. Several Short Empire flying boats operated from Cleveland Bay for the crucial period in 1941-1942. The American Troop Carrier units later dominated. The Pacific Air Wing of the USATC established maintenance hangar facilities for C-54 Douglas 'Skymaster' 4 engined transports on GARBUTT in mid 1944, after November 1943 activation of the unit. Late 1944 aerial obliques in the Arch Fraley Collection show an igloo hangar erected on a N-S axis, about 1 000′ north of the present day terminal. 1947 vertical photography shows a 'Butler' type hangar replacing this igloo. It is quite likely that this 160′ x 130′ 'Butler' was in place by 19 July 1945 and that it came from the twelve dismantled at STOCKROUTE ready for shipping north in January 1945. By July the USATC had 'moved out' and the RAAF sought, probably unsuccessfully, to acquire this hangar to serve as Transport Squadron workshops. What happened to this Butler hangar I do not know. However on this site since 1965, has stood the sole remaining RAF Type 'A' (1917 style) hangar, which Rod Voller's camouflage men busily painted March 1942, at the corner of Duckworth Street and Ingham Road.

Although GARBUTT had been fully handed over to the RAAF for purposes of maintenance and construction by October of 1943, as late as 7 April 1945, the US reserved its rights to maintain operational control. This situation apparently was sufficient to cause the RAAF to review its plans at that time for relocation of the NEI Equipment and Personnel Pool and the RAAF's B-25 Training Flight. Both of these elements had been slated for relocation to CHARTERS TOWERS or possibly even GARBUTT. By 31 July 1945, two weeks prior to the cessation of hostilities, GARBUTT was in use by RAAF personnel undergoing conversion to P-51 Mustang aircraft. There was the real problem of congestion and for that reason some activity was shed to nearby fields such as ROSS RIVER and BOHLE RIVER.

GARBUTT was always planned to be operated by the RAAF after the war. The airfield continues to be a dual-use defence forces and civil aviation facility. It is an important International Gateway to Queensland's tourist industry.

GARBUTT *(27/7/1991)*
Inside the Butler prefabricated steel framed hangar still in use at GARBUTT today. Notice that this structure utilised permanent metal side and roof cladding in contrast to slinging a canvas 'tent inner' which was an option called on in the case of the several Butler structures erected on the old STOCKROUTE strip. *[Author]*

STOCKROUTE *(31/3/1942?)*
Three months into the Pacific war, this oblique (PA 545) photographed from an RAAF aircraft shows the freshly graded STOCKROUTE Airstrip. A developed Duckworth Street extends southward angling across through the bushland of what is now the Townsville suburb of Vincent. On this section of the taxiway a 24 Squadron Wirraway is passing what looks like a parked 24 Squadron Lockheed Hudson. Along Duckworth Street where it reaches northward to Ingham Road on the southern boundary of GARBUTT, the two prominent hangar buildings associated with US Project No 1 can be seen along with sundry accommodation buildings. *[NLA (RAAF Official)]*

STOCKROUTE

'...Where the hell was that?'

This was my involuntary response to a 22 April 1985 letter from the RAAF's CPE, Laverton, Victoria. I had written 11 April, for 'a complete list of RAAF airfields and bases during that WW2 period, their locations, the dates of relevant aerial photographs of that period, their availability and cost.' This STOCKROUTE mention came along with some twenty others, geared simply to a CPE 'reference number-367' and a date of 22 July 1943. I asked for more information. I received no reply. Circumstances meant that I waited until a private opportunity arose to visit Melbourne in January 1987. After quite some hassle in even getting permission to visit CPE, I at last inspected some envelopes containing negatives only. When I saw the STOCKROUTE negative (see Appendix 4) I was elated - there were dozens of aircraft actually visible! The RAAF attendant did oblige by checking, presumably only from some present day gazetteer. He informed me that he couldn't find a 'Stockroute' as such, but that maybe the photo related to an entry for 'Stockyard' at lat/long 27°38'S and 152°04'E. Against this I scribbled - 'maybe it's OAKEY??' Again, so much for asking the RAAF.

Not until my later perusal of 'key diagrams' obtained from the very helpful AUSLIG staff officer of the time Mr Warren Webb, did STOCKROUTE fall into place. Of course Rod Cardell did not have this problem identifying it! In his recently published book ***Wings Around Us*** he tells of his boyhood days during WW2 living only meters from what was urgently cleared and titled STOCK ROUTE.

Project No 2 seems always to have been the major commitment that Maj-Gen Lewis Brereton's November 1941 visit to Australia produced. Mentioned in Supplement No 3 to the Commonwealth Government's War Cabinet Agendum 334/41, it was the 'Main Depot' component, estimated then at £2.8M, some fourteen times the value of Project No 1. The latter, located on the southwestern corner of GARBUTT, was intended as an erection facility for fighter aircraft. Tediously slow as the Australian Government's 'system' was, at least when war in the Pacific came, its two major hangar structures at GARBUTT and several identical 121' x 115' units ear-marked for the component of Project No 2 at EAGLE FARM were 'in the pipeline'. Typical of the RAAF's heritage, these units were essentially the RAF Type 'A' hangar (1917 style) phased out in England around 1935.

Harry Parker, then a fitter 2E with the RAAF's 24 Squadron, recalled that squadron's advent to the strip. Harry dated this as early in 1942 -

'...When the Americans took over Garbutt, 24 Squadron moved to Aitkenvale (Stockroute) and used a road for a runway which was also used for Army movements which wasn't very satisfactory. One of our Wirraways hit a tree dodging an Army truck...'

STOCKROUTE (4/12/1942)

'South end of south hangar of Air Force Project No 1 4th Air Depot at an airfield in Townsville Australia'. That is the caption on this photo from the USAF pre-1954 collection. Two such hangars of this design actually preceded any of the nailed wooden arch igloo type. This style was utilised also at EAGLE FARM and CHARLEVILLE airfields. Note the variegated camouflage pattern.

[Neg A 29890 USAF Photo Collection NASM Smithsonian Institution]

Harry was involved in mending the 'dent'. This he did by shaping a piece of wood as a filler block, gluing it into the leading edge and fabric taping over it.

Jim Trench, a freshly inducted RAAF Pilot Officer had arrived at GARBUTT in February of 1942. Records now in Australian Archives mention a telephone message on 22 March 1942, from P-O Trench NEA apparently to DWB Melbourne, which amongst other things reports -

'...Garbutt aerodrome (Townsville) - dispersal strips - Stockroute strip finished; Bohle River strip will be finished on 25 March 1942...'

As Rod Cardell has recounted, March 1942 also saw elements of the USAAF's 22 BG (Medium) with its B-26 Martin Marauders begin to utilise aircraft dispersal clearings in the vicinity of STOCKROUTE. At the time the Group CO was Lt-Col Millard L Haskin.

Walter P Maiersperger arrived in Australia as Flight Leader of 'A' Flight, 19 BS, 22 BG(M) on 22 March 1942. This was the now historic first landfall of B-26 Martin Marauders at ARCHERFIELD, Brisbane's municipal airport of the time. Capt Maiersperger and his No 3 wingman 2nd-Lt Robert R Hatch had left the No 2 wingman, 2nd-Lt Chris Herron behind in New Caledonia with a burst tyre. Maiersperger's log shows 40-1495 as the aircraft he brought over the Pacific and on to Townsville. It is likely that Hatch flew 40-1496 'Dixie' over the Pacific to become the Marauder the 9 year old Rod Cardell most loved and associated with the STOCKROUTE dispersal. Maiersperger noted arrival in Townsville, in his log as -

'...Mar(ch), Sat(urday) 3/28/42, Martin B-26, 40-1495, P & W R-2800 (engines), Amberley to Townsville, 1 landing, Disperse into woods, Bumgarner (Lt W C) taxys into pole - wrecks wing tip, Krell (Lt Walter) taxys into tree - wrecks wing tip...'

The 9 year old Cardell was right in his considering possible reasons for the -

'...two crew members sitting casually on top of each fuselage, with their legs dangling through a hatch in the roof... it was possible they were there as lookouts... during taxiing and parking.'

The 25 year old Capt Maiersperger, who was a graduate engineer and had been the 19 BS Engineering Officer, as well as a pilot, was moved into 'squadron operations' a few days after. His log entry for 2 April 1942 shows a 2 hour 'local' flight in 40-1496, '5 landings - Trans(ition) for Jay Zeamer'. Such 'transition' meant testing the co-pilot for advancement to pilot-in-charge. Reflecting on this entry, Walter told me he 'failed' Zeamer who was his co-pilot in the USA, believing he did not have the necessary maturity. Either Walter's judgement was too harsh or Jay Zeamer changed, for Walter explained that Zeamer -

'...was (later) transferred to a B-17 outfit, where he won the Congressional Medal of Honour after losing half his crew and suffering severe combat injuries himself.'

Walter Maiersperger recalled his earliest quarters near the STOCKROUTE strip. His experience tends to balance some more colourful accounts -

'...Officers did not live in hotels, but in tents scattered in the woods near their airplanes' dispersal areas. There is record that some livelier types kept rooms in some hotels which they utilised for 'recreation' to suit them when allowed out of the dispersal area...

My 19th Squadron Operations tent was located on a concrete slab, originally the floor of a slaughter house. There was a drainage channel so that the floor could be hosed down. At the low end there was a pit to which the slop drained. One day I saw an Iguana lizard which had fallen into the pit and couldn't get out. I shot him with my .45, the first animal that I had ever killed in my life. A few days later a 7' black and yellow snake was roosted on a tree branch where our path led, I dispatched it the same way. I've never forgotten these killings and am ever more thankful that I was never ordered to bomb or strafe anyone.'

Such accounts go some way towards explaining the

desultory small arms fire Rod Cardell remembers late afternoon on that day of arrival - 'rifle shots' that he put down to 'letting off steam'.

On 21 June 1942, a month after his transfer from CO of the 22 BG, to Chief of Staff HQ AAF NEA, Lt-Col Haskin issued a directive, stemming from a recent visit by General Brett. This directive scaled down fresh airstrip construction along the Townsville to Cloncurry route and set various minimum standards for buildings associated with new aerodrome works. In respect of STOCKROUTE, it confirmed that it was then to be 'gravelled'. This aimed at controlling the dust nuisance. Clearly if bitumen had been available, it would have immediately followed. US Drawings show it sealed by 1 October 1942.

One of the lasting memories Walter Maiersperger has of his flying out of STOCKROUTE was the period May to mid July 1942. A Lockheed 12 (C-40) twin engined passenger aircraft had gravitated to the NEA HQ from the Dutch. Maiersperger's log lists several interesting flights from the 'Dalrymple Strip' in this C-40, serial '21302'. He may well have been pleased that the week-end flight of 11-12 July 1942 was his last in this aircraft. Australian 'Signal Corps Telegraph (line) Inspectors' requested 'the altitude be kept below 500′ all the way' in their aerial inspection IRON RANGE/Portland Roads to COEN and southward through WROTHAM PARK to Mt Surprise -

'...I kept to about 500′ knowing the C-40 did not have feathering props and would not hold altitude on one engine. They were in jeopardy on that flight but didn't realise it - I did!'

Mr R W Voller was Chief Camouflage Officer for North Queensland in the Department of Home Security's civilian organisation. Rod wrote, referring to his diary -

'...On 4 July 1942 I went to Aitkenvale, Stockroute, Armstrong's paddock and Garbutt with P-O Roy Grounds, now my contact on camouflage in RAAF to see arch structures - these being mainly for camouflage of aircraft in dispersal bays...'

The following extract from ***The Corps of Engineers:The War Against Japan*** by Karl C Dod, in addressing problems of logistics, touches on 'warehouses' -

'...A fourth (change in warehouse design) and among the most widely used, was the igloo warehouse, modelled on a hangar designed by Mr Brizay, a French engineer who had come to work for the Base Section engineer in Brisbane after fleeing from Malaya. Brizay's hangar consisted of arches constructed of scrap lumber, held together by purlins and covered with camouflage netting... adding a concrete floor and a corrugated iron or fibrolite roof, and closing the ends, a large and roomy warehouse could be quickly provided.'

Rod Cardell's research shows the arrival of the first elements of 4 ADG at Mt Louisa, Townsville as 2 October 1942. The CO was Lt-Col Victor E Bertrandias 'one time crew chief of the famous WW1 aviator, Eddie Rickenbacker, and later a Vice President of Douglas Aircraft Corporation'. In fact, Vic Bertrandias had been here before - on 29 April 1936 he test flew the first DC-2 Airliner 'Bungana' delivered to Holymans Airways in Melbourne.

The first mention of this Depot construction work in the AA holding 'General Index Cards' of the then Department of Interior, Works and Services Branch (Qld), is Job No USM135, 'registered' on 8 October 1942. It is described as 'StockRoute - Erection Heavy Bomber Hangar'. Scores of other such 'jobs' follow in the Index Cards and while it is true most were organised by the AWC, the US forces were always quick to expedite their work using their own 'troop labour'. From inception, USASOS advised RAAF HQ that copies of plans were so limited, only layout 'Plan A 100' was transmitted. This 13 October 1942 letter added-

'...one copy of detailed plans is transmitted to AWC, if they are to be the constructing agency...'

AWC Minutes dated 26 October 1942, record US Col. Robinson as 'not agreeable' to the Contractor (T F Woollam) that AWC had sent north to do the job. Nor was he happy with Contractor Dunne as 'his work at Charleville was too slow'. US troop labour was involved in this early part.

George Richard Graf remembered Bertrandias from the times 'Sally' was at the 'Depot'. B-17E 41-2633 was named 'Sally' after repair and reassignment to Gen Kenney's HQ.

STOCKROUTE *(1/12/1942)*
Construction crews are here busily working on what was designated Repair Hangar No 6 erected on the land west of Duckworth Street. On the extreme left, the southern tip of Repair Hangar No 1 is just visible. Both R1 and R4 (on right edge in background) were constructed close by the western side of Duckworth Street and both had heavy lift aircraft gantry cranes set up at their NW corners. This view clearly demonstrates the three pin arch nature of this type of structure. The two identical arch elements of each frame are 'hinged' and securely bolted to concrete foundations and meet at their top or ridge. The design pictured here preserves the hinged nature of that junction at the top with only nominal framing erected above to provide positive drainage to the roof sheeting. A 1986 aerial photograph shows the concrete floor slabs of R6 and R4 still existing at that time. The structures of course have been removed.
[Neg A 29887 USAF Photo Collection NASM Smithsonian Institution]

STOCKROUTE *(4/8/1942)*
This photograph shows early work on nailed wooden arch aircraft hideout structures several of which were erected and camouflaged in and around STOCKROUTE. This view shows the three dimensional simulation of the crowns of individual trees by elevating netted and scrimmed sections above the nominal surface of the hideout structure itself. *[RAAF Official, courtesy R Voller]*

STOCKROUTE *(12/12/1942)*
This oblique shows the airstrip running across the centre of the photograph with GARBUTT in the right background. The work of construction on warehouses in the immediate foreground and the workshop hangars of the centre left is clearly evident. What looks like a RAAF DC2 transport aircraft is on the strip in the centre of the photograph. The isolated small dwelling just left of centre is unique in that this was and continued to be the family home of Rod Cardell, who as a boy enjoyed a life style in close proximity to this bustling activity of the early Pacific War period.
[Neg SC 166034 US National Archives]

STOCKROUTE *(1944-45?)*
Both the date of the photo and identity of this B-17F are unknown though possibly the aircraft could be identified by reference to a comprehensive catalogue of 'nose art'. The site is just east of Repair Hangar No 5. *[A Dean via D Vincent]*

STOCKROUTE *ca(-/3/44)*
Looking southwest along the centre of the now discarded strip, seven of approximately twelve Butler steel framed hangar/warehouses are here pictured in the course of erection. *[W Maiersperger ex USAAF]*

Dick Graf was radio operator in its crew and tells me that Bertrandias was known to many as -

> *'..."Vicious Vic"... when he needed carpenters for the erection of buildings at Garbutt 4th Air Depot, he simply lined up his men, gave them all hammers and nails and those without hit thumbs after some time, were the carpenters!!'*

By mid 1943, 4 ADG had been joined by both 12 and 15 ADGs, and the facility was termed Depot No 2. Dick recalls-

> *'..."Sally" was at 4th Air Depot in August 1943 for the fitting of an extra compartment behind the Radio Operator for MacArthur to sit while on board...'*

A heavy component of Depot No 2 activity was modification of aircraft, particularly in armament. Capt Paul I 'Pappy' Gunn, a US airman with a colourful reputation and an heroic record, was involved in this work, almost regardless of where it was being carried out. Air Depot 'modifications' were not confined to Townsville. Charters Towers and Brisbane certainly figured in such work and the credit for development of the very effective B-25 'Strafer' is somewhat contentious. Walter Maiersperger writes of some 4 ADG work as -

> *'...In late 1943... "Pappy" was also in the "modification business", in Hangar 2 (Eagle Farm) with a pick-up crew of volunteers who probably had the ideas...When one of his modifications killed Edward L Larner, one of my class mates, the son of Gen H "Hap" Arnold's friend,..."Pappy" was relieved of his "mod" business...'*

Maiersperger writes of Lt-Col Bertrandias, then CO of the Advanced Echelon of the 5AF ASC, bringing to Depot #2, a Maj Jake Shuster. Shuster arrived around July 1943 from Wright Field (USA) to be Bertrandias' 'resident expert Engineering Officer'. Maiersperger knew Shuster from the 1941 service test programme on the B-26 and thought him a bit of a 'bull in a china shop'. Walter writes-

> *'...The rage was to add (still) more nose guns to the A-20 and B-25, which already had six .50 cal in place of a bombardier nose. Shuster mounted two more blister guns (.50 cal) on each side of the B-25 fuselage under the wing. On test over the bay, the muzzle blast loosened the leading edge of the wing. When he attempted to pull out of the strafing run, the leading edge peeled off...'*

STOCKROUTE *ca(1943)*
Open air engine change by RAAF 36 Squadron men. Personnel are unidentified except for the man in the centre believed to be McFadden by name. The photo is taken from the cockpit of a C-47, looking southeast across the middle of the strip. The Igloo warehouse in left background is No 1 in the 4th Air Depot complex. Note the tubular steel lifting frame for chain hoisting of the engine. Such 'frames' are evident in photos at MAREEBA and WOODSTOCK. *[L Baker ex RAAF]*

All four on board were killed and Walter says Bertrandias then wrote to Gen Kenney pointing out that (Maiersperger's) Special Projects Section at EAGLE FARM 'had no Wright Field expert(s) and therefore should be dissolved!' Walter was detached to other duty but the rest were posted to Depot No 2.

Rod Cardell feels the loss of a modified B-25 being test flown on 23 September 1943, by Maj R P Duncan was the only such loss he located in his research. Maiersperger didn't date the loss he described, except to say it occurred 'over the bay'. While the much touted ***Details of Crashed and Missing Aircraft in Australia*** (AWM 54, item 85/9/9 - ex CRS A2663, 'opened' to 31 December 1948) is fraught with errors, it carries only one listing which fits this loss -

> *'...Index No 331, 18°58'S, 145°37'E, Area - Phillips Reef (Rattlesnake Is) Qld, Date unknown, Aircraft - USAC Mitchell B-25 Serial unknown, Remarks - wreck submerged on Phillips Reef...'*

Typically for A2663, that lat/long puts 'Rattlesnake Is' 70km inland! The same A2663 is still relevant today, as instanced by the finding, early August 1994, of 'Beautiful Betsy', a WW2 bomber which disappeared 26 February 1945, enroute Darwin to Brisbane with eight on board. National Park Rangers found the wreckage near Rockhampton. A2663 lists this as 'Index No 9 USAC B-24D 42-40387'. (Strangely, this aircraft was used to trial the parachutists' exit slide described in the LEYBURN segment, page 184).

Early in October 1943, with the majority of Depot No 2 warehouses and all of the Repair hangars operating, news of the final expansion appeared. A cypher message from 3 DWO Townsville to RAAF HQ Melbourne reads -

> *'...AMERICAN AUTHORITIES NEGOTIATING FOR TRANSFER 36 SQUADRON TO AITKENVALE STRIP. THEY PROPOSE ERECT WORKSHOPS ON REPEAT ON STOCK ROUTE STRIP ELIMINATING LANDING FIELD...'*

Reasons given were elimination of the safety hazard involving vehicular traffic in close proximity to aircraft movements and the long troublesome dust. Propeller blown dust intruded to the priority 'clean air' buildings of the nearby engine overhaul, instrument repair and machine shops. While it seems a detached flight from 36 Squadron went to the WEIR strip for a while, the end of STOCKROUTE strip as such was in sight.

Engine overhaul meant follow-up engine testing.

STOCKROUTE (24/4/1944)
Part of KD 1026 this photograph shows development since 22/7/1943. Repair Hangar No 9 has been added complete with an aircraft heavy lift gantry, which itself has been relocated from Repair Hangar No 4 (see photograph 22/7/1943 in Appendix 4). STOCKROUTE strip ceased to function on 29/11/1943 and as shown in this photo its valuable hardstand surface was ideal for the speedy erection of several Butler hangars complete with canvas tent inners. A similar Butler hangar has also appeared on the short entrance taxiway to Hangar No 43 north along Duckworth Street. Again the shadows aid interpretation, e.g. the exposed steel frames of the Butler hangars give individual shadows helping to identify their real nature. The heavy lift aircraft gantries are also identifiable by their shadows[AUSLIG]

February 1944 saw building start on seven engine test stands and two associated water cooling systems just off the southwestern side of the now disused strip.

Depot No 2 was soon complete - able not just to assemble, modify and repair the airframes, but also to overhaul engines at a rate around 300 per month. The huge advantage so gained by the AAF, stemmed from the Depot's location '600 miles from Port Moresby'. Tactical 'island hopping' meant little if 'the Depot' couldn't follow. Early in 1945 the four thousand strong technical heart of Depot No 2 packed its equipment and moved north to the Philippines.

Not surprisingly the STOCKROUTE airstrip was rejected by the Department of Civil Aviation as having any requirement in the post war scene in April of 1946. While the associated igloo structures remained, several were disposed of for relocation and up until the mid 1970s at least one, just west of Duckworth Street, was in use for structural steel fabrication work. The present scene shows none of these buildings remain, although some concrete floor slab-works may.

STOCKROUTE (26/7/1991)
Now known as Dalrymple Road this view is from the southwestern end of the original STOCKROUTE strip looking northeast to Castle Hill. The Dalrymple Road reserve remains an amply wide artery serving both dual carriageway road and electricity transmission towerlines. [Author]

BOHLE RIVER

'...Garbutt Aerodrome (Townsville) - dispersal strips - Stockroute Strip finished; Bohle River Strip will be finished on 25 March 1942...'

This is part of an archive record of P-O Jim Trench's phone message from NEA to DWB on 22 March 1942. The search for a dispersal strip in the BOHLE area had been on for some few weeks. On only 5 March 1942 a report by the RAAF's Sqn Ldr Towers mentions that a possible site a couple of miles further west of BOHLE had been considered but rejected for various reasons, including the need for extensive new road access. That site was termed ALICE RIVER and like many others has figured briefly in archive records, before fading as one of the many on which no work was commenced.

P-0 John Keays had been involved in this quest for dispersal strips for GARBUTT. His 14 April 1942 report points out that ALICE RIVER, among its other criticisms, was deemed by American advisers as 'too close to GARBUTT for the role of a dispersal strip'. Instructions to Keays had been that he should find four further fighter fields each with dispersal strips. By April, he was of the opinion that -

'...Giru, Upper Ross, Cardington and Alice River will be included in the official list of requirements.'

As history now shows, US Engineer Units commenced work on both GIRU and UPPER ROSS, but CARDINGTON never got past the 'pegged centre line' stage and ALICE RIVER was completely discarded.
While Jim Trench's phone message said that BOHLE

BOHLE RIVER (27/7/1942)
This right oblique Trimet photograph shows the early stage development of BOHLE RIVER strip (bottom centre). No dispersals have yet been constructed. Away on the left hand edge of the photo the Ross River itself can be seen winding northward to Townsville. The weir that is clearly visible is near what was developed as AITKENVALE or the WEIR STRIP. The three intersecting clearings of the UPPER ROSS airfield project which did not come to effective fruition appear back up the valley. The prominent 'R' painted on the starboard engine nacelle of the P-38F served to correctly identify the negative and avoid printing in reverse. [AUSLIG]

would be 'ready', it should be realised that such a state of readiness was the bare strip, cleared and rolled, with possibly a thin gravel addition. This was the state in which BOHLE would have been used by US fighter units such as the 40th Pursuit Squadron with P-39 Airacobras on their arrival in the Townsville area during April 1942.

It has not proved easy to document the earliest use by aircraft of some of these strips. A message dated 21 April 1942 to Air Board from NEA helps to suggest which strips were providing much needed dispersal duty for the dozens of newly arriving aircraft -

'...Request requisitions £500 each for camouflage works at following: Bohle River, Mt St John Zoo, Aitkenvale Weir, Civil Aerodrome, Antil Plain, Woodstock, Reid River, Macrossan, Fanning River, Torrens Creek. Work chiefly connected with hideouts...'

The Department of Home Security's Camouflage Report for the week 13-20 May 1942 includes -

*'...**Bohle River** - 4 painters and an apprentice painting tents at this strip and expect to have all complete by May 23. Photographs of the strip and surrounding country have been obtained and dispersal points and general camouflage is being planned. A squadron of aircraft has arrived and they are asking for cover.'*

Bob Gow and Norm Poppleton were both long term members of the RAAF's 100 Sqn. Bob recalled two periods at BOHLE RIVER - the first time only a stopover en route to CAIRNS, early May 1942 -

'...eight people crammed into a Beaufort outfitted for a crew of four - tool boxes, kit bags with minimum clothing, plus aircraft spares wherever they could be stowed... one aircraft had a push bike in the bomb bay!'

It took until 27 May 1942 for BOHLE to figure in the Allied Works Council Minutes and when it does first show, the reference is clearly to construction work for a 3.7" anti-aircraft battery. I've not managed to make contact with any veterans who are able to say that they even remember seeing such 'fixed anti aircraft defences' around BOHLE strip. Camouflage reports refer to a 'Bohle River A-A Station' and it may well be this was located further downstream along Bohle River towards the coast.

In my striving to learn what airstrips such as BOHLE were like in those early developing periods, I have become aware of the evolution in camp buildings. The permanent type buildings in vogue during the period prior to the entry of the Japanese into WW2, had to be discarded and rightly so. The influx of American Forces and aircraft to these many airfields, simply meant that resources and time were not available to continue in that vein. Moreover there were people, even in those pressed times, able to see that these airstrips were temporary installations and that the buildings

BOHLE RIVER (21/7/1943)
This oblique shows the dispersals to the east of the airstrip. Bohle River itself appears briefly in the bottom right corner and re-appears on the right edge on its way to the sea. Roughly circular 'rings' near the dispersals are thought likely to be fire breaks around ammunition and/or fuel dumps. [NLA (RAAF Official)]

BOHLE RIVER (26/7/1943?)
This vertical is one of twelve frames of KD 828, which cover the airstrip and its dispersal area. Note the conflict of dates which is not uncommon in these sorts of records. At this point in time it appears impossible to know which date is correct. The date here ascribed as 26 July is that which appears on the edges/margins of the first and last frames in Run 2. 26 July, also appears in the RAAF Master Negative Register. The photograph clearly shows a number of twin engine aircraft, one proceeding north on the runway. These aircraft are probably Beaufort Torpedo Bombers of No 8 Squadron RAAF. [AUSLIG]

should be so tailored.

Some drawings survive to show the location of mess buildings, ablutions and latrine buildings. Obviously all 'accommodation' was in tents. The skillion (one-way sloping) roofs didn't get such frills as guttering and down pipes. In fact the very plan view of such buildings clearly shows efforts to stagger the edges and break up the outline of the galvanised or other, roof sheeting. Combined with some chosen paint colouring of the roof, this served to blend such a building into its surroundings.

As lesser standard buildings came to apply at various fields, so also the US authorities, from their point of view, quite rightly deemed that lesser standards should apply to some runways. Archive documentation clearly shows that even in such relatively urgent times the RAAF authorities were looking to the future and trying to decide which installations they would require developed to more permanent standards. Dated 8 September 1942, the RAAF advised to AWC, a list of dromes in north Queensland -

'...to which RAAF standards of construction should apply. Aitkenvale, Breddan, Bohle River, Garbutt, Fanning, Ross River, Stockroute. Balfes Creek aerodrome was originally constructed under USA direction as the second bombardment field in Charters Towers area. It is now under consideration as the site for an RAAF Aircraft Repair Depot but you will be definitely advised when a decision is reached.'

Around this time, the RAAF's 30 Squadron, with Beaufighters, occupied BOHLE for a few weeks before going on to New Guinea. 100 Sqn was back at BOHLE RIVER about the middle of September 1942. Camped at the airfield, it was preparing for a move to Gurney Field, Milne Bay. Bob Gow, a Fitter-armourer, was in a party of thirty ground crew flown out in a 'converted B-17E (41-2497) 'MacMac' on 30 October 1942'.

Gordon Howe, a veteran of RAAF's 8 Sqn described its several weeks stay mid 1943, before proceeding to Good-

BOHLE RIVER *ca(-/6-1945)*
Here both date and photographer are uncertain but not so the distinctive South Pinnacle mountains at the end of the strip. The RAAF man on the wing of Wirraway '617' is LAC Cook (initials unknown). Kittyhawks of the RAAF's 86 Sqn in the background are believed to have been at BOHLE RIVER between February and August of 1945 when that squadron underwent conversion to Mustangs. The duty pilot tower shows just right of the radio mast. [C A Mason ex RAAF]

enough Island. Alan Payton was a Wireless Air Gunner (WAG) in that Torpedo/Bomber Squadron. His period at BOHLE embraced the July 1943 vertical (KD 828) photo coverage surviving to this day.

BOHLE RIVER like others went through the saga of being assessed as needing elaborate aircraft dispersal extending to bomb splinter proof revetments. It appears none of that particular work was actually done. Possibly 'top cover concealment' on some of the aircraft dispersals was done, but the July 1943 aerials show the prepared dispersal bays quite open and devoid of such cover.

Half way through 1944 the tenor of life at BOHLE RIVER can be gleaned from a RAAF review of unoccupied service buildings -

'..Bohle River - is still used as a Squadron Camp and staging area. It is usually occupied for this purpose for approx 6 months of the year.'

The RAAF's 86 Squadron was at BOHLE late in the Pacific War. Dated 31 July 1945, the AOC NEA wrote to DWO Townsville on the subject of runway maintenance -

'...it is the intention of this HQ to use both Ross River and Bohle River for operations of fighter crews undergoing conversion courses to Mustang aircraft. This will obviate congestion on Garbutt field. As far as can be foreseen, at present, these courses will take at least 8 months and possibly up to 12 months. It is further estimated each squadron will make a maximum of 240 takeoffs and the same number of landings per week. Its not considered provision of hardstandings and running up bays is warranted... Suggested therefore any resealing work considered necessary should be completed as early as possible.'

Post war, BOHLE was retained in use by the RAAF at least until 1950, when records show the 'marking of runway (with) letter "Z" ', a sure sign of its lapsing. 1986 aerial photos still showed most of the winding taxiways, and the runway is still noisy - during drag car meetings!

BOHLE RIVER *(27/7/1991)*
The strip is now used for drag car racing. The South Pinnacle mountain skyline to the southwest would be familiar to WW2 pilots with 'time' at BOHLE. [Author]

ROSS RIVER

'...Old Civil Drome Ross River - clearing and grubbing plus gravel skin complete - ready for occupation 9 April 1942.'

This brief reference appears in a lengthy report dated 14 April 1942 in RAAF files under the signature of P-O J Keays. Obviously ROSS RIVER being the Civil Drome of the day warranted immediate attention to bring it to a serviceable state for Air Force purposes. The US Engineer Forces, small as they were in number, had arrived in the Townsville area in March and it is possible that some of their direct effort was applied to ROSS RIVER. Certainly, archive documentation refers to MRC involvement and later of course, AWC mention of further development. However the initial hasty runway construction at ROSS RIVER may have been undertaken directly by American Forces.

A report covering the inspection of several aerodromes in the Townsville area by P-O J W Campbell dated 28 May 1942 includes the following reference -

'..Ross River Strip - the 75° landing strip has been rough graded and consolidated, gravelling in progress at eastern end 700' - balance complete. Surface in good order... Buildings - administrative, kitchen, mess hut, latrines and ablution buildings in progress and now almost complete... MRC at present boring for water-depth to date 70'. Three tanks with stands have been erected.'

Air Board Agenda No 4043/1942 under the signature of Air Vice Marshall G Jones Chief of Air Staff, paved the way for War Cabinet approval of the initial dispersal and camouflage works -

'...In accordance with the policy laid down Air Board memo No 5, it is necessary to provide hideouts and other dispersal works for the dispersal of aircraft at... Ross River, a dispersal field for RAAF Station Townsville... Investigations are in hand to evolve a satisfactory scheme to combine concealment of the aircraft with protection but details are not yet available ...Tentative cost supplied by AOC, NEA is £5 000,

ROSS RIVER *(27/7/1942)*
*Part of an extensive '**coastal trimming**' programme this Trimet right hand oblique (ROB) photo, neatly shows the RAAF strip freshly constructed, very close to the site of the civil landing ground. While ROSS RIVER strip appears centre bottom, the rest of the frame itself is interesting in that taken clockwise from 12 o'clock, the three intersecting strips of UPPER ROSS can be seen through the whirling propeller, the AITKENVALE (2nd WEIR STRIP) is seen about 3 o'clock close to the bend in Ross River, ANTIL PLAINS shows up at 9 o'clock with WOODSTOCK at 9.30 o'clock and REID RIVER a little further south in the distance.* [AUSLIG]

...work has been given A1 priority.'

Notation at the bottom indicates War Cabinet Approval came on 20 June 1942. While ROSS RIVER is included in record correspondence dealing with arch hideout structures, I've not found evidence of any proceeding there.

With the urgent need to work on many airfields simultaneously, the approaching 'end of year' wet season spelled more concern. By 3 August 1942, F-Lt John Keays was the RAAF's NE Area Works Officer. A copy of several pages from an extensive undated review by the Area Works Officer survives in AA -

'..During the early portion of the year aerodrome construction works were concentrated on the clearing and rough grading of strips so that they were useable under dry weather conditions. Some gravel was added to fill pot holes and depressions and to partially stabilise areas that would be unsafe after small rainfall, but it was not considered that this gravelling would do more than provide an improved sub-grade for the construction of the final all weather gravelled pavement. Soil conditions generally throughout the area are such that in the wet season, aircraft will not be able to operate on other than fully constructed metal or gravel pavements and as the wet season will commence in 4-5 months, it is essential that the whole question of runway construction should be reviewed. The clearing and rough preparation of strips was carried out by organisations supplied by MRC and by the USA Corps of Engineers but with the opening up of further fields to the north the USA organisations have been gradually withdrawn leaving MRC to complete all fields south of Coen...'

The review continued listing the seven strips which had by then been gravelled. ROSS RIVER and STOCKROUTE were included but those still to be done numbered eighteen. Where gravelling was complete -

'...a considerable amount of earthworks will be necessary to complete the flanks. Little drainage work has been carried out and no dispersal works have been constructed, all of which will require plant and labour...'

The outcome was a refinement of priorities for the weather proof sealing of various airstrips and importantly the importation of further manpower and equipment from projects down south.

A report by Herman G Cox, Capt (US) Corps of Engineers, Air Engineer (Works) Officer dated 8 August 1942 addressed to Lt-Col W T Abbott Air Engineer HQ USAAS covers many of these projects, and in respect of ROSS RIVER indicates -

'...4) Ross River - this strip is primed with tar for entire length and bitumen water-tight seal is now underway, and should be completed by 15 August. This plant and equipment will be moved to Antil Plains where the 90° strip will be primed and sealed.

5) A conference was held with Main Roads and following completion dates were agreed on: Mareeba 15 October, Ross River 15 August, Antil Plains (90° strip only) 1 September, Woodstock 1 November, Antil Plains 15 December, Reid River 1 November, Breddan (1 strip only) 22 August, Torrens Creek 1 November.'

The dispersal works at ROSS RIVER may have included one of the early cases of using 'wire mesh' as a pavement substitute or ground surface reinforcement. The following mention appears in a RAAF file, dated 19 August 1942 -

'...Forwarded are copies of Plan 42/43/542 and schedules... covering... dispersal works... at Ross River. The work described in the schedule is to be charged to requisition DWB 1387/41-42 previously forwarded to you for an amount of £5 000 for the construction of splinter proof pens or concealed hideouts and other dispersal works.'

Plan 542 carries the signatures of both Rod Voller, for Home Security Camouflage and John Keays as Area Works Officer. A note reads -

'...access to hideouts from main taxiways to be provided by wire mesh tracks.'

This was ARC welded wire mesh of an appropriate wire gauge and pattern laid directly on the ground and staked in place. In theory, the advantage was to be no tell-tale gravel offshoot to the camouflaged dispersal bay! Such finesse soon proved unnecessary and my guess is that vehicle tyres would have been spiked before too many aircraft tyres were risked! This wire mesh should not be confused with Pierced Steel Plank (PSP) or Marston matting which became widely used for runways.

Harry Parker was a Fitter 2A (Airframe) posted to the RAAF's 24 Squadron early in 1942. His notes to me included -

'...24 Sqn left Stockroute and went to a landing strip adjacent the meat works. We lived in tents with bedding filled with local grass which I consider was Mitchell or Spear grass. I refused to use the grass and so had to put up with my bed consisting of hessian for a mattress. We also had a shower rigged up with bore water which was very black and smelly. We used a hand pump to bring the water to the surface. I had relations in Townsville where I had a shower about 4 times a week. There was a bitumen landing strip there thanks to the Americans. About 200 goats used to pass the camp in the morning and return about 5 p.m. They cleaned up a lot of uneaten food, etc...'

Somehow Harry spent Christmas 1942 in Townsville before his posting south. Whether through exaggeration or not, Harry told me the Turkey that Christmas was actually Emu!

ROSS RIVER *(23/7/1943)*

One of a pair of known obliques taken by RAAF Survey Flight's De Havilland 84 Dragon A34-82, this view shows the southward development of dispersal taxiways and hard standings. Aircraft can be seen parked on some of these hard standings. One appears on the western 'alert hard stand' at the end of the strip. The operations building of the day was located just south of this western end and can in fact be seen in this photograph. *[CPE (RAAF Official)]*

ROSS RIVER *(25/7/1991)*

This shot from the lookout at Mt Stuart shows the site of the old ROSS RIVER airstrip. It is not easy to see evidence of the strip but sufficient of the geographic features common in the 23/7/1943 RAAF Official oblique can be picked out. This site is just north of the Army's present day Lavarack Barracks but aircraft do not now use the old landing ground. The Meatworks remains but a large sporting complex now encroaches on the old airfield. *[Author]*

Strange how food, its beauty, its sameness, or maybe the simple lack of it, sticks in the memory. Jim Legge was a RAAF pilot with 7 Sqn and Jim recalls a period flying out of ROSS RIVER during December 1942 and January 1943. This was the particular Christmas when his CO, Sqn Ldr K R Parsons detailed him to 'fly to AUGUSTUS DOWNS to pick up the pigs for Christmas'.

I asked Jim about RAAF Beaufort A9-132 pictured below warming up at ROSS RIVER. Jim's careful reference to his logs showed -

'...We did fly it from 28 January 1943 when we took it from Ross River to Horn Is until 14 February 1943 when we handed it in at Amberley on our way south on leave...It does not appear again in my log book after 14 February 1943.'

In March of 1943 ROSS RIVER, like many other northern airfield sites came in for fresh allocation of aircraft dispersal works including splinter proof pens. Eighteen medium bomber pens were set down for construction at ROSS RIVER but after high level contact between General MacArthur and Prime Minister Curtin, none of that work proceeded.

A DWO Memo of 8 May 1943 to the Department of the Interior Townsville, is interesting in its reference to fencing of the airfield. Fearing accidents, a request had recently been made -

'...No action has yet been taken... (on this work)...and several accidents have since occurred due to stock on the runway. As it is now necessary to place an additional squadron of fighters on this strip, it is essential that immediate action be taken to put this work in hand.'

It's not yet clear to me which fighter 'squadron' is referred to here. Certainly Jim Legge's 7 Sqn were based back at ROSS RIVER for several months of 1943, but that involved twin-engined Beauforts on seaward convoy and anti-submarine patrols.

ROSS RIVER served as home to the RAAF's 84 Sqn late in the war. That squadron and 86 Sqn at BOHLE RIVER, were the two selected for conversion from Kittyhawks to the new high performance North American Mustangs, using P-51K models acquired under land-lease.

With its review of airfield needs post war, the RAAF decided not to retain ROSS RIVER. By April of 1946 Civil Aviation authorities had advised they had no continuing need for ROSS RIVER.

ROSS RIVER *(1943?)*
It has been difficult to date this photo of a RAAF 7 Squadron Beaufort A9-132 warming up by the Operations Building near the western end of ROSS RIVER airstrip. This typifies the nature of such buildings erected in the first few months of 1942. Less effort was applied to those erected at following airstrip developments. e.g. COOKTOWN MISSION. [North Queensland Air Force Museum (RAAF Official)]

AITKENVALE

'This house where we are, in Silky Oak Street would be right in the middle of the runway.'

The late Dan Gleeson OBE, former long term Mayor of Thuringowa City had a long association with this district. As a young man he was employed by the MRC as an owner/driver 'truckie' and was associated with the urgent construction of several of the airfields in the Townsville area. In fact one of the weirs nearby on the Ross River carries his family's name. Dan recalled the period early in 1942 when particular work by MRC was drawing to a close at CAIRNS. He remembered the MRC engineer O F 'Pines' Anderson suggesting that Dan should 'return to Townsville as you'll be needed down there pretty soon'.

A signal from NEA to Air Board on 22 March 1942 addressed progress on several strips in the Townsville area-

'...Antil Plains - 2 strips complete, additional strips and access roads ready 27 March. Woodstock - 1 strip finished 2 more on 27 March. Bohle River - 1 strip ready 24 March. Stockroute - ready 24 March. Weir Site ready 28 March. Location of aerodrome and dispersal strips Reid River in progress...'

At this time, just a few weeks after the first Japanese air raid on Darwin, W Cdr Lerew was busy there. Lerew was due back to Townsville soon and the signal continued -

'...Following instructions W Cdr Lerew; Fighter Strips graded only 75' wide, timber cleared 300'.'

Progress on such defence works certainly came at a price in terms of disruption to the lives and businesses of the inhabitants of the area. But then what else is war? Many had already vacated their homes in North Queensland. In the vicinity of AITKENVALE strip it is of interest to note archive documents relating to the 'hiring' of premises for use as RAAF 76 Squadron HQ and personnel accommodation. A particular letter dated 14 April 1942 relates to the plight of the Gleeson family. Signed by Mr G Roberts of Roberts, Leu and North, Solicitors and notary public of Townsville, it is addressed to the Hon. A W Fadden PC MHR Canberra -

'...Dear Artie,

I am acting for Gleeson family, who as you know, live

AITKENVALE *(-/7/1941)*
Prior to war in the Pacific. Note the two weirs. *[AUSLIG]*

at Upper Ross River, the father having selected the property somewhere around 1870, and made their home there ever since. The father of course, is dead.
Yesterday, the Air Force served notice on them, to get out of the house, within 24 hours, and to get rid of their cattle and fowls etc. However, I adjusted that, to give them a reasonable time to vacate, and made satisfactory arrangements, under the circumstances; but what I am writing to you about, is to say that it does not seem right to me for the Air Force to turn old inhabitants out of their home. They of course, say it is a danger zone, but if these people are prepared to take the risk, why should their homes and their businesses be upset altogether, for the Air Force to live in houses, when the whole of the Military Department have to live in tents. It does not seem right to me, that such preference should be given to the Air Force, although I know, they do wonderfully good work; but they live in palaces compared to what the soldiers do, and why, to comply with this, should old inhabitants be put out of their homes?
I am making arrangements, in this case, to vacate the home subject to satisfactory arrangements being made, so far as compensation is concerned, and for the people to live in the hay shed, adjoining the property, so that they may still carry on their business. It is however a bit rough, at their time of life, to have to move and live in a hay shed, to give up their comfortable home to the Air Force, who could be quartered in tents.
Another grievance that people have here, is the terrible delay in arranging compensation. People are asked to get out of their homes, at a moments notice, and when they get to somebody, supposed to have charge of the fixing of compensation, they are told they know nothing about it at all, and this is happening almost every day in this place, and it should not be.
If you were to be put out of your home at a moments notice, I feel sure that you would like to know what rental you were going to get before you committed yourself to finding another home, because you would have to base your rental for a new home, on what you were receiving from the Authorities.
There are streets of houses being taken here, and not even a word yet, said about compensation. If only a little bit of the red tape that the army seems to have to go through, could be got rid of, it certainly would assist matters materially.'

There are seventy odd entries on the 'Minute Sheet(s)' of this 'hiring file' which extends to 1947.

P-O John Keays' report of the same date, 14 April 1942 notes progress as -

'...2nd Weir Site - clearing and grubbing complete, skin of gravel applied, strip in use as from 9 April 1942.'

It appears this initial use of the WEIR STRIP was by the RAAF's 76 Squadron, freshly moved there from its formation at ARCHERFIELD.

Greg McGregor was a RAAF Fitter Armourer with 76 Sqn at the WEIR. He wrote that in April to July 1942 -

'...the strip was gravel, not sealed. The Kittyhawks were dispersed into trees on the north side, roughly midway along the strip... armament section was in that vicinity and the squadron camp was a little further north.'

He recalled the Weir State School located just south of the southern end of the strip being taken over by the RAAF and used as an Orderly Room and a Medical Centre.

In April of 1942 the heat was on to conceal or camouflage the dispersal of aircraft. A message from NEA HQ to Air Board, Melbourne dated 21 April 1942 requested '£500 each for camouflage works' at a string of airfields round Townsville, including AITKENVALE WEIR. Rod Voller, in charge of the Dept of Home Security's Camouflage Section in NEA, recalls from his diary notes -

'...two US officers; Hughes and Murphy came to Townsville and had discussions with F-Lt Dawkins and myself on 16 May re the use of these (nailed wooden) arch structures in relation to the general policy of RAAF. We all went to the Aitkenvale strip to decide positions for these arch structures to hide aircraft.'

On 20 May 1942, Rod's notes refer to 'painting of tents...(and)...5 labourers employed on erection of netting for aircraft'. Again on 21 June 1942, he refers to a visit 'to see the arch, with Gp Capt Thomas, CO at GARBUTT, who gave his approval'. During June 1942 Air Board approved an allocation of £5 000 for the provision of 'hideouts and other dispersal works'. This was based on a

AITKENVALE *(16/5/1942)*
This rather tattered frame is the vertical of a Trimet Run starting approximately 30 miles north of Cairns and running down the coast to Townsville. The AITKENVALE STRIP can be seen (top left) adjacent the weir on the Ross River. The rest of the photograph is interesting in that it shows clearly the ROSS RIVER development (bottom centre) at that point without dispersal taxiways etc. and STOCKROUTE also is just shown at the top edge of the photograph. *[AUSLIG]*

AITKENVALE (21/7/1943)
One of a pair of obliques from systematic coverage of such airstrips in mid-1943, this view to the southwest shows the dispersal taxiways. [NLA (RAAF Official)]

tentative estimate supplied by the AOC, NEA. While I've not found any further corroboration, it seems likely some arch hideouts served for a time at the WEIR.

John Short was another 76 Squadron armourer. He reported to the unit in May 1942, '...on attachment from the Base Squadron at Garbutt.' 76 Squadron, the second RAAF squadron to be equipped with P-40E Kittyhawks -

'...was operating there in an Operational Training capacity, preparatory to going to New Guinea which turned out to be Milne Bay. In fact I notice I have "Fall River" entered in my original RAAF pay book (Fall River was a code name for the base at Milne Bay).

We sailed for there on 15 July on the MV Schwartenhondt (Dutch for black dog) Squadron Leader "Ali" Brooker RAF was CO at the Weir Strip but Peter Turnbull assumed command to go to New Guinea. There was quite a compliment of experienced pilots - ex Spitfire boys from 452 Squadron (Pat Tainton, "Blue" Truscott and "Bardie" Wawn) - Peter Turnbull was from 3 Squadron; there was Peter Ash from 21 Squadron, George Pagan, George Inkster and "Slam" Sullivan are but a few names which come to memory.

Though we must have had some rookies from SFTS schools I can't name any so you can see we had quite an experienced bunch to face the Nips...

There is a very strong bond between former 76 Squadron members - the Squadron earned the name of "hydraulic Squadron" at Townsville 'cause it would "lift anything" I told you about acquiring "after hours transport" - well, when the Squadron had its final inspection before embarking, the inspecting officer Air Cmdre Lukis made a comment about us not being worthy of that name, missing simple items such as laces in sandshoes, when a tow truck of the US Army pulled out with a ¾ ton "Blitz" Wagon that we had acquired and on which we had stencilled a fake RAAF number on the engine bonnet - "AF 111 111" - I know, because I made the stencil!'

As with many veterans who have searched their memories to help me in this project, John added -

'...I hope I have given you some information of value. I was too much of a newcomer at that time, then only about 8 months in the RAAF, still a little naive - hadn't then taken my first strong drink and I steered a straight course as to rules and regulations. In fact I had only left home once prior to that when I was about eight with my mother on a holiday to grandparents, so you can see I really was green then.'

In August 1942 a formal design for the 'dispersals' was drawn up by F-O Bruce sent to the area by Air Board. His approach to AITKENVALE was similar to others in that it too featured the 'wire mesh access tracks' between gravel taxiway and 'concealed hideouts'. Bruce's visit was prompted by W Cdr Towers' July inspection, when it became clear funds allocated to dispersal taxiways and hard standings were being consumed in arch hideouts!

Air Board agendum No 4299/1942 dated 6 October 1942, addressed the needs of AITKENVALE strip after 76 Sqn had departed -

'...On Air Board Agendum 4018 approval was given for the dust laying treatment of the natural surface runways, used as dispersal strips for RAAF station Townsville... The natural surface runway at the Aitkenvale dispersal strip has been subjected to almost continuous use as a fighter operational strip and as a result has been cut up to such an extent as to make it impossible to satisfactorily treat the surface without forming and gravelling... Unless gravelled and sealed, the existing runway will not carry aircraft in wet weather, and damage to aircraft is likely...'

The 'serviceability date' comments on RAAF Landing Ground (Scale 1"=1000') Drawing 42-43/1014B serves as some indication of progress on this work -

'...15/2/1943 - gravelling of taxiways 50% complete, dispersal bays not started, drainage in progress, runways sealed 100%, dispersals U/C (under construction?) (from US plan), 9 hideouts.'

While the presence of straying cattle represented a hazard to many little used airstrips, resulting in the need to fence them, in the case of AITKENVALE the situation was somewhat different! During 1943, when the strip was not actually in use, it became necessary to temporarily reduce the strip's fenced width to thus widen the riverbank 'stockroute'. This was the route along which mobs of cattle were driven to the Ross River and Alligator Creek Meatworks. To avoid loss of meat quality, it was advisable to remove the constriction in the route. As a result temporary fences were erected right against the southern edge of the runway. This temporary fencing was realigned in company with other restoration works at the

AITKENVALE [1992]
The SW-NE Runway and associated dispersals superimposed on an extract from a current street directory. [Universal Press P/L]

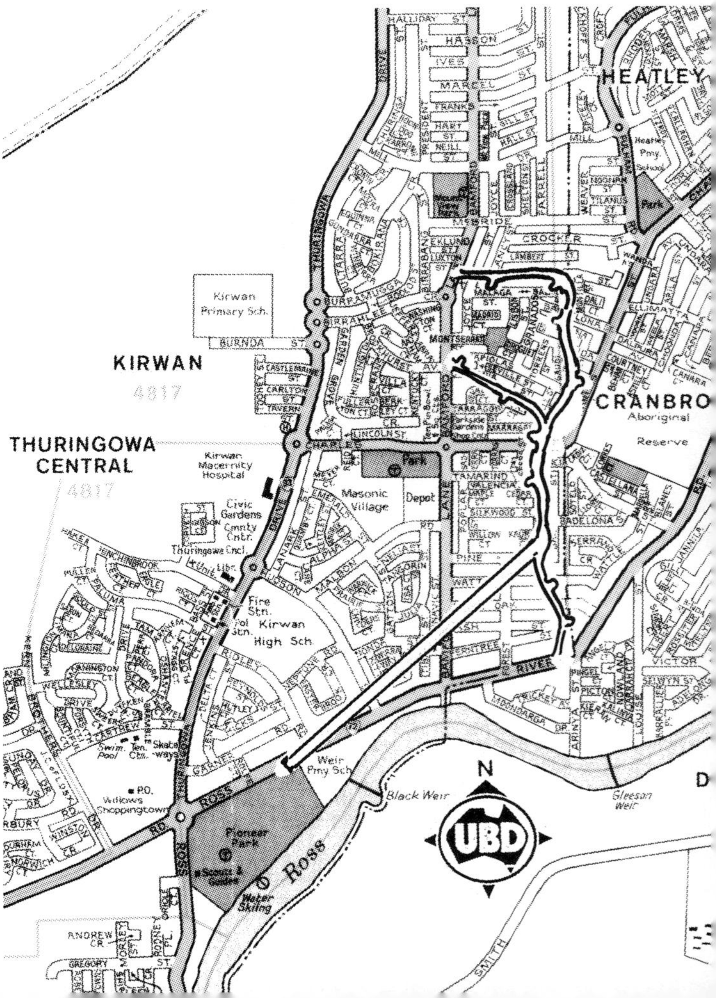

airfield during September of 1943. Such instruction to the Dept of the Interior, Townsville ended with the statement -

'...it is intended to make use of this runway within the next few weeks'.

The 'use' then intended for AITKENVALE soon became clear with the American Forces' move to turn the old STOCKROUTE runway over for the speedy erection of some eleven Butler hangar/warehouses. Consequently the RAAF's 36 Squadron required an alternative strip, if only for the usual training on 'circuits and bumps'. 36 Sqn moved to a more permanent base at GARBUTT around February 1944, which appears to have left AITKENVALE more or less deserted. Certainly archival documents report the following in March 1945 -

'...Reference is made to the maintenance of Aitkenvale strip which is now not in use and has no Unit based on the Aerodrome Area. A recent inspection revealed that cattle were grazing on the area and doing great damage to the bitumen sealed portion of the strip. Droppings from these cattle now make necessary... (sweeping of the runway)... and bitumen surface patching. Fencing around the strip is in very bad condition... (with) vegetation growing on edges... some shrubs have reached a height of 6' '

No further use was made of the WEIR strip.

Today, just southeast of the Kirwan Primary School, there is a small residential development, adjacent an Hotel. With street names of popular Australian beers, can it be entirely coincidental that this was the 1943 site of the RAAF Sqn Camp?

ANTIL PLAINS *(5/1992)*
Walter Harvey (ex USAAF), seated back in the cockpit of the P-39 in which he departed ANTIL PLAINS for HORN ISLAND on 1 May 1942. This was Walter's first visit back to ANTIL PLAINS and he and his wife were able to visit the Mossman cane farm to see the aircraft being carefully restored by CART (Cairns Aircraft Recovery Team). Walter's comment when he climbed back into the cabin was ***'Yep, its just the way I left it, both fuel gauges with the needles against the empty pegs!'*** *[Ian Mullins]*

ANTIL PLAINS

'...I had no real personal interest in Antil Plains except that Walter Harvey departed there on 1 May 1942. I picked up when I heard from Nick Watling, who just recently got the information from Charlie Falletta (Charlie was the pilot of the other (Syd Beck's P-39). Charlie, who has kept contact with Walter, has been a fairly frequent visitor to Oz over the last few years, and nearly always contacts Nick when here. This time (only a few days ago), Charlie told Nick that Walter would be coming to Oz for the Coral Sea shindig, so Nick subsequently told me - what does that make it - third hand info?... So it occurred to me that if I do contact Walter while he's in Townsville, he might just express a wish to have a look at Antil Plains,... so I wanted to be ahead of the game.'

Such was Ian Mullins' approach to me for 'information on ANTIL PLAINS'. Ian and Nick Watling have laboured for years on the meticulous restoration of the P-39F 41-7215 that Walter Harvey put down in the 'turkey bush' near Charlie Falletta's on 1 May 1942 up near the tip of Cape York.

Early in April 1942, while new aircraft were arriving in Townsville, Falletta and five other US pilots flew to Port Moresby to gain front-line experience alongside the RAAF's 75 Squadron. Falletta flew seventeen combat missions in five days destroying three enemy aircraft with probable damage to others. On April 11, Falletta's party returned to Townsville to put this experience to work in the much needed training of young pilots fresh from school.

With refuelling stops planned at COOKTOWN and HORN ISLAND, eleven Airacobras left Townsville on 27 April. With no radio navigational aids and only elementary maps, they hardly needed the further hazard of low visibility wet weather. All eleven came down on various beaches after failing to find HORN ISLAND. Lt Falletta was due to lead the remaining six a few days later. On 1 May they set out from ANTIL PLAINS, only to meet the same fate. From this six, Falletta, Harvey and three other pilots survived.

The area around ANTIL PLAINS had been systematically photographed in 1938. A drawing in AA, titled ***RAAF Landing Ground Antil Plains showing area required for Air Force purposes*** bears an initiation date of 9 November 1941. Hence it appears this area was under consideration for an Air Force airfield since sometime around the outbreak of WW2 at least. Early March 1942, the RAAF's Sqn Ldr G J Towers investigated various sites, in company with two US personnel, Mr Pettus, a civilian, on the staff of Sverdrup and Parcel, engineering consultants, and Major Urwin (Irwin? Francis T), and Mr Mathison of the MRC. Towers' report dated 5 March 1942, described this site as being -

'...On the western side of the Townsville-Charters Towers railway line and road, and on the northern side of Antill Creek. This site is an excellent one. The clearing is light and a full sized drome could be rapidly constructed... Only a thin skin of gravel would be required for runways. The Main Roads Engineer assures me that ample gravel supplies would be available nearby. The water supply could be obtained by sinking spears into Antill Creek...'

A signal from RAAF HQ Melbourne, to RAAF HQ, NEA, on 18 March 1942 read -

'...Aerodromes Townsville area. Approval given construct Woodstock and Cardington as USA Pursuit Aerodromes and work may proceed accordingly. New site Antill Plains and present aerodrome Garbutt

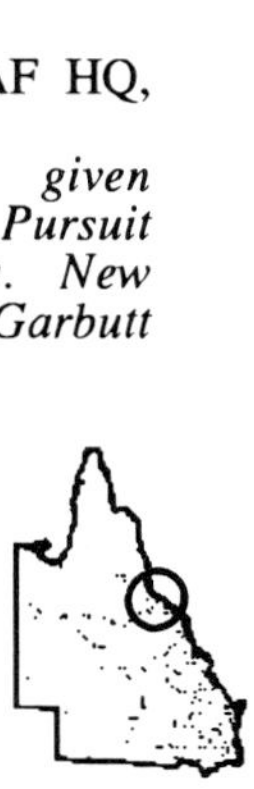

selected as RAAF Fighter Aerodromes. P-O Keays DWB now Brisbane will design aerodromes and arrange detail work in conjunction Doig and Mathison. Advise Doig.'

P-O Jim Trench, then a newly inducted RAAF Works Officer in NEA, telephoned airstrip construction progress to Air Board on 22 March 1942. In part this recorded -

'...Antil Plains - 2 strips complete (cleared only) third strip ready in 1 weeks time.'

Early in April, John Keays and Jim Trench had, in the company of Eric Doig (Dept of Interior), and Jack Mathison MRC, checked progress on the ground. Keays' reports of 8 and 9 April 1942 embraced the US 46 Engineers' work at WOODSTOCK and REID RIVER. It took in MRC work at several other fields out Charters Towers way too. In respect of ANTIL PLAINS he said -

'...Four strips cleared and grubbed, NE-SW strip in occupation, other strips levelled, too dusty for use. Initial requirement is a skin of gravel.'

Early in this project I found mention of more than the two ANTIL PLAINS strips shown on the 1976 topographic mapping, but could not establish where these others were. It took the finding of the 16 May 1942 photography titled ***Panoramic of all strips in Townsville Area*** and later AA searching to clarify the locations.

By 14 April, Keays added of the NE-SW strip, that -

'...after clearing and grubbing (it was) found to be satisfactory for use without grading. It carries a good growth of grass but once this goes the surface will powder and become very dusty.'

Jim Dudley as a recruit fresh out of basic training, was one of hundreds of US 'replacements' shipped to Australia on the *Queen Mary*. In great secrecy, his contingent slipped out of Boston harbour on 18 February 1942. The journey via Trinidad, Rio de Janeiro and Capetown to Freemantle must have been much like others the two Cunard *Queens* logged except for 'the fire'. Jim described

ANTIL PLAINS (-/6/1942?)

This is a good example of misleading marking of photography of the WW2 period. The frame covers that part of the Townsville to Charters Towers road and rail, crossing Antill Creek with Antil Plains Railway Station of the day in nearly the middle of the photograph. By June 1942, the 144° strip, just SE of the Railway Station and the 170° in the road reserve, had been graded and abandoned. Therefore this photography is most likely June of 1941. [AUSLIG]

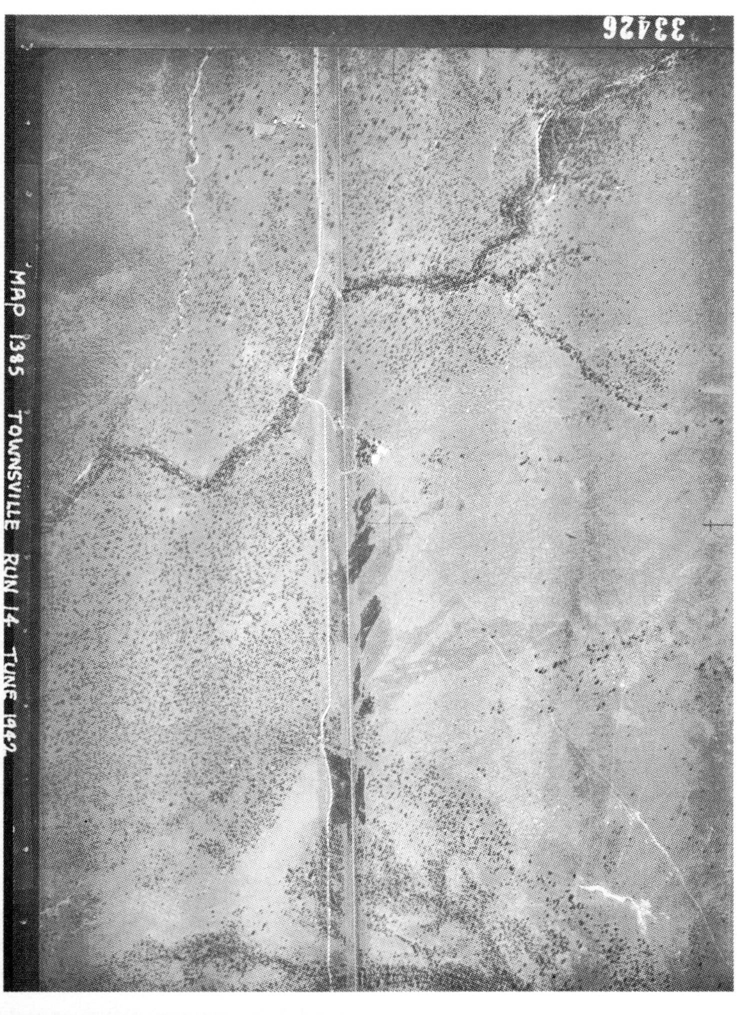

an 8 hour period east of Capetown, during which a fire 'up front in the officers' quarters' actually brought the ship to a standstill. Although the cause was never explained to Jim, the ship's stopping for even 'a couple of hours', was most unusual.

Jim's early days in Australia were spent in Victoria. His Queensland period started in earnest with 'the Army calling out the next twelve names' on the Townsville Railway Station platform. 'Our names all started with "D"! Mid-April we were on a truck to ANTIL PLAINS.' Somewhat disillusioned, Jim Dudley took some settling into Army life at ANTIL PLAINS.

'...well they just used us for anything... We dug slit trenches for weeks... I wasn't too fond of digging slit trenches. You go down about 30" with soil and then you get rock - and that last 18" was tough. You had to pick it and break it. I didn't like that... I screwed up a

ANTIL PLAINS (16/5/1942)

This is a vertical of a Trimet set titled ***'Panoramic of all strips in Townsville area'****, flown only a little over a week after the important Battle of the Coral Sea. The finding of this photograph was a high point in research for ANTIL PLAINS airfield in that it provided the first proof of strips being graded in the road reserve itself and east of the road and rail. By this date elements of the US 8 Fighter Group had transited through ANTIL PLAINS on their fateful progress northward to Port Moresby. Both Charles Falletta and Walter Harvey in their P-39 Airacobras, had been forced down due to bad weather on Cape York. Both veterans consider themselves fortunate in that their particular Airacobras were recovered by interested parties in the 1970's and are undergoing restoration.* [NLA (RAAF Official)]

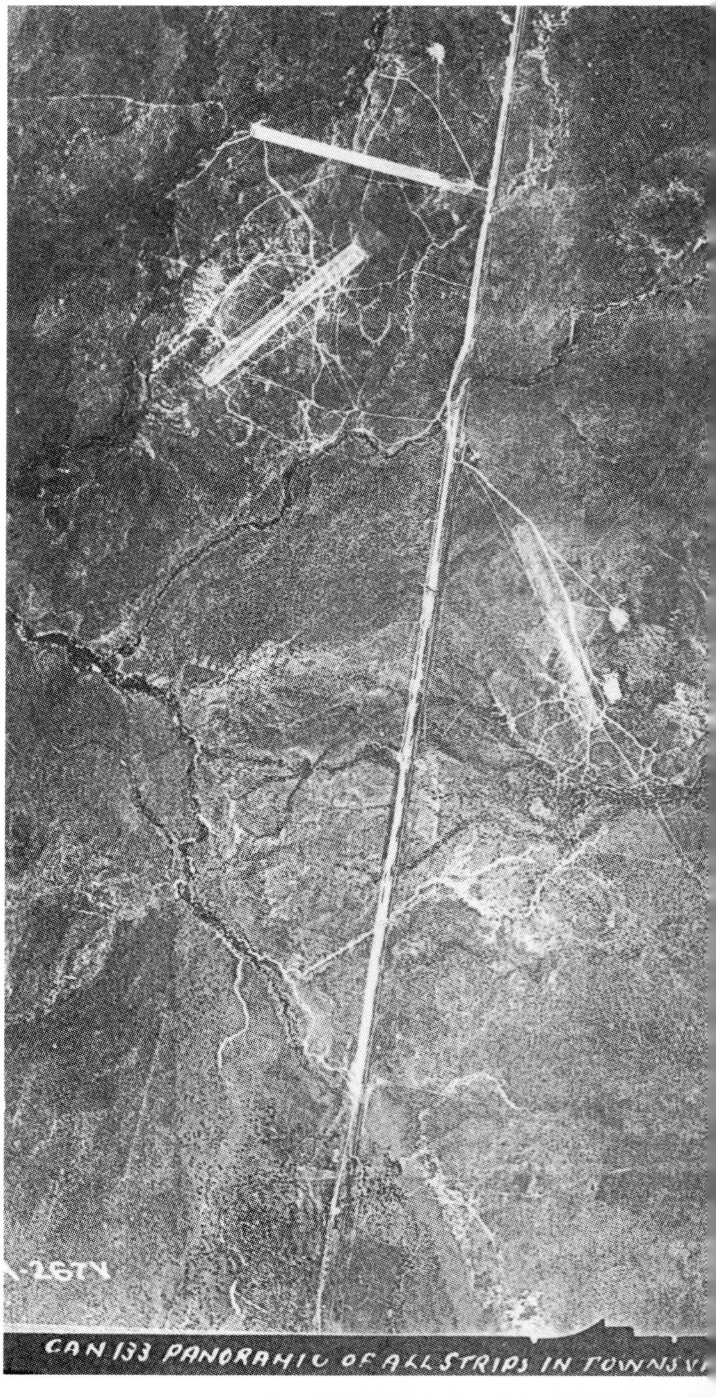

bit, after I found out that I couldn't be a flying cadet which I signed to be, and I didn't really have any faith in anybody telling me the truth, so I kind of did what I wanted to. The result was they put me on a detail pumping water... about a mile back in the woods... a little put-put pump... there was animals and birds and fantastic environment. It was very interesting to me because I like outdoors and there was a Russian man with a truck who was hauling water for the Australian Government. He was getting his water by taking a 5 gallon bucket, dipping it, taking it, and putting it in the truck. Take him about an hour and a half to fill his tank. Shortly after, I met him and he couldn't communicate except smile and make gestures, but I found a "T" on the top system where it come off the pump and I rigged the "T" with a plug and put a piece of hose on it and within a few days... he could pull up and I'd pump his truck full of water and he was happy and I was happy. We got to be real buddies. He started the first day to take off as soon as I filled his truck, which would take about 15 minutes and I persuaded him not to hurry because it would generally take him 1½ hours with a bucket. So he got the spirit of what I was trying to tell him, so we had some long fellowship. We couldn't talk but we just enjoyed being together. So anyway that was the best duty I had really. After my cycle of punishment was over and they put me back in the routine of things, I screwed up again and here we go, I got two weeks that time. Loved every bit of it...'

The two strips west of the road and railway soon became the preferred pair. However, an inspection on 25 May 1942 by P-O J W Campbell shows the spread of early use of the four. The 90° runway was by then gravelled full length and being widened from 110′ to 150′ -

'...no excavation work has been done on the runway - it has been built up with gravel filling to the correct profile. The 35° landing strip has been cleared, grubbed and consolidated to natural surface full width 450′ x 4500′ long. Appears to be very wet at the SE end. The runway is in use. The administrative building is near completion by the MRC. The Americans are camped in the area between the 90° and the 35° landing strip.

The 144° landing strip has been cleared, grubbed and consolidated to natural surface. Clearing width 350′ by length 5300′. Runway is in use. The runway parallel to the road and railway has been roughly graded and rolled, but has now been abandoned...'

Robert 'Bluey' Field still lives only a few kilometres south of ANTIL PLAINS. As a young man he remembers all of these strips and even 'a few fighters landing on the road, but not often'. By contrast, Jim Trench couldn't remember the road strip, though he did recall how -

'...a gentlemen named Pettus, attached to the American Army with the honorary rank of Colonel, was keen on strips on roadways.'

David Pratt was a young pilot, one of about six RAAF personnel attached to the US 33 BS of the 22 BG (M) flying Martin B-26 Marauders. Dave was uncertain of when he arrived at ANTIL PLAINS but it seems to have been late in May 1942. He was emphatic in his statement that 'the only other aircraft I saw during my couple of months there, was a RAAF Hudson which came in one day for a while'. It appears the US 8 FG elements had moved from ANTIL PLAINS by this time.

During June 1942, the role for ANTIL PLAINS became clearer. As a result of a tour of inspection by General Brett, and subsequent discussion with the AOC NEA, a programme of airfield construction evolved. Starting with Milne Bay (New Guinea), then PORTLAND ROADS and MAREEBA, ANTIL PLAINS was listed as (j) in an alphabetical treatment of priority. Out toward Cloncurry several projects were curtailed. The requirement at ANTIL PLAINS was spelled out 21 June 1942, as -

ANTIL PLAINS (31/3/1942)

*PA569 - this is one of an important group of obliques, held in the National Library of Australia. They are edge marked as **WOODSTOCK and ANTEL PLAINS runway** and this particular frame is notable in that it shows the graded 144° strip to the east of the road and railway and the strip in the road reserve proper. In fact this strip shown is one of two or three such urgently prepared strips in the road reserve which were constructed in the belief they would blend with the road itself in any reconnaissance photography gained by enemy aircraft. The Antil Plains Railway building appears at the bottom left edge of the photo.* [NLA (RAAF Official)]

'...accommodation for 2 fighter squadrons to consist of:
1) 2 parent aerodromes each with 2 dispersal strips...
2) camp facilities for 500 men at each parent aerodrome. All construction necessary to accomplish this programme has been designated a US responsibility and the control thereof will be exercised by HQ USAFIA... The camp facilities listed above will be based on the use of tent camps supplemented by semi-permanent kitchens, mess halls, latrines and showers. It is essential that kitchens and mess halls be screened....'

The absence of drainage at ANTIL PLAINS and the consequential effect of rain on aircraft movement mid 1942 is a matter raised by both Dave Pratt and Jim Trench. Jim may not agree philosophically but does understand the principle of providing the much needed working runways first. Gambling on the dry time of year lasting until resources allowed work on drainage, was the path Allied planners had taken. Jim remembered 'some very heavy rain (resulting in) all their (US) bombers (being) bogged where they stood'. Dave Pratt was confident that in his time, only one B-26 suffered any damage in attempting to taxi over suspect ground in the dispersal area at ANTIL PLAINS. Walter Maiersperger, ex 22 BG and freshly posted as Engineering Officer to the US 35 Service Group at CHARTERS TOWERS, recalled a 'broken B-26 nose wheel strut' at ANTIL PLAINS. Walter's log shows his initial inspection of such damage on 4 August 1942, and his later, 31 August visit to 'test hop' the repaired aircraft, 40-1491. Jim Dudley also recalled -

'...we got some heavy water and the airplanes actually sank about 30"... they had to use air bags to pick 'em back up when they dried out... they put an air bag under each wing and they had a "Cletrac" a tracked vehicle with a pump (air compressor) on it... they simply pumped it up and put something under it, and they'd let it sit there til the ground got hard enough to

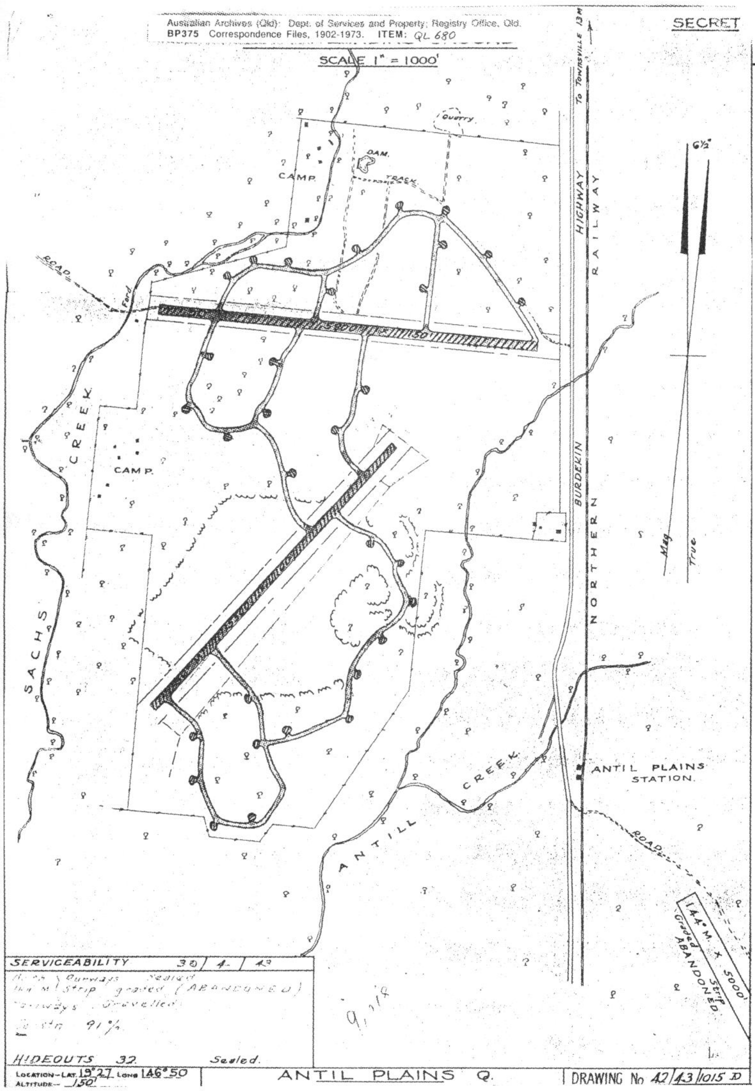

ANTIL PLAINS *(30/4/1943)*
This is the ***'serviceability/status date'*** *on Drawing 42-43 1015D, one of a set of drawings surviving in Australian Archives Brisbane Region. Of interest is the varied spelling of the airfield compared to the creek and adjacent railway station, and of course that on the oblique photo margin for PA 569. [AA BP 375 Item QL 680]*

transport it and all. In one case they thought it was hard enough... Lt Robinson, I believe he was the one who was taxiing it around to sort of get a feel for the cow pasture and see how good it would be and he hit a soft spot and nosed over...'

Lt R W Robinson went on to become CO of the 33 BS. Even that early in the Pacific War, Robinson must have left an impression on Maiersperger, already posted away from 22 BG. In Walter's log entry for this taxiing incident, he added 'Robbie did it!' (see photo page 269).

ANTIL PLAINS *ca(-/6/-1942)*
Here Martin B-26 Marauder '01421' throws up a cloud of dust taxiing out from its makeshift dispersal on the western side of runway 36°. Dave Pratt was a RAAF pilot assigned to this US 33 Bomb Squadron. Dave's period at ANTIL PLAINS extended to July and he speaks of it as ***'only a grassed strip that we used'****. He also recalls a heavy rain period which* ***'closed the place down for several days - there was no drainage'****. In fact Dave showed me a couple of snapshots of a Marauder, nose wheel collapsed and part bogged as a result of, as Dave put it,* ***'an unwise move to taxi from the dispersals'****. [D Pratt ex RAAF]*

Bitumen was scarce and safe delivery of further stocks was rendered uncertain by enemy submarine activity. An 8 August 1942 progress report addressed ROSS RIVER and indicated the 'next move' as -

'...plant and equipment will be moved (from Ross River) to Antil Plains where the 90° strip will be primed and sealed (by September 1)...'

Other airfields were to follow before the rest of ANTIL PLAINS was due for bitumen sealing 'by 15 December 1942'. Jim Trench felt ANTIL PLAINS was one of 'those airfields' where -

'...The strips were smoothed over with a grader and sealed. The whole lot failed. The bitumen seal peeled off like a carpet. A type of construction called "Armour Plating" was adopted. A layer of coarse screenings was spread over the gravel and rolled in, leaving a rough surface to key into the seal coat and prevent the seal peeling off...'

The RAAF's F-Lt D Farran took over from Jim Trench when Jim was posted to New Guinea around September of 1942. While I've not found any mention of such failure(s), there is under Farran's signature a memo to Dept of Interior dated 16 October 1942 which reads in part -

'...Owing to the large number of aircraft operating from this aerodrome, it is requested that the following matters be treated as urgent and completed prior to commencement of work on the second runway. Works required are:...
1) Completion of taxiways as previously requested
2) Completion of 12 hardstandings nearest the runway as shown by red crosses as marked on plan enclosed
3) Completion of hardstandings on sites adjacent to the western end of the runways on sites to be indicated on ground by an officer of this Works Division
4) The sealing of the loose 3/4" screenings near the western end of the strip
5) The laying of a light coat of gravel 3" loose thickness to act as a dust layer along the flight strip as indicated on plan...
It is requested these matters be treated as urgent, as this work is required for operational use.'

It seems possible 'item 4' related to pavement treated by 'Armour Plating'.

Which squadron(s) made up this 'large number of aircraft operating from ANTIL PLAINS' at this time is not clear. The Marauders had moved over to WOODSTOCK earlier. The airfield was being seriously considered for upgrading and use as a US Fighter replacement training school. The AWC Minutes show their receipt of advice of A1 priority for this upgrading on 5 January 1943 with a set time for completion of 3 months. However, on 21 June 1943, Rear Echelon USAFFE advised DWB Melbourne of cancellation of relevant works at ANTIL PLAINS in favour of establishment at CHARTERS TOWERS.

By October of 1943 the record shows fencing only being done by six men under control of MRC. The same report shows the original fighter field project as 91% complete.

A review of unoccupied service buildings in the Townsville area dated 12 June 1944 gives some side light on the airfield -

'...Antil Plains - constructed for USA 5th Air Force has been unoccupied for 18 months. When the USA Squadrons moved, buildings were stripped of moveable fittings, such as shower roses, taps, electric light switches, etc. and it is not known whether this was done by the American personnel or stolen by others. The buildings were not guarded for some time after the departure of the Americans, due to the shortage of RAAF personnel in the area, except for weekly visits by patrolling motorcyclist from 23 OBU. In January 1944 permission was given to No 1 L of C (lines of communication) to occupy the buildings for a Coastal Artillery School. These personnel have now vacated the site. It is understood that 23 OBU, subsequent to this,

moved two buildings to Garbutt, as stated in W Cdr Hepburn's report, and also that 80 Squadron (RAAF P-40s) were given authority from NEA HQ to remove a pump chlorinating plant and tanks to New Guinea. As there seems no possibility of the remaining camp buildings being used in future, it is proposed to demolish them and reuse the materials in the construction of the proposed camp at Stuart for 103 Mobile Fighter Control Unit...'

The RAAF reviewed unoccupied dromes in Queensland in June of 1944. ANTIL PLAINS was one of thirteen confirmed by DCA as having no value to that Department, certainly not to the point of it acquiring the land.

Around April of 1945 the RAAF was considering relocating the NEI Replacement Pool and the RAAF's B-25 Training Flight to the CHARTERS TOWERS area on the basis that the Training Flight would be able to use such fields as ANTIL PLAINS and WOODSTOCK still retained under US control. Although by then the RAAF had 'taken over' CHARTERS TOWERS Airfield from the Americans, the US Forces had reserved the right to reoccupy it and others including ANTIL PLAINS on '30 days notice' if needed. This RAAF proposal lapsed, but when the war ended the RAAF was still considering the retention of ANTIL PLAINS in its post war requirements.

Today the airfield is bisected by an E-W fence line which places the old 90° runway and the dispersal area north of it in private property. To the south of the fence line the old airfield is embraced by the Townsville and Thuringowa City Water Supply Board's huge Ross River Dam. This dam is presently developed to its Stage 2 level and the Charters Towers road and rail have been relocated eastward to skirt the dam reserve. While there is a proposed Stage 3, it is generally considered doubtful it will proceed in view of the recent connection by pipe line to the Burdekin Dam supply source. The present Stage 2 full storage level reaches to the southern end of the old 36° runway and a flood stage sees water covering about half of that runway length.

Ian Mullins prepared a brief plan and statement of ANTIL during its WW2 period to give to Walter Harvey when he revisited the strip in May of 1992. On this plan Ian added -

'...just inside the airfield gate on the old road there is currently a radio controlled model aeroplane club aerodrome. They complain from time to time of low flying P-39s.'

ANTIL PLAINS (25/7/1991)

Looking east at a point half way along the southern edge of the 90° runway. John Stephenson, an Officer of the Townsville City Council-Thuringowa Water Supply Board is orienting a layout drawing and showing Jennifer, the author's wife, where things were. Note the myriad of ant hills, the abundance of which must surely have given further latitude for varied spelling of this historic airfield. [Author]

GIRU

'...See, he's still got his cattle here and he flies down and looks at them. He uses that little strip all the time now...'

Such was Tom Leon's comment about the remnant of the E-W 108° strip at GIRU. Tom is a cane farmer living just to the southeast. He was sixteen when troops arrived and commenced work on the airstrips. Tom was speaking of another long term resident of the district, Jim Burry, when he referred to Jim's present use of 'that little strip'.

Archive records show that on 1 April 1942 possession was taken of property owned by R S Burry for a USAAC aerodrome. A 14 April review by P-O J Keays noted -

'...Giru - three runways located in form of a triangle so that approach is through gaps in the surrounding hills. Surrounding country is subject to flooding but actual site is above flood. Timber cover is fair, and dispersal area adequate. Water available from shallow wells in any portion of area. W Cdr Lerew to inspect from operational aspects.'

The arrival in the Townsville area of the US 46 Engineer Regiment in March, was closely followed by elements of two US Engineer Battalions (Separate). These initially provided manual labour for projects rather than serving as fully integrated engineering units. Col. US Army (ret'd) Burton B Bruce wrote me that he remained CO of the 91 Engineer Battalion (Sep) the whole three years it served in the SWPA. During that time the unit was reorganised to full EGS function. The 91st, comprising some 1 200 coloured troops disembarked at Townsville on 11 April 1942. The US Archive sourced ***History of the 91st Engineers*** details its early days at WOODSTOCK assisting the 46 Engineers. 17 April 1942, the 91st 'marched the 17 miles' to GIRU to start work on a new airfield. Their 29 April journal entry includes -

'...still no news of the transportation and heavy equipment. Most of the work is being done by hand and the use of dynamite.'

It carried on, 30 April 1942, with an account of 'the alert' -

'...upon news that a Japanese fleet was in the South Pacific ocean, headed this way. Troops were dispatched immediately as outposts. Co "D" was sent down the Haughton River to within about 4 miles of its mouth.'

While a field hospital was being set up at WOODSTOCK the journal added, on 1 May 1942, some classic hearsay -

'...Report from an enlisted man returning from Townsville Qld, established the fact that Japanese reconnaissance planes were over Townsville and at least one of them was shot down...'

Blacked out ships were seen in Bowling Green Bay and, inevitably, quite a quantity of 30 calibre ammunition was expended on the 'periscope and conning tower' of what turned out to be a floating log on the incoming tide. The ***History of the 91 Engineers*** records this period as the 'Battle of Mosquito Bend'.

During May, a horse drawn mower appeared to help those cutting grass with machetes, and some local farmers' trucks were added to the hand-pulled wooden beer box wheel barrows. The Unit's journal triumphantly records the following milestone on 26 July 1942, one day before it is directed north to start work on JACKY JACKY strip -

'...first plane to land on the aerodrome at Giru landed today on No 1 Runway. Col. Mills, 46 Engineer Regiment (and a small party) were the passengers. Their plane was a light two engined job... and the pilot an Australian flying officer...'

Soon, both equipment and strategic positions improved. Work on GIRU lapsed. During 1943 the project was formally abandoned by the US and handed over to the RAAF. NEA Headquarters requested that 'the gravel

runway' 41° be retained for emergency purposes. Both the other strips were fully cleared, with '108°' partly gravelled.

Like many urgent wartime construction projects the job was not finished until the paper work was done. This often involved the reconciliation of claims by land holders for compensation and recovery of property. In this case a claim lodged by property owner and local Butcher, Robert Stockdale Burry read in part as claiming for -

'...a) filling in trenches, six wells, gun pits and two chinamen; b) pair of wooden gates; c) loss of six head of cattle; d) fencing erected; e) droving expenses; f) timber destroyed. Total Claim £223/5/0.'

A written assessment of this claim gave as background -

'...some 1 500-2 000 men were camped in the area but vacated about the end of July 1942... (Mr Burry) had difficulty getting fences replaced... He constructed a "lightning fence" at his own expense and since 1 October 1942 has had partial use of the area. However in July 1943 American personnel came back into the property to remove many tons of earthenware piping. The grass had grown so high around the pipes that it was impossible for the men to work, so they set fire to the grass with the object of burning off around the pipes. The fire apparently got out of control and burnt out the grass on the whole of Mr Burry's property...'

A figure of £147/15/0 was negotiated and recommended by the Hirings Officer.

The 41° strip was promptly incorporated into the main road between Giru and Woodstock to the west. The US publication, ***Engineers of the Southwest Pacific 1941-1945*** indicates that an existing pre-war aerodrome at GIRU was extended and developed during the war. The small Camping Reserve at the north tip of the airfield triangle was used by the Giru Race Club as a racetrack. It likely did see use as a pre-war landing ground. Nowadays, the only semblance of aerodrome activity is at Jim Burry's 'little strip' where he flies over to 'check out his cattle'.

GIRU (27/7/1942)
Part of a Trimet coastal trimming flight, this frame shows the classic triangular layout of GIRU airstrips. From 20 000' the airfields of WOODSTOCK and REID RIVER can be seen to the southwest just behind the propeller and in front of it respectively. In the foreground, is the mangrove etched foreshore of Bowling Green Bay, where US Engineer Troops were hastily dug-in around May 1 1942 in anticipation of a Japanese landing. [AUSLIG]

WOODSTOCK

'...Crabbe was the name of the people where they built the first strip... The next one was along the Charters Towers Road, that's the Flinders Highway and then the third strip was on the highway between Giru and Woodstock...'

Robert 'Bluey' Field was a lad of seventeen in 1942 when this airfield was begun. 'Bluey' still lives a little north of here and this was his brief description of the three airstrip complex, named WOODSTOCK after the nearby railway station.

As with ANTIL PLAINS airfield, WOODSTOCK was a site under consideration as early as October of 1941. The 'initiation date' of 30 October 1941 appears on an unnumbered drawing titled ***RAAF Landing Ground Woodstock Queensland*** in the 'Hiring of Site' file. The drawing has obviously been added to as development on the field took place. That date alone testifies to its consideration prior to the Japanese entering the war. In a report to Air Board dated 6 March 1942, Air Cmdre F W Lukis, AOC NE Area reviewed the WOODSTOCK site in conjunction with others. He stated a need for three fighter dromes apart from the Townsville RAAF Station itself, each to be positioned to respond to the likely direction of enemy attack. He knew the Operations people didn't like sites being 'too close together' and concluded by saying -

'...If pilots are well trained and ready for War before being sent to the Squadrons stationed at these places, it does not seem that any undue risks would be run of the respective turning circles cutting each other with consequent danger of (aerial) collision...'

In the prominent US historical treatise ***The Corps of Engineers-The War Against Japan***, Karl C Dod wrote -

'...After two weeks of combat training near Melbourne the 46th Engineers left by rail for Northern Queensland, an area which might at any time become a battle zone. To protect themselves against enemy air attack the men mounted machine guns on the flat cars of their trains. During the trip they passed train load after train load of refugees headed southward. On 13 March the first of the troops reached Woodstock, a town 20 miles southwest of Townsville...'

Lt-Col Albert G Matthews was CO of the regiment and as Dod continues -

'...On 18 March the entire regiment began the work of clearing and grading three runways for a giant airfield at Woodstock. Matthews ordered round the clock operations scheduling three eight hour shifts a day...'

P-O H V (Jim) Trench is on record as reporting by telephone on 22 March 1942 to the RAAF's Directorate of Works and Buildings, Melbourne -

'...Woodstock aerodrome - 1 strip complete (cleared only) 2 others ready in 1 weeks time.'

From this evidence it appears that the 'first aircraft to land' would have done so on the 70° strip on Crabbe's property.

WOODSTOCK ca(2/4/1942)
*Brice Allen's caption is **'P-39's kicking up dust 4/1942'**. It is believed the photo is taken on the 104° strip, where the early arrivals from both 35 and 36 Pursuit Sqns landed. The US 46 Engineers' history mentions first use of the airfield as; **'A plane carrying Lt-Col A G Matthews and Major William H Mills landed on the first runway on March 22, four days after construction was started.'** Progress records indicate the 70° was in use first.* [W (Brice) Allen ex US Engineers]

WOODSTOCK *(20/5/1942)*
US Army Chaplain Henry Chambers stands on the seating round the Country Womens' Association (CWA) hall at WOODSTOCK to thank all for the social evening. Veterans of the 46 Engineers' HQs and Service Company tell me an organisation called 'The Comforts Fund' (Australian Comforts Fund?) sponsored a train load of girls from Townsville for the dance. Les Turner tells me there was some ill feeling about 'that dance' because no Australian boys were invited. The photo portrait of Mrs Clara E Haberecht, a foundation member of the Woodstock CWA, looked down from the galvanised iron clad wall that night - and it still does!
[W (Brice) Allen ex US Engineers]

Heimo Raski was a member of the US 46 Engineers. After his return visit to WOODSTOCK a few years ago, he wrote recalling the -

'...Australian civilians who worked along with our troops. I remember at least two father and son teams with their crawler tractors and small scrapers... they (and others) "were assigned and on rations"...'

P-O John Keays' 14 April 1942 report to Air Board included a tabulation of some twenty-nine strips listing their type, priority etc., works required and a final column headed 'action taken'! Against WOODSTOCK, there is the comment '...most of clearing and grubbing done by MRC. Balance carried out by US Engineer Corps...'

Brice Allen was a young lieutenant assigned to the 46 Engineers' H & S Co, responsible for operation of the motor pool and heavy equipment. That company's CO Capt Harmon C Broyles accumulated several photographs of the 46 Engineers' work in Queensland, and Brice Allen has furnished some from Broyles' collection. Brice tells me Capt Broyles and Chaplain Chambers arranged the 20 May dance.

Keays' 14 April 1942 report on his assignments in selection of aerodromes in the Townsville and Charters Towers area said -

'...US Corps of Engineers is presently gravelling the strips...(which) are located along Charters Towers and Ayr roads with a third strip at right angles to Charters Towers road. The width of clearing for the strips makes them obvious from the air. Road traffic has been seriously inconvenienced. Traffic on Charters Towers road is heavy and at present is fenced off the road and forced to traverse a very dusty and rough side track...'

Keays was in touch with the overall progress of these airfield construction activities and was vitally concerned with the prospect of achieving the minimal gravel surfacing by the required dates. With the severe shortages of equipment and manpower becoming evident, Keays could see the whole gravelling programme was in jeopardy if extra plant could not be transferred from southern Queensland or interstate.

Very minimal camp facilities were in place for the arrival during April of the P-39 Bell Airacobra fighter aircraft of the US 8 and 35 Fighter Groups. WOODSTOCK was for the squadrons of those units only a brief pause before moving on to Port Moresby. Their transit was fraught with weather losses on the way. Their combat experience was the bitter sort - where a young pilot often needed luck to stay long enough to learn. July 1942 saw them return to the Townsville area generally for re-equipping with P-38 Lightning aircraft.

Like ANTIL PLAINS, WOODSTOCK proved to be an area where aircraft could not move under wet conditions unless their taxiways were at least gravel paved. Les Turner describes those taxiways as 'simply gravel placed on top of the natural ground, but eventually bitumen surfaced'. He recalls the dispersal areas as being camouflaged by the simple draping of camouflage netting over cables, supported often times by the natural tree growth in the area.

P-O Campbell reported to Air Board on his inspections of several fields, 25 May 1942. Here he said -

'...Woodstock... The 102° landing strip has been cleared, rough graded and consolidated... Runway is in use. The 173° landing strip is cleared for width 500' generally. Surface fair - runway in use... Western edge of runway is about 60' off centre line of main road. The watercourses at either ends of this runway have been covered over with round timber box culverts. The 68° landing strip is cleared for 500' width... Runway joins the 173° runway and about 500' of it will be ineffective as it does not give necessary approach clearance of 1 in 30 from the main telephone lines along the Railway... Reported this runway was originally set out bearing 75° in order to avoid hills in the foreground. The US Engineer Corps however, altered it to a bearing of 68°... runway in use... Administrative, kitchen, ablutions and latrine buildings are nearing completion... A (water) well has been sunk near the gravel pit...'

In June, US Anti Aircraft units were positioned at various centres in north Queensland. Records show that

WOODSTOCK *(16/6/1945)*
This vertical is part of the systematic coverage for mapping 'Manton' carried out by RAAF Survey Flight. In the bottom right hand corner is the wartime farm residence of Les Turner. The initial airstrip construction on the Charters Towers road for runway 175° was similar to work done a little further north for the earliest strips at ANTIL PLAINS. After the war, the northernmost one of these three strips (104°), became an improved section of the road east to Giru and Ayr. *[AUSLIG]*

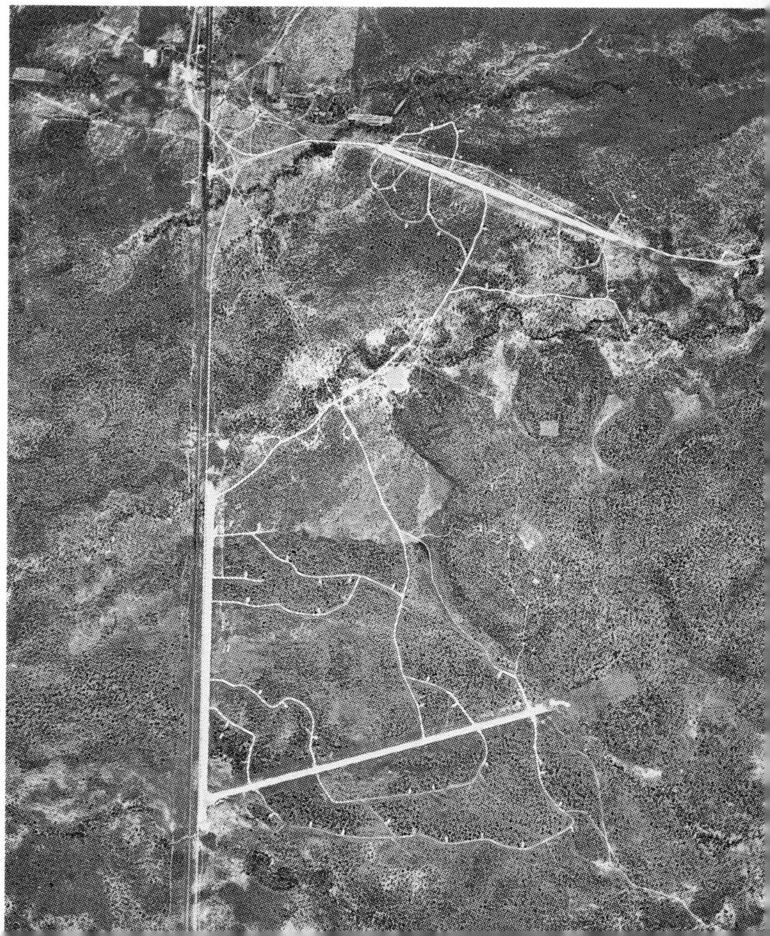

WOODSTOCK (-/8/1942?)
Parked south of the 70° runway '01493' was photographed bearing evidence of makeshift repairs to shrapnel damage to its rear end. [D Pratt ex RAAF]

HQ 3rd Battalion with Batteries D, H, L and M and 1 Platoon each from Batteries A and E of the 94th Coastal Artillery (Anti Aircraft - A-A) defended the airstrips at WOODSTOCK and REID RIVER.

Also during June a review of the operational aerodrome requirements in NEA resulted in the Commander, AAF, directing that WOODSTOCK and REID RIVER were collectively to provide accommodation for one complete medium bombardment group. The directive carried a note-

'...The group at these two airfields will later be moved to the tip of Cape York Peninsula, but it is desired to retain these fields for the use of squadrons that will be formed later.'

The 22 BG (M) was that group. HQ and HQ Sqn together with 19 and 33 Sqns occupied WOODSTOCK. During the first couple of months, aircraft operated from Port Moresby, returning to WOODSTOCK during stand-down. To meet these requirements, WOODSTOCK and REID RIVER were to have camp facilities for 750 and 500 men respectively. This 21 June 1942 directive listed the work at WOODSTOCK alphabetically at 'f' ahead of that for 'j', ANTIL PLAINS. The following general provision covered the minimal standards of camp buildings and the timing of their erection-

'...On fields already constructed and lacking the necessary camp facilities, it is desired that the necessary construction be provided as soon as possible. It is anticipated this will require in some cases, that engineer troops be used for camp construction and that this will delay somewhat the programme of runway construction. It is therefore recommended that civilian labour be utilised wherever possible to replace engineer troops on camp construction...'

It was very likely in this context that Les Turner's older brother became involved 'helping to build some of the buildings on the WOODSTOCK Airfield'. Les was eleven years old and recalls his brother had 'an American Uniform' and was 'paid by them' though Les cannot remember whether that was in currency or some 'other form'.

Graham B J Robertson was a young RAAF co-pilot assigned late May 1942 to the 19 BS, 22 BG. Robertson was paired with Walter Krell on the 9 June 1942 raid to Lae in which the US Congressman Lyndon B Johnson flew as an observer in a 19 BS B-26. P-O Robertson's diary described his impression of WOODSTOCK mid July 1942-

'...arrived back from Sydney (torpedo school at Nowra) to find the camp all changed. Showers installed, an Officers Club where there is electric light, crap table, chess, ping pong, bar etc. and it is rather good. Two wirelesses belch all the swing you can stand and the news come through regularly and causes good amusement to all concerned (up here)...'

Camouflage treatment of aircraft hideouts seems not to have gathered any real momentum at WOODSTOCK. Cable strung netting is probably the most sophisticated treatment used, however minimally, beyond soft nets manually thrown over the machines. By 10 August 1942, a report by Lt Longland an Australian camouflage officer passing through WOODSTOCK, referred to -

'...along flight strip parallel to railway line... noticed bombers parked out in the open in bright moonlight...'

Jim Dudley was a member of the US 33 BS ground echelon, moving with it from ANTIL PLAINS to WOODSTOCK. During Jim's recent visit back there he reminisced with Les Turner about an incident near the junction of the 70° strip with the one 'on the road' involving a B-26. Jim said the pilot was -

'...pre-flighting it, you know,... brakes on and give the engines full power and be sure they can get the right vacuum... then the nose wheel broke and it fell over and burnt - it was destroyed - they got the crew out, but there was one guy that almost didn't get out...'

Remarkably, Les filled in some background. He believes the B-26 was 'The Fly'n Parson', named for the father of Capt Fred Nichols, its usual pilot. Les' sixteen year old brother, 'who worked for the US', was in the nose of the aircraft with a parachute on, but with little or no instruction on 'how to get out in a hurry'. Les agrees there would have been a 'great to do' if young Jim Turner had died. However, the experience can't have been too sobering for young Jim because as Les says -

'...soon as he was out he tore off down the strip on his motor bike to see when the next plane was going up!'

Certainly in the early part of this construction programme General Casey, Chief Engineer US Forces, had quickly concluded that drainage was to carry a lower priority than that of providing operative runways. After late wet season rains which bogged aircraft at some of these near Townsville strips, it became clear that drainage was imperative. Early August 1942, F-Lt John Keays as NEA Works Officer, followed up his earlier concerns for the gravelling programme. In his plea for augmentation of equipment in NEA, he wrote -

'...Plant and labour is available in the south and could be transferred to this area. Four heavy bomber fields are being constructed west of Brisbane, each one with a complete and separate construction organisation. Construction of the parent fields should be almost complete and the organisations now constructing

WOODSTOCK (-/10/1942)
Officers of the 19 Bomb Squadron 22 BG (M) photographed in front of 'Officer's Club' WOODSTOCK in October 1942. Lt Bob Thompson, who gave this photo to Les Turner is in the rear, five from the left. While names don't appear on the photograph, I believe Charles Eberly is third from the right in that rear rank and possibly Frank Allen is in the front row, extreme left. Walter Krell and his RAAF co-pilot Graham Robertson missed this group photo. Krell was hospitalised after trying in vain to rescue Robertson from their burning B-26, 13 September 1942 at IRON RANGE. [R Thompson ex USAAF via L Turner]

WOODSTOCK (-/8/1942?)
Here an engine is being uncrated by use of a chain block slung off a demountable hoist frame. Clearly seen in the foreground is the PSP, Marston matting laid out to facilitate access to the B-26 Marauders. David Pratt was one of a few RAAF pilots assigned to the 22 BG when its own personnel were redistributed to aid in the formation or strengthening of other units. [D Pratt ex RAAF]

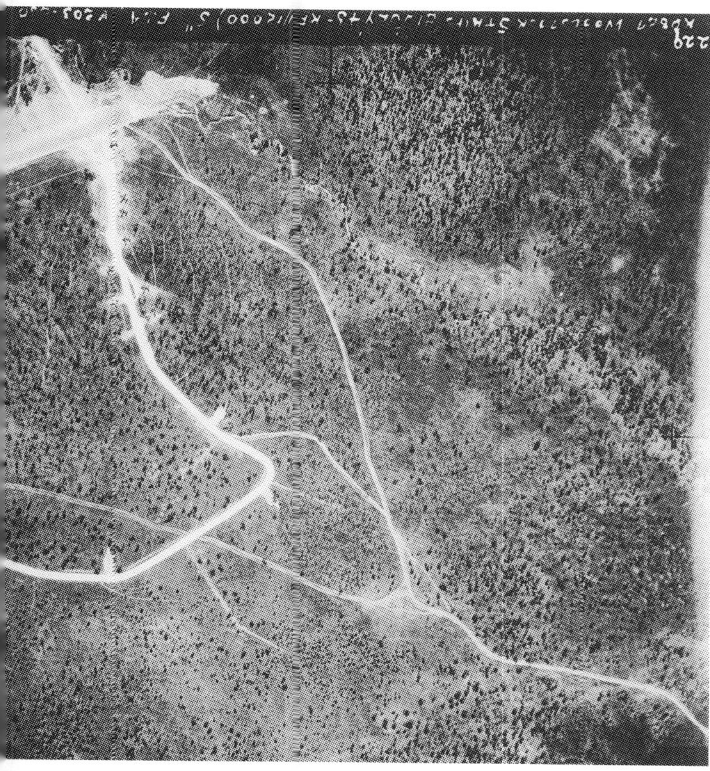

WOODSTOCK (21/7/1943)
Frame V 229, part of 'KD624', is one of several surviving prints in the National Library of Australia. The companion obliques are also available in company with an interesting set of some twelve photographs taken on 31 March 1942. This particular frame shows a number of twin-engined B-25 Mitchells with some B-26 Marauders. As a boy living nearby at the time, Les Turner, recalls vividly those childhood years of going to and from school under the noses of such aircraft. Les's home was out of this photo along the track leading to the lower right. [NLA (RAAF Official)]

(associated) satellite dispersal strips. It is contended that the construction of parent field in this area should (now) take priority...'

Col. Willis E Teale, Chief Engineer USASOS in a letter to AWC dated 3 August 1942 set out in detail, priorities for the bitumen sealing of some thirteen runways in the Townsville vicinity. All three at WOODSTOCK figured, with target dates set in November of 1942. It took until February of 1943 for the last of the runways to be sealed at WOODSTOCK and by then the operations buildings and camp accommodation for 1 300 men were available .

When the 22 BG with its B-26 Marauders came back to WOODSTOCK in February of 1943 it was with only thirty two of their original fifty six aircraft. These were given a thorough repair and overhaul. When they were returned to service, it was on allotment to only the 19 BS, who proudly flew them with a clean metal finish and bearing the unofficial name of the *Silver Fleet.*

Col. Jarred V Crabb commanded the US 345 BG (M) when it arrived in the area from AMBERLEY around 16 May 1943. While the 22 BG's B-26 aircraft had been shipped to Hawaii before flying the rest of the Pacific, the 345 BG's B-25s flew all the way from California. HQ squadron and the 498 BS occupied WOODSTOCK, while the remainder were distributed between REID RIVER and CHARTERS TOWERS. This was a new group without the heavy sprinkling of pre-Pacific war Air Corps men. During their few weeks training period at WOODSTOCK, the newcomers learned from 22 BG, themselves re-equipping with the B-25. The 498 BS trained for 'bombing from 8 000′ ' but they were to wreak havoc amongst the Japanese with the new format 'strafer' aircraft. The Group was known as the 'Air Apaches'. I wonder if Col. Crabb was ever aware of the WOODSTOCK family by like name, from whose property they flew.

By mid 1943 with consolidation of the strategic position in New Guinea, the US Forces sought to make clear at which airfields in northeastern Australia they intended to retain responsibility for construction and maintenance. WOODSTOCK is one of six such fields. By October of 1943 three of these, WOODSTOCK included, were passed over to the RAAF for these matters.

Records indicate that the 22 BG had departed WOODSTOCK by October of 1943. Certainly Richard Graf a member of the crew of B-17 'Sally' assigned to

WOODSTOCK (24/7/1991)
Compare this recent aerial to the direct vertical V 229 taken almost forty-eight years ago. This shot picks up the eastern end of the original 70° runway with the winding creek reaching back towards Les Turner's old home. There is now only the suggestion of the taxiways on which the Mitchells were once parked. [Author]

WOODSTOCK (-/10/1942)
With eight missions to its credit and 'Bugs Bunny' on its nose, this B-26 Marauder is photographed with a crew. Centre of this group is Lt Bob Thompson of the 19 BS, 22 BG(M). So far it has not been possible to establish the date of the photo beyond that shown here. Bob Thompson proudly gave it and another to his young Australian friend Les Turner. [R Thompson ex USAAF via L Turner]

WOODSTOCK (17/11/1992)
US veteran Jim Dudley (left) and local identity Robert 'Bluey' Field stand on the foundation of Jim's old 'Sergeants Corporals and Privates Mess', southeast of the present day Donnington Air Park strip. [L Turner]

WOODSTOCK *(1942?)*
*This photograph has written on rear face - '**Capt Len Klosterman 33rd Bomb Squadron, Navigator, grounded himself in early rough period - became Transport Officer 33rd Bomb Squadron, 22 BG(M)**'. Clearly the photo was taken at WOODSTOCK shown by the unmistakable skyline of Mt Elliot National Park. Note the wheel ruts on the ground, clear evidence of 'boggy when wet' conditions.*
[W Gaylor ex USAAF via S Beck]

General Kenney for immediate transport requirements recalls, in effect the only time he knows of 'Sally' dropping into WOODSTOCK. Dick thinks it was -

'...About the middle of 1943 when everything had moved into New Guinea, ...Kenney just decided to have a look and see what was going on at these three airfields, because nobody really knew. So on one of our trips to Moresby, we did a little detour and we landed at Charters Towers, Reid River, Woodstock and Torrens Creek just to see what was going on. Except for Charters Towers we found three bare airstrips, with nothing else, absolutely nothing... At one of them, and I've forgotten which, a herd of sheep were grazing... we had to buzz that runway several times at low altitude before we could safely make a landing... whatever the Americans had put in there initially had gone completely and there was absolutely nothing...'

Unoccupied buildings at several airfields came under RAAF review in a report dated 12 June 1944 -

'...Woodstock; has not been used for 12 months except for emergency landings. The USAFIA Base Section No 2 has intimated they intend to demolish the whole of the buildings for re-erection elsewhere. These buildings were erected on three £5 000 requisitions issued by the Base Section...'

After the war, the RAAF did intend to utilise WOODSTOCK as evidenced by a review in January of 1946. While it is not the intention here to cover in any detail the period since the war, it is fact that the RAAF certainly has no requirement over WOODSTOCK today. The Commonwealth Government, through the CSIRO, has long held the land on which the only survivor, the old 70° strip was located. Moves to sell that land have brought objection from heritage minded citizens.

A small private airfield is located south of the old 104° runway and east of the WOODSTOCK railway station. Known as Donnington Air Park it is a popular general aviation venue for light aircraft and gliding interests.

WOODSTOCK *(7/5/1992)*
Although the three wartime strips do not now function as such, there is an active strip known as Donnington Air Park, located in their midst. Ray Smith is pictured here by a locally owned Cessna 150 with the unmistakable profile of Mt Elliot National Park in the distance. *[R Smith]*

REID RIVER

'...(h) an area on the south side of Reid River... 34½ miles SSE of Garbutt... eastern side of Townsville/Charters Towers road and railway line, adjacent to the siding at Reid River. The (site) offers sufficient area for a full sized aerodrome but has a depression through the centre draining towards the northeast, which could however, be satisfactorily dealt with. The big disadvantage of this site in my opinion is the proximity of a high mountain range on the north at a distance of about 3 miles... Considerable drainage would be required to satisfactorily handle the depression referred to.'

This extract is from a comprehensive report dated 5 March 1942 by Sqn Ldr Towers of RAAF HQ which went on to recommend the development of both WOODSTOCK and CARDINGTON as 'the two US Pursuit Squadron aerodromes'. CARDINGTON was only a couple of miles south of the REID RIVER site. Towers' report concluded with the rider that if CARDINGTON was chosen, its dispersal strips could be provided -

'...at Reid River (and) either at Mingela or on the road between Mingela and Cardington.'

RAAF HQ's approval was signalled on 18 March 1942. Eleven days later, Companys B, C and F of the US 46 Engineers moved from their WOODSTOCK activity to commence work on REID RIVER. Even with the site's detractions, the 46 Engineers appear to have exercised their discretion, launching straight into the full three strips at REID RIVER. The CARDINGTON sites then lapsed, probably as the need vanished. Certainly the site 'pegged' on Lux Foote's boyhood home would have meant the added work of a bridge over the Haughton River.

In my recent endeavours to find and obtain any published or formal history of the US 46 Engineers' operations here in 1942, I am informed by the US Army Military History Institute, Carlisle Barracks, Pennsylvania, that 'none is on file there'. From the US Army's viewpoint, I am fortunate to have contributions from 46 EGS Regiment veterans, Harry Besancon, Brice Allen, Wally Wanielista and others, who all appear to have drawn 'from the files of John A McFee', another '46' veteran. The

REID RIVER *(21/7/1943)*
'P 5', is one of two obliques recorded as being taken on this date. Prints of both survive in the National Library Collection but to date no companion vertical prints nor any record of their existence has been found by the author. Comparing the fretwork as shown on these photographs to the lack of such taxiway evidence on the current day mapping (see Appendix 1) will demonstrate the desirability of seeking out such information before visiting such an airfield site. Like many others this site is now private property and it is essential that permission be gained before entering. The northernmost (87°) runway is now used infrequently by the owner with cattle interests in the area. *[CPE (RAAF Official)]*

REID RIVER (?/6/1942)
The 46 Engineers' 'Workshop' boasted some paltry shade cover over basic benches. On the right of the photograph the 'oxy-cutter' at work is believed to be Ralph Stevens. [W (Brice) Allen ex US Engineers]

oft quoted official histories ***Engineers of the Southwest Pacific 1941-1945*** by Hugh J Casey and ***The Corps of Engineers:The War Against Japan*** by Karl C Dod, provide a broad coverage. Dod, by footnote, page 125, clearly indicates some ***History of 46 Engineers GS Regiment 1942*** existed so I am disappointed my enquiries have not produced it.

P-O John Keays' 14 April 1942 progress report to RAAF HQ is available in Australian Archives. It states -

'...Reid River: This site has been constructed by the US Corps of Engineers, in preference to the selected site at Cardington. Runways have been located clear of hills and site has been approved by operational personnel. Grading will be necessary and site will have to be gravelled before use...'

REID RIVER (15/6/1945)
Part of Frame 3673 in the systematic aerial mapping for 'Mingela' this extract shows the layout of REID RIVER at the end of hostilities. With its proximity to water, the main camp area, was understandably, along the northern section between the river itself and the airfield. [AUSLIG]

Karl C Dod referred to the Unit's initiative, quoting Albert G Matthews' post-war correspondence -

'...Matthews described the construction of these airstrips as "field improvisation at its... best" The Officers of the 46th were largely on their own and "did what seemed good to them and in most cases their engineering common sense was the primary and single qualification for the work". Construction was held to the simplest standards . Because equipment was short , dispersal taxiways, hard stands and revetments were omitted . As the long dry season was just beginning, drainage was dispensed with. Since the engineers had no asphalt or tar, they surfaced the runways with rotted granite, semi-decomposed shale or sometimes merely with gravel and clay. The fields were rough and crude, but planes could land on them.'

Keays' lengthy 14 April 1942 review shows he was a party to constructive evolution of design standards -

'...Width of clearing and pavement: strips are being cleared for width of 300' only and it has been suggested width of pavement for fighter strip should be decreased from 150'. A direction in this regard is required. I consider 300' will be insufficient for bombers and suggest 400 or 450' of clearing with the full 150' of pavement (for bombers).'

This is the time by which the first P-39 arrivals at WOODSTOCK had landed. Rod Voller, then the Dept of Home Security's man in charge of camouflage at Townsville, cannot find specific reference to first aircraft (P-39s?) at REID RIVER. His diary notes blanket the matter and he simply says -

'...these outlying strips were used as soon as they were formed without any bitumen surfacing. They were very dusty and visible! (from the air)...'

The thirteen page history from the files of John A McFee tells of early aircraft arrival and of a diversion -

'... During the latter part of April, the 18 BS (redesignated 408 BS on 22 April 1942) of the 22 BG (B-26) arrived at Reid River and began operation from strip No 1. These bombers would fly to Port Moresby to refuel and load up and strike at Lae and Rabaul. Within weeks a second squadron arrived using strip No 1. On May 3 1942, operations were temporarily stopped when the regiment was put on the alert for a supposed enemy invasion of Northern Australia, possibly in the vicinity of Townsville. Company F returned from their outpost overlooking the town of Giru during the alert and resumed work on strip No 1. Company's D E and H & S moved to Reid River (from Woodstock over the period 5-9 May 1942) and commenced work on dispersal roads and parking bays. Lt-Col W H Mills assumed command of the regiment on May 9 1942 relieving Col. A G Matthews.'

In June of 1942 Anti Aircraft Troops and equipment were deployed to the various airfields and other points in Townsville area. In respect of REID RIVER these troops appear to have been from -

'...HQs 3 Bn with batteries D, H, L and M and one platoon each from batteries A and E of the 94 CA (A-A) defended the airstrips at Reid River and Woodstock...'

The disposition of these batteries is not known and it appears their weapons were nothing heavier than 0.5" calibre machine guns.

While the 22 BG squadrons stayed for many months, the 46 Engineer elements had all departed REID RIVER by 7 July 1942. The plan then for REID RIVER was advancement to bitumen surfacing on all three strips and the erection of two squadron camps. Both were located between the 87° runway and the river. A large CCC camp developed west of the railway siding.

During August, renewed concern for the adequacy of several of these North Queensland airstrips in the next wet season arose. By then most of the work was being undertaken by the Allied Works Council through the Main Roads Commission and the Forces of the Civil

Construction Corps, because the US Engineer organisations had moved either to Port Moresby or into Cape York area. It is also a fact that significant work was then progressing on airfields west of Brisbane and it became a matter of deciding priority. The outcome was curtailment in the south so that renewed effort could be applied in the north.

The RAAF 'Aerodrome Works File' for CHARTERS TOWERS records the dissatisfaction of DWB with work underway by MRC at CHARTERS TOWERS and REID RIVER, mid November 1942. Agreement reached at a recent conference between AWC/MRC and DWB officers, meant '1' and '2' (operational) runways were to be opened up, in turn, to allow culvert installation. Apparently MRC people on the job diverted from this plan incurring -

'...the construction of 4 500′ of drain which will be about 5′ deep... (involving) an entirely unwarranted expenditure of funds and also tie up plant, (which) cannot be agreed to...'

No resolution appears 'on file', but drawings of REID RIVER show miles of open drains and no culverts under the runways! Perhaps the 22 BG and/or the US 5 AF had the final say!

Both the 2 BS and the 408 Bomb Squadrons converted to B-25 Mitchells during 1943. For a few weeks in May and June they shared the airfield with two squadrons of the 345 BG (M), a fresh unit en-route to New Guinea.

In May 1943 the HQ Allied Air Forces SWPA assigned to the RAAF, responsibility for construction and maintenance of, all airfields in 'Base Section 2', except those at CHARTERS TOWERS, REID RIVER, GARBUTT, WOODSTOCK, MAREEBA and IRON RANGE. They were to remain under the control of USAFFE. By October, three more of these, GARBUTT, WOODSTOCK and REID RIVER were also assigned to the RAAF.

After the 22 BG departed, the US Army maintained some guarding presence, sufficient it appears to minimise the vandalism and/or pilfering which occurred at other unoccupied airfields. By April of 1945, the RAAF was considering relocation of some NEI Airforce units and the RAAF's B-25 training flight to the Charters Towers area. REID RIVER was not as attractive as CHARTERS TOWERS to the RAAF because it lacked 'accommodation and storage facilities'. This move was eventually shelved because of the USAAF's adherence to its reserved right to re-occupy 'on 30 days notice'.

REID RIVER is one of the few airfields which the RAAF did propose early in 1946 to retain post-war. However, no further construction work of any relevance ever proceeded at REID RIVER and certainly the RAAF divested itself of the holding many years ago.

REID RIVER (24/7/1991)
*Loren Wheeler, caretaker of this property, is photographed showing Jennifer one of the old US camp building foundations that still remains. The prominent white post is an historic marker bearing a recently attached durable plate reading **'2nd Bomb Sqn Mess Hall established April 1942'**. This camp area is near the eastern end of runway 87°, between it and the river.* [Author]

FANNING

'...Contractor T F Woollam says "can't get timber, and (his) men refuse to go to Fanning". Decision of AWC - as they are men of the Civil Construction Corps, they will go where they are told!'

This is the essence of Allied Works Council Minutes around 15 December 1942. Work on clearing and levelling of strips at FANNING was commenced by the MRC, early April 1942. At the outset the assigned role for FANNING was 'repair strips' in contrast to other sites being developed as pursuit (fighter aircraft) strips and bomber strips.

P-O John Keays attached to the RAAF's DWB Melbourne, was heavily involved in the siting of this and other airfield needs. His comprehensive report of 14 April 1942, refers to exploratory survey work underway to prove a 1000′ eastward extension of the 'strip at the Christian Kruck mine', this extension being 'required to compensate for altitude and allow use of the strips by heavier aircraft in case of emergency'. Keays included a tabulation in which FANNING figures as 'Repair fields for Garbutt and Antill (sic) fighter fields'. 'Action taken' shows as -

'...clearing and grubbing (by MRC) is complete, levelling in hand (priority - low in comparison with fighter and dispersal fields).'

FANNING seems to have had an identity problem. With mistaken layout of its 'northern' strip to the tune of some 30°, its 70° cleared strip struggled on until eventually it was scrapped. P-O J W Campbell appealed to Air Board for co-ordination of strip identification in his construction progress report dated 25 May 1942 -

'...During the recent emergency rush for selection of sites and preparation of aerodrome strips,there appears to have been a considerable amount of duplication of effort and overlapping in instructions to surveyors, and of multiple control of the construction works generally and various officers giving different instructions.

FANNING (31/1/1943)
*This is the **'serviceability/status date'** on Drawing 42-43/591C. The comments re serviceability indicate that at that time the 95° runway was sealed, that 'dispersal works gravelling' was complete but that drainage had not then been commenced. Significant drainage work was done at this field and ample evidence remains to this day. The two indicated **'workshop'** locations, appear now only as bitumen sealed areas. The 'D' amendment to this drawing carries **'serviceability date of 30 November 1943'** and shows camp buildings enclosed by the strip and circuit roads, but not the two Workshop Buildings. Notice the bracketed name **'Cardington'** appearing above the airfield title. The more correctly termed CARDINGTON site(s) were about 25 km NE of here, closer to REID RIVER.* [AA BP 375 Item QL 680]

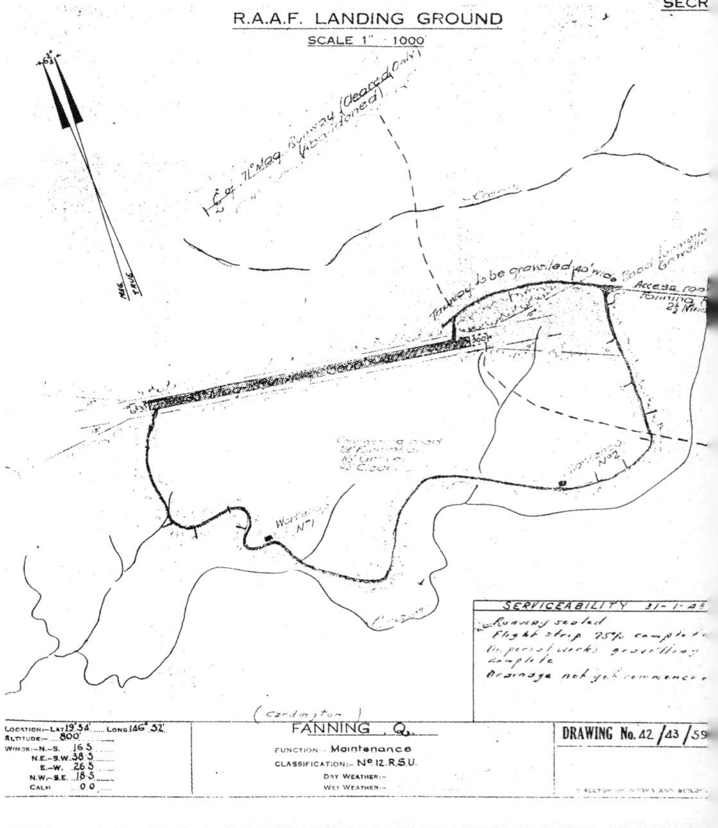

The USA Engineer Corps describes the strips or runways as A-A B-B and C-C, or as No 1, No 2 etc. The RAAF usually describe them as NE/SW or as 90° etc... (obviously) a standard method (is desirable). The two Fanning dispersal strips about 1 mile apart, are called by surveyors as Virginia Park No 1 and No 2, and No 2 is also called Christian Kruck in construction plans. Similarly other strips are called after the property name. To avoid confusion in the filing system, it is desirable that a standard method be adopted for naming landing strips. (For example its suggested these be called) Fanning 70° and Fanning 94°. Fanning is the nearest railway station.'

In my efforts to establish whether in fact the southern runway, FANNING 94°, ever received fighter or other aircraft for however short a period, I queried Rod Voller, then the Dept of Home Security's Camouflage Officer for NEA. While he visited there on 30 May 1942, both Rod's diary and his memory stop there. He feels FANNING probably did see some of the 'early arrivals'. Remarkably, I've not made contact with any veteran able to testify to any aircraft use of FANNING other than (surely!) some undocumented emergency or 'communications' use.

John Keays, by 1 August 1942, was posted to NEA as Area Works Officer. His review of progress achieved and work outstanding said -

'...The clearing and rough preparation of strips was carried out by organisations supplied by MRC and by the USA Corps of Engineers but with the opening up of the further fields to the north the USA organisations have been gradually withdrawn leaving MRC to complete all fields south of Coen. The original programme called for the formation of gravelling gangs that would operate independently of the clearing gangs and although this procedure has been followed, the progress shown by these gravelling gangs has been very poor. Gravelling was commenced at Charters Towers in March and in five months the following runways have been gravelled: Charters Towers 7°, length 7000′; Charters Towers 46° length 6000′; Ross River 5200′; Stockroute 3000′; Antil Plains 90° length 5700′; Fanning 4800′; Breddan 58° length 6800′. Whilst some gravel has been applied to other runways these are the only runways that could be considered for use in the wet weather conditions that can be anticipated next January and February.'

Indeed, on 3 August 1942, the US authorities are on record advising AWC of their priorities for sealing various operational aerodromes in Townsville and vicinity. 'FANNING No 1 runway' was listed seventh in a line up of thirteen. However, on 13 September 1942, the RAAF

FANNING *(23/7/1991)*
Taken from near the western end of the old runway 95°, this view shows some of the problem that livestock represented to the use of many of these airstrips during WW2. This field is located on a private holding known as Virginia Park. [Author]

halted construction on two aerodromes pending review of its operational needs. Fanning was one of them. A further month passed before the RAAF decided to form 6 ARD at BREDDAN and to shift 12 RSU from there to FANNING. So DWB was urged, 13 October 1942, to complete -

'... the buildings (at Fanning)... as quickly as possible.'

This was where contractor Woollam came onto the scene. His organisation was then stretched between work in MACROSSAN and Townsville. MACROSSAN being fairly close to FANNING, it was deemed sensible that his organisation be assigned this building work at FANNING. AWC Minutes show that Woollam was then the -

'...only large contractor from the south who had yet been given work (in the north).'

The work was allotted on a 'cost plus fixed fee' basis -

'...3% profit and 1½% overhead and plant.'

Throughout January of 1943 work proceeded on this site for No 12 RSU. Mid February 1943 a letter from contractor T F Woollam to the Allied Works Council suggests he was aware of a down-scaling of this role -

'...I wish to report that my project at Macrossan is 95% complete. Fanning has developed into tent floors, latrines and a few mess buildings which are well in hand. Townsville, including the extra 14 buildings to original contract is nearing completion and should be

FANNING *(14/6/1945)*
This extract from systematic mapping photography discloses a straight access road passing a prominent gravel pit halfway between the airstrip and the Charters Towers road. For reasons unknown the land owners have allowed this access to lapse. Note the abandoned 71° strip. [AUSLIG]

in 3 weeks. I have 130 men employed and offer my services for further work. I have been away from my home in Brisbane for the last 12 months and carried out defence work at Charleville, Rockhampton, Macrossan, Townsville and Fanning. I would appreciate a change, if it is at all possible, in the vicinity of Brisbane or Darling Downs...'

The RAAF decided to locate only one RSU near its developing Aircraft Repair Depot at BREDDAN. Hence 12 RSU went to the more advanced MACROSSAN site and the RSU role for FANNING lapsed. So this strip, one of several the RAAF had built to higher standards than the US authorities stipulated for their aerodrome needs, was suddenly relegated to an ELG role. With the completion of the water supply reticulation, FANNING reached 'standard operational base' category, and settled into limbo. Fencing to exclude cattle was considered in November 1943, but no action was taken as it was then doubtful the strip would be further maintained.

For a period in 1944 the RAAF's No 1 Signals Task Unit was in occupation here. Unfortunately no contact has been made with any veteran of that unit and it is not clear to me what role the unit did serve. By mid 1944, the Civil Aviation Department did not see a post-war need for FANNING and the RAAF eventually decided not to acquire the land.

A visit to the site in 1991 showed 'no buildings' and no remaining evidence of water storage tanks, although the 4" diameter asbestos cement reticulation piping was noted exposed in some areas.

FANNING *(23/7/1991)*
One of several junction pits in an extensive collector drainage pattern along the northeastern fringe of runway 95°. Note the exposed crushed rock backfill of the connecting lines making for reasonably effective 'soakaway' of surface water. [Author]

BREDDAN

'...Site A - this is situated on the Northern (Inland) Highway 9 miles north of Charters Towers, between Sandy Creek and Mitchells Creek...'

In July 1941, the RAAF began investigations to locate an airfield 'west of Townsville', in the Charters Towers area. The Civil Aviation Dept was also seeking to find and develop a site close in, but the RAAF searched -

'...all the country within a ten mile radius of the City of Charters Towers, as far as the roads which for the most part are only bush tracks, would allow...'

F-Lt J W Hurst reported on this search carried out over 30-31 July 1941 in company with Mr A Ellis, Dept of Air, Mr N Roeszler, Defence Liaison Melbourne and Mr A C Crawford, Dept of Interior Brisbane. Hurst considered two CAD sites in the northern suburbs but said-

'...these are at present unserviceable for aircraft and do not offer any prospect of development as first class landing grounds but could be improved for immediate use if required for medium sized aircraft'.

Hurst's report recommended site A -

'...(The Northern Inland Highway) is unconstructed (here)... and can be relocated without difficulty. Water supply is available (by pumping about 2½ miles) from the Charters Towers City water supply dam on the Burdekin River... (Electricity supply)... would be (by extension from Charters Towers by the private company Charters Towers Electric Supply)... in the event of a waterborne system of sewage disposal being required, full treatment would be required as the site drains into the catchment area of the Burdekin River from which the town water supply is obtained... Charters Towers is unsewered so arrangements could be made with a local contractor for clearance of pans if the pan system is adopted.'

Sites B and C, both to the east near Sellheim, were considered but deemed less attractive. The CAD's 'town site' was eventually chosen. When war with Japan came, the BREDDAN site was revived. It reappears in RAAF files as part of P-O John Keays' report of 8 April 1942 on fighter strip sites in the Townsville area and out toward Cloncurry. It said -

BREDDAN *(9/5/1945)*
This extract from systematic mapping, Run 5A Frame 1867, shows development near the end of WW2. Notice how Runway 13° was originally a straight section of the 'Northern (Inland) Highway', construction of which skirted the western end of Runway 60°. Notice also how at the eastern end of Runway 60°, advantage was taken of the cleared splay area to develop a sports oval. [AUSLIG]

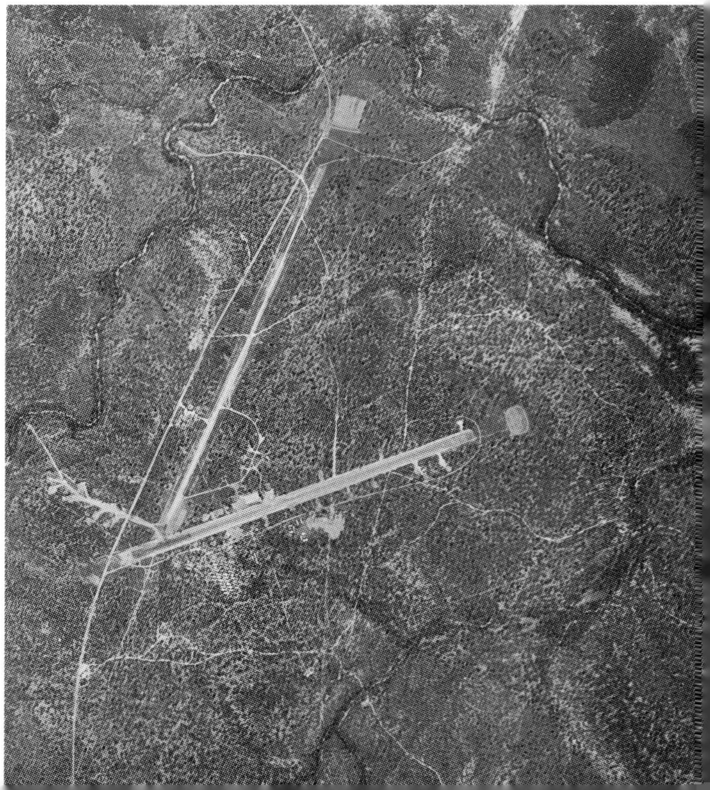

BREDDAN (20/6/1986)
While only part is visible, it will be seen that this is a substantial reinforced concrete structure surviving to present day. Its role was a Squadron Aid Post presumably deserving a more durable construction because of its medical role. It is located about 500 metres north of the eastern end of Runway 60°. The exploring party was Don Johnson on left, Jennifer Marks in the front seat with daughter Andrea and friend Leigh in the back and the author on the right. [Author]

'...Work will be commenced tomorrow on the clearing of a strip at Breddan which has been selected as a dispersal strip for Charters Towers. The second dispersal strip for Charters Towers has been located at Southern Cross...'

Work started on April 10. Four days later Keays said -

'...an E-W strip is being constructed at Breddan as first dispersal strip for Charters Towers, replacing the original strip selected at the cemetery site. Construction is to be in accordance with the original plans for the complete field so that if necessary it may be increased to a full field.'

Whether BREDDAN saw any of the early US fighter arrivals is doubtful. A-24 Dauntless dive bombers of the US 3 BG reached CHARTERS TOWERS late March 1942, before BREDDAN strip was started, so it is likely some of them visited BREDDAN in its basic state.

April and May were crucial months. Confusion reigned. At the highest levels of Allied command, MacArthur and the US Navy Force in the Coral Sea shared, less than fully, each other's thoughts on the imminent Japanese threat to Port Moresby. In North Eastern Area, RAAF and US officers struggled to achieve cohesive direction in providing urgently needed airfields. P-O Keays wrote to Air Board 17 May 1942, pleading for clarification. He had left Melbourne 1 May 1942 with RAAF requests spelled out as -

'...Interceptor Fighter Sqn:- Garbutt with dispersals at Bohle, Ross River and Maintenance Echelon at Fanning. Dive Bomber and GR/Torpedo Sqn:- at Charters Towers with dispersals at Breddan and Southern Cross and Maintenance Echelon - Balfe Creek.
RSU at Thalanga; ARU, RSU and T(LAND) Sqn at Prairie with dispersals at Eurunga and Penrice and Maintenance Echelon at West Prairie.'

US requirements were spelled out by a Schedule of Air Projects given to the RAAF by 19 May 1942. On 26 May 1942, after discussions presided over by the AOC NEA, directions were issued. Under a heading of CHARTERS TOWERS these included -

'...The heavy bombardment field is to be transferred to Breddan and it is requested that the Torpedo Workshop go in the vicinity between Charters Towers and Breddan. When the Bombardment Squadrons vacate Charters Towers the aerodrome will then be used for No 12 RSU, for the Base Section Pool and the Air Base Group. Camouflage works will be required...'

Further reassessment of airfield needs followed the American victory at Midway, early June 1942. Much of the programme west of CHARTERS TOWERS was abandoned. For BREDDAN the plan became one of completing both strips to gravelled stage and the erection of camp facilities.

Rod Voller was in charge of the Dept of Home Security's Camouflage Section in NEA. Rod's diary references show his earliest visit to BREDDAN on 29 May 1942. He was back there on 1 July 1942 for 'site selection for 'arch covers for aircraft'. Such arched hideouts certainly were built at CHARTERS TOWERS and elsewhere, but I've found no evidence of any at BREDDAN. An interesting drawing dated 19 July 1942 survives in AA. Its only title is ***Commonwealth of Australia, Department of Home Security, Camouflage Section - Suggested Dispersal of Aircraft Project.*** By legend, the 35 **sites chosen for immediate use on edges of runways** are shown and these appear on 'drawing 1259' (see Appendix 3) as clearings with 'irregular inclusions' which I interpret as other than arched hideouts. The 19 July drawing shows both 'workshops' built over by Sandy Creek on the western boundary, the MT (Motor Transport) workshop near the southern entry and only one camp site in the NE quarter of the airfield. A US Base Section 2 (Townsville) progress report dated 31 July 1942, mentioned -

'...Breddan... to be occupied by US Sqns and 12 RSU.'

On 30 July 1942, in Hawaii, two squadrons of the US 38 BG(M), the 71st and the 405th were preparing for their flight across the Pacific. John 'Hank' Henry's ***History of the 38th Bombardment Group (M)*** says -

'...After some weeks of rest, sightseeing and minor adjustments to their planes, the two squadrons took to the air proceeding to Christmas, Fiji and New Caledonia, arriving their final destination, Breddon (sic) Airdrome, Queensland, Australia, August 22. Upon landing they found their ground echelon waiting for them. The B-25s were armed, conditioned for combat and operational training was started.'

Much had happened since the ground echelon had sailed for Australia on 31 January 1942. They arrived in Melbourne with no real prospect of their air echelon following promptly. The men were dispersed to several centres according to need. At AMBERLEY and EAGLE FARM they assembled P-39 and P-40 fighters. Their 69 BS and 70 BS, equipped with B-26 Marauders, had been diverted first to the US action at Midway and then to New Caledonia and the Fiji Islands. It had also been decided to re-equip the 38 BG with B-25s so these two squadrons eventually went to another Group.

It is not clear when the 60° Runway was bitumen sealed. Progress reports of early August indicate it was programmed for completion by 22 August, the date by which most of the 38 BG's two squadrons of Mitchells had arrived. The Allied Works Council Minutes don't make any mention of activity at BREDDAN until 15 September 1942 and that relates to confusion as to which Department would be responsible for construction of 'USA camp at Breddan'. However, Lt Longland, a Camouflage Officer attached to HQ Queensland Lines of Communication Area, had toured north Queensland extensively late July 1942, and reported on, among other things, the appearance of BREDDAN flight strip from the air. His report indicated -

'...there are approximately forty buildings which urgently need camouflage painting'.

The two squadrons of the 38 BG had moved on to Townsville by the beginning of October by which time BREDDAN was being considered as the future site for the RAAF's No 6 ARD with 12 RSU to go to FANNING. On this basis the familiar winding dispersal taxiways were deemed unnecessary at BREDDAN.

By December 1942, the designation of the ARD unit for BREDDAN had changed from 6 to 13, which ultimately came into operation at the airfield. An impressive array of workshop, hangar, engineering and other buildings

BREDDAN *(20/6/1986)*
This shot again does not really do the building justice. It is one of the reinforced concrete 'igloo' shaped structures believed to have been stores for aerial torpedos. At the time of this visit the property owner uses them for private storage. It was only with the kind permission of the owner that inspection access to this airfield was gained. Current day access is doubtful as the present owner may be an explosive manufacturing or distributing concern serving mining needs in the district. [Author]

developed over the following months. Prefabricated steel Bellman hangars, to 1938 dated drawings, were valuable items in the RAAF's stores scene. Archive records tell a remarkable tale of four such hangars, delivered early in 1942 to Kingaroy. Seems some or all of this lot were reconsigned, by train, to Hughenden. During May, telegrams, cypher messages and other documents tell of parts off-loaded at Cloncurry, being eventually returned to Hughenden. There they awaited finalisation of location of what was to become No 8 Stores Depot MACROSSAN. Two units got erected there, but the others were railed to Townsville, apparently destined for Pt Moresby. Again, changes saw them returned to BREDDAN, 'less shortages'. By war's end it seems somewhere between two and six Bellman hangars were erected at BREDDAN.

On the taxiway running westward across the road towards Sandy Creek, clearly there were some workshops. Remaining concrete foundations suggest 'igloos' yet the 9 May 1945 vertical photos from 17 000′ indicate shadows typical of a gable structures like a Bellman. Unfortunately I've not managed sufficient contact with veterans of 13 ARD to clarify this matter. Similarly I've not located on-ground or other relevant photos. Surely there would be some in the extensive Department of Information photo holding frozen in AWM Canberra (refer Appendix 4).

To complicate this question of type of building, the following extract appears from an AWC Works file memo dated 22 December 1942, the subject of which is No 13 ARD Breddan:American Type Hangars -

'...it is now proposed to include in this project four in number 95′ x 45′ galvanised iron covered steel workshop buildings which have been reconstructed in Melbourne from material available from distressed cargoes...'

The constructing authority responsible for most of the work at BREDDAN was the Department of Public Works. Early in 1943 the record shows that the personnel of the RAAF's No 13 ARD were due to arrive on the site when significant building works were outstanding. Doubtless it happened elsewhere too, but at least BREDDAN is one case where the men of the incoming unit were given the task of erecting significant buildings as virtually their first task at the base. The AWC minutes record in early March 1943, a request by the Department of Air to agree to a proposal for service personnel to erect the 94′ x 45′ American type hangars. It was proposed the men of an advance party of 13 ARD would erect these 'hangars' and sundry other buildings, including some latrines. The AWC agreed so presumably 13 ARD personnel did this work.

The construction of the Torpedo Maintenance Unit's camp, workshops and store buildings, located east of the airfield, appears to have proceeded quietly because little evidence shows up in the AWC Minutes or elsewhere in archive material that I've located. Certainly the robust reinforced concrete igloo style torpedo stores, some nine in number, were, when I last visited in 1986, surviving monuments to the work of presumably, the Public Works Department.

Most veterans I've reached who remember BREDDAN, recall it as somewhere they visited during their time with 'Test and Ferry flight'. Bill Henderson, a long term resident of Charters Towers, has researched BREDDAN over several years. He has been instrumental in the construction and dedication of a memorial to the men and women who served at the base. He is in touch with several and plans to see their story is told in greater detail. To set this scene he once told me he believed BREDDAN was -

'..A typical behind the line base, used for the repair and maintenance of aircraft,... It was occupied by men and women who felt the war was passing them by. Morale was low - in fact, requests for transfer were very high. Barring the odd night in town, a dance, a sporting fixture and a visit by some 'Who's Who', it would have been a very mundane place to spend a war.'

Bill recently showed me the Memorial plaque in the front lawn of the Charters Towers RSL. Cast in bronze, the plaque has been mounted near a genuine pre-cast concrete aircraft tie-down block retrieved from BREDDAN itself.

MACROSSAN *(15/7/1943)*
*This is the **'serviceability date'** appearing on Drawing 42-43/1085C. Of note is the fact that the two **workshops** shown on the northern loop taxiway were built. Only the concrete foundations remain. The 69° runway was the only one developed and used and it remains serviceable for the Defence Forces of today. MACROSSAN is now a facility utilised by the Australian Army's 2nd Field Supply Battalion, headquartered in Townsville. It is therefore Defence Department property and permission to enter is required. [AA BP 375 Item QL 680]*

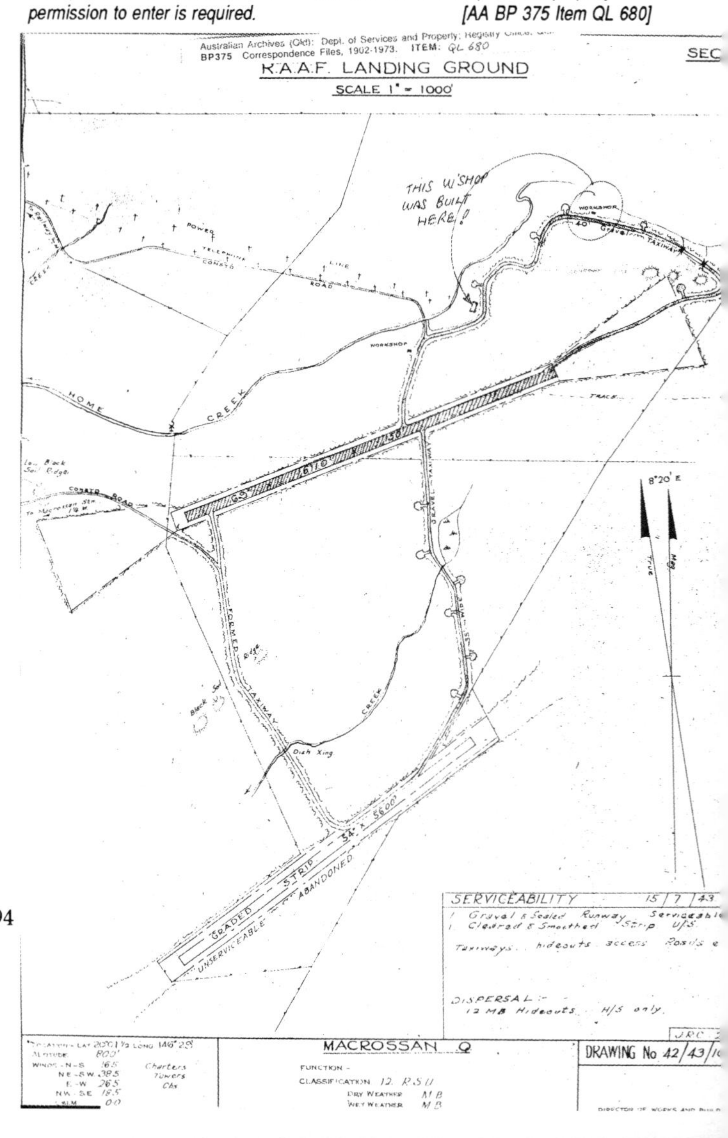

MACROSSAN

'...Concern was expressed by the Director General at... slowness of progress at Macrossan... It is apparent... in... the camp section, there have been delays due to the fact that designs have been changed by RAAF. In respect of the two big stores - Nos 50 and 51...drawings were supplied on 8 April, and there is no reason, other than...shortage of carpenters etc. why the work should not have been further advanced. (For) one building, all material is on the site and additional manpower is required to expedite the work...Allied Works would be justified in reducing the pace on certain storage proposals adjacent to Townsville... to ensure... 100 more men were engaged on Macrossan, which is an Aircraft Repair proposal and therefore justifies the highest of priority.

...Special consideration should be given to... construction of railway siding... supply of round posts... 92 (in No.) 60' rounds are required, and orders should be placed at an early date to secure delivery... Supply of the light cranes, in order that workshops may function as soon as buildings are completed... An electric drill has been ordered, but delivery not yet completed... Work can be expedited if an electric crane with a 75' jib could be provided... (It is) understood that a steam crane of this type was formerly used on the Agriculture and Stock Building, Brisbane, and should now be available.

...Please examine position... of Macrossan personally, and take any steps... necessary to ensure... completion of at least one of the two major buildings at an early date, and also... the camp, so that this may be available simultaneously with completion of the first major Workshop.'

This is the gist of a letter of 18 October 1943 from the Assistant Deputy Director General, Allied Works Queensland to that organisation's Supervising Engineer in Townsville. It concerns the RAAF's No 8 Stores Depot which was then urgently required so that Unit could vacate unsatisfactory operating premises in Townsville. This Stores Depot took a long time to 'settle'. Its need was envisaged during 1942 and various Bellman hangars (prefabricated steel framed structures) were railed to various parts of Queensland as planning stumbled along - to KINGAROY, then Hughenden, then Charters Towers area in general and finally to this MACROSSAN site.

MACROSSAN (13/6/1945)
This composite photo comes from prints 1876 and 1877 of Run 1A. The RAAF's No 8 Stores Depot buildings are grouped in the north west corner and the two airstrips of 12 RSU are away to the southeast. Of the two (300' x 230') large igloo shaped buildings, the western one remains to this day. The other was burned down in the mid 1970s. They were served directly by a spur rail line passing through the southern ends of both buildings. [AUSLIG]

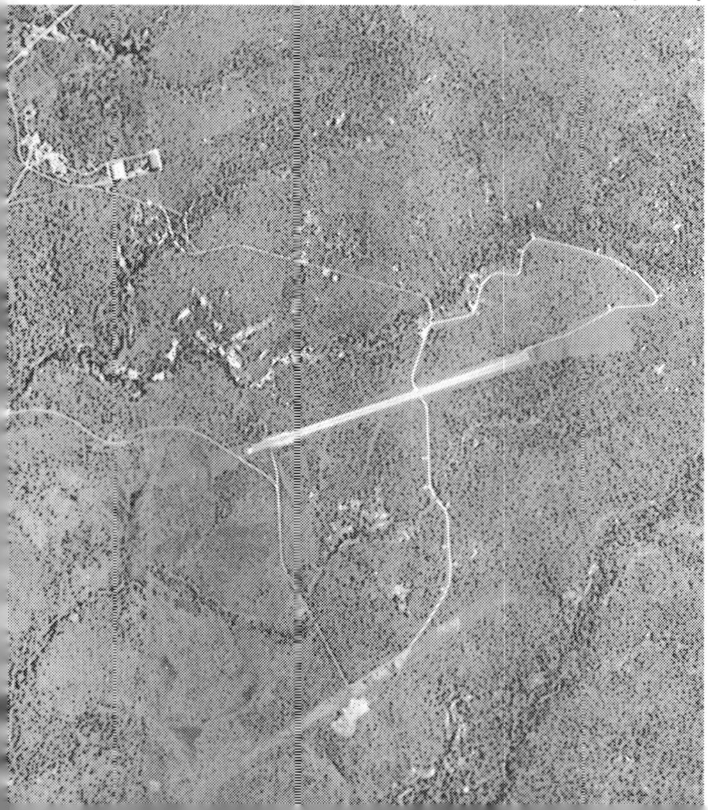

The airfield site itself is one of several identified and urgently commenced in the early months of 1942 to accommodate US Pursuit (Fighter) aircraft. Its proximity to a major river, the Burdekin, ensured its water supply for attendant camp establishments. As a report by P-O John Keays dated 8 April 1942 mentions -

'...Main Roads (Commission) clearing and levelling gang has completed four strips at Antil Plains, the clearing and levelling of two repair strips at Macrossan and clearing of two repair strips at Fanning.'

The RAAF's P-O Campbell visited MACROSSAN on 13 May 1942, some few days after the Battle of the Coral Sea and the transit through Townsville area of the US 8 Fighter Group aircraft. His report reads briefly -

'...Macrossan strip - these two strips are on flat country approx 1 mile apart. The 70° landing strip known as No 1, is rough graded, site depressions filled and rolled, full width 300' length 5 300'. Now being graded to proper levels, width 200' with slopes of 1 in 100 from centre line outwards... 2 500' of boxing completed, but in rough state. The 56° landing strip is similar to above but proper grading work has not commenced.'

Along with the other strips around Townsville, MACROSSAN shared in various requisitions for aircraft dispersal and camouflage works. Rod Voller was in charge of camouflage activities for the Department of Home Security in Townsville. He tells me he visited MACROSSAN on 30 May 1942, the same day as his earliest visit to FANNING. While Rod feels these fields were likely used in the receipt of the US fighter aircraft then accumulating round Townsville, he cannot confirm that MACROSSAN was so used.

As a result of a visit in June 1942, by General Brett, then Commander of US Air Forces in Australia, the urgent and somewhat haphazard construction programme for aerodromes along the Townsville/Cloncurry route was curtailed to some extent. However, for MACROSSAN, the instruction was that both strips were to be completed and gravelled and that a camp in line with the then established minimum building policy was to proceed.

The first mention of MACROSSAN that I found in the AWC Minutes came early in October of 1942 showing an RSU was to be transferred there from BALFES CREEK. Further, that the Department of Works had just finished a similar job at BREDDAN, and now a contractor, T F Woollam was to do such work at MACROSSAN. On 13 October the AWC Minutes mention -

'...Following decision by Air Board in relation to the location of maintenance units, the following aerodrome(s) are being constructed to RAAF standards (as distinct from USA standards)...(list includes) Macrossan (No 10 RSU)...as a single strip...'

Woollam was apparently one of the earliest, if not the first major contractor from southern Queensland brought into the north Queensland AWC activities. His immediate difficulties were the usual shortages of materials and equipment and manpower deficiencies made better (or was it worse), by the drafting of CCC men from further south. Generally, such draftees were disgruntled at being sent away from home to 'rough it' in often poor 'camp' conditions with the 'flies and the heat'. Add to this the numerous disruptions caused by the 'RAAF changing its mind(s)...' Woollam was 'given' a similar RSU and camp job at nearby FANNING.

While the cantilever style workshop buildings at FANNING weren't constructed, they do appear to have been completed at MACROSSAN. It seems Woollam was responsible only for the 'amended' concrete floors of these

MACROSSAN (23/7/1991)
Taken from the northern side looking to the south, this shows the remaining reinforced concrete foundations of the eastern one of the two workshops. The building featured a cantilever style roof capable of covering the front half of aircraft under maintenance. The isolated plinths in the foreground carried bracing members to the structure. Wrought iron tie down rings are still evident on the concrete slab work and presumably were available to tie down aircraft either against high wind disturbance or during some attitudes of maintenance. [Author]

two buildings, on the northern loop taxiway. The design was somehow changed and the buildings, as appearing in plan form on drawing 43-44/1261 endorsed as 'considered correct to 3 December 1943', were finally erected by the Public Works Department. Interestingly, contractor Woollam's progress report of 2 February 1943 shows against 'Buildings Nos 34 and 35 - Aircraft Workshop - building to be omitted, foundations to be laid and concrete floors to be laid for new plan'. Unfortunately, I've not managed to find any photograph of these structures nor any effective contact with personnel who served at the site.

By January of 1943 records indicate that the need for aircraft dispersal facilities for MACROSSAN in the RSU role, had been discarded. However, by then about 70% of such work was done.

Around 22 January 1943 the site for No 8 Stores Depot was 'in vicinity of Charters Towers'. Contractor Woollam wanted a change of work scene, and apparently it suited the AWC to allot the major work of the Depot to the Public Works Dept. It was 'established' in the Charters Towers area and its local officers were dissatisfied when a contractor from southern Queensland had been introduced back in November 1942.

The MRC was allotted site works for the Stores Depot on 2 April 1943. This included earthworks for the mile long spur line which the Queensland Government Railways (QGR) were to lay. While 12 RSU was preparing for its several month's occupation of the camp that Woollam built near the airfield, QGR and the RAAF were finessing the goods handling scenario for this spur line. As the AWC Minutes of 19 April 1943 say -

MACROSSAN (23/7/1991)
The northern end wall of the remaining large store building (referred to as No 50 during WW2) is seen with one of several WW2 period Bellman hangars in the right background. [Author]

'...shunting on the siding will be carried out from Macrossan and by "up" trains only, and delay to main line trains is anticipated...'

From the viewpoint of the RAAF's prosecution of its War, this 'delay' was nothing compared to all those to follow in the final emergence of 8 Stores Depot. AA files indicate a sophisticated project clearly intended to be there, post-war. They still contain quite detailed specifications with reference to appropriate drawings. In the case of the two large General Purpose Stores, buildings '50' and '51', the drawings are geared to 1941-42 financial year and they have an air of permanency. This contrasts with the many acres of storage and workshop space created at Townsville for the large US Air Depot. These recognised only the present war need. Perhaps behind the scenes, this difference in perspective prejudiced the scheduled completion of 8 Stores Depot.

Joe Mills was an Observer with the RAAF's 7 Sqn during 1943. Joe's diaries mention two occasions of flying into MACROSSAN. These were 31 August and 7 September 1943. Both trips involved delivery of Beauforts to the RSU. Joe recalls the second visit most -

'...It was the only way we could get leave - that was to ferry a kite south to Amberley or elsewhere. This time it was a 100 Sqn kite, called "Black Demon" - it was "Pranger" Harrison's kite. We picked it up at Breddan and we had to call in at Macrossan to collect some "records" for the plane. Well, something went wrong on the short trip over. We were losing height and we had no flaps and we were landing from W to E, pretty fast and the port undercarriage snapped and we spun around. We all got out OK and waited for someone to show up. We had finished up near a Ryan Trainer that had pranged only a day or so before and we walked over to inspect it...'

The airfield itself according to the Works and Services Branch property cards holding in AA, 'scored' a duty pilot tower around September of 1943. Apparently, it was not started by 1 October 1943 due to 'lack of a job number'. No such Tower appears on drawing 43-44/1216 so it may not have proceeded. According to RAAF records, 84 and 86 Interceptor Fighter Squadrons were here for a few months each during 1944.

MACROSSAN (24/7/1991)
Interior view of WW2 named store building No 50. Note particularly the heavy timber columns paired with gantry crane support columns. These column timbers were rough hewn from solid trunks showing to this day the broad-axe work of their shaping. The trusses are all timber with metal gusset plates. [Author]

BOWEN

'...Yes, J C Mathison, a very fine gentleman, a very big tall fella; he rang me up one day not long after I'd got settled here and he said, "Oh George, we want you to prepare a temporary slipway, to slip a Catalina, because it has to be slipped tomorrow at 2 o'clock". I said, you're talking to Bowen people here - there's nothing here to build a bloody thing like that. He said, "Oh well, do the best you can".'

This is how George Davidson recalls what was the 5 May 1942 urgent request for a slipway at BOWEN. George was then recently appointed Overseer to the Wangaratta Shire, which covered the area surrounding the small Bowen Town Council. While the requirement for this work was in Bowen itself, at that time the Wangaratta Shire and Bowen Town Council were working as one. This temporary slipway was built just on the northern side of the surviving town jetty.

The association of RAAF amphibious aircraft with BOWEN reaches back to around 1927 when Flight 101 operated Supermarine Seagull 111 bi-planes to provide aerial photography over parts of the Great Barrier Reef. The RAAF's 9 Fleet Cooperation (FC) Squadron operated a detachment of Seagull V Walrus aircraft here in the early years of WW2. With Japan's entry into the war, the suitability of Port Denison as an alighting area for the Catalinas came very much to the fore.

The AWC Minutes first mention BOWEN in terms of a 26 June 1942 requisition for a 'Flying Boat Base - A1 priority'. George Davidson's organisation did many jobs associated with the war effort. Most vividly he recalls the two other slipways he built. The first of these, a structure something like 1400' long, was prefabricated out of fairly heavy timbers, bolted together in sections and manhandled into place. The majority of it was under water. This slipway appears to have been that referred to in the General Index Cards, as '3 August 1942; Temporary Slipway near Datum Rock', at the Port end of Brisbane Street. George recalls the structure suffering badly from marine borers. The other more permanent design provided for concrete to below low water. It came later with the development of the

BOWEN *(26/6/1940)*
This oblique (FPZ 24 (18)) shows BOWEN eighteen months before the outbreak of the Pacific war. Viewed from the southeast the strip had not yet been lengthened and the road south from Bowen appears in the foreground of this photo. The railway alignment crossing the Don River in top left is essentially the same rail location of today. The small farmhouse in the centre foreground is on about the spot that the wartime building complex was later developed. *[CPE (RAAF Official)]*

BOWEN FLYING BOAT BASE *(24/6/1945)*
An extract from Run 1 Frame 4046, this enlargement shows the FLYING BOAT BASE in the centre with the pier, bottom right. The two heavy shadows on the eastern edge of the FLYING BOAT BASE are those cast by the twin cantilever workshop buildings erected for maintenance on the Catalina Flying Boats. Two 'Cats' are parked on the apron. At the pier, 'shape and shadow' make the vessel on the right a warship. *[AUSLIG]*

Flying Boat Repair Depot south of the pier. Commenced around October 1943, its construction tested the ingenuity of its builders. Though deteriorated through time and service, part remains to this day.

The conventional airfield was started by Bowen Town and the Wangaratta Shire Councils in 1939-40 with the aid of Commonwealth and State Government advances. DCA took over the site mid 1940 and set about its development. Grading was undertaken during 1941 as the field was to be one of the RAAF's Advanced Operational Bases (AOB) for 24 Sqn based at Townsville. By October of 1941, 24 Sqn still saw BOWEN as substandard to its requirements. Viewed in the context of other AOBs from COOKTOWN south to MACKAY, the plea was for attention to the -

> *'...poor water supply, no drainage facilities and inadequate showers and latrines. Storage racks, ventilation and petrol dump covers are also required. At present petrol is stored in a hangar and this will be required for its original purpose shortly...'*

Archive records of the construction of various RAAF hutments at Queensland AOBs during 1942, show 30 June 1942 progress on these items at BOWEN -

> *'...additional buildings and services - 92% complete - nearly finished. 12 splinter proof pens - work 40% complete. Obstruction barriers - work 80% complete...'*

Splinter proof traverse walls were erected around important installations such as the W/T hut and the W/R buildings at the airfield in June of 1942. These usually took the form of two timber sheeted walls tied together and filled between with earth.

BOWEN like other airfields on the Queensland coast, provided brief rest and refuelling stops for fighter planes transiting northwards to New Guinea. On occasions it served as a relief ground for Townsville. P-O Graham B J Robertson was a RAAF co-pilot assigned to the US 22 BG (M) flying B-26 Marauders. He was killed in a landing accident at IRON RANGE which US pilot Lt Walter A Krell survived. Graham's diary entry for 5 July 1942, describes a hurried exit from Port Moresby after a spell of 'ops' -

'...About twenty minutes out from Townsville we struck very bad weather at 13,000' and started to let down through it. When the altimeter registered 150' above sea level we found we were over land so turned onto a course of 90° to take us out to sea again. On that turn, taken at about 100' above the trees at 220mph, we passed directly between two hills and never even saw them. Being lost, turned South, following the coastline at about 100' and eventually struck Bowen, where we landed on an emergency runway with about fifty gallons of gas in each main tank and one hour's daylight left. We were forced to stay three days there and in the same clothes we had left Townsville in, so we were all very 'high' (as in body odour!).'

Air Board Agenda 4266 dated 4 September 1942 advised on the provision of an 'operations building' -

'...Approval was given on Air Board Agenda 4081/42 for the construction of a temporary hutted camp at Bowen Operational Base... For the efficient operation of aircraft from this aerodrome, it is essential to provide an operational building, and a site has been selected adjacent to the campsite. In pursuance of... policy... it is proposed to construct a semi-underground reinforced concrete splinter proof building, designed to withstand blast, bomb splinters, front gun attack, and to be specially adapted for ease of camouflage.'

Drawings survive detailing this building with its characteristic semi cylindrical shape. So too does Rod Voller's report of his visit in August 1942 to plan camouflage treatment for the camp. This embraced the 108' x 21' concrete operations building then sited on the NW corner of the camp area. The war progressed and policy changed. Work was deferred in February 1943. By year's end, the 'operations' role at BOWEN was integrated with a system of Key Communications Centres, the buildings being furnished in timber.

George Davidson well remembers the early work put into constructing splinter proof pens at the airfield. During November 1942, closer attention was directed to better dispersal on the airfield. F-O Jim Trench is on record concluding that 'dispersals should be on hard standings within the aerodrome boundaries', because of several constraints such as tidal marshland to the southeast and railways on the northern and western boundaries. Rod Voller's diary notes of 18 January 1943 support this -

'...with F-O Grounds we saw F-Lt Farran at RAAF HQ Townsville re the splinter proofing of "ready flights". He showed us an Air Force directive stating that all

BOWEN *(20/7/1944)*
'P 132' (oblique) looks to east southeast, with the rail bridge (left) and the road bridge of the day in the right foreground. It is easy to see development since 1940. Prominent are the seven or eight bomb splinter proof aircraft dispersal pens. These were considered necessary because of the paucity of forest area nearby in which aircraft could be hidden. *[NLA (RAAF Official)]*

BOWEN *(24/6/1945)*
This extract from Frame 4048 of Run 1 of the systematic aerial photography for mapping purposes, shows the airfield layout in plan with the main runway extended to the southeast. The earlier entry to Bowen from the south at this stage skirts the end of the runway and a narrow track can be seen leading southward directly below the airfield building complex near the intersection of runways. This track extends only a few hundred metres before an apparent dead end spur runs off to the southwest. Along that spur two substantial reinforced concrete bomb stores were built. These exist to this day. *[AUSLIG]*

aircraft dispersal positions on airfields on Cape York Peninsula as far south as Charters Towers and up to 300 miles inland from the coast were to be splinter proofed. This meant reconsideration of camouflage policy. F-O Grounds advised that all high RAAF Officers were of the opinion that overhead cover of some sort for aircraft was essential. In the meantime Works Dept had prepared plans for splinter proofs at such places as Bowen.'

One such drawing is undated and titled 'Camouflage to splinter proof aircraft bays:AOB Bowen'. It shows earth heaped against a sloping inner log wall and the scrimmed or garnished netting spread over these entire structures at a level of approximately 20' above ground. Systematic aerial photography over the airfield in June 1943 shows most of these top covers in place though somewhat ineffective. An aircraft is clearly visible through one!

Runway obstructions, such as heavy timber hurdles, so urgently arranged early in 1942 are shown on records to have been removed during February of 1943.

An interesting mention occurs in the RAAF files for BOWEN, Buildings and Services, dated March of 1943. It relates to a proposal, for an 'AML Teacher'. It's not clear to me precisely what the function of such a building is. Surviving drawings, dated 30 March 1942 indicate a form

BOWEN *(21/7/1991)*
In the rear of this residential property where in 1991 a dog boarding kennel operated, this surviving reinforced concrete 'bomb store' appears. Note the educt vents on the roof. Built around June 1943 the building was mounded over with earth to disguise its existence and for further protection. *[Author]*

BOWEN (21/7/1991)
This dilapidated structure is all that remains of the WW2 building complex. It was a passenger waiting room and flight administration centre. [Author]

of projector, approximately 15′ above a floor, able to project a cone of light to the floor, apparently in some role of instruction for bomb aiming. This is described elsewhere as a Bomber Teacher, in some cases as a 'half unit' and in others as a 'full unit'. A file reference dated 26 March 1943 indicates confusion in 3 DWO Townsville. The Officer sent to BOWEN found neither the COs of 11 or 20 Squadrons at the F/B Base, nor the CO of 9 (FC) Sqn at the airfield knew anything of the AML Teacher assignment. A handwritten note on his report says -

'...Bombing Teacher must be on Flying Boat Base, i.e. where crews are operating from...'

After all that, a December 1943 advice says the -

'...Bombing Teacher Mark 3 at present held at Bowen is not required... and instructions for its issue to Cairns will be given when the building (there) is ready...'

Owen Callard was a RAAF F-Sgt in 'stores section', posted there 13 August 1942. He recalls Datum Rock -

'...where the Catalinas came in for short overhauls. When they landed the ground crew would have their bathers on rolling out landing wheels (motor car tyres weighted down with water I think) to be attached to the hull both sides. Then the Cats would come up on the beach where there was a work hut...My time for 10 months was in the town itself and from various small buildings...eventually, 11 and 20 Squadron took over the picture theatre and we transferred our equipment. While I was there I was sharing a room in the Hotel Denison requisitioned by the RAAF. I remember opposite the Hotel was a ladies lingerie shop which was run by Mrs Caruthers I think, as my mate and I got to know her and usually were invited there for Sunday tea. Also we nearly bought the shop out to send undies to our wives. Being in the "stores" we didn't get out much during working days and when we did go down to the beach it was for a swim...
I am reminded of the Christmas I was there...it started raining before Christmas day and lasted until New Year's day, so much that the Burdekin River was in flood and two trains were held up at Bowen Station...'

BOWEN FLYING BOAT BASE (1945)
This low altitude oblique shows the extensive pavement accessed by the slipway disappearing below water level. Two Catalinas are evident, one positioned under the awning of a maintenance building. Of the buildings showing on the right of the huge parking area, some portions still remain to this day, incorporated into a Sailing Clubhouse. [RAAF Museum Point Cook (RAAF Official)]

In October 1943 some consideration was given to the redeployment of six of the several nailed wooden arch aircraft hideouts constructed at MAREEBA. The thinking was to relocate two to BOWEN as workshops for 9 Squadron. Walrus 'pusher engined' fabric covered bi-plane amphibians were being operated from the airfield, three of these quite clearly visible in the 18 June 1943 vertical photos. DWB cast doubt on any envisaged cost saving and certainly on gaining US and RAAF approval. The AOC, NEA by letter of 1 October 1943 said -

'...It is assumed that 9 Squadron are expected to make use of facilities provided for Nos 11 and 20 (Sqns). Admittedly, these facilities are inadequate for present commitments at this centre, but it is known that action is now being taken at RAAF HQ to provide additional workshop facilities at the waterfront. It is understood (such) development will eventually lead to establishment of Bowen as a Flying Boat Repair Depot...'

General Index Cards from files of the Department of Interior's Works and Services Branch Queensland, now held in AA, show an entry, 'concrete slipway - £16 000' against the date of 11 October 1943. AWC Minutes of 20 October 1943 indicate MRC to 'do the slip and the buildings to go to Dept of Public Works'.

During 1944, a detachment of 1 APU was in camp near the airfield. I'm not clear what it did, but in August 1944 it was followed by a Chemical Research Unit. The latter helped conduct trials at the Gunyara Camp near PROSERPINE. My delving into that WW2 period landing ground shows that, while it was used infrequently by communications and transport aircraft, DWB replied to 3 DWO Townsville, on 13 July 1944 saying it had been -

'...decided to operate aircraft carrying out trials for the Australian Field Experimental Section from the existing RAAF drome at Bowen. Therefore it will not be necessary to consider any construction at Proserpine site.'

Requisitions for 'additional Buildings and Services' including the two cantilever awning maintenance structures at the Flying Boat Base were noted as received in the AWC Minutes of 24 April 1944. Time for completion was twelve weeks, but this wasn't met. A photo, apparently dated 24 October 1944, appears on page 126 of Geoffrey Pentland's publication ***RAAF Camouflage and Markings 1939-45*** Volume 2. One of two Maintenance structures is shown in an early stage of erection.

The Catalinas are long gone and records show that one of the 'nose hangars' went to Rose Bay in Sydney. Port Denison Sailing Club makes good use of what's left. The airfield has been improved over the years and is now operated by Bowen Shire Council.

BOWEN FLYING BOAT BASE (21/7/1991)
The Port Denison Sailing Club uses the paved area associated with the old FLYING BOAT BASE and the launching ramp. The rear portion of this new building, incorporates part of a wartime building. [Author]

CHARTERS TOWERS

'...I joined the RAAF on 1 February 1942 without doing the normal School of Administration and was attached to the Department of Works and Building (DWB) at RAAF HQ, situated in the "Barracks", St Kilda Road, Melbourne - aerodrome section. At that stage I had never seen an aerodrome or had anything to do with one. Rank - Pilot Officer. The first job I was given was to design a brand new runway system for a proposed aerodrome at Charters Towers, construction of which was to start immediately. I was to work in conjunction with another engineer who joined the Air Force a day after me and who certainly knew no more about aerodromes than I did!'

So started Mr H V (Jim) Trench's written notes to me on his time in NEA Townsville. Jim described the airfield site

CHARTERS TOWERS (5/7/1942)
Early in the period of US occupation of this airfield, the 'Operations Tower' was cunningly erected between these two dense foliage trees as a measure of camouflage. B-17 Flying Fortress aircraft on at least one occasion are recorded to have 'buzzed' this tower causing the hapless occupant to jump out! Viewed from the north, the gabled dwelling behind the tent on the right is one of several flanking runway 9°. These were promptly taken over for airfield purposes. The truck facing the runway appears to be carrying fire extinguishers. [A Fraley ex USAAF]

CHARTERS TOWERS (24/5/1945)
This extract from Frame 1649 Run 2 of block photography flown for mapping purposes by RAAF Survey Flight, shows the intersecting strips of CHARTERS TOWERS airfield. The east west 'scar' on the left is, the 'cemetery strip' which served as standby during various construction phases on the main airfield. Centre right is the Racecourse, further developed and used to this day. [AUSLIG]

and his understanding that it had been selected by -
'...a gentlemen named Pettus who was attached to the American Army with the honorary rank of Colonel'.
A Mr Pettus figures in documentation around Christmas 1941. As a civilian with Sverdrup and Parcel, Engineering Consultants to the US forces, he had something to do with contour survey of the 'Town site'. The gentleman seems etched in Jim Trench's memory as, with a touch of cynicism, Jim says this site was chosen 'for camouflage purposes as the three 7 000′ runways, 150′ wide, would blend in with the 12′ wide town streets and be invisible from the air!'.

Jim Trench reached CHARTERS TOWERS around 4 February 1942. Drawings and schedules were sent to Townsville with a covering letter of that date saying -
'...this work is one of great urgency... requested that the NE-SW runway be graded first, followed by the N-S... minor details can be arranged on site between your Resident Engineer and this Department's Area Aerodrome Officer (P-O Trench).'
Because the plans failed to arrive, Jim says that together -
'...with a MRC engineer, we redesigned the whole thing based on what I could remember of the standards we had adopted in Melbourne. Construction went ahead... (later) the original plans and those prepared in Charters Towers proved to be almost identical...'
Reg Jennings was that MRC engineer and though he and Jim have probably not met since those war years, each recalls now the other's involvement.

The ***MRC History*** published in 1949 mentions in respect of CHARTERS TOWERS aerodrome, that -
'...In September 1941, sites... to accommodate heavy bombers and combat planes of the RAAF were examined in the vicinity... Work started in February 1942 with construction of temporary landing strips, two at the old football grounds and one near the cemetery for use while the drome proper was under construction.'
RAAF files in AA reach back to July 1941 and cover arrangements for those early inspections. One of the sites termed Sandy Creek, was later developed as BREDDAN.

Minute No 10 dated 10 January 1942 on a RAAF file gives a brief progress statement to that date -
'...telegram from NA (North Area) advising no plans or schedules relative to requisition 305 or 306 held - flight strip Cemetery site graded. Town site road runway 1500 yards graded also short runway at most SW-NE graded about 1000 yards. Urgent survey being completed by Main Roads.'

Air Cmdre Lukis, AOC of what was now North East Area, wrote to Air Board Melbourne on 7 February 1942 -
'..Reference... works on the townsite aerodrome at Charters Towers... considered that existing facilities should be utilised rather than proceeding with the erection of new buildings... submitted that all houses in the vicinity of the aerodrome should be taken over

immediately, in order that servicing of flying personnel may be housed as inconspicuously as possible... Reference... proposed AOB buildings it is considered that erection of these at such a late stage would not be satisfactory, owing to the fact that Fibrolite roofs are most conspicuous, and the formal shape of the buildings merely makes a better target for enemy action... requested... immediate arrangements be made to have the runways gravelled, and also to see to the erection of splinter-proof dispersal areas. This work is very important when it is considered that this is the only inland aerodrome which will afford strategic support to the main aerodrome at Townsville.'

While these preliminaries commenced in January, work on the 56° strip started in earnest on Monday 16 February. According to the MRC History, basic earthworks and gravelling were complete seventeen days later.

Alexa Gallop, in her biography of the MRC's Rollo D Gallop, ***Bush Engineer,*** drew on RDG's notes and recollections. Her treatment extends beyond the MRC History's oft quoted passage about the 10 p.m. Saturday (14 February 1942) phone instruction to mobilise -

'...At the time, Mr Engineer Reg Jennings was stationed at Mingela, supervising two road schemes in progress from Reid River. One Sunday morning he received a telephone call from District office advising that two special trains were being loaded at Townsville with men and machines for Charters Towers in order to build an aerodrome for the American Air Force, due to arrive in four weeks time... With the help of RDG a hurried trip was made to the Towers to prepare for the arrival of the special trains. Even though it was Sunday evening a cafe was commandeered and arrangements made to supply meals for up to one hundred men on their arrival. A building was leased to provide sleeping accommodation. The old Mining Exchange building was rented to serve as the Project Office. Buses were hired to transport the men and by 3 a.m. the first train arrived... Overseer Bill Crittenden took charge out at the site of the first airstrip (56°) and put the machines to work as they arrived...'

Through Alexa's biography, RDG identifies the involvement of several of the Local Authority and Shire Engineers - Frank Morris from Cairns City 'eased the building programme', Alex Black from Hinchinbrook Shire controlled a 'vast road building programme', and the construction of INVERLEIGH aerodrome, S Cochrane of Beaudesert Shire was 'given the construction of TORRENS CREEK drome and fifteen ancillary strips', Bob Hope from

CHARTERS TOWERS *(1942-43)*
This more permanent control tower replaced that built between the trees. Both were located in the southwest quadrant of the airstrips. *[A Fraley ex USAAF]*

CHARTERS TOWERS *(ca 14/9/1942)*
Navigator Jack Woodward (left) and pilot Jack McMasters, two of the three man crew of an RAAF 22 Sqn Boston DB-7B. Taken shortly after arrival from Richmond, NSW, Jack assures me the 'things that look like beer bottles' on the ground under the port wing are in fact part of their flying gear! *[J Woodward ex RAAF]*

a NSW Shire was 'placed in charge of BREDDAN, and later JACKY JACKY'. Overseer Bill Crittenden was later 'transferred with men and machines' to IRON RANGE, working under Clem Wilson's leadership.

Common belief is that all this work at CHARTERS TOWERS was being done for the American Air Force. The RAAF was however busily developing its '73 Squadron Plan'. Air Cmdre G Jones, in a document of 11 March 1942, addressed the need for an orderly approach to site selection of dromes. While Sqn numbers were not fully assigned, the accompanying tabulation proposed locations of the full array by Unit/Type. Noted as 'already in existence', 24 DB (Dive Bomber) Sqn was due to be 're-armed' and located in the Charters Towers area. Stated as 'in existence', 100 GR/T Sqn was slated there too. Jones added -

'...Wherever possible, use is to be made of existing roads and main highways as the basis on which to construct landing strips, the normal traffic being diverted or handled otherwise as appears suitable. It is again emphasised that this work is to be pressed on with the utmost speed, as any delays may have a vital influence on the defence of the country.'

A file Minute of 17 March 1942 shows Jones directed -

'...site at Charters Towers should be developed immediately for occupation by operational units. In accordance with our siting of new units under the 73 Sqn plan we will require this site for No 100 G.R/T. Sqn and No 24 D.B.Sqn. Will you therefore proceed with all possible speed to have this work put in hand in accordance with specifications laid down. Torpedo need not be included for the present.'

History shows 100 Sqn didn't come up to Queensland until May and 24 Sqn stayed at STOCKROUTE until June. All four flying squadrons of the US 3 BG (L) are recorded as stationed at CHARTERS TOWERS by 17 March 1942, the day General Douglas MacArthur landed at Batchelor NT, on his way to Melbourne. This was 'teething time' for the USAAF and the RAAF's working alliance. A signal from NEA dated 21 March 1942 advised 'one runway 4 700′ ready for use'.

'First use' of the 56° runway is uncertain. It ranges from an unspecified aircraft landing 'on the first runway sixteen days after commencement' (i.e. 4 March) to an 'A-24 flown by Major 'Big Jim' Davies, on 25 March 1942'. The first appears in Australian Parliamentary papers, the 7th Progress Report of the Joint Committee on War Expenditure, 27 November 1944. The second date is mentioned by Noel Tunny in his 1991 publication ***Gateway to Victory.*** The journal of the US 435 BS, 19 BG kept by Lt Walter H Johnson records on 21 March 1942 -

'...at ten minutes to one, Townsville had its first air raid warning. One lone Jap, four motored, bi-winged plane, probably taking pictures, was sighted. Men dashed to their ships and others to slit trenches. Three B-17s took off headed for Charters Towers inland 80 miles. The runway there is reported ready. When the alarm

was over Major Carmichael contacted the planes and instructed a return...'

Again on 23 March, Johnson notes all planes were *'...dispersed to Charters Towers and Cloncurry...'*

James A Simpson was with the US 47/479 Service Squadron, part of the 35 Air Base Group (later re-designated Service Group). As chronicler of a 1985 Reunion book, he sheds further light on this aspect. Soon after berthing in Brisbane on the USAT 'Ancon' on 25 February 1942, men from the convoy arrived in Charters Towers, on a US troop train, 2.30 a.m. 10 March 1942 -

'...The 47th Materiel Sqn (later redesignated Service Squadron) and the units of the 3 BG followed on succeeding trains... Upon arrival, troops were carried by Australian Army trucks, procured by Captain Kane, as Advance Echelon Commander from nearby Camp Selheim, to the campsite some 5½ miles outside of town. Since no tents were erected, everyone stretched out on the ground awaiting the dawn, and the birth of a new experience. It was a rude awakening, mostly caused by ants... Areas were selected for each unit, and by supper time... filled with tents... The Australian workmen (Main Roads Commission) had begun to lay water mains to the various camp areas...The runways had not been completed... On Saturday March 21, the first planes landed. They were A-24s manned by pilots of the 27 BG who had originally come to Australia to pick up their planes to fly them back to the Philippines and resume the fight against the Japs. There were at least a dozen ships in the flight led by Lt-Col John H Davies, whom we had known in Savannah... These men were assigned to the 3 BG and Col. Davies assumed command... Within a few weeks a stream of B-25s arrived... (They) bore the insignia of the Netherlands Government, and had probably been originally meant to be used by the Dutch. However, these and other B-25s were soon... flown by the 13 and 90 BS... the 8 BS and the 89 BS flew the A-24s. Soon A-20s arrived and the 89 BS flew this type of plane....'

Noel Tunny's ***Fight-Back from the North*** notes the arrival of the first batch of B-25s as 'inauspicious' with the Corinda ridge hump in the 56° runway claiming damage to three aircraft in landing. Tunny also says of the 3 BG's A-20s, the first got to CHARTERS TOWERS in June. Others followed after their assembly and weapon reconfiguration were completed at AMBERLEY, 'under guidance of "Pappy" Gunn and Robert Ruegg'.

Perusal of the RAAF Operational Record Books (Form A50) for NEA HQ (Townsville) show an early reference to CHARTERS TOWERS as -

'..April 5 - CT 1 (origin of sortie Charters Towers - Sortie No 1 for that date) all serviceable aircraft (B-25's) 3rd BG proceed Moresby attack Gasmata 6/4 return to Cooktown. Five out of six aircraft arrived, airborne 0415Z/5.'

P-O J Keays in a wide ranging report on the selection of sites for new aerodromes, Townsville area, dated 14 April 1942, records CHARTERS TOWERS as -

'...Whole scheme provides for four runways involving heavy earthworks with metalling and pens, bitumen surfacing 150' wide...One strip (is) completed and base course metal laid; this strip in full operational use by all types of aircraft. Another strip very well advanced. Some doubt expressed by US senior officers as to advisability of completing full scheme.'

On 20 May 1942, CHARTERS TOWERS came in for some mention in the camouflage reports. Personnel on a Committee, meeting in Brisbane, were listed by names -

'...Minutes of 27th meeting of SDCC Qld - those attending R A McInnis (Chairman), R P Cummings (Civilian Member), J D McColl (Dept Interior), Lt-Com V W Bowden (Navy), Lt A T Longland (Army), A B Leven (State) and Dr W H Bryan (Specialist Advisor), W Cdr Hartwright was unable to attend...'

Money for 'camouflage treatment, including tents' was recorded. Mr G Rae was mentioned as the man handling CHARTERS TOWERS matters and, further that -

'...600 gallons of paint have been sent to this drome.'

Dale Kruger came to Australia in 1942 with a group of US Army 'casuals'. While his memory of detail is faint, he recalls eventually 'training it' to CHARTERS TOWERS on assignment to the 350 Service Sqn (Provisional) early September 1942. Dale wrote of the experimentation on added 'front guns' for the medium bombers. Simpson's Reunion book relates how the 30 cal. -

'...guns in each side of the nose section of the A-20s weren't proving to be sufficiently effective. After some experimentation by the Engineering Sections of the (35 Service) Group and the 3 BG, T-Sgt Victor J Mitchell of the 47 Materiel Sqn was instructed to draw up a plan for the installation of four 50 cal machine guns in the nose section of this plane on May 1, 1942... he presented his drawing and an A-20 was given him... By working day and night he overcame the many obstacles... his installation... proved very satisfactory... Mitchell was later awarded the Legion of Merit for his work... Experimentation was also carried on with 50 cal machine guns in the nose of the B-25, both in Charters Towers and in the various Air Depots in Australia and in the States... (leading to) the B-25 "Strafer".'

Dated 10 June 1942 there is in RAAF files an interesting 'note' by General Brett on a visit to northern areas. Bearing on CHARTERS TOWERS it states -

'...an immediate need for an Engineering Officer to be detailed to work the 35th Group under Col. Davies. Check on the detail of 20 men out of the 89th at Archerfield that were placed on erection, but now have been placed on other various jobs...'

Walter Maiersperger's log shows his posting from HQs AAF US Air Command No 2 in Townsville, to the 35 Air Service Group in CHARTERS TOWERS, as Engineering Officer (Eng O) effective 20 July 1942. Walter recalled Paul I 'Pappy' Gunn as one who claimed -

'...to have been responsible for converting the B-25 bombardier nose into a gunnery attack nose. He may have been a Navy enlisted man during WWI. At some time he became a pilot and was an executive in the Philippine Airlines before Pearl Harbor. One day he showed up in Brisbane and became a Major and rated pilot overnight. He was attached to the 3 BG... but flew missions wherever it suited him and seemed impervious to the control of local commanders in New Guinea or anywhere else. His modus operandi was to fly one or more missions and then fly back to Brisbane and report to Gen Kenney. Nobody else knew what he reported but naturally, all commanders seemed afraid of him, though they played up to him as if they weren't. As the 3 BG was based on Charters Towers at the same time, I caught some of "Pappy's" sarcasm and ridicule from time to time. He served as Gen Kenney's point

CHARTERS TOWERS *(29/7/1943)*

'P 21' is an oblique of the airfield, viewed from the northeast. The town of Charters Towers can be seen on the left hand side of the photograph roughly on the extension of the north-south runway 9°. *[CPE (RAAF Official)]*

man... He was very brave, took the fight to the enemy and was very lucky in not being killed until the war was over, by trying to fly "contact" in IFR weather in the PI. He took the President of the PI with him in that crash. A flying school classmate of mine, Maj-Gen Bob Ruegg, ret., survived the rout from the PI and (was) the lone survivor of the A-24 attack to stop the Japanese landings at Buna. In the Spring he claimed to have... made the gunnery conversion in Brisbane. The first I knew of this conversion was when these B-25s arrived at my 35 Service Group in Charters Towers with ruptured hydraulic lines to the gun chargers. My men educated me to the fact that soft copper tubing had been used instead of hard copper tubing, hence the ruptures. My men replaced all the tubing and I wrote an unsatisfactory report (UR), to HQ 5th AFSC in Brisbane, who sent it to the newly formed 81 Air Depot at Eagle Farm for reply by endorsement and correction. This made me an enemy of its Eng O, Frederick G Hoffman, Class of 39C, two classes ahead of me. In 1947 he became my deputy as chief of the Special Projects Section, Engineering Division at Wright Field and my good friend until he died about two years ago. Fred claimed to have originated this conversion idea in the nose of the B-25... I believe him. He was a graduate engineer, a former Wright Field power plant engineer, extremely well versed in armaments and politics. He had been the Eng O at Nichols Field, PI and the last transport, a C-33 (early version of the C-47) to leave the PI with a double load of US Army nurses, at night to Darwin. Fred claimed he used soft tubing because that was all that was available, and the conversion was needed urgently to stop an enemy landing. Besides, the soft tubing did the invasion job so what was the UR for anyway? He wanted credit for his work, which he never got. In his book, Gen Kenney gives the credit to "Pappy" Gunn.'

The 'igloo' structure was part of the mid 1942 scene and onwards at CHARTERS TOWERS. Rod Voller, then in charge of Camouflage activities in Townsville for the Dept of Home Security, had earlier met the 'igloo' designer, Emil Brizay. Rod's notes suggest -

'..Brizzay (sic) did not appear in the (NE) Area to my knowledge. I went to Charters Towers with Hughes (US Army) on 17 June to see about using the "shelters"...'

Used initially as hideouts they became more of a sunshade over maintenance operations on aircraft. Some were developed further to full-blown roofed hangars.

Air Vice Marshal Bostock, Chief of Staff in a 19 June 1942 memo to the Chief of Air Staff RAAF and the Commanding General USAAS, reported further on 'operational aerodrome requirements in NEA'. For CHARTERS TOWERS it was to be -

'...one light bombardment group. Note: this unit will be moved when Portland Roads site is finished, and then this field may be available for other than operational use. It is requested that you should indicate your requirements in regard to aerodromes for maintenance depots, RSUs etc., utilising ... either existing grounds, or grounds already in course of preparation.'

This foreshadowed clearly the cessation of what had been until now the urgent identification and development of numerous airfield sites in north Queensland.

An excerpt from ***Anti Aircraft Artillery Activities of the Pacific War*** published in 1946 by GHQ US Army mentions some start on such defence -

'...in June, Regimental HQs with HQs 2nd Bn Batteries E (less 1 platoon), F and G of the 94th CA (A-A), and with the Australian 32nd Light A-A Battery, and 64th A-A Company RAE attached, established a defence of the airstrips and repair facilities at Charters Towers...'

In my search for hard evidence of the presence and location of anti-aircraft defences on CHARTERS TOWERS I found drawing T331 in AA Accession J-153. Dated 29 June 1942 it is titled - *US Camps 94th CA/A-A (Coastal Artillery/Anti-Aircraft). Relative positions of batteries Charters Towers (not to scale).* South of the field in a rectangle marked simply 'Park', a group of hutments is figured as '2nd Battery and HQ'. North of the airfield, batteries 'E' and 'F' are shown. Continued searching in RAAF files turned up a 12 July 1942 drawing headed *Commonwealth of Australia - Dept of Home Security (Camouflage Section) - Suggested Dispersal of Aircraft, Project 105.* Siting for sixty-six aircraft from fighters to medium and heavy bombers is shown, mainly north of the field. It also shows 'F' battery, on the right of the 5 chain road north, near Sheepstation Creek. Current 1:100 000 topo mapping (8157) Grid Reference is 245869.

An added bonus appeared in the signature on this drawing; G Collings, Senior Camoufleur, Section 5. Certainly Rod Voller recalls Geoffrey Collings and so does Walter Maiersperger -

'...he showed up when I was stationed at Charters Towers and we hit it off. As the operational units gradually moved north he did too, while I was sent to Sydney. To my surprise, he showed up there and had an exhibition of his paintings. He asked me to make the opening address. The request floored me. In my view A-R-T is overrated. However, I liked Collings' style of painting, and he was not ARTY. So I agreed... Collings was 'in' with Sydney's society. The Lord Mayor's daughter attended and others of her social set, all of whom I met at the reception afterwards. I can still see their faces as they looked at me and wondered what this Yank airman had to do with ART!'

Though I've not yet found the paintings, if indeed they still exist, Walter's speech survives. Clearly the paintings related to IRON RANGE. I've been fortunate in locating Geoffrey Collings. While he remembers all this, his WW2 period graduation to a new 'love' in documentary cinematography sits more prominently in his memory.

By 21 June 1942 instructions for airfield development at CHARTERS TOWERS resulting from a recent visit by Gen Brett and discussion with the AOC, NEA, were

CHARTERS TOWERS *(29/7/1943)*

This is Frame V 374, one of a systematic coverage under KD 832. Noticeable are lines of aircraft near top right and mid right on the south side of runway 56°. The latter are close to one of the nailed wooden arch hideouts. Camouflage net covered, it served as concealment for fighter servicing. The narrow building immediately south was an Armoury. On the western edge of runway 9°, two dense foliage trees blend to one. They shrouded the first 'Operations Tower'. The 1943 'Control Tower' is further towards the runway intersection. The northern one of two Igloo Hangars in bottom right, remained for many years post-war. The RFDS used it until it was destroyed in a freak storm, just a few years after its listing by the National Trust in 1984. *[NLA (RAAF Official)]*

issued. They were simply to 'seal one runway and complete gravelling of the second'. It appears sealing of the '45°' runway did not take place until around 31 July 1942. A US Progress Report of that date mentions -

'...7° runway, gravel- condition good; 45° runway, tar seal-not yet useable. US Squadron operating.'

In this respect the two strips at CHARTERS TOWERS were further advanced than many others in the Townsville area. In mid 1942 alarm bells were rung in respect of the need to get some twenty strips gravelled and thoroughly useable by the next wet season. Resources from other parts of Queensland and outside were brought to the area to augment an organisation struggling to achieve this goal.

At some stage after the bitumen sealing of the main runways at Charters Towers, Reg Jennings recalls mention in some American newspaper of an aerodrome 'paved with gold'. The story turned round the MRC having crushed local gold mine mullock heaps for screenings. The mine owners moved to sue (presumably the Government) for unauthorised use. In the end, sampling indicated insufficient value to warrant 'retrieval'. This is all quite apart from the sure fact that the Supreme Command would not for a moment have allowed such an interruption to the operational use of this airfield!

The following excerpt from a report dated 10 August 1942 by Lt A T Longland, an Army Camouflage Officer, shows a realisation of the construction difficulties associated with access to camouflaged hideouts -

'...hideouts and dispersal areas or roads, not proceeding as OC Aerodrome states that completion of runways is first priority and there is no labour available for other secondary work. It is believed that Camouflage adviser, Mr Hughes of the US Army is to supply pre-scrimmed netting for covering of arch hideouts...believed it has been laid down that all hideouts must be situated at least ½ mile distant from runways and at least ¼ mile apart, thus 20 or 30 hideouts would cover a large area of country which must necessarily be well timbered. In addition... many miles of roadway, at least 100'in width will be required to give access to hideouts and roads must necessarily be of sound construction sufficient to carry heavy rolling loads... It is believed that Messrs Hornibrook has a commission to build 20 hideouts at Charters Towers drome. These are in course of construction and it is understood that one has been completed. Camoufleur Adams is giving attention to the siting of such hideouts but the position as regards access for (their) future use... is not at all clear...'

F-O J W Campbell's 11 December 1942 'Dispersal Taxiways' drawing 42-43/27A shows the great relaxation of these constraints which did occur.

Lt Longland addressed several other sites, including -

'...No 223 Australian Light A-A Charters Towers - good work apparently done by (Camoufleurs) Messrs Stubbs, Adams and Dorey. A variety of covers and frames have been tried out over the A-A gun pits situated around the drome with a good measure of success. Arrangements will be made to forward paint for 46 tents in Area.'

Rod Voller visited CHARTERS TOWERS again with Mr Hughes on 24 September 1942. His diary notes that he saw a 'heavy bomber arch structure' at the airfield.

CHARTERS TOWERS *(ca 1943)*
Looking approximately southwest through a nailed wooden arch hideout being utilised for fighter servicing. The building in the background is shown on various drawings as 'Armoury'. [US Photographer Halbert via A Fraley ex USAAF]

CHARTERS TOWERS *(ca 1944)*
With the Armoury building evident on the right hand side the 'arch hideout' has by this time been replaced by a prefabricated steel 'Butler' hangar frame complete with canvas tent inner. [A Fraley ex USAAF]

Donald Englund was with the US 319 BS, 90 BG flying B-24s. His recollection of the airfield is undated -

'...we spent several days there while the air crew of Capt Holmes practised night take-off's and landings. We were able to go see a movie and while at the movie we sat in the open air and were situated so that we were in the flight path of our plane and we could see them go overhead to the airstrip. It was a chance to eat food in a restaurant, and the fellows who wanted a drink found a local lady who could supply their needs at a price that caused one of the fellows to dab a few drops of gin behind the ears. I wasn't a drinker, but I guess we all thought being there was worth the price...'

This might have been after the 17 November night take-off accident at IRON RANGE. Donald added that -

'...Holmes' crew must have benefited from the experience since they always made it safely home from the missions they flew.'

Dated 16 December 1942 Minute No 68 on a RAAF CHARTERS TOWERS file, is brief. It simply states -

'...Letter from 3 DWO To Messrs Chesterfield & Jenkins, Roads Com. at Charters Towers, advising trees should not be cut down which is contrary to Camouflage Regulations, and (that the contractor's) tents were the only white ones in the area.'

Chesterfield & Jenkins were principals in a construction company of that name. They were responsible for a number of airfield and other defence works over much of Queensland and remained active for some time after the war. It's interesting to note their apparent flouting of the camouflage requirements of the day in respect of their 'white tents'. I first met 'Chesty' in the early sixties on a job outside Maryborough, Queensland. He rolled up to a bitumen job in a Rolls Royce!

On 19 July 1943 an American request was made for some £20 000 worth of further paving and bitumen sealing works to taxiways, 'fighter areas' etc at the aerodrome. The RAAF file shows a 20 July 1943 Air Board Minute which tends to highlight some 'problems' associated with RAAF work as distinct from US requirements -

'...The Board understands that the commencement of any new works for US Forces, in practice, depresses the construction of RAAF works in the same area. Approval has been given to the construction of an ARD at Breddan and a Stores Depot at Macrossan... Board is concerned that the work it is now recommending (for US at Charters Towers) may militate against similar works at Breddan and Macrossan for the RAAF.'

Air Board's approval to this A1 priority work was surely a foregone conclusion. Still it recommended that -

'...this matter be represented to the appropriate authority with a view to ensuring that the works of the

ARD Breddan and the Stores Depot Macrossan are not adversely affected by this new project.'

As an indicator of the cost dissection nightmares already arising, Air Board's recommendation added -

'...In view of the fact that Project No US 19-Charters Towers Aerodrome Qld - is now a sole US project, the Board recommends that the expenditure on this project be transferred to sole US account.'

The base served throughout most of 1943 as a vital US Training and Replacement Centre. Withdrawal was signalled with US HQ FEAF's letter of 19 June 1944 -

'...The Fifth Air Force intends to withdraw completely from Charters Towers. The airdrome, however is being classified as "reserve" airdrome, which means that it may be used at a later date for the reception of additional Air Force Units. This... will mainly consist of staging, and will in no way interfere with daily operations of RAAF units. If it is...required, your HQ will be so advised at least 30 days in advance. All facilities left by HQ FEASC except building G15 (Water Purification building) and miscellaneous transformers and wood tent floors, may be utilised by RAAF.'

The RAAF resumed responsibility for construction and maintenance of the Airfield on 20 May 1944 and DWB naturally asked direction for the future. Gp Capt Walters, then DASP, replied, 4 September 1944, saying -

'...there is distinct probability of its being occupied, during 1945, in the event of British Forces being based in this country. The Prime Minister has declared and it has been published in the press, that such forces are to be expected here...'

The *Marching In/Out Certificate* on file, shows date of occupation by RAAF of CHARTERS TOWERS aerodrome and bomb dump area as 11 December 1944. It also shows these areas being vacated by the 35 Service Group (USA) on 10 March 1944 and the Chemical Warfare Supply (USA) on 7 December 1944.

Around April of 1945 when the RAAF was looking to locate the Netherlands East Indies (NEI) Equipment and Personnel Pool and its own RAAF B-25 Training Squadron at this airfield, it was made clear by US authorities that they still reserved the right to re-occupy 'on 30 day notice'. Instead, the NEI Pool went to BUNDABERG. However, the US reservation persisted as this 14 August 1945 File Minute indicates -

'..DWB TO DASP and DO. Your attention is invited to enclosure 149A wherein it is advised that Charters Towers aerodrome will probably be used by USAAC for temporary emergency use by B-29 aircraft...'

While it appears at least one B-29 Superfortress aircraft visited Townsville on 30 October 1945, it's probable that none ever visited CHARTERS TOWERS.

January 1946 the Dept of Air advised DCA that the RAAF planned to assume responsibility for CHARTERS TOWERS, post war. That has long since lapsed and today it is operated by the Charters Towers City Council.

CHARTERS TOWERS *(22/4/1991)*

Located halfway along the 5?° runway and on the northern edge, this large circular concrete pad of the Compass Swinging Platform is nearly all that remains of the wartime structures. Like the minute hand of a giant clock, a wooden member was advanced to cardinal point settings. With both the aircraft's main wheels breasting the wooden member, compensatory adjustments were made to the onboard compass. Pictured on the right having shown the author around the area, is Bill Henderson a long term resident and military historian of Charters Towers. [Author]

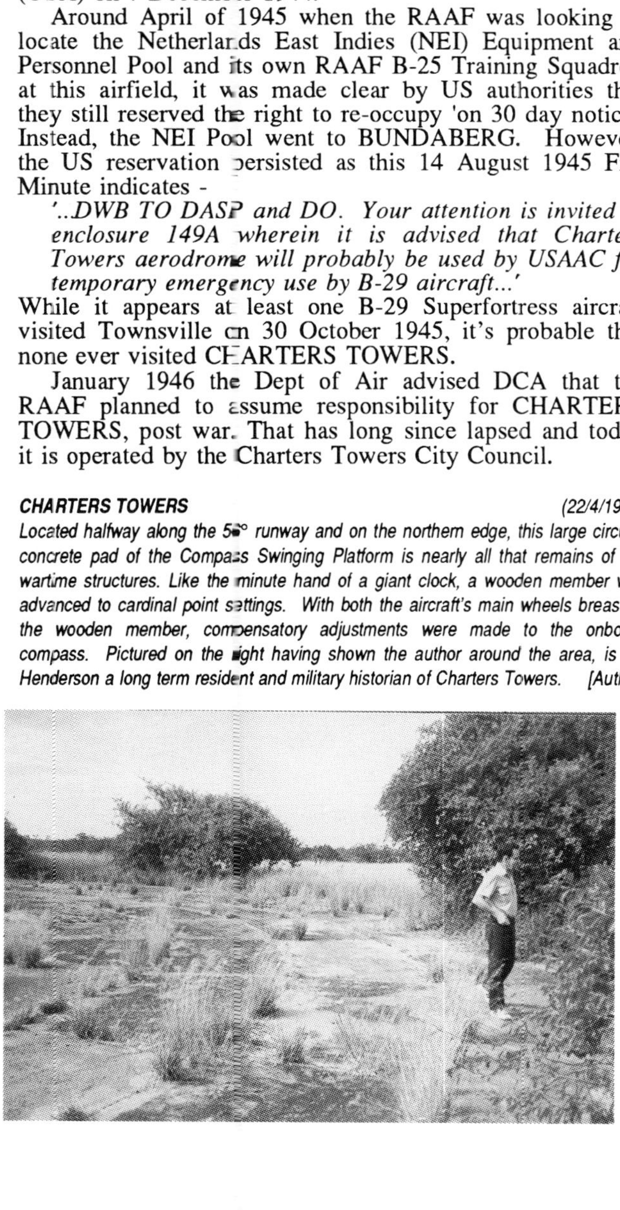

BALFES CREEK

'...Seems 1943 was a quiet year for this strip, both "on the files" and on the ground.'

This was my own reflection after reviewing assembled archive and other mentions of BALFES CREEK. This site was requisitioned on 1 April 1942, the same day as GIRU. Here P-Os Jim Trench and John Keays came together probably for the first time in their RAAF "Works Officer" roles. Before the war and for years after they were colleagues in a prominent consulting engineering and surveying firm. Keays' 8 April 1942 report to Air Board ranged over several sites 'west of Charters Towers' -

'...inspection made...in conjunction with P-O Trench and Messrs Doig and Mathison. A tentative site for a Dive Bomber field has been located at Balfes Creek immediately to the northeast of the railway siding. Investigations in regard to water supply were not entirely satisfactory and arrangements were made with Mr Doig for one of his officers to (investigate further)... The site at Cotonvale is satisfactory, but it is suggested that as the best access is from the siding at Karoon, the name be changed accordingly. The original site selected at Prairie was rather restricted in area and a better and considerably larger site was located at Strathglass. It is proposed that the dispersal strip be placed on the original site. The 5 miles of road immediately to the east of Eurunga is ideal for a road dispersal strip and strips are also available by Penrice Bore and to the east of Bullock Creek. Preliminary surveys have been completed at Balfes Creek and surveyors are now operating on the other sites. The Main Roads clearing gang expect to complete all works on the north and east of Charters Towers by end of this week and will then proceed to the bomber sites.'

In April and May of 1942, confusion reigned in respect of just what purpose which fields were eventually to be put. There was also the problem of how to get them built in time considering the staggering shortage of both equipment and men. By 14 April 1942, John Keays was aware of altered requirements. Dive Bomber fields were halved to just one - CHARTERS TOWERS. He asked if BALFES CREEK was still wanted, adding -

'...strips have been located and pegged to the east of the creek in the form of a triangle...Taxiways will connect

BALFES CREEK *(1975)*

Recent 1:100 000 topographic mapping shows no indication of this airfield. Apart from the 'strip on the road' the remainder is private property. [AUSLIG]

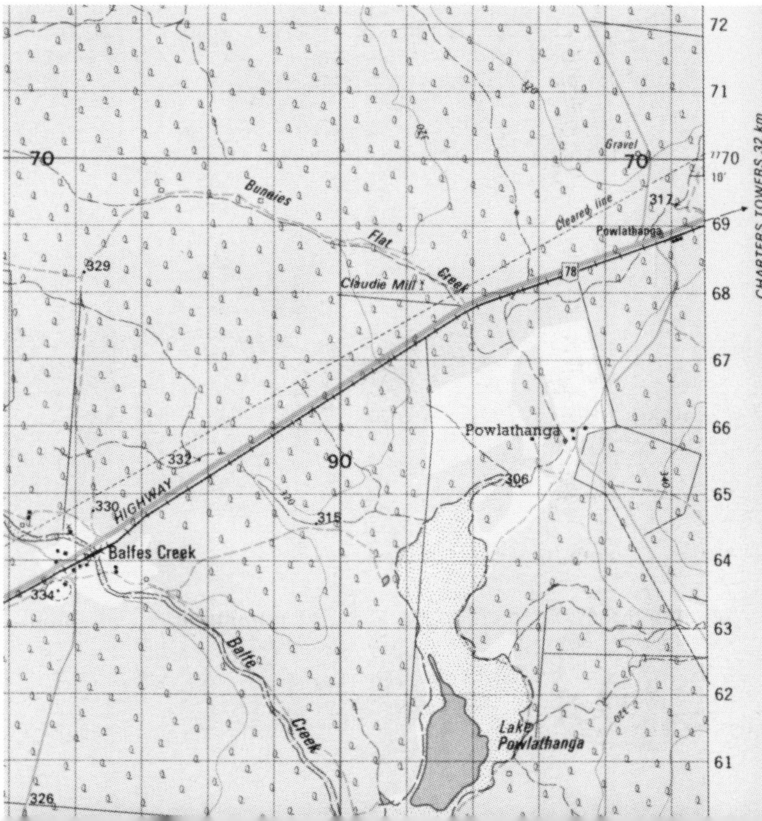

the ends... (Surface) will be solid in dry weather but spewey after heavy rain. Timber cover is satisfactory... water supply is from well in Balfes Creek and further well is available approx 1 mile north of the area. The Railways draw from two spear points in the creek bed... The site is on the east side of the Creek and the (railway) siding on the west side so that it will be necessary to provide a Creek crossing. It is suggested that the existing railway bridges be decked so they may be used by road traffic...'

The tabulation accompanying this report by Keays mentions by name, two dispersal strips for the BALFES CREEK field. 'TARANGIE' and '4 miles east'. The former was cleared but the latter, referred to as THALANGA, was not. Instead the POWLATHANGA site north of the railway, was cleared.

Walter Wanielista was with Company 'E' of the US 46 Engineers when it was ordered from REID RIVER to BALFES CREEK on 11 May 1942. They were only there five weeks, during which work proceeded on at least three strips. Walter recalled -

'...Strip No 1 was surveyed by ("E" Co. CO) Capt John Olsovsky and myself. Ant hills were blasted and levelled with our only D-4 bulldozer. The strip was cleared... and with an Australian tractor pulled grader, readied for aircraft...'

Camped south of the town, Wally described it as 'small' -

'...but it did have a fine country store with all sorts of goods. When our orders arrived to move (on) we purchased much of the goods at the store...'

'E' Company went to Milne Bay, New Guinea from here. The 46 Engineers left just the No 1 strip 'readied' and during their final week, Wally says Gil Schumacker, his 'platoon sergeant, who was the last to leave, recalls vividly, a few B-26s and B-25s landing on it'.

Six weeks after the US Engineers departed the AWC Minutes mention 'a strip at Powlathanga being built as a dispersal for Balfes Creek'. It then joined the confusion of what if for any purpose it was needed!

A status drawing 'No 14' included in the US 46 Engineers' material forwarded by veterans, suggests the 52° strip was gravelled October 26 and sealed November 20. If this is so, BALFES CREEK may well have been one of the strips Jim Trench remembers where 'the bitumen seal peeled off like a carpet'.

As 1942 closed the Allied Works Council were formally advised that BALFES CREEK was to be completed 'only to gravelling stage' until a final decision was made. Its nominated role had changed from dive bomber squadron site to RSU site to a now unassigned role and this is where 1943 came and went leaving little further activity or progress. During 1944 further mention appears only in respect of reconciliation of costs and what to do with another one of these superfluous airstrip assets.

Eventually both the RAAF and DCA concluded they had no need for what was in effect, a gravel runway parallel and close to the highway.

BALFES CREEK *(26/5/1945)*
This extract (Frame 4306) shows a field typical of many only just cleared before they were abandoned. In this case only one runway, just north of and parallel to the road and rail, was advanced to a good gravel standard. The loop taxiways towards each corner of this triangular layout are evident in the photo. *[AUSLIG]*

BALFES CREEK *(22/7/1991)*
Looking along the old gravel strip, its proximity to the highway is evidenced by passing traffic and the telephone pole line. In the foreground describing the old strip to Jennifer is the late Don Johnson, author and historian of Charters Towers. Don felt that 'as a kid, he could recall seeing Airacobras on this strip'. *[Author]*

CLONCURRY *(31/10/1942)*
The Central Hotel viewed from the verandah of the Post Office Hotel. Frank Allen, a member of the US 19 BS, 22 BG (M) took this photo on a stopover en route to Darwin. By this time, the field had returned to its role as a vital link in the ferry route across northern Australia. *[F Allen ex USAAF via Jeanne Newell]*

CLONCURRY *(18/8/1944)*
Looking west to aircraft parked along the southern edge of Runway 124°. Boomerangs of the RAAF's 83 Sqn were in transit from the Northern Territory to southern Queensland, accompanied by a 13 Sqn Ventura. Behind the nearest fighter is the gabled outline of the historic QANTAS Hangar. *[D Oliver ex RAAF]*

CLONCURRY

'...By the time we got there, the runway was lit up and we landed with no drama.'

George R (Dick) Graf was a Radio Operator in a B-17 aircraft of the US 22 BS, 7 BG on assignment to a 'recon' Squadron destined on 16 May 1942, to become the 435 BS, 19 BG. Early in 1942 CLONCURRY was used as a dispersal field. Since the war, Dick has lived many years in Australia and certainly he could not help but chuckle in telling me of one night landing there -

'...In those early days there were no US Air Force radio ground stations at all so we used the RAAF ground stations or when necessary the 'aeradio' stations of the Australian Civil Airways...the operators of these stations were very good indeed. I used them... whenever possible. Anyway, we were heading for Cloncurry at night. I called them on the radio several times and got no answer. After a while Longreach 'aeradio' came on the air and said "I think Cloncurry has closed down for the night. Do you want to land there?" I answered "Yes of course but we can't do it unless there are some runway lights laid out." So Longreach said "Well, I have the operator's home address so I'll send him a Telegram. Even though its late, the Post Office will deliver it to him and then he'll go down and get things organised at the airfield." And it all happened just as he'd said. By the time we got there, the runway was lit up and we landed with no drama.'

Dick Graf's description of the Unit's early contact with CLONCURRY sets a few other scenes -

'...The standard mission in those days was, leave Townsville, 4 hour flight to Moresby, land just at dark to be safe, and refuel. You had your bombs with you because there weren't many to spare at Moresby; so we always flew up fully loaded. Fly your mission, get back to Moresby, middle of the day or early afternoon. Put in just enough gas to get to Horn Island which was safer than Moresby, because sure as anything the Japanese would be following us. Finish refuelling at Horn Island and then back to Townsville.

Pretty soon some high ranking people remembered that a lot of aircraft had been lost on the ground in Java and there wasn't much in the way of defence at Townsville. There weren't many bombers, so let's be really safe; we'll keep the bombers at Cloncurry. So what happened was - let's take a B-17 that's just arrived at Townsville from Moresby, via Rabaul. A skeleton crew takes the aircraft out to Cloncurry and stays with it there until its needed again. They fly the aircraft back to Townsville and the combat crew takes over. It's refuelled, goes up to Moresby, does the mission, back to Moresby, refuel, back to Horn Island, back to Townsville, over to the skeleton crew which takes it back to Cloncurry. Well Cloncurry is a nice airfield, but not much of a town and the poor aircraft were being worn out flying back and forth. So this nonsense kept on for a few weeks and then stopped.'

The daily journal written by 1st Lt Walter H Johnson of the 435 BS, 19 BG has been preserved in US Archives. Johnson arrived at ARCHERFIELD, Brisbane on 19 February 1942, the day Darwin was first bombed. There were then five B-17s serviceable at ARCHERFIELD out of the twelve which set out to cross the Pacific from Hawaii on 9 February 1942. Storms in the Coral Sea resulted in only two aircraft reaching Townsville directly as planned. The five joined them on 20 February. Johnson's next journal entry was for -

'...February 21, 1942 Saturday. Townsville to Cloncurry, Australia - And now we seemed on the retreat, for before dawn Maj Carmichael left the field in a Hudson bomber. He radioed from Cloncurry for us to follow him inland, which we did, all except one ship. The other 22nd ships had come in, all except Capt Swenson, who had motor trouble in the Fiji. Lt Roberts and crew stayed behind in Townsville awaiting Swenson and trying to dig the tail wheel of their ship from the mud. The edge of the runway is soft, and his tail wheel got a little outside the edge when he was turning.

All the Americans had been agreeably surprised with the livability of Townsville, its tropical air housing, its successful camouflage. Food was excellent, and the enlisted men did not complain too much about the canvas tied between two-by-fours they slept on.

As we flew over the nearly 400 miles of almost uninhabited area of Australia (still free kangaroo country) and landed at Cloncurry we became convinced we'd never see Townsville again. At Cloncurry we took over the eight hotels of the town, hired drivers and trucks to haul us about. Our planes were two miles from town, two miles of winding trail-like road.

Cloncurry is a hot dry place of a million flies per cubic foot of air. The men who kept out of the bar rooms found comfort and rest in shows, 6 to 12 months old, American made. The theatres were four walls, with a ceiling of sky. Seats were canvas, outdoor type, half bed and half chair. Lt Morris was made OD and two men per ship stood guard thereat. We figured we'd come to stay and we figured Cloncurry didn't offer much the Japs would be interested in...'

Here Dick Graf's memories struck one of those sour notes which don't easily fade. About a week after -

'...we flew a mission from Cloncurry, the only time we actually did that I believe. We took off late at night and flew a mission to the Lae, Salamaua area...'

While the details and the mission results may have blurred, the homecoming seems clear -

'...We didn't go straight back to Cloncurry... they sent us to Townsville instead. We had left all our possessions at Cloncurry. We flew... out of Townsville a few times and then were sent (back) to collect our possessions because we were tired of flying and sleeping in the same clothes we'd had for some time. We found we'd all been checked out of the hotels and all our baggage was stacked on timber at one side of the airfield in one big area, supposedly being guarded by MPs. However, they hadn't guarded it very well. A lot had been rifled and a fair number of items stolen. One of the unhappy things that happens in war.'

CLONCURRY was a key aerodrome on the Darwin-Sydney air route. Its wartime importance was automatic. A RAAF Aerodrome Works file commenced two days after Pearl Harbor, bears some testament to the effects of geographic isolation in 'getting things done there'. A file Minute dated 22 December 1941 reads -

'...signal from F-Lt James, Longreach... advising Contractor Taurir awarded Cooktown Job. Civil Aviation will forego in favour Cloncurry. Alternative arrangement Cloncurry difficult owing shortage plant, also requesting this Dept confer with Liaison and advise Comworks, Brisbane...'

Surely 'Taurir' in this typed Minute, shows someone (the typist?) misread the handwritten name of Tanner! Tanner & Kenny was a firm of earthmoving contractors in North Queensland during the war with work on CAIRNS airfield well to its credit. The diary of Walter Edward Tanner, details furnished through his son John Tanner, shows T & K Constructions left Yungaburra, via Mt Garnet, for CLONCURRY with a road convoy of plant to 'continue construction of an airfield' on 10 January 1942. The firm had done earlier work there during 1941. W E Tanner's diary records 5 January 1942 as the 'handover date' of a completed contract on the COOKTOWN airstrip.

Those same 'Minutes pinned inside the front of the file, changed from typed to handwritten a couple of days before

Darwin was bombed. Runway extensions and upgrading, temporary 'hutted camp(s)' and associated items of water supply are typical of the urgent work being organised at CLONCURRY. This first handwritten 'Minute' summoned various forces, concluding with -

'...Proceed urgently 24 hours per day.'

The RAAF/Dept of Air took control of CLONCURRY on 6 January 1942. P-O Keays is first mentioned on 22 February as replacing F-Lt James. Keays was due in Brisbane 25 February with full -

'...authority (to) design and direct aerodrome works.'

Consultations in Brisbane, visits to the site and discussions during March 1942 appear to have achieved coordination. Yet a following Minute around 1 April 1942 says -

'...Signal from General Brett - stating that Dept of Interior Brisbane interfering with work. Also requesting suitable water supply for Air Force Camp.'

On his return to Melbourne in late April, John Keays reported on this situation in CLONCURRY. While he personally had not been there since early March -

'...P-O Trench (NEA) inspected on 28 March 1942 and Mr V G Crawford (Dept Interior Brisbane)... on 17 April 1942. P-O Trench advised that little work had been carried out on buildings since my inspection as the contractor had been engaged on erection of huts for US temporary camp. I understand from Dept of Interior, Brisbane, that little progress has been made since then and that US temporary camp has now grown to 3 000 personnel. Contractor is asking the Dept of Interior to assist in obtaining further supplies of material for camp on hill as materials have been used on US works...'

By 17 April 1942, all four runway extensions were gravelled. Most of the western perimeter taxiway was formed and was being gravelled. Keays continued -

'...I have no knowledge of any US request that work should not be started on one of the runways, and I know that no request to this effect had been lodged with Dept of Interior Brisbane when I left Brisbane on 21 April 1942. Dept of Interior... considers Cloncurry to be a RAAF job and is not prepared to alter plans etc. without direction from this HQ. Works in progress in Queensland have been inspected by numerous officers attached to service and civilian organisations and entirely conflicting instructions given to the officers in charge of construction....

Keays supported an orderly chain of command. On the water supply matter he wrote -

'...I have no definite information...but was informed in Brisbane that there was some friction when MRC engineer acting under instruction from US Officers proceeded to pull up water main from the well by the river to the tanks on the hill and to use it to connect the town supply to the temporary US camp. Mr Crawford... was to investigate this matter.'

John Keays' last paragraph is interesting, as it bears on an alleged US request not to start work on a runway -

'...When I was in Townsville early this month I was advised by Mr Doig that Charters Towers was to be restricted to two runways. Mr Mathison was aware of this decision and I have no idea as to how works were started on the third runway.'

It is entirely possible that events at CHARTERS TOWERS were being confused with those at CLONCURRY. An American presence was being established at both centres simultaneously. US operational aircraft with tricycle undercarriage were soon to dominate making cross wind runway movements more usual and the third leg of the traditional triangular runway layout less needed.

Earle W Curtis was another veteran of the US 19 BG again as a transferee. In a taped response to an enquiry by my friend the late Don Johnson, he explained -

'...On 8 April 1942 our crew and three others from 11 BG 7th Air Force which was stationed at Hickam Field at Hawaii, were assigned to four B-17s with the new ball turret which replaced the ineffective remote

F-4 CLONCURRY

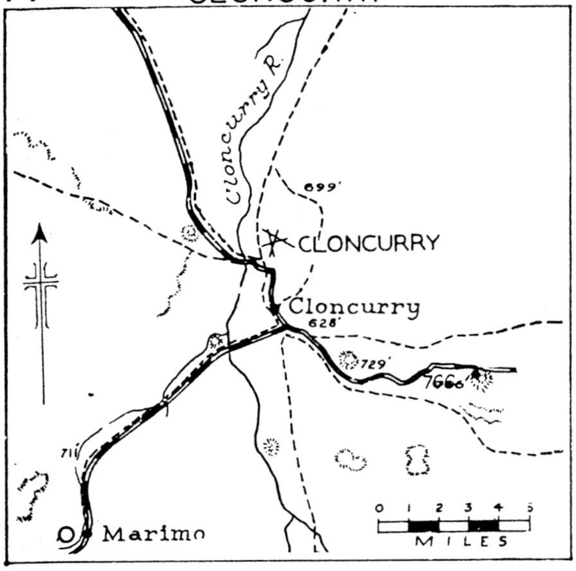

Location: Approx. 2 miles N. of Cloncurry & immediately E. of Cloncurry River.

Dimensions: 015° - 5090' - Sealed.
064° - 5090' - Sealed.
124° - 7000' - Sealed.
164° - 5100' - Sealed.

Surrounding Country: Open country with patches of light scrub & timber along river.

Obstructions: Control buildings & hangar (40') masts (70') on E. side. Tanks 75' 600 yds. S.E. from Control Tower. Obstruction lights on all above. 4 masts 30' 200 yds. inside Western boundary. Hill 70' 700 yds. N.E. 1 Mast (180') & 2 masts (185') 700 yds. from Northern boundary. Hill 500' 5 m. S.W. Tower 70' 800' S. of control tower - unlighted.

Radio Aids: Aeradio, HF/DF, Homer,ASV, QGH approach

Accommodation: Not available.

Servicing Facilities: Refuelling & minor repairs.

Meteorological: Forecasting Met. Section.

Lighting: Electric flare paths. Rotating beacon.

Remarks: Suitable H/B all weather.

Control: Civil Aviation.

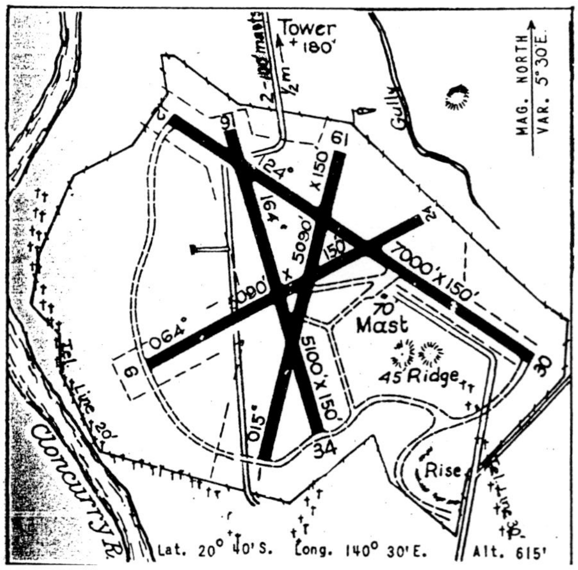

CLONCURRY *(10/1945)*
The RAAF Aerodrome Guide (Publication No 694) shows this detail for Cloncurry. This is typical of the presentation given sixteen prominent airfields in Queensland at that time. *[RAAF Museum Point Cook]*

turret. We flew to Australia arriving in Cloncurry on 18 April 1942. Our crew was assigned to the 30 BS and began flying reconnaissance missions from that field, our first being to Kavieng. On 4 May we were transferred to the 435 BS at Garbutt field in Townsville. Sure is a long train ride from Cloncurry to Townsville!'

Lawrence J Moore was yet another transferee into the reforming 19 BG. In one of those confusing coincidences of the numeric identification of USAAF Squadrons and Units, Lawrence was a member of the 11 BS, 7 BG on ship to the Philippines when Pearl Harbor was attacked. He gravitated to the 28 BS, 19 BG after time in Java, and the subsequent retreat to Broome in March of 1942.

In my efforts to prompt accurate recollections, I've distributed various questionnaires to veterans. That approach has proved useful and the response sometimes humorous. Though my questions on 'airfield defence' and the 'local wildlife' were not directly related, at least Wallace Fields of the US 435 BS wrote that -

'...Most of the places from which we operated were well protected by flies, particularly Cloncurry.'

Three weeks after the Coral Sea Battle, some order crept into the flurry of new airfield work. A 28 May 1942 report to Air Board by the AOC, NEA mentions -

'...All works at Cloncurry have been stopped, with the exception of the temporary camp. The only work left to be done there is to seal the runways, and it is considered that this should be of very low priority.'

A letter from the Director General of Civil Aviation dated 17 September 1942, to the Dept of Air, Melbourne, shows what a few more months did bring. On the subject of maintenance of the landing area, it read, in part -

'...It is now understood that the original proposal for the establishment of a large RAAF Station at Cloncurry is not being proceeded with. On 6 September 1942 our (DCA) Officer-in-Charge Cloncurry aerodrome, advised as follows: "since Cloncurry ceased to be an active operational base, only the following service units have been maintained here.

1) One USAAC Lieutenant whose main duties are in connection with American Transport aircraft.

2) A Meteorological Section consisting of approx twelve American observers and forecasters, one Australian Officer and one Sergeant.

3) A nucleus ground crew to refuel transports and service aircraft.

4) A RAAF signal section to operate point to point traffic with two Cypher Officers.

5) A RAAF detachment to operate the H/F Direction finding equipment.

6) Two RAAF photographers and one stores section aircraftsman.

The net result is that there is no effective control over the aerodrome or the services to aircraft. The aerodrome itself is not administered either by the American officer or by the RAAF, and it is with considerable difficulty that any attention is secured to any requirement or regulation". A separate communication will be forwarded to you regarding the aspect of operational control at Cloncurry aerodrome, but, in view of the above and the non-establishment of your large Station at Cloncurry, it is suggested that the maintenance of the landing area and its development could revert to this Department. On 25 August 1942 we received advice that the RAAF had no maintenance staff or equipment at Cloncurry and that the Queensland MRC... (presently) completing certain construction works at that aerodrome, will remove its equipment and plant upon completion of this in about a month's time.'

Present day Australians will recognise a little of the long winded written language they've come to expect of Government Departments.

The Aerodrome Works file Minute of 27 May 1943 refers to a memo to NEA from DWO Townsville requesting action be taken to have 'machine gun pits refilled and resited.' While the same file shows these gun pits were apparently ordered filled in around 22 June 1943, slit trenches and what very likely are some of these gun pits still appear in the August 1943 aerial photos.

As a focal point in the vast cattle and sheep lands of the Gulf of Carpentaria and the Barkley Tableland, CLONCURRY figured in the growth of the Royal Flying Doctor Service. Hudson (later Sir Hudson) Fysh, the great co-founder of QANTAS rightly said of this airfield site, fifty years after its development, that 'it was the only aerodrome on the Darwin to Sydney air route, still in the same position'.

CLONCURRY *(21/7/1988)*
The Cloncurry Shire Council owns and operates this airfield. It has changed little from the layout indicated in the RAAF Aerodrome Guide of 1945. The John Oxley Library, Brisbane, holds an historic panorama of CLONCURRY aerodrome taken around 1937. On the handwritten description is the following: ***'...Back in 1919 Hudson Fysh and McGinness (later principals in QANTAS Airways) sat on this hill (45' high ridge) and said, "This is where the aerodrome is going to be"'. In 1969 Hudson Fysh said it was the only aerodrome on that route that is still in the same position! '*** *[Department of Lands, Qld]*

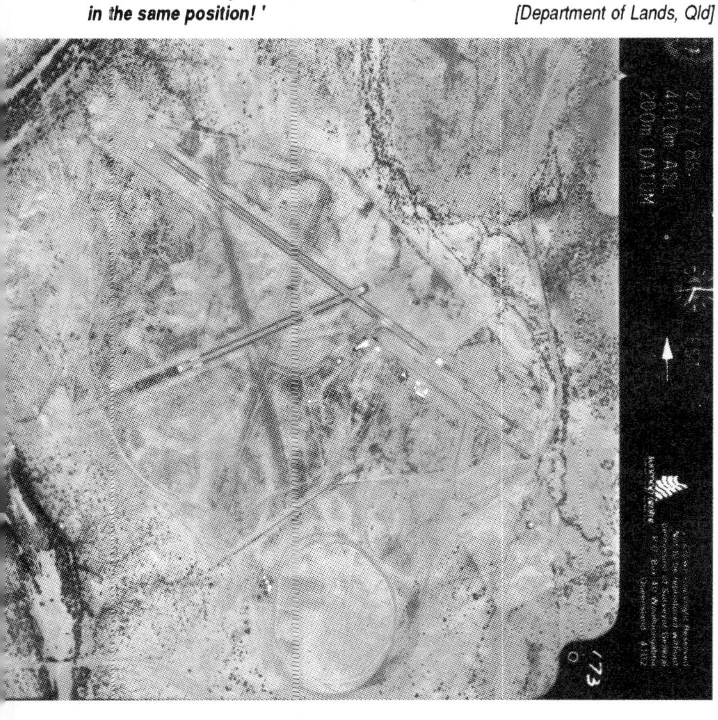

TORRENS CREEK *(1945?)*
This oblique taken from the northeast shows the three airstrips with the township of Torrens Creek in the foreground. The 76° runway, roughly parallel to the railway, appears with its ends white (gravel surfaced) and the middle length dark. The middle was overlaid for a time with 'pierced steel plank' matting by US Engineer Troops in 1942. This use of PSP appears to be the first of significance in the southwest Pacific theatre. The matting was removed as the war effort moved on. *[AA CRS A705/1 Item 171/93/532]*

TORRENS CREEK

'...Sunday, June 7: Waited all AM for plane. Left at 1: PM. Saw ant hill strips, Charters Towers landed at 3: PM. At Torrence (sic) Creek. Com. Off. Woodstock of Ft Worth sent for Walter B King - he has 700 to treat, learned more medicine in last 2 months. Dengue, flies, etc. Aussies know nothing of sanitation. Takes 6 hours to get to hospital. Last mail I got was in March. Think I will go to Moresby. Glad to see plane... first we have seen. No town group of perhaps 300. Bushy country 150 miles northwest of Townsville. Made my own hospital. Runway made of steel mat - cut like biscuits. Walter B, big, fat, blonde, good sun - looked fine - had just treated boy from Waco - negro on elm st. Arivee Charter Tower. Saw plane land on belly. Toured post - excellent. Moreale, camouflage, etc. Air repair depot 1 300 men.'

Such is the raw typed text of a 7 June 1942 entry in a diary obtained from the Lyndon Baines Johnson Library, Austin, Texas. Congressman Johnson was newly commissioned in the US Navy when he appeared in Townsville on a brief fact-finding tour of the SWPA. LBJ's trip over Lae on 9 June in the nose of a B-26 Marauder gained him notoriety, a Medal and a swift posting away from 'active service'. His return from Port Moresby via Darwin has been documented. Through a faulty compass adjustment, the party's B-17D made an unscheduled landing on 'Carisbrooke', a sheep property near Winton. These things I knew. What I didn't know of until the 'diary' arrived was LBJ's visit to TORRENS CREEK.

I was fortunate to have received historical detail of the US 46 Engineers' period in Queensland from a few veterans. They proudly credited this to one other of their ranks, John A McFee of Arlington, Virginia. A list of officers therein included Capt Walter B King Jr in the Medical detachment. The correspondence that followed brought a better understanding of LBJ's diary. Despite a long family connection, Walter did not know of its existence, preserved only 90 miles from his current home.

Walter set about 'translating' the diary -

'...the Battalion CO at Torrens Creek was Maj (later Col.) Herman Blackwood, from Ft Worth. Woodstock of course was an airfield back near Townsville...'

It was then clear that many of the short statements represented what the young Lt King had told Congressman (later President of the USA) Lyndon Johnson. No one was more surprised than Walter when he was called to the strip after the C-47 landed. Heading the party was the first General he'd ever seen, who exclaimed -

'...Lieutenant, I don't know who the hell you are but this Staff plane detoured out here so Commander Johnson could see you.'

Walter walked over to the tall man in the Navy uniform and introduced himself. The Congressman said -

'...Walter, I'm Lyndon Johnson. Your Aunt asked me to look you up. How are things going?...'

That's when Walter told him about -

'...no mail, poor food, difficulty with supplies etc.'

Food is a universal subject. Opinions on its attributes will often be as many as the participants to the discussion. The 46 Engineers' Company A was first dispatched west to TORRENS CREEK on 22 March 1942. The Unit's history included the following *Situation and Progress Report* dated 25 March 1942. To Lt Col. A G Matthews, it said in part -

'...1) Arrived on time and unloaded equipment at once. Made reconnaissance of area with Mr Coleman while unloading, selected job site and camp site at one end about ¼ mile from water. Spent day setting up camp, established water supply, putting in center line, and making reconnaissance of alternate sites. Started work Tuesday, (clearing etc)...Steel arrived Tuesday night but will not be necessary or used on any sites selected. Hard, sandy, soil well compacted with small amount of clay. Mr Cook, inspector for Pettus, will see you and give details of sites, soil, etc. Cleared and grubbed about 600′ today... Had to blow most of them (stumps)...
2) Have received no supplies of any sort. Initial supply of rations exhausted at noon today. Require daily rations delivery, 30 drums of gasoline per week, all oil and grease originally ordered, medium tractor if available, and 500 additional pounds of Gelignite. Have procured gasoline locally but must replace it.
3) Called Maj Irwin this noon and received promise of immediate shipment of rations. Request Lt Vernor determine what is wrong with ration problem and straighten it out at once. Also request proportionate share of candy and cigarettes, if any, be sent regularly.
4) Time for shower and fresh beef steaks. WE'RE HAVIN' A WHALE OF A GOOD TIME.
(Signed) ***Leonard L Haseman****, 1st Lt 46 Engr Regt (GS) Commanding CO "A".'*

A Coy was joined by B on 4 April and C on 27 April 1942. On the PSP or Marston matting, the history notes -

'...in five days a section 2500′ long by 100′ wide was laid and ready for use. This set a new record for US Army Engineer Troops laying steel matting, per sq yd per hour.'

The comment by Lt Haseman, suggests this steel mat laying may have been more in the nature of an exercise, than real need. Quite likely it was a 'belt to braces' move to ensure an all weather strip during the confrontation building in the Coral Sea. After all, this site was to be the reserve ammunition depot, well inland from Townsville.

I've not found any hard evidence of concerted aircraft use of the freshly laid mat. I do find, in Walter P Maiersperger's log, an entry for May 4, 1942, where he flew a party comprising (at least) 'Col. Porter, QMC, Col. Matthews C of E, and Maj Hirsch, Ord' in the C-40 twin-engined Lockheed 12 from the STOCKROUTE strip. They landed here before flying on to CLONCURRY.

The earliest Archive mention I've found of TORRENS CREEK is a 17 March 1942 report to Air Board by Air Cmdre Lukis, AOC NEA on the subject of *Proposed heavy bomber aerodrome, Charters Towers-Hughenden District -*

'...(On) instructions, P-O Trench proceeded to Torrens Creek to assist in the (site) selection.'

Others were involved too. E J 'Eddie' Connellan, pre-war founder of Inland Airways which later bore his name, was busy flying a small exploration party in search of aerodrome sites. For this the RAAF impressed a privately owned Beechcraft 17 'stagger wing', VH-UXP. Connellan's strength was his knowledge of and ability to recognise soil types by colour and vegetation. In his 1992 posthumously published autobiography ***Failure of Triumph***, he claimed TORRENS CREEK as 'one of his'.

Jim Trench's telephoned advice back to DWB on 22 March 1942 reads in part -

'...Antil Plains - two strips complete (cleared only) third strip ready in one weeks time.
Woodstock aerodrome - one strip complete (cleared only) two others ready in one weeks time.
Hughenden (Torrens Creek) aerodrome - USA Works Battalion starting on 23 March 1942 to lay a portable mesh runway which will be completed in one week.'

A report by P-O J Keays dated 14 April 1942 described TORRENS CREEK as 'an excellent site', which was -

'...being built by US Engineer Corps. E-W strip cleared and levelled 7000′, steel mat being laid 4000′ x 100′ and clearing commenced on another strip.'

This suggests that until then no effective use had been made of the one only (E-W steel matted) strip. Some historical accounts indicate that B-17 Flying Fortresses were based there for some unstated period immediately after construction of the strip 'in February of 1942'. A report of 29 April by P-O Keays, generally aimed at clarifying which sites were to proceed, mentions -

'...Dispersal strips (sites) are available... Eurunga could be extended to a full field. It is however only 7 miles (West) from the heavy bomber field being constructed by US Corps of Engineers at Torrens Creek.'

EURUNGA is typical of several sites identified, possibly surveyed, assigned a RAAF 'Landing Ground Drawing number', in this case, '42-43/590', yet never to progress further. By 28 May 1942, the AOC's directions on works in a priority list then current, were -

'...Priority No 6, Torrens Creek: only works to be completed are - paving, sealing and camouflage work.'

The 46 Engineers pulled out of TORRENS CREEK on 10 June 1942, just three days after LBJ's visit. A week later, they were at IRON RANGE. Brice Allen was Regimental Motor Officer in the 46 Engineers. His work meant contact with the civilian vehicle dealers in the hunt for spare parts. Scarcity of equipment and parts ensured the utilisation of even the obsolescent. Brice wrote of a real steam roller the 46th had left at TORRENS CREEK.It had been used in Townsville area before being railed west -

'...My instructions were to load this steam roller on... a rail wagon for return to Townsville. After an uneventful train ride I arrived in Torrens Creek to find a Negro Company there to secure and maintain the airstrips. I could not find a man among the Negro Company that had any experience or knowledge of how to operate steam powered equipment. With some help I fired up the steam roller and drove it to the loading ramp...'

Brice told of the smooth metal drive wheels spinning on the inclined ramp, and some innovative rigging of truck winches. Working with unfamiliar controls, he concluded -

'...Thank God the stop was made before dropping off the end...!'

General Brett's visit during June resulted in a 21 June 1942 directive -

'All works west of Torrens Creek are to be curtailed as well as proposed sites at Pentland and Homestead. The ARU-RSU layout as originally planned for Jardine Valley (Skull Creek) will be shifted to Torrens Creek.'

This directive referred to the remaining US Engineer Troops involved and said of outstanding work-

'...With few exceptions, MRC will complete...'

Maiersperger was back at TORRENS CREEK 1 August 1942 'checking off (a) Capt Horgan (for) strange field landing' in one of the early B-17Es out of CHARTERS TOWERS. Walter can remember the visit, but only that he met up with a class mate he'd not seen since graduation in May 1940.

Elements of the US 43 BG (H) were stationed at TORRENS CREEK over periods in the second half of 1942. Its 403 BS History colourfully relates the arrival by train of that squadron's ground echelon on 27 August 1942. Apparently by then the 63rd and the 65th Squadrons were in camp there. According to the 63 BS history -

'...On August 14 1942, B-17s with full combat crews began to arrive at Torrens Creek for the 63rd squadron... the first in the 43rd BG to be activated. Within a week, most of the 63rd personnel had been flown either to Townsville or to Mareeba, to train with the 19th BG for actual combat operations. The first combat crew with the 63rd was Lt Murphy, Capt Scott, Lt Reeder, Capt Kenneth McCullar and Lt Sogaard.'

TORRENS CREEK (22/7/1991)
On the middle strip (141°) looking to the southeast. While the clearing is still evident and some gravel remains this strip is in no way useable by fixed wing aircraft. Flinders Shire Council maintains the 76° strip in current use. [Author]

With the departure of the 43 BG, TORRENS CREEK settled back to a token number of US personnel and the big munitions dump 10 km south of the airfield.

AWC Minutes first mention activity at TORRENS CREEK in October 1942. These covered such matters as Air Board's decision to apply the lesser 'USA Standards' of construction to the two strips then scheduled for gravelling and sealing - an indication the RAAF saw no post-war need there. By 22 October the priority of this work compared to 'Townsville' was being urgently reviewed. A week later all three runways appear in an amended 'hit list' of strips to be sealed before the start of the wet season.

During 1943, the munitions dump moved to Kurukan, north of Townsville. By mid 1944, TORRENS CREEK was listed as one of the many 'unoccupied aerodromes'. In a 12 June 1944 survey of unoccupied service buildings the following mention is made -

'...All buildings have been demolished and removed by USAFIA. Steel matting strip has also been removed.'

September 1943 vertical photography from 6 000' suggests the steel mat was then still in place. PSP was a vital tool in the equipment establishment of the airfield construction units in the southwest Pacific. Hugh J Casey's ***Engineers of the Southwest Pacific 1941-1945*** states -

'...the first pierced steel mat produced in quantity was used for surfacing a runway at Torrens Creek.'

Certainly the 'locals' seem to have treasured what sections were 'left behind'. It is not uncommon to find the gate to a station property in the area adorned with a piece each side, often painted white to show up in the headlights at night.

MACKAY (27/7/1942)
This is a vertical (1-800V) of a Trimet coastal trimming run down most of the Queensland coast. By July of 1942 a dispersal loop had been constructed at the southeastern end of runway 140° and several bomb splinter proof pens constructed for the protection of dispersed aircraft. Other earth mounded pens can be seen in the southwestern corner of the airfield. [AUSLIG]

MACKAY

'...At Finch Hatton (west of Mackay) we were asked to go turkey hunting and we rented horses to ride and two young men took us out to the farm of the Andus family where we were given a nice lunch. After lunch the boys took us on a hunt. It was fun, but we didn't get a bird... It was a great experience and I hope the people liked us as much as we did them. Especially the Andus family and their young nephew, who with their son, took us hunting.'

Donald Englund was one of several USAAF veterans with such recollections of very welcome leave periods spent at Mackay. His 'R and R' there was probably around October 1943 when he came down from New Guinea.

Situated on the coast, MACKAY always was an obvious rest and refuelling stop for north-south air traffic. The aerodrome was started by the local Council, but it was not until 1940 that major development came. The Commonwealth Government extended the area, funded construction of the unsealed gravel runway 140°, and DCA took over the operation from Mackay City Council.

MACKAY was slated as one of the AOBs for the RAAF's 24 Sqn in 1941 yet upgrading works progressed with no more vigour here than at others. Some detail of this upgrading is evident in drawing 41-42/556 which superseded an earlier drawing from the financial year 1940-41. It shows firm positions for twelve dispersal points and a detailed schedule describing elements of the work, such as -

'...Pyrotechnic store, workshop, accommodation hut, a W/T receiving station, W/T transmitting station, two machine gun posts, one bomb dump and six individual 1000 gallon underground petrol tanks.'

The machine gun emplacements were to be -

'...two of type 2 - drainage through spaces of the flooring with pipe to drum or cask sump, to be kept bailed out.'

In a high rainfall lush cane growing area such as Mackay, it must surely have been fun keeping such pits 'bailed out'. Doubtless those on the airfield at the time were extremely grateful never to have needed them.

A week before Pearl Harbor, 24 Squadron was still pleading for attention to the basics of accommodation at MACKAY. RAAF files now in AA carry little of MACKAY over the period October 1941 to March 1942. Clearly the emphasis was elsewhere. However on 19 March 1942 plans and specifications for a second runway were tabled with 'request to expedite approvals'. The Queensland MRC history shows that this work was commenced in April by the Mackay City Council acting under the direction of the MRC.

The aerodrome came in for camouflage attention. On 14 May 1942 SDCC reports mention -

'...on 2 May the Chairman discussed the position regarding this AOB with the City Engineer of Mackay. He has been in charge of construction and it has now been converted into a full sized landing field. A number of splinter proof pens have been constructed under instructions from Air Board. The City Engineer was able, in consultation with a representative of the RAAF to greatly improve the siting of these. They are now mostly placed under large paper bark trees adjoining the field but Lt Longland reported that in passing over the drome on his recent trip to Townsville he noticed that several of them are built right out in the open.'

The term 'hutment' appears to cover various small buildings from accommodation huts to stores and offices. In monthly progress reports of the time, it is recorded that by 30 June 1942 the provision of additional buildings and services was 92% complete - 'nearly finished', and work on

MACKAY *(17/5/1991)*
This is an extract from recent systematic coverage. Here the new terminal is five months away from its opening. [Department of Lands, Qld]

twelve splinter proof pens was 40% complete. 'Obstruction barriers' were put at 80% complete. These barriers were probably heavy timber railings resting on timber frames, or fuel drums filled with earth or concrete. They had to be capable of being manhandled in and out of position on the airstrip. The RAAF's 1" = 1000′ drawing 42-43/1330B carries a serviceability date of 5 May 1943 and notes -

'...two gravel runways - useable - obstructed but landing lanes left open.'

Air traffic increased with ferry activity and particularly in 1943 when the area became a centre for R and R visits by American servicemen from New Guinea. While further improvement of the airfield was constantly sought through that period, there was always more important work elsewhere. On 14 June 1943 an American request is on record to RAAF HQ for -

'...Control tower at Mackay - Air Traffic through Mackay now amounts to more than 100 aircraft arriving and departing each week. The field control

MACKAY *(2/6/1944)*
*Walter Maiersperger, pilot of this US Stinson L-5 'Sentinel' 298828 en route to Townsville, provided this caption - **'Flew 700 miles in 6:10 one day in this little plane - counting two stops for gas and an hour for lunch, it was 8:20 from start to finish. Those clouds held rain, both behind me and in front of me.'** In the background is the Duty Pilot/Control Tower. Not so easily seen are the timber frame, galvanised iron sheeted earth filled 'traverse walls', erected round the wireless equipment hut behind the Airways Rotating Light (ARL) in the centre of the photo. [W Maiersperger ex USAAF]*

people have been requesting a control tower... and I believe one is urgently needed there. I understand that the authority to build this control tower has to go through RAAF HQ. Request that steps be taken to install one of your standard steel Control towers... at Mackay. If this is not possible, then I request permission for the US Army to construct some sort of control tower at the Airdrome.'

MACKAY was 'not controlled by the RAAF'. DCA therefore had to agree to the erection of such a control tower. By 16 August 1943 DCA had responded -

'...It is now desired to advise that this Department proposes to proceed almost immediately with the erection of such a tower. The building will be similar to that erected at Rockhampton aerodrome. Plans for the building have been prepared and this Department proposes to install therein radio telephonic equipment.'

By September the heat was coming off, with respect to camouflaging various things. Dated 29 September 1943, a memo from 3 DWO Townsville to DWB Melbourne on the matter of bomb stores at both MACKAY and BOWEN -

'...It is proposed to omit fencing around these stores as they are all situated on Air Force land, omit camouflage and earth cover except at Bowen where earth cover is complete. If you concur with these amendments, it is requested that a signal be dispatched, repeated Allied Works Brisbane, so that instructions can be issued before the work is commenced.'

The bomb store buildings at MACKAY appear to have been semi cylindrical reinforced concrete structures built at ground level as at BOWEN. Progress on the item is shown to have been 90% complete by 1 October 1943.

For Dale Kruger an ex USAAF ground echelon member of the 90 BG (H), Mackay sat well in his memory -

'...Most of our leaves started after we arrived in Port Moresby, New Guinea. I went down to Mackay, Queensland two times for ten days each time. Once I flew down on a C-47 and another time on a B-24; one leave in 1943 and one in 1944. They were good times.'

'Good times' as these were, they were marred 14 June 1943 with the crash of a B-17 transport, shortly after take-off. Loaded with men returning from R & R, the aircraft crashed into a bridge over Bakers Creek.

On 6 October 1991, the Mackay Port Authority formally opened its new airport terminal. Positioned in the southern quadrant of the crossed airstrips it is located generally where, early in 1942, hasty efforts were made to provide protective splinter proof revetments for aircraft in the defence of this country.

MT HEDLOW *(-/6/1944)*
Extract from 1:63 360 (1 inch = 1 mile) Military Topo Survey Map. This field was only 'discovered' by research at AA. *[Department of Defence]*

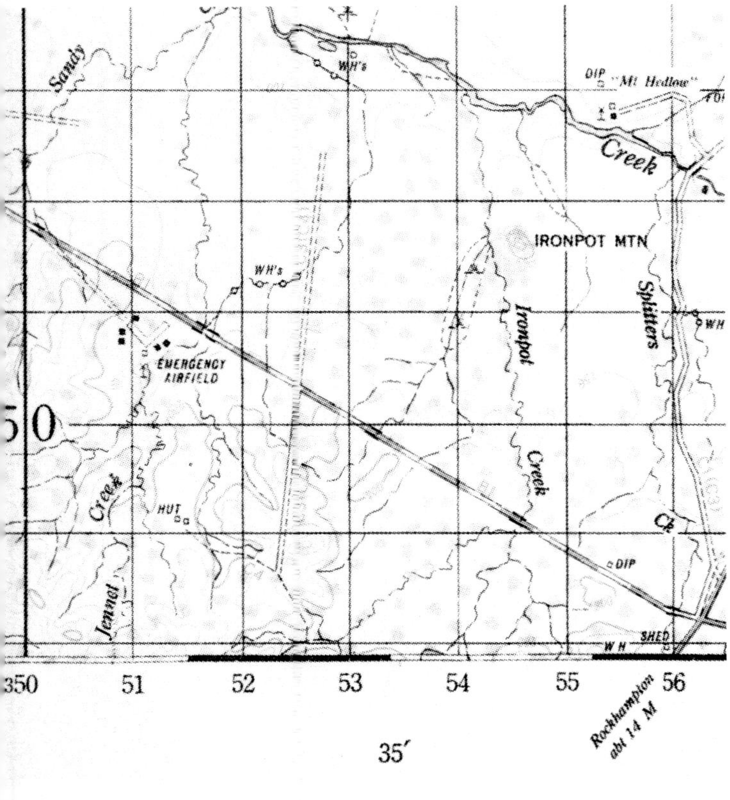

MT HEDLOW

'...Painting of three hangars at Mt Hedlow drome.'

One of my better days during a period of consistent Archive study, produced the following excerpt from a weekly report to the Camouflage Committee in Brisbane -

'...Division 4 - Rockhampton - USM 14 Nerimba Camp - the painters are engaged on the three new buildings. Parkhurst buildings have been inspected but are not ready for painting. Artillery camp - painting of three hangars at Mt Hedlow drome has been completed. Two others are nearing completion. The painting and colours are very satisfactory. These buildings could have been sited much better had Mr Rae been consulted by US Engineers.'

That report was dated 21 March 1943. Immediately I began a search for 1944 mapping and any aerial photographs that might show detail of this small grass surfaced drome. I located 'Key Diagram No. 560' at the Department of Lands photo library in Brisbane but they unfortunately did not hold the photos. The Key Diagram indicated photography dated 25 November 1944! This would have been ideal to pick up early 1943 construction. Alas, my check with the NLA print holding revealed that

MT HEDLOW *(15/1/1944)*
This is the compilation date on Plan CR-CA-15, one of numerous carefully prepared record plans surviving in AA. These plans carry the signature of John C Parker Jnr, Captain CE, Resident Engineer Base Area Command 3 and certainly are a credit to someone in that organisation at the time. The holding extends to several clear photos of military installations in this Rockhampton sector. *[AA BP 198/1]*

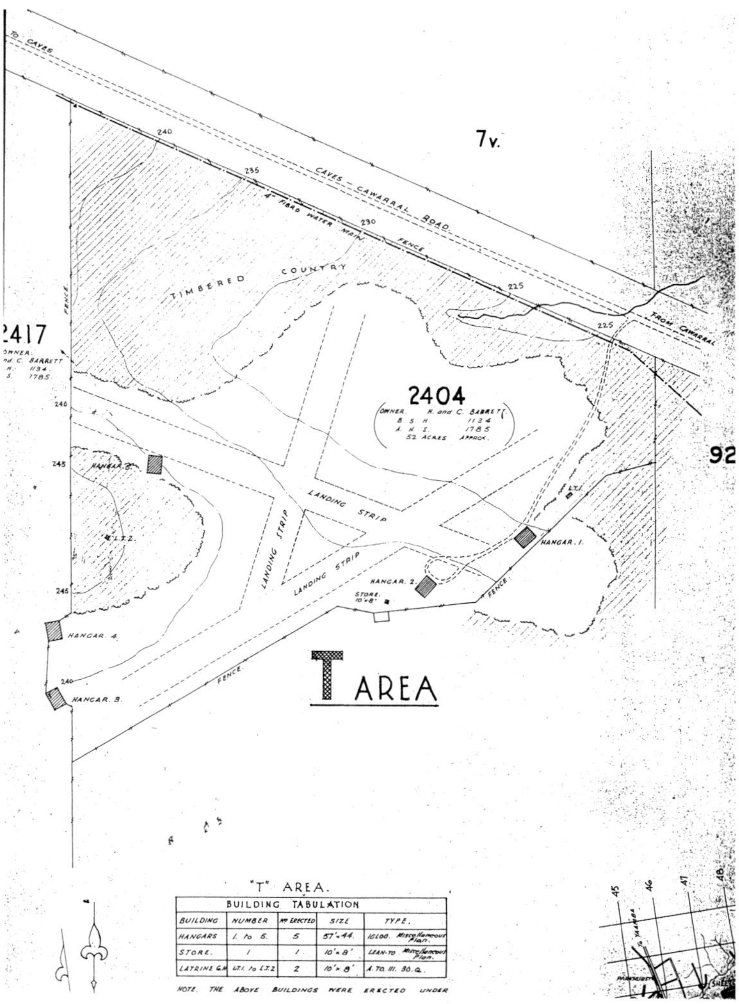

MT HEDLOW (22/5/1992)
Ron MacDonald (ex RAAF), a long term Rockhampton resident, accompanied us to see what was left of MT HEDLOW. As the sign on the gate shows this is private property and time did not permit our seeking permission to enter. From the gate it was clear that the '57' x 44' Igloo hangars' were nowhere to be seen. [Author]

the true date on 'edge of photos' was 25 November 1941, well prior to the construction. That was just another example of misleading dating on photographic records.

AA Brisbane hold a truly magnificent item in BP 198/1. It is a stoutly bound folder of mostly construction layout plans for WW2 US Army camps and other facilities in the Rockhampton area. The signature on these plans is that of a Capt John C Parker Jnr. The same officer was involved in work at CHARLEVILLE in 1942 and I suspect also at IRON RANGE around July 1943.

Sure enough, one landing ground appeared in Area 'T'. The five hangars, each 57' x 44' were listed as Igloo type. From available information they were probably rolled sheet (corrugated?) steel on maybe a light tubular steel frame. Complete with the full triangle of grass strips, the field catered specifically to light 'spotter' aircraft, such as Aeronca L3 or Piper L4. Their use was justified in areas of relative air superiority. The US 24 and 41 Infantry Divisions trained here together with several Field Artillery Battalions.

Light aircraft provided the 'eyes' for the artillery units and a range of communication use. Wherever large ground formations, particularly Artillery units, trained in Queensland it is likely 'spotter' aircraft were involved. The smaller types were far less reliant on airstrip preparation. Cleared paddocks and straight country roads were often enough. Hence accurate documenting of such 'landing grounds' now is almost impossible. The foresight of someone, maybe John C Parker Jr back in 1944, has allowed MT HEDLOW a firm place in this review.

ROCKHAMPTON (19/7/1991)
This current day Staff Amenities building housed Aeradio equipment and personnel during the war years. In the foreground is Roy Barney, long term member of Rockhampton airfield ground staff. Roy is moving off to his vehicle before showing me various locations by the runways where demolition access shafts were constructed during the war. [Author]

ROCKHAMPTON

'...I'm not real sure whether it was a Liberator or a B-54 or whatever - it was a bomber type aircraft that overshot the runway at Rockhampton. It finished in Lion Creek. The word soon spread amongst the boys of the town - I would have been about eleven or twelve - that there were souvenirs for the taking. We got on our bikes and all rode there. We started attacking this dammed plane. I remember one guy was even trying to get the half steering wheel off the joystick column. We were having a great time, when suddenly we were spotted from the American complex and a whole truckload of armed guards arrived, chased us all away. We stood round in awe at these guys with their sub machine guns, very impressive indeed. So that sort of ruined the souvenir hunting. Then as we watched for a while, one guy, one of these Americans, said to them all - "chow time". They all got in the truck and drove back to their huts again; left us to it. So we immediately started again, attacking steering wheels and things, and suddenly someone over there at the chow hut spotted us again. But they were too busy eating to come back, so some guy came out and fired a rifle in the air and that was enough for us!'

For Graham Wilson of Gladstone, this was his first recollection of ROCKHAMPTON airfield, along with -

'...the frequent arrival of fighter aircraft to refuel and head on northwards'.

A letter from P H 'Skip' Moody, Manager, Air Taxi P/L to the Minister for Air dated 16 May 1939, a few months before the outbreak of war in Europe, serves to set the scene at ROCKHAMPTON aerodrome. It reads in part -

'...This aerodrome is controlled by the writer and Capt H L Fraser who have a 16 year lease from Rockhampton City Council. The aerodrome was originally established by Capt Fraser at considerable cost in clearing and surface preparation, no financial assistance being granted by either the Commonwealth or the City Council. This aerodrome has advantages over many others on the coast now being prepared by the Commonwealth at considerable cost. Its principal advantage being its strategic position which precludes it from enemy naval raider attack. It is situated 16 miles from the coast, which in turn is flanked by the Keppel Shoals...In view of the establishment of RAAF squadrons at Brisbane, Townsville and Port Moresby, it is reasonable to assume that considerable air traffic between these stations will result. Rockhampton is half-way between Brisbane and Townsville... As it has recently been equipped with wireless and meteorological facilities it is probable that more and more use will be made of this ground by RAAF aircraft.'

Mr Moody listed items of existing facilities, not the least important of which was underground storage capacity of 7 000 gallons embracing 'all grades of aviation fuel'. He outlined two schemes of financing further work at the drome. 'Proposal A' involved the Commonwealth taking over the drome. The 'alternative B' amounted to an annual subsidy which would allow free use of the drome by all RAAF aircraft and civil aircraft operating under Air Mail contracts. He concluded -

'...In the course of our management here, we encounter a certain amount of opposition to paying landing fees, not pronounced perhaps, but nevertheless it is sometimes apparent, even from RAAF officers. We cannot avoid this, we have made quite an extensive financial investment and must necessarily see that a return is forthcoming.'

Moody's established drome was referred to as CONNOR PARK. In November 1939, DCA and the RAAF considered alternative sites, resolving 'in the present emergency', to proceed with improvements linked to unemployment relief schemes, at CONNOR PARK. The next two years saw attention to its development as an AOB with 'bomb dump' etc., apparently with more expedition than evident further north. By January of 1942, with the outbreak of war in the Pacific, the push really was on to provide Ferry Station facilities at ROCKHAMPTON airfield and elsewhere. A report by F-Lt P James applies -

'...Existing "Civil" and "AOB" aerodrome is capable of extension and development as required, subject to the eastern approach and flood position being regarded as satisfactory... Part of the aerodrome was flooded from the Fitzroy River in 1918, but a report by the Chief Engineer of Public Works, Queensland, indicates that such a flood may be expected once in 100 years. An alternate site known as Bob's Creek, is situated 14 miles south of Rockhampton near Archer Railway siding... soil is particularly good... practically cleared and on good contours, but the road between Rockhampton and the site is subject to much more frequent flooding than is the Rockhampton aerodrome, and services are not available within easy distance...'

F-Lt James attached upgrading plans and estimates and proposed sites for added buildings. He noted a campsite -

'...having good natural cover, and adjacent Lion Creek for the disposal of sewerage effluent, is on allotment 24.'

ROCKHAMPTON (-/1/1946)

This is an extract from 1:63 360 (1 inch = 1 mile) Military Survey Mapping. This map shows the airfield in contrast to others produced in 1942-43 which purposefully excluded any indication of airfields. [Department of Defence]

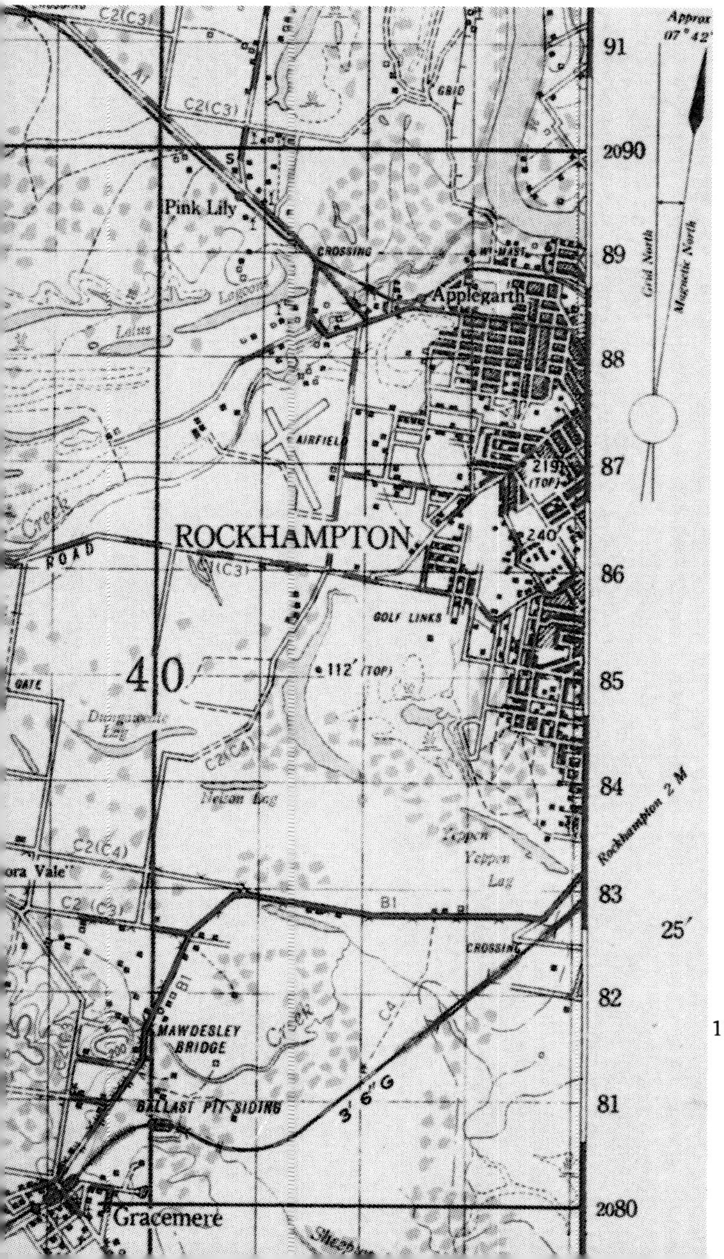

The first sustained movement of new fighter aircraft northward, occurred over the next few weeks. This would have been the time Graham Wilson remembered seeing numbers of Kittyhawks staging through ROCKHAMPTON. Barney Davies flew one of the first ten Kittyhawks of the RAAF's 75 Sqn north from Bankstown. Barney called in -

'...on 8 March 1942 at Rockhampton... the small lot of ground staff had just refuelled 30-40 US P-39 Airacobras - no help from their own pilots - and were pretty exhausted. We got stuck in and helped do our own Kittyhawks - I think out of 44 gallon drums.'

While defence strategy called for airfield build-up, there was also the need for contingency plans. Here movable 'obstacles' were fashioned and preparations made to mine the runways. A 30 June 1942 report shows -

'...provision of additional buildings and services - work 92% complete; 12 splinter proof pens - work 40% complete; obstruction barriers - work 80% complete.'

The Dept of Home Security's camouflage people were active here too. A 27 June 1942 report mentions -

'...Rockhampton Aerodrome: construction of runways and general groundwork is nearly completed, no buildings have been camouflaged. Positions for splinter proofs have been selected but they are in the open and have no concealment. This could have been obtained as large trees are not far from (the chosen sites)... A village is being erected on the outskirts of the drome in continuance of the residential section of the city. It is being handled with a great deal of thought... Roads are being formed and Mr Rae has suggested better track discipline on the part of the contractors. Mr Besly has been put in charge of the work on this drome.'

This village site is likely F-Lt James' 'allotment 24'. Mr O J Besly, Camoufleur, reported further, 26 July 1942 -

'...The toning down of the aeradio station would be completed at the end of the week. The painting of the drome buildings has commenced. Ashes were ordered by Dept of Interior to spread over sawdust paths which had been put down previously by the Army guard as sentry paths. The colour schemes for buildings (etc.)... in the village scheme are also being supervised.'

As early as July 1942, Rockhampton city was identified by USAAF crews as an attractive leave destination. Walter H Johnson's diary of the 435 BS, 19 BG mentions, for July 20 1942 -

'...Ezzard and Simmons and crew pick up vacationers at Rockhampton, at the same time checking off the crews in LB 30 and Radar operation...'

AWC Minutes of 19 August 1942 note 'advice' from Col. Sverdrup, Chief Construction Engineer USASOS, to the effect that mining of runways should cease. A DCA letter to the Dept of the Army, on 16 September reported -

'...shafts for demolition purposes had been excavated at the sides of, and extended under the runways at Rockhampton Aerodrome. Our officer reported that work had been held in abeyance since 26 August 1942.'

They were apparently 'marked and lit', but DCA urged -

'...that these open shafts be either filled in or made safe by timberings, and the shaft openings covered over to permit of the safe passage over them of heavy aircraft.'

Two weeks later the same file shows 'work suspended on demolition of runways'.

Early in 1942 the US committed a large ground force to the SWPA. The 41 Infantry Division landed in the south around May. MacArthur soon had them north bound, initially for acclimatisation and training, and later New Guinea. Rockhampton became the training area, with camps springing up along the Yeppoon road. Artillery training called for aerial spotting and the RAAF's No 4 Army Co-op Sqn was one unit involved. Jack Williamson was then at KINGAROY with 4 Sqn. Around 30 August 1942 a flight of five Wirraways -

'...went up to Rockhampton and we cooperated with the Americans and their artillery. They hadn't fired with air to ground direction... they fired their 105mm guns...'

Jack described the stay as 'a couple of weeks only', during which the aircraft operated from ROCKHAMPTON. Interestingly, during Jack's return to KINGAROY, engine trouble caused his unscheduled landing at MUNDUBBERA and an overnight stay at the 'local Pub'. The departure from that 'all-over grass field with only a tin shed and a windsock' was by unceremonial road tow. One 3 ton truck took the wings and the other in the recovery party, hitched up the tail wheel and towed the aircraft on its own main wheels, back to KINGAROY!

The airfield remained under shared but dubiously effective control. Mounting traffic taxed facilities at the ground. When DCA voiced concern over 'electrical interference' to Wireless equipment, the RAAF investigated. Sqn Ldr H V Davies reported for DWO Brisbane, dated 29 October 1942 saying, in part -

'...Rockhampton Council own the original aerodrome site, such additional land required to develop the aerodrome to its present boundaries has been taken by the RAAF. A small area (approx 2 acres) of Rockhampton Council's land has been leased to Civil Aviation who have erected their own Meteorological and Wireless communication building with aerials, also an emergency power house...'

Davies found RAAF and Civil Aviation were trying to operate their separate radio facilities in this one inadequate building. He found recently installed High Tension lines were causing the other technically based 'interference'. Davies' answers were 'undergrounding the H/T' and a RAAF 'Comet' barracks hut nearby which could better house the RAAF W/T section. He also saw the need to reconcile 'control matters' -

'...At present, neither the RAAF Commanding Officer nor the Officer in Charge of Civil Aviation has received directions as to who is in charge of the aerodrome...'

The problem went deeper. Allied service aircraft were not -

'...tuned into Civil Aviation's W/T wavelength, which is the only present means of communication with aircraft in the air as no Duty Pilots tower has been provided for visual control...it is suggested that a Duty Pilots tower, together with necessary equipment, be provided so that RAAF have control of all aircraft...'

Colin Munro was another RAAF veteran with experience at ROCKHAMPTON. He was with 5 Army Co-op Sqn based at TOOWOOMBA, early in 1943 -

'...The Squadron at that time was in three flights and I was OC B Flight I think. I received instructions in the morning to take a flight to Rockhampton to train with the American 41 Division. We were to provide some air support for training purposes...it took us about 4 hours to pack up a whole flight and put it on the road to Rockhampton. The aircraft took off then and of course arrived well ahead of the transport...'

ROCKHAMPTON (27/10/1943)
Frame V 1045 is one of the several overlapping verticals in KD 915. A single C-47 Dakota Transport is on the tarmac with others difficult to identify. ROCKHAMPTON was on the ferry route to New Guinea. [NLA (RAAF Official)]

Again, their stay was a matter of weeks only. Colin speaks of 'landing on and operating from paddocks'. Very likely he used strips like MT HEDLOW and the one near Nerimba that I've not been able to locate.

By May 1943 camouflage reports show something of the tenor of life at ROCKHAMPTON. On 3 May -

'...visited Rockhampton AOB and contacted CO, F-O Bernard. Buildings on drome are well dispersed and toned down. Runways are gravelled. Under joint control of RAAF, DCA and US. Busiest airport south of Garbutt... refuelling base for US transport.'

Continuing work on camouflage had begun to recede -

'...the wiring has been completed on pyrotechnics and detonator stores building and garnishing in dark earth and khaki green will commence on Monday 10. Fences erected to simulate farm properties were being removed to allow increase of tarmac space. At Mr Rae's suggestion the CO has agreed to move only those that are absolutely necessary. Sqn Ldr Murphy visited with camoufleur Wyeth and informed Mr Rae that nothing had yet been done regarding placements of gun pits...'

This need for small anti-aircraft machine gun pits had almost passed as a memo dated 16 June 1943 from DCA to Dept of Air notes -

'...existing gun pit in front of the RAAF building fouls one of the proposed taxiways and permission to remove this pit is requested. If deemed necessary, a replacement pit will be created... suggested this pit should be removed in any case as its close proximity to the apron is considered a potential danger to aircraft.'

Early bitumen sealing of gravel runways at ROCKHAMPTON met with varied success. The airfield's Ferry Station role gave it not the highest of priorities and the background reluctance of Dept of Air (RAAF) to increase its stake showed in a RAAF file minute (DASP to DWB) dated 15 July 1943 -

'...I don't think we need encumber ourselves with responsibility for maintenance of Rockhampton aerodrome. I presume that CAD is still responsible for maintenance at Parafield and Alice Springs and Cloncurry where we have RAAF units. As CAD still exists it might as well have some responsibilities. In relation to priorities we can support application by civil...'

August 1943 saw the completion of further upgrading work at the aerodrome. It also saw the commencement of construction of a Duty Pilot tower. Its heavy timber framed construction took 3 months. By 22 December 1943 it was basically finished awaiting equipment installation. Archive documents show the local builder was Robert Cousins and that the same gentleman erected the

ROCKHAMPTON (14/1/1991)
Part of some project photography flown to record serious flooding at the time, two frames have been merged by scanning techniques to achieve this photo. Developing Regional Tourism has allowed the recent commissioning of a modern passenger terminal building seen in the centre of this photograph. The problem of the airfield's siting in the Yeppen Flood Plain of the Fitzroy River has been with it from inception. [Department of Lands, Qld]

'galvanised iron' hangars at MT HEDLOW.

Percy Yates was a navigator in the RAAF's 71 Reserve Sqn operating from LOWOOD. Their Ansons were engaged in convoy escort and anti submarine patrol. It was probably in 1943 that he recalled staging -

'...in at Rockhampton to refuel after a morning sortie to take over an evening job before staging overnight at Bundaberg. Unknown to us while taxiing through long grass at Rocky, a grain of seed found its way into the pitot head blocking the supply of air to the speedo. We found on lift-off that the speedo wasn't working, but as it was a daylight sortie we carried on... However, the patrol stretched out longer than expected and by the time we headed for Bundaberg, it was dark. The pilot had never been to Bundy before and with a fairly heavily loaded kite an approach to a strange field with no air speed indicator was rather awkward. To add to the joy, when we arrived there was night flying training going on with aircraft circling all over the place. We dived into the approach to ensure sufficient air speed for safety, and as the skipper concentrated on the approach he had me poised over the flap lever to pump down full flap as soon as he crossed the fence...'

They landed and managed to brake short of the far fence -

'...On later finding why the instrument wasn't working we said some rude things about the way the grass was let grow at Rocky airfield.'

Reconstruction and sealing of both runways was long overdue by April 1944. Work had been executed piecemeal, with the typical constraints of less than 'A1' priority. A DASP to DWO Minute of 22 April 1944 resolved -

'...in view of the current and future necessity of the aerodrome for Service and commercial purposes, it would appear desirable to make a complete job of the reconstruction and sealing.'

The RAAF's ***Aerodrome Guide***, January 1946 addressed sixteen prominent airfields of this State with diagrams and listed detail. 'Control', either RAAF or Civil Aviation, is the last listed on all save ROCKHAMPTON. There, no mention is made! The Commonwealth Government owned and operated the Airport after the War. During the 1980s, the Government moved to reduce this responsibility. Negotiations with Rockhampton City Council (RCC) were protracted. Agreement reached led to RCC taking over the new Terminal operation in February 1989, and the whole Airport in September 1991.

Runway upgrading in 1987 brought echoes from WW2. Areas of suspicious subsidence appeared. Sure enough, after checking archive records, and some exploratory drilling, remnants of the 1942 demolition tunnels were discovered and hopefully put to rest. Also echoing from the war years, the builder of the new Terminal was T F Woollam and Son P/L.

LONGREACH (-/7/1942)
'Ready! Point! Don't fire! Pistol inspection for US Airmen'. *This is essentially the caption accompanying this photograph when it was first published by National Geographic in January 1943. The location and details of the unit involved were censored and excluded from the article. However, Australians would surely recognise the faded name on the hangar gable.* [National Geographic]

LONGREACH

'...BASEBALL GAME - FIRST SOFT BALL GAME IN WEST. All Longreach turned out to see the first soft ball game in the West when a large and enthusiastic crowd attended the show grounds last Sunday afternoon. The exhibition of a game new to Westerners was appreciated and the afternoon was also a financial success, the gate takings being £35, while the refreshment booth provided an additional £11. The proceeds were for the benefit of... (Australian Comforts Fund?).'
[The Longreach Leader, June 6 1942]

This newspaper report was cut short and the reader left wondering who benefited! Presumably, strict press censorship meant the US 19 BG personnel who no doubt staged this game on 31 May 1942, were not mentioned as having so recently arrived in Longreach. The worthy organisation that did benefit financially, was probably the Australian Comforts Fund or similar war charity.

It has been said Australia grew up on the sheep's back. Wool was for years the dominant export of this land. Longreach was, and in many ways still is synonymous with 'wool'. The motor truck grew to replace the bullock wagons feeding the railheads, and the isolation helped breed resourceful automotive mechanics. The transport revolution that was aviation impacted early on Longreach. Only poor roads linked the western ends of the rail lines from the coast and their connection by air beckoned to the QANTAS founders. The countryside was flat and forgiving in emergency landing. There were capable engine repairers and as a bonus, the whole region lay on the Great Circle, Darwin south eastward to Sydney.

In his book ***Qantas Rising*** on the earliest days of the airline, Hudson (later Sir Hudson) Fysh refers to a bleak period in the fledgling company's finances. They decided -

'...to take to the sky and visit the smaller towns which had not yet seen an aeroplane and do some joy riding in order to keep solvent. McGinness was to go west and north in the Avro taking Baird with him as mechanic, while I was to go south with Herb Avery.'

The time was May 1921. Herb Avery was, even then a respected automotive mechanic in Longreach and Herb Avery was my maternal Grandfather.

It is not surprising that LONGREACH as an airfield should figure prominently on any inland ferry route from the populous southeast of the country to Darwin. The Americans certainly realised this prior to the entry of the Japanese into WW2 in the Pacific.

RAAF records surviving in AA contain drawings dating to January 1942 relating to a 'temporary hutted camp' at LONGREACH. One shows such a camp sited close to the

LONGREACH (15/9/1989)
The US pistol inspection group has long gone, but the National Estate registered QANTAS hangar remains. [Author]

LONGREACH (-/7/1942)
National Geographic photographer Howell Walker has confirmed what really is the obvious, in that this 'briefing session' was actually being conducted within the QANTAS hangar at LONGREACH. Now on the National Estate Register, the building's architecture is obviously that of the early 1930s. The site of this briefing was the northeastern corner of the hangar. [National Geographic]

racecourse on the northwestern boundary of the field. It also shows an E-W runway to be developed taking off from the intersection of two existing short runways. The MRC History records a start on such construction on 9 February 1942. On 16 March 1942 an inspection of these works was carried out by P-O Keays.

A review of items registered on the General Index Cards of the then Works and Services Branch of the Department of Interior shows five lodgements by the 'USM' (US Military?) at LONGREACH airfield. Most relate to extensions of runways, but one other reads 'US accommodation - hiring of buildings - Town Hall, Masonic Hall and Oddfellows Hall'. These are all registered in the period 8 June through to 1 July 1942.

American records show the date of arrival of units of 19 BG at LONGREACH as around 5 May 1942. A RAAF signal of 4 May 1942 to Air Board from NEA reads -

*'...**Longreach**. Wilson of Main Roads advises work on 3 000' runway almost complete. Unless approval forthcoming will be necessary to put men off and plant which may be needed in additional extensions to 6 000'. This extension necessary. US Officers at Longreach have also requested taxiways and parking areas. Unit already in occupation. Please advise.'*

The MRC man mentioned is the late C H (Clem) Wilson who later served several months at IRON RANGE in charge of the AWC organisation there.

Late May the strategic situation had improved to the point where the Allied Air Forces could think of moving units closer towards the coast and away from the 'safety' of LONGREACH. The AOC's report to Air Board dated 28 May 1942 reviewed planning for all airfields in NEA. In respect of LONGREACH it said -

'...the 2 000 yard runway has been completed at Longreach and it is not proposed at this juncture to proceed... with any other construction. Works in this district are to be carried out only after aerodromes in the north of York Peninsula have been completed.'

Interestingly the General Index Cards for 'Longreach RAAF' show an entry dated September of 1947 relating to 'acquisition (of land) for underground Ops. Building £30'. That suggests such a building was constructed, though where precisely I've not been able to learn.

After a visit by Gen Brett and his discussions with the AOC NEA, a 21 June 1942 directive curtailed aerodrome work along the Townsville-Cloncurry route. It addressed LONGREACH saying it -

'...will revert to ferry-route status when the units (19 Group) now occupying... are moved to Mareeba...'

The SDCC had LONGREACH within its Division 4, centred on ROCKHAMPTON. The earliest relevant mention I've found in camouflage reports show -

'...camoufleur V Corlett arrived in Longreach on 24 June. He contacted the American CO who received him well and gave him charge of all camouflage work. No hideouts are being constructed there being no small aircraft on the drome. B-17s are the only planes arriving there at present. There is very little timber, and it is almost impossible to hide the big planes...'

Vic Corlett found by 30 July 1942 that the Americans 'did not want any camouflage done at Longreach'.

Lawrence Moore was an Electrical/Electronics specialist involved in aircraft maintenance work. He was with the 11 BS, 7 BG. They had gravitated here after the retreat from Java via Broome. Lawrence transferred to the 28 BS, 19 BG during the time in LONGREACH. He recalls being billeted in the Longreach Shire Hall -

'... my bunk was on the stage! and the Mess Hall was in the basement. At Longreach Airfield we had one corner of a hangar for setting up our workshop.'

I'd asked Larry how he'd 'found the locals' -

'...There were some refugees in from Darwin... to get away from the Jap bombings. Overall the "locals" treated us as well or better than the "locals" back here in America. Several of the fellows married local girls. We were frequently invited into homes for dining. They were specially glad to see us as most of the local men of service age were away in the Middle East.'

The National Geographic Magazine sent Howell Walker to Australia in 1941 to write back on 'the making of an Anzac'. After Pearl Harbor, he headed for Darwin sensing a story of its impending bombing. Even with War Correspondent status he recalls, thankfully, the Authorities at Alice Springs denying him permission to 'cycle north to Birdum'. He reached Darwin, got to ride his bike and fell off, 'near a pond in the Botanical Gardens'. With a leg in plaster, weeks of summer heat and 'still no air raid', Howell gave up and left for the south on 18 February 1942 -

'...While I was on the train, the raid happened and, you know, a lot of people beat us to Birdum!'

Howell then set about a story on Sydney. Right on cue, the Jap midget subs attacked there, 31 May 1942. With his leg still in plaster, he met the Sydney girl he was later to marry, decided to join the US Navy and got an invite to LONGREACH to go on a 'bombing mission to New Guinea', all in quick time. Howell recalled that -

'...goats were something of a problem at Longreach. When they weren't cluttering up the airfield, they swarmed through the business section of the town, licking away at cinema hoardings and other glue pasted posters... Like Ralph and Rachel Dietz, Sheila and I met and married in Australia in 1942.'

Ralph Dietz is one of those 19 BG airmen who met and married a Longreach girl, Rachel Curtis. Like a few others

LONGREACH (-/7/1942)
B-17E '12462' makes an imposing backdrop to this USAAC unit parade. The National Geographic photo/journalist Howell Walker, tells me it was pure luck that the elevated concrete water tower at Longreach is just visible on the skyline in this photo to right of the QANTAS hangar. [National Geographic]

LONGREACH (15/9/1989)
In place of the 4 engined B-17 we now see a typical twin engined commuter aircraft on the ground at LONGREACH. Both the QANTAS hangar and the elevated concrete water tower (just to its right on the skyline) confirm this scene as where that parade took place 'fifty years ago'. [Author]

LONGREACH ca(-/6/1942)
I'm not sure how many P-40 Kittyhawk fighters pancaked on LONGREACH Airfield during May to July of 1942 but this photo shows at least one. In the background are the QANTAS Hangar and two US B-17E bombers. [R Dietz ex USAAF]

in recent times they have revisited Longreach.

During this study I have asked veterans with whom I've made contact, for their recollection of any 'notable event' that links them in memory to the particular airfield in question. Lawrence Moore answered this way -

> *'...there was an event that happened in Longreach that still remains an unsolved mystery, at least as far as I know. It happened one night when it was very stormy with much rain and lightning - it was just pouring. There was a safe in our Squadron Orderly Room where many of the men had put their recent pay for safe keeping. There was over £3,000 in it. Someone stole it - carried it across the stage (the "orderly room" was in fact in the Longreach Shire Hall) and out of the building, or tossed it out of the window to a waiting vehicle below, and drove off with it. As far as I have heard no one ever found the safe. In July we moved to Mareeba and we were preparing to come home to US in December 1942. Two medics in the Squadron stole a staff car and headed for Longreach. The driver went to sleep and ran off the road into a tree and cracked up the car and badly injured both the men. Their excuse for stealing the car was to go see their girl friends! Had they wanted to stay with them they could have asked to be transferred to the other group that replaced us (43 BG). I don't know if there was ever any questioning of them about the theft of the safe, but it looked very suspicious with all of what had happened. Who knows? maybe that safe with £3,000 Aussie pounds is still buried some place near Longreach...'*

LONGREACH (-/7/1942)
Photographed on the steps of their billet, the Longreach Shire Hall, these three US airmen are probably in process of 'checking out'. The moulded lightweight 'pith' helmets on issue to many of these servicemen at the time, shows they were expected to be serving in the tropical Philippine Islands. [L Moore ex USAAF]

LONGREACH (-/7/1942)
Not all the USAAF men flew when it came to shifting to another base. This group were treated to the Queensland Railways journey eastward to the coast. On the right is a typical group of local bystanders. [L Moore ex USAAF]

The Longreach Leader for Saturday 30 May 1942 reported widespread, drought breaking yet 'stormy' rain but I could find no mention of Larry's stolen safe, even for weeks after. My enquiries to the Queensland Police Museum in Brisbane have also been unsuccessful. The matter may have been kept entirely within US Service Police jurisdiction. So as Larry says, maybe such a safe still lies discarded or buried somewhere near Longreach.

Angela Moffat's book titled ***The Longreach Story*** covers briefly the wartime period and the USAAF in Longreach. It supports the presence of very capable automotive workshops and repair facilities in Longreach and notes that in several instances urgent aircraft repairs were made possible virtually on the airfield. Mrs Moffat's story touches on reports of aircraft flying low in the district and 'machine gunning the countryside'. While no doubt there are two sides to such reports it does remind me of a close personal link with my own Grandfather, Herb Avery.

While I'm not sure how he acquired it, he did at war's end have a serviceable .50 calibre Browning machine gun. I believe my Uncles, both now dead, told me that it had been 'given' to him by some US Air Force person or persons from 'up CLONCURRY way'. Considered opinion from some of Herb's remaining contemporaries, is it was 'left' with him in an unserviceable state. With Herb's love of firearms, it was certain to be mended! I believe my Uncles told me that at some stage they had it bolted on the tray of a utility and it's not too difficult to imagine the scene out kangaroo shooting with such a weapon.

However, sticking to the facts as I know them, after my Grandfather died in 1953 I asked my Grandmother to release the 'gun' to me. I set about visiting the three or four addresses around Brisbane where separate components of the weapon had been held by Herb's friends. They were happy to release their parts to me. And so it was with some trepidation that I made the overnight train journey back to Maryborough after attendance at some School Cadet camp in Brisbane. Imagine the task of transporting the largest item, namely the body with barrel attached, about 6′ in length. Wrapped in blankets and tied with string, I shouldered it past the ticket inspector and onto the 'Rocky Mail' train. We lived a few blocks from the Railway Station and the 2 a.m. arrival meant a couple of trips on foot to get it all home! I spent weeks restoring it, but with no ammunition, there was no chance of 'testing'.

LONGREACH (23/9/1943)
Frame V 1015 is one of several in KD 878. The QANTAS hangar is the prominent building on the airfield. By this time both the 87°and the 40° runways had been lengthened. *[NLA (RAAF Official)]*

When I finished school and faced the reality of study years away at University I offered the weapon, complete with its ammunition belt box, to the Maryborough Air Training Corps Unit on the basis that they might 'like to look after it for me and maybe use it in some instructional mode for their Cadets'. Years after, on my return to Maryborough, I approached the ATC Officer in Charge to recover the weapon to my own safe keeping. Needless to say, no one knew anything at all about it!

I have tried since to find out what did happen to it, but to no avail. Firearms collectors embrace all walks of life - from Electricians to Commonwealth Bank Managers. Somewhere there is the proud, but quiet, holder of this .50 calibre machine gun, which very likely came from the 19 BG's stay in LONGREACH.

LONGREACH (22/4/1990)
This photo is one of a survey to record the extent of flooding. It shows some changes in the last fifty years. The main runways have survived and been upgraded. The QANTAS hangar still dominates the building area. The Longreach Shire Council owns and operates the airport today. To the southeast across the road and rail is the Australian Stockman's Hall of Fame complex, proudly displaying the heritage of outback Australia. *[Department of Lands, Qld]*

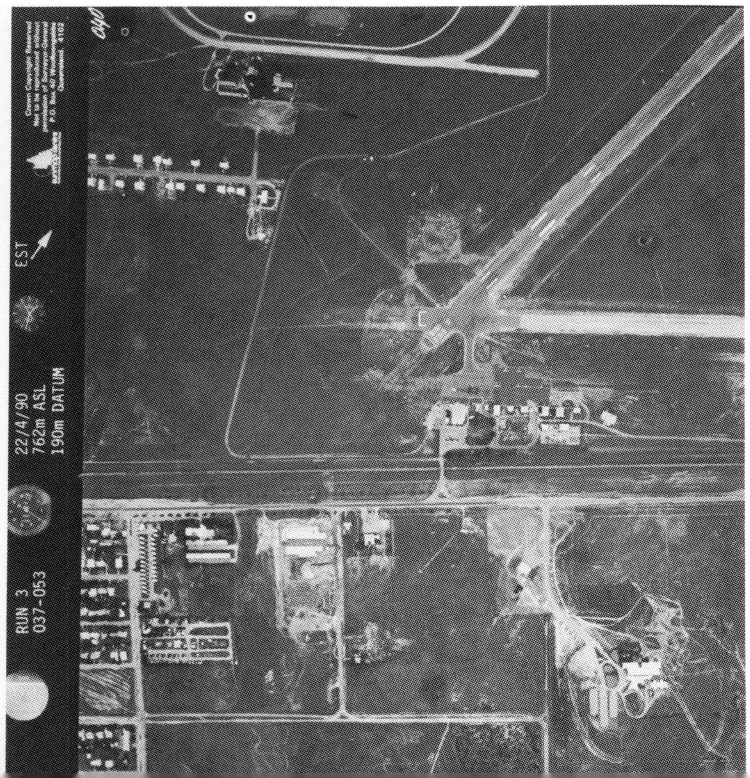

BARCALDINE

'...They were a pretty harmless lot .'

The late Clem H Wilson remembered the American, predominantly negro, unit there in 1942. Archive documents show the 585 Engineer Company (Dump Truck) (CLD), comprising at least 120 men, reached Barcaldine May 1942. Clem, at the time Engineer in Charge of the MRC's Central Western District, told me the townspeople felt rather sorry for the Unit. They camped at the Showgrounds before moving to the 'old wool scour' a few miles west of the town. They weren't there long.

Isabel Hoch, in her ***Barcaldine 1846-1986*** history published by Barcaldine Shire Council (BSC), refers to a Council minute recording full cooperation promised to-

'...P-O Trent and party of US Air Corps, April 1942.'

By 22 April, P-O H V 'Jim' Trench accompanied Clem Wilson, and Mr Andrew, an 'architect with Dept of Int', on an inspection of various 'sites'. They investigated alternatives for WINTON and LONGREACH and sites for two Heavy Bomber fields, at Barcaldine and Alpha. Surely Jim Trench was the 'P-O Trent' inscribed in the BSC Minute book.

When I first interviewed Clem Wilson in 1991, he could not recall any wartime development of the present airfield site. As testimony to the frailty of most human memory I later found ample Archive evidence to show Clem fostered the new BARCALDINE strip site. A signal to Col. Matthews CE US HQ Townsville from 'Cox, Comworks Brisbane' dated 19 May 1942 referred to -

'...(your) Company of Engineers at Barcaldine. Could Wilson make use of (them) clearing strip at aerodrome site? Plane recently made emergency landing at (claypan NW of) Barcaldine, with great difficulty.'

The RAAF's ***Aerodrome Guide*** of January 1946 mentions only a 'natural surface' area ½ mile NW of the town. However, BSC applied for a license for, presumably, this 'new aerodrome' in May 1945. DCA granted licence '486' effective 28 June 1945. The ***MRC History***, published in 1949, summarised BARCALDINE-

'...1 mile east of the township a strip of 4 000' x 500' was cleared by a detachment of USA engineers using coloured labour. The strip was...left unfinished. The MRC completed the formation and gravelling, erected Army reserve storage and an oil depot.'

BARCALDINE (11/8/1951)
This is an extract from systematic mapping photography. It shows the established airfield of today in the southeastern quarter. This is the site on which work was commenced by coloured US Troops in 1942. The triangular area north west of the town was the claypan landing ground of the early 1940s. *[AUSLIG]*

BLACKALL

'...I didn't even know there was an airstrip there!'

This is the comment I received when I spoke to someone in a house just at the northern end of the old Blackall civil aerodrome, and it's not surprising. It is quite easy to find a relative newcomer to a district.

The old civil field south of the river at Blackall was the only one there during the early years of the war. With the decision to construct major Air Depot facilities on the New South Wales-Victoria border, the need arose for an adequate Inland Ferry Route. Queensland stations on this route to CLONCURRY were to be CUNNAMULLA, CHARLEVILLE, BLACKALL, LONGREACH and WINTON. Some consideration was given to Quilpie but it was discarded. The civil drome was far too small so a fresh site west of Blackall was adopted.

The ***MRC History*** notes that work on the new drome -

'...continued from August 1942 to March 1943...'

The SDCC fortnightly reports touch on this work in September of 1942 -

'...Mr Wilson (MRC) has asked for a camoufleur to accompany him for a few days on tour of inspection in his area...'

Mr P Handfield was detailed this task. His report describes the two strips on a clay pan in black soil country -

'...Tree cover is poor and concealment will be difficult.'

No provision was authorised for camouflage work but Handfield did advise -

'...the merging of barren clay pans into the edge of the strip to help break the straight edges.'

Presumably the finished field was used 'as required' by aircraft 'on ferry'. No 'occupation' is recorded to my knowledge and the RAAF's ***Aerodrome Guide*** of January 1946 lists it only as an 'RAAF ELG'. BLACKALL was one of these new WW2 airfields that DCA did take over for post war use. It has survived as the present aerodrome, owned and operated by the Blackall Shire Council.

It is interesting to note the following extract from a periodic visit by RAAF Airfield Inspection Officers dated around June 1943-

'...Sheep in considerable numbers were running in the fenced area at the time of the visit. These sheep appear to have been placed there, as the fences are well constructed and stock proof. They could cause trouble to landing aircraft and should not be there.'

BLACKALL *(31/7/1952)*
This extract from mapping photography shows the relationship of the old (cross strips) just south of the township, to the triangular field initiated as one of the airfields on the Inland Ferry Route. *[AUSLIG]*

BUNDABERG

'...In the words of the present CO of No 12 Elementary Flying Training School at Bundaberg, "the ground cannot be condemned too strongly".'

Such was the tenor of an inspection report dated 6 November 1941 from the CO of No 2 Training Group to Air Board in Melbourne. The principal physical detraction to the landing area and the associated ELG, was -

'...thick timber on three boundaries and buildings on the fourth...The surrounding country is heavily timbered for many miles with the exception of cane fields some miles to the east...aircraft operating (there) will be piloted by trainees, and... in large numbers. In order to overcome the menace of the timber to (some) extent, it is recommended that clearing should be undertaken (for) at least 300 yards in all directions... Extreme caution will (be needed) in night flying where pupils tend to underestimate their height of approach, and it is known from experience that trees can prove most deceptive in judging their height, even to experienced pilots. An ELG is being provided at Moorlands, distant 14 miles by air. The intervening country is heavily timbered without even a small break - most dangerous in the event of an engine failure... It is considered entirely unsuitable as any sort of training ground. In the words of the present CO of No 12 EFTS at Bundaberg, "the ground cannot be condemned too strongly"...'

The Training Group CO drew no comfort from -

'...the only other cleared field is at Childers, 24 miles by air... with heavy timber between it and Bundaberg...'

He criticised most the decision to operate both this EFTS and the already committed SFTS from BUNDABERG -

'...for (these) reasons... it is recommended that every endeavour be made to expedite work at Lowood so 12 EFTS can be moved there during the first week in January, and that work of clearing obstructions at Bundaberg should be given immediate attention.'

This site was first developed as an airfield in the mid 1930s. Cyril Broome recalled its construction was started by Bundaberg City Council, partly as an unemployment relief scheme. He said it continued -

'...at least until the Council was ***told*** *to stop!'.*

The early 1940 expansion of the RAAF led to establishment of Flying Training Schools throughout the country using existing aerodromes wherever possible. This Bundaberg site, already named Hinkler Airport after its famous aviator son, was the object of considerable local and political pressure to be selected for an SFTS. A DWB detailed Minute on RAAF files in AA aimed to clarify some development cost estimates. Dated 11 January 1941, it recognised sites for -

'...SFTSs are now most difficult to find and Bundaberg is considered the most suitable site in Queensland. It is understood that it is the Government's Policy to locate 8 SFTS in Queensland. In addition, it is an AOB which must be developed in any case for strategic reasons...and it seems desirable to make the one expenditure fit two purposes.'

Bundaberg was already an important civil drome, and likely to expand so after the war. DWB concluded on the need for 'firing ranges' and 'practice bombing ranges' -

'...These considerations make it impossible to establish SFTSs at such existing well established aerodromes as Mascot, Archerfield, Parafield, Essendon or Maylands.'

Camouflage considerations were applied mid 1941, when the RAAF modified its earlier specification to the SFTS constructing authority, the Dept of the Interior -

'...a clean-cut straight line of timber would be particularly easily observed from the air by a possible enemy, and it is thought that an irregular line would be

somewhat more difficult to pick out...'

Work proceeded on the buildings and on gravelling the runway. The MOORLANDS Relief Landing Ground (RLG) was scheduled to be available by September. On 9 October 1941 the Business Administration Board endorsed a Department of Air proposal for preparation of a second RLG at Bingera, on a site north of the Gin Gin to Bundaberg road and between it and the Mt Perry Railway Line, some distance west of BUNDABERG. That site needed extensive clearing and it appears this was enough to stall any further move until a better site was found at MEADOWVALE, 7 miles northwest of BUNDABERG.

While LOWOOD was being prepared for 12 EFTS that unit was temporarily located at BUNDABERG, effective 16 October 1941. Vic Waddell was one of the early participants in 12 EFTS's commissioning. He recalled a -

'...shortage of tools and our WOE (Warrant Officer Engineering) bought up what was available in the local stores. Being assigned to maintenance, we spent many hours making and case-hardening spanners.'

Vic said this free and easy attitude ended -

'...7 December 1941 when even our Tiger Moths were engaged in coastal reconnaissance...'

That was also when he 'volunteered' for the last time! -

'...The Orderly Sergeant came looking for two lads to ride Harley Davidson motorcycles... I spent the whole night on guard duty on the strip which had been covered with 44 gallon drums to prevent a night invasion.'

Cyril Broome arrived at BUNDABERG on 2 January 1942. As a Flying Instructor working under the Chief Flying Instructor, Sqn Ldr H F Boston, Cyril was there 'to familiarise instructors with the Anson aircraft'. Of the first twenty seven Avro Ansons to arrive during January 1942, ten came from 3 SFTS AMBERLEY. Cyril Broome described a -

'..."scare day" when the armament section was in confusion - all pilots were told to get the aircraft off to Dalby... The armaments people wanted to make available Lewis guns but didn't have magazines for them at Bundaberg - plenty were available at Archerfield - they (Archerfield) said that if someone came down from Bundaberg, they could get them...'

Cyril contributed to the 'confusion' by exercising initiative. He 'fixed up' an unserviceable Anson, took it to ARCHERFIELD and collected the magazines. That didn't stop the Armament Officer at BUNDABERG wanting to censure Cyril for the 'unauthorised flight'! 8 SFTS Operations Record Book (ORB) shows this period as 1-5 February. ACH Townsville had slipped up in the relaying of communications associated with the procedural identification of the Aircraft Carrier USS *Langley* and its escorts approaching Brisbane. Indeed, RAAF Station BUNDABERG was not the only airfield placed on alert at that time!

Neville Halligan and Frank Lawrence were both members of 19 P (Pilot) Course of trainees, the first intake of 8 SFTS. That course was one of the few where most of the trainees had done both their Elementary and Service Flying training at the one station. They arrived early in February when heavy rain led to severe flooding in the Burnett River. With no possibility of flying operations, the Air Force helped in community flood relief and evacuation work. A strong bond formed and prospered for many years.

It was through my contact with both Neville and Frank that each realised the other was 'still around'. Both had gained 'their wings' and as Sergeant Pilots been assigned to duty in Europe. Frank flew Lancasters with distinction in the RAAF's 460 Sqn. Neville was posted to an RAF unit. Frank studied a group photo and picked Neville as a bloke who used to - and still does! - excel at tennis. Neville recalled a solo cross-country to ROCKHAMPTON when bad weather made the return difficult -

'...in 10/10 cloud, down to 50' wet through and lost, I reached the Burnett River and followed it out to the mouth... I finally landed, wondered why the aircraft took so long to pull up and then looked at the wind sock. The Chief Flying Instructor came out to meet me, and ticked me off for landing "down wind". Maybe he remembered I was only 19 'cause he said, "I suppose you were a bit scared up there - you'll be pleased to know your fellow pupil (who was doing the same trip) had to land at the satellite (Meadowvale) aerodrome, so you're one up on him"... I'm nearly sure Frank Lawrence, was that other pilot!.'

Harry Edwards did his Initial Training in Tamworth New South Wales. He joined 21 Course at 8 SFTS in March 1942. He was to return to 8 SFTS, in July 1943, this time as an Instructor where he remained until 30 September 1944. Harry remembered MEADOWVALE RLG but not MOORLANDS. He added -

'...many a welcome let down at Meadowvale was made (sanctioned or otherwise) to gather mushrooms...'

Harry was unable to date the time of bad weather forcing -

'...about 12 US Bell Airacobras to overnight there... the young American pilots had so few "hours in their machines" that several aborted their take-off, and had to go back to try again... the Colonel leading the group was circling the field - waiting. He had to land again to top up his own fuel!'

8 SFTS ORB shows this date as 27 April 1942 and the flight as one under the command of a Lt Greasley. Presumably the aircraft belonged to 8 or 35 Fighter Group pressing north to get to Port Moresby where fighter aircraft were greatly needed.

BUNDABERG had rightly attracted the care and attention of the State Defence Camouflage Committee in its conscientious work. Their officers spent much effort in policing the infringement of National Security Regulations which legally prevented land holders 'interfering' with natural tree growth on their own land. The Minutes of the 28th Meeting of the SDCC Queensland include the following references -

'...On 28 April 1942 Mr Wyeth reported that the fight to save the trees was still being waged. He has now arranged with the contractor for the splinter proof pens, that only the trees blazed by our camoufleur are to be removed... A Main Roads Surveyor has commenced setting out the taxiways and approaches to the hideouts but on 1 May the contractor for the splinter proofs had been instructed by the Dept of Interior to stop work. Enquiry... confirmed... the contract was to be terminated and no more splinter proofs constructed.'

Camouflage Reports of 20 May 1942 include an update on work at MOORLANDS, where the scheme -

'...is simple but will be very effective when completed. It involves mowing and burning off to a pattern giving the appearance of cultivated fields with haystacks.'

F-Lt John F Keays is recorded inspecting BUNDABERG aerodrome on 4 July 1942. He wrote -

'...inspected... with Mr P Moller of Commonwealth Works Department... details of dispersal scheme were discussed with Mr Hewitt of MRC... layout... is good,

BUNDABERG *(ca 28/4/42)*

One of those photos that servicemen were not supposed to take! With camera hidden somewhere near his knee the photographer has snapped this shot of a US Airacobra staging northward through Bundaberg. The Duty Pilot Tower is partly obscured by the propeller blade. Harry Edwards had an 'arrangement' with a local pharmacy where the odd film was processed and stashed outside in a rubbish bin to avoid periodic checks of such premises by Service Police! [H Edwards ex RAAF]

use having been made of existing roads for the greater part of the taxiway system... Station is at present being used for training only and the scheme appears rather elaborate for such conditions... A direction is required as to relative urgency of the works compared with others away from the coast belt.'

Amongst other works under way, Keays mentioned -

'...MRC has commenced the mining of the runways. Construction of slit trenches by MRC is complete...'

Such preparation for demolition of runways has proved to be a colourful episode. Archive records show it as an Army initiative. Few of the RAAF veterans I've contacted even knew of the effort. Cyril Broome recalled when 'explosives' were so scarce that Ansons due to go 'on patrol' had to wait until the others returned so that the 2 x 250 lbs bombs could be transferred! Harry Edwards knew of the 'tunnels' under the sole runway, but viewed their need cynically -

'...the runway was thin gravel and was only used by visiting aircraft, civil included... the trainees were told to keep off it because their propellers would blow it away as dust!'

Harry described a locked glass cabinet on the wall of the HQ building for which there was a key duly entrusted to a responsible officer on duty. He understood that if demolition was ordered, the drill would have been -

'...break glass, press button and - boom!!'

8 SFTS ORB carries this entry for 23 June 1942 -

'...US B-26 Marauder aircraft (No 40-1638, Lt J W Oliver) crash landed on the drome after becoming unserviceable whilst airborne. The Marauder, after repairs, departed on 17 July 1942.'

BUNDABERG (6/12/1944)

*Caption from the Australian War Memorial's Collection reads - **'RAAF Base Bundaberg Queensland, aerial oblique photograph taken by war photographer William Robinson from an Anson aircraft based at Bundaberg'**. Viewed from the north west, the foreground includes most of the aircraft dispersal and hideout works. One can relate the sites of these dispersals to present day development. Veterans remember 'route marches' round these roads past farms with fresh fruit but not much in the way of planes in the hideouts!! The prominent narrow clearing in centre left was a rifle range.* [AWM Neg P 279/64/18]

BUNDABERG (late 1942 early 1943)

This is very likely an extract from a 'RAAF Official' photograph. Its date must be prior to April 1943 because of development work known to have taken place by then, not appearing here. The significance of the numbered black dots is not clear. They may be just gravel/soil depth sample locations! They should not be aircraft revetments because of their number and proximity to the 140° runway; e.g. No 69 near the northern tip. They may relate to anti aircraft machine gun pits, although other records indicate further pits not marked here. At the extreme bottom of the building group the steel assembly for a Bellman hangar can be seen in progress. The heavy black shape near the large hangar 'Building No 161' is the approx 10 a.m. shadow cast by elevated water tanks. [W Maitland via D Vincent]

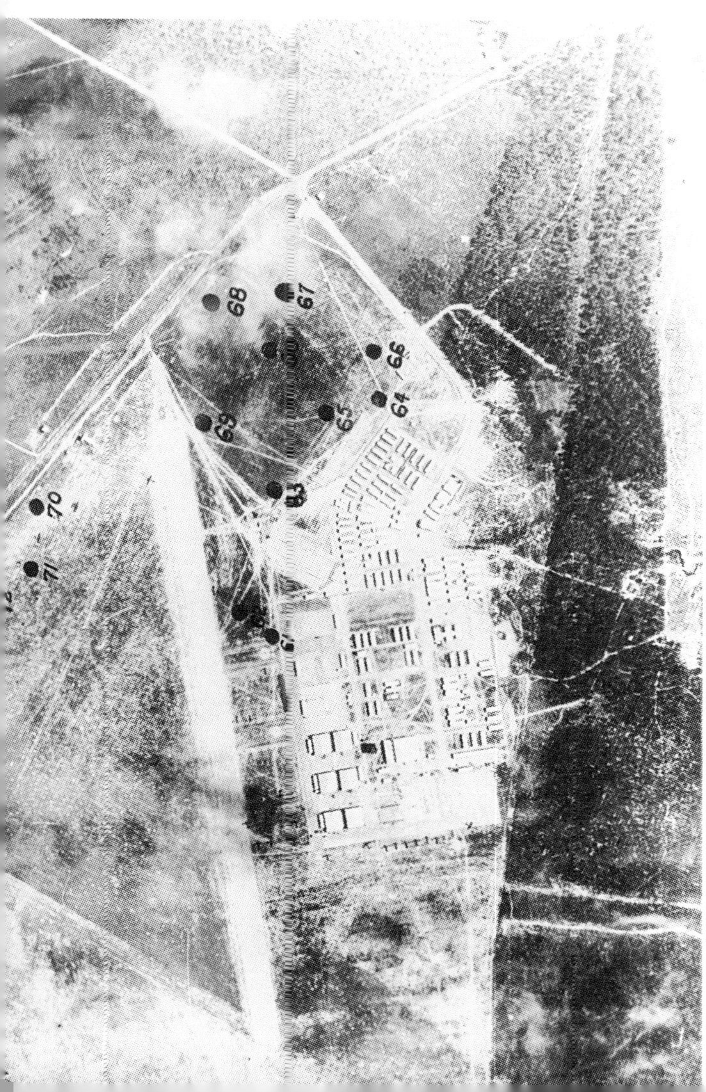

Harry Edwards described it landing parallel to the Childers/ Bundaberg road and skidding to a halt near a bomb storage building on the western end of the building area -

'...RAAF personnel were kept away from the aircraft by an American with a .45 automatic in his hand.'

Harry's confident the aircraft's Norden bomb-sight was the reason for the zealous security.

Harry and Cyril Broome, both described the practice bombing and gunnery ranges. Cyril took me to inspect a surviving concrete observation 'blockhouse' south of the drome. Harry Edwards, who claimed an A in chemistry, told me of the fine aluminium powder filled 'sea marker' bombs and the one containing Stannic Chloride, which produced dense white smoke. He produced an unused 'empty' 8½ lb practice bomb! As a 'souvenir' he'd -

'...secreted it inside a hollowed out 2lb loaf of bread, parcelled it up and mailed it to Mum in Armidale!'

In Australia at the outbreak of WW2, opposition to the employment of women in the Defence Forces in other than nursing roles was very strong. By early 1941 the need to rapidly increase numbers for the RAAF eroded some of this opposition. On 25 March 1941 the formation of the Women's Australian Auxiliary Air Force (WAAAF) was announced with an initial establishment of 250. A narrow field of job opportunities was cited with short engagement periods up to twelve months. The pay rates did not exceed two thirds of the male rate for the same duties. The changes to follow can be gauged by WAAAF numbers, which peaked around 18 000 by the end of 1944. Many barriers were broken, but the pay difference remained.

On 10 April 1942 Flight Officer Hawthorne arrived at BUNDABERG as the first WAAAF Officer to be stationed there. By 31 July 1942 there were 120 WAAAFs there compared to a RAAF strength of some 900. Accommodation at RAAF BUNDABERG amounted to the fencing off of a cluster of existing huts in the close sited building group, with of course some alterations and minor additions. AA records show these arrangements and those at a few other stations in Queensland, in contrast to the situation at AMBERLEY where some of the WAAAF sleeping huts were of 'aircraft packing case construction'! Such was the pressure on building services during 1942.

By the end of 1942, 8 SFTS claimed success in producing nearly 300 qualified pilots with another 200 in training. While the battle for New Guinea occupied the minds of the tacticians, at least questions were being asked about any sustained need to blow up the runway at BUNDABERG. A memo to HQ East Area, Point Piper NSW dated 2 February 1943 warned -

'...there exists at Bundaberg Station, demolition tunnels similar to those at Maryborough. There is evidence of erosion... outside the timbering of one tunnel at the southern end of the runway and it is considered that this tunnel should be filled in to obviate any danger... It is consequently proposed to make arrangements accordingly unless it is considered that there is any special circumstance that would render it necessary to retain the existing facility for demolition. Please advise...'

A few weeks later instructions came to fill in the tunnels. This coincided with planning to enlarge the eastern hardstand and raised questions about the five machine gun emplacements there. While it was time to fill in the demolition tunnels, it seems from notations on drawings in AA, these five gun pits were replaced by six elsewhere!

Abe (Joe) Howsan was a Fitter 2E whose role was aircraft engine maintenance at BUNDABERG. John Galvin was CO of the RAAF's 71 Reserve Sqn, to which Joe was attached. Anson aircraft were here being pressed into service, on what amounted to an 'operational basis'. As John Galvin wrote recently to Joe -

'...(we) did tests on the bomb lifting capacity of the Ansons, which as you know were clapped out with years of pupil training... I remember you being with me on the final test when we lifted the 2 x 250 lb bombs, your job being to check the engines which were way overboosted... I used to feel sorry for all the engineering staff... at that time in 1942 we had to have 3 fully operational aircraft on standby in the No 71 Reserve Squadron continuously as well as maintaining full advanced flying training facilities. I distinctly remember the very late hours in the workshops hangar to keep up the serviceability. There was a lot of strain placed on ground staff and in particular the NCOs.'

Without shifting station Joe Howsan was posted from 8 SFTS to 66 Reserve Squadron, which, in company with a detachment of 32 Sqn, was to take over the coastal patrol work from 71. Joe's health is still troubled by the close working conditions in maintaining serviceability of the Ansons there. He says -

'...that whilst the spray painters in our hangars were equipped with protection masks and issued with daily drinks of fresh milk as additional protection, we fitters were given no such consideration.'

In March 1944 the Divisional Works Office in Brisbane in a memo to DWB Melbourne, reviewed proposals for extending runway 140° at BUNDABERG. It seems one stated need was for 'continuous all weather use by Ventura aircraft'. Apart from increasing gravel thickness, there was even mention of bitumen sealing the surface! However, already the RAAF did not see a post-war role here and

BUNDABERG *(-/1/1943)*

This is an extract from the 1:63 360 (1 inch = 1 mile) Military Survey Map. The airfield is shown only by a black 'dot' symbol at its centre. [Department of Defence]

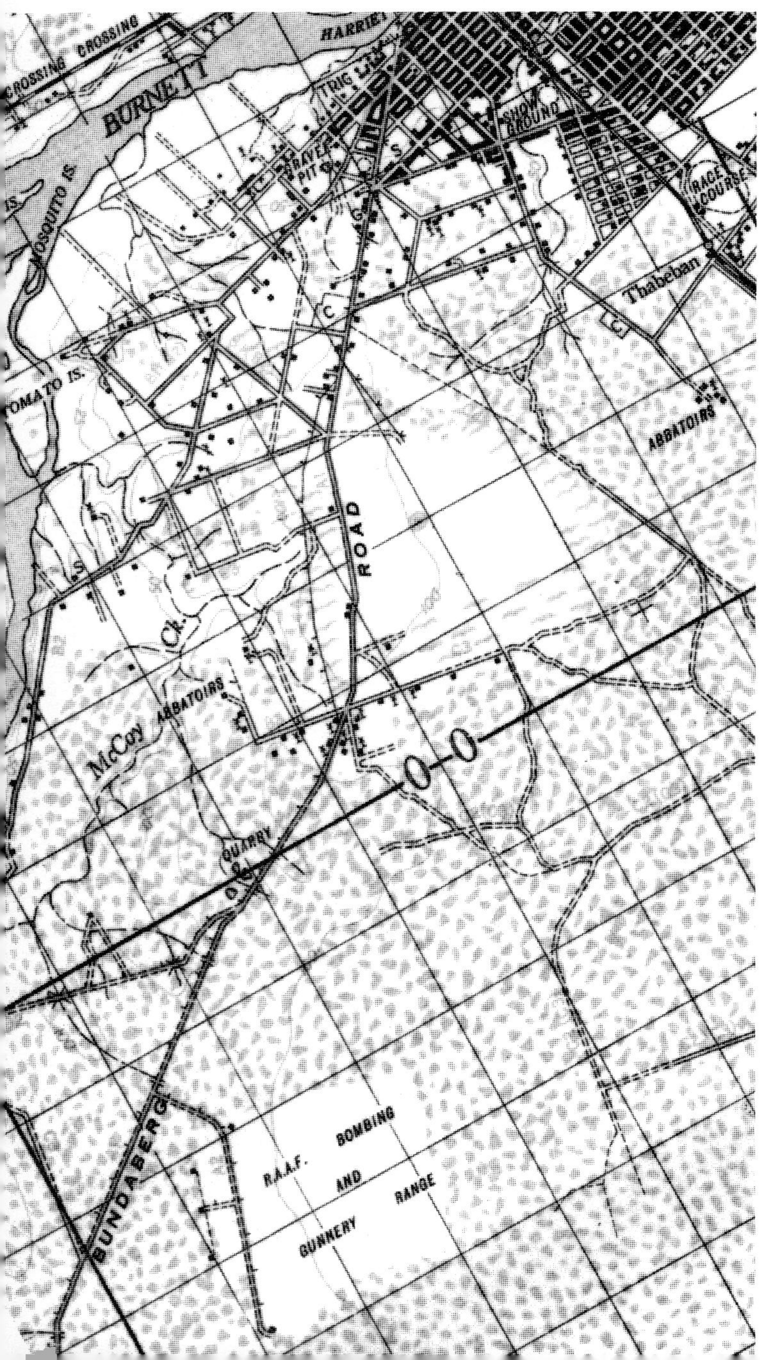

BUNDABERG *(1991)*

This extract from current UBD Directory shows the Sugar Land Shopping Town development located in the area that once served for aircraft dispersal. Hideout positions have been 'mapped' from Drawing 42-43/2877, dated 7 December, 1942. [Universal Press P/L]

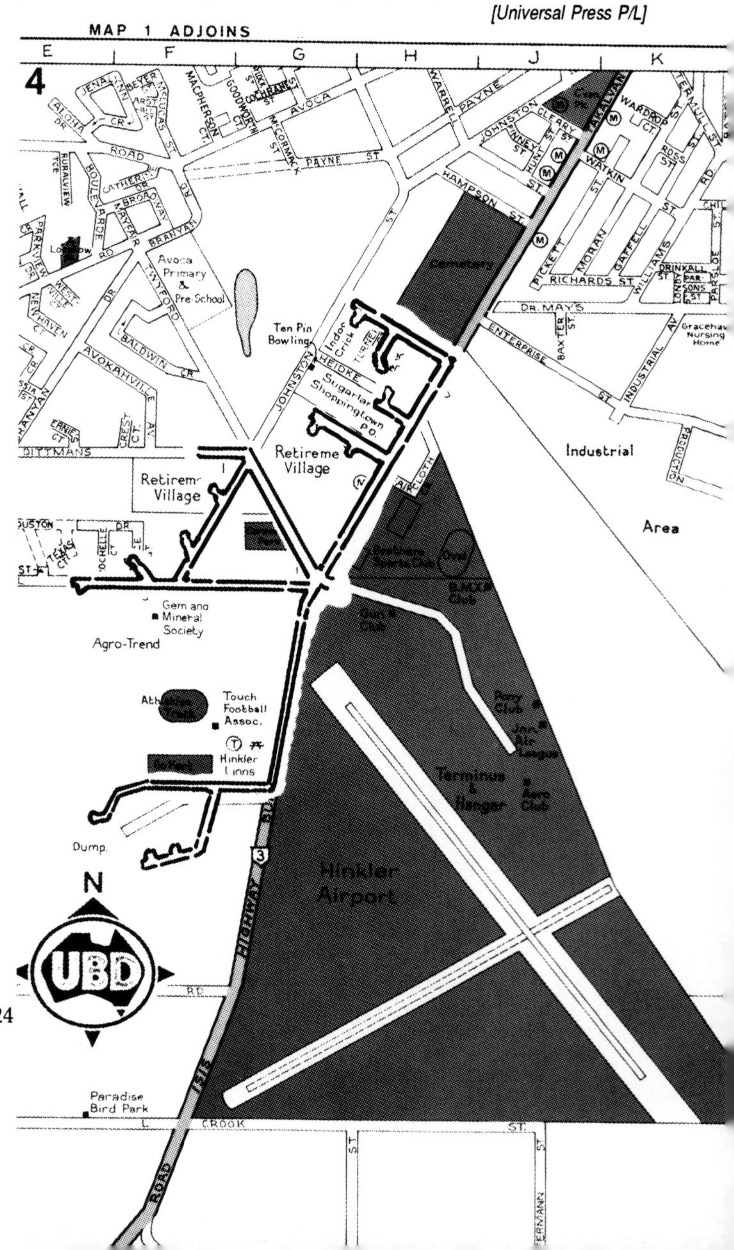

13 Sqn with its Lockheed Venturas, moved further north to the sealed runway of COOKTOWN.

Joe Howsan went back to 8 SFTS from 66 Reserve Sqn on 22 December 1943 and was posted away from BUNDABERG on 3 October 1944. By then he'd spent about three years there, mostly on or close to an 'operational' basis. His war service discharge read, in part -

'...that the above named airman did not serve outside Australia in the present war... (and) that the above named airman did not serve for a period of 3 months or more in NT north of parallel 14.5°S latitude between 19 February 1942 and 12 November 1943.'

There must have been something magical about the latter date. Surely the former is tied to the initial bombing of the Australian mainland at Darwin. War Service Pension matters sometimes 'hinge' on such facts.

The improvement of runways at BUNDABERG was clearly 'on hold' with respect to what developments might take place to serve the Royal Navy elements arriving in the Pacific after the Allies' victory in Europe. There was also a need to provide facilities for the Royal Netherlands East Indies Units particularly the Air Force.

On 12 December 1944 all flying training with 8 SFTS, ceased. Ironically on the same day, at Bundaberg cemetery, a burial service was held for the two victims of its most recent Anson crash. Flying training had thus finished just three years and one month after 19 (P) Course began with 12 EFTS at BUNDABERG on 14 November 1941.

Amongst the interesting aircraft to visit BUNDABERG during the war years, must surely be mentioned Avro Lancaster 'Q-Queenie' on 22 April 1944. Almost a year later on 11 April 1945, Lancaster 'G-George' visited. Both were open for a degree of public inspection. Both were involved in Victory Loan raising publicity and both served to show the Bundaberg community the very fruit of theirs and the RAAF's local effort of several years. 'George' survives today on static display, in the AWM Canberra.

Once it was decided the RNAS would use MARYBOROUGH, the way was cleared for BUNDABERG to be made available for the Netherlands East Indies Air Force. An advance party of that organisation visited on 25 June 1945, to finalise arrangements for the handover of the station. On 17 August 1945 Lt Col. B J Fiedeldij, Director of (NEI) Military Air Force wrote to the Department of Air -

'...It may be expected that with the surrender of Japan, airfields in NEI, will be able to be reoccupied under reasonable, serviceable conditions. This being so it would appear the need for extension and sealing of Bundaberg may not now be necessary.'

It took some months before the NEI Units moved on to Biak in Dutch New Guinea. With the handover of wartime RAAF assets to DCA, runway 140° was finally slated for refurbishment and sealing. Hinkler Airport resumed its peacetime role.

In my efforts to reach veterans of all units at these various Queensland airfields, in the case of the NEI at BUNDABERG, only Huub Beckers responded -

'...I joined the RAF as soon as the first US Army liberated the town of Massfricht, (Holland where I was born and lived) in September 1944. I passed the medical exam and selection board for pilot/navigator in February 1945 (I was just under 18 at that time). I was shipped via Belgium to Wolverhampton in England in April 1945 for the initial military drill, with a group of about 100 in Wolverhampton. A party of about 350 of us were then sent to Australia in three groups. 100 were destined to be trained for air crew and 250 for ground duties. We were meant to join the Pacific War after being trained by the RAAF instructors. The 100 potential air crew were all stationed at Bradfield Park, St Ives, Sydney...Towards the end of this course the atomic bomb fell on Hiroshima and Nagasaki...and we were then told, there would be a delay in flying training as the Australian authorities did not see any need for such training. We were split into two groups of fifty, one group went to a camp near Casino and the other to which I belonged, went to Bundaberg. There we were told that there was a great shortage in Indonesia of weather observers and forecasters and we had to study for this. About twenty refused. Out of the remaining thirty, fifteen succeeded as weather forecasters in May 1946 and 15 as weather observers in December 1945. I was one of the forecasters and was flown to Batavia at the end of May 1946 to train Indonesians to become weather observers and also had to take over the forecaster duties from the RAF at Kemajoran Airport Batavia (now Jakarta)... Whilst at Bundaberg we were part of the Netherlands East Indies - Personnel and Equipment Pool (NEI-PEP) which had been established at Bundaberg since 6 June 1945 under Lt Col. Asjes who handed over command in September 1945 to Lt Col. Van Haselen.'

Huub described the 'huts' and most other facilities. Of the Course he added -

'...The Australian, New Zealand and Dutch teachers were all very able and knowledgeable. We flew only every now and then to become acquainted with the cloud formations and what a typhoon would look like - from a safe distance.'

BUNDABERG *(7/11/1992)*
Leading it 'by the nose', Phil Ainsworth tows out this early development example of the Jabiru. Aimed to fill a gap in international markets, this now dual-certified (Australian Ultralights Federation (AUF) and the orthodox Civil Aviation Authority's VH) locally manufactured plane uses structural composite plastic construction. In the left background is part of the WW2 workshop hangar '161', used now by regional commuter airline maintenance sections. [Author]

MARYBOROUGH *(6/12/1944)*
One of a series of photos by wartime photographer William Robinson taken from an Anson based at BUNDABERG, this oblique shows the near peak wartime development of the airfield. Four Bellman Hangars front the tarmac. [AWM Neg P279/64/15]

MARYBOROUGH

'...I was a Wireless Air Gunner and trained here at No 3 Wireless Air Gunners School (WAGS) on No 20 course, from 13 November 1941 for six months, flying in Tiger Moths, Wackett Trainers and DC2s. Exercises were of 1 hour duration and we just flew round and round the town. The DC2 carrying 8 or 10 wireless operators, did a 3 hour flight down to Redcliffe and back. As this (December 1941 to January 1942) was the time of the Japanese invasion scare, wooden trestles were placed at intervals along the runways to prevent any planes landing and we were issued with rifles, ammunition and "iron rations" which we had to keep with us at all times. We were told that, if the Japs landed, the Station would be abandoned and we were to make our way south, independently. There was no ack ack defence at the Station, and that "Brisbane Line" was no myth!'

This is how Alan Payton first recalled his period at MARYBOROUGH. Alan went on to serve in the RAAF's 8 Torpedo Bomber Sqn.

MARYBOROUGH *(14/7/1940)*
This extract from mapping photography is marred by cloud shadow but it serves to show the transition from the small 'all over field' within the round coursing track to the enlarged airfield which became the RAAF's 3 WAGS and ANS. [AUSLIG]

The airfield in use before the outbreak of WW2, was in no way suited to the operation of modern military aircraft and the RAAF first took steps to secure land adjoining it in December 1940. Planning at that stage was for the setting up of No 3 WAGS and No 3 Air Navigation School (ANS). Drawing 40-41/502 now in AA, had its origin then. It remains a classic record of the layout of a RAAF Station in an environment considered free of enemy action. On 26 December 1941, just after Japan entered the war, the Camouflage authorities decided to -

'...seek toning down of airfield buildings especially Maryborough and Bundaberg - these dromes can be located by their mass of white buildings, long before the flying field itself can be distinguished.'

The MRC was responsible for construction of the three runways which continued during 1941 with bitumen sealing achieved by year's end. Here they found a somewhat unique problem with small freshwater crayfish burrowing up into the pavement from the water table below. A March 1944 explanation of 'cost over runs' mentioned that -

'..."yabby" holes under the ground to be filled necessitated additional filling material to make good the many depressions formed by the rollers...'

Alan Payton recalled when he arrived at MARYBOROUGH the 'Station was still being built'. Archive records contain an exchange of letters between the Works and Services Branch of the Department of Interior Brisbane and building contractor T H Dennis. Picking up the tempo, H W Barker, the Department's Works Director in Queensland, instructed working the men -

'...10 hours per day for six days a week as provided for in your contract and also to work through the holiday period at Christmas...'

Application to the Industrial Court and its approval for overtime beyond relevant Awards was needed. The Dept was quite prepared to pay such overtime but -

'...you will be responsible for supply of artificial lighting, if such is required.'

Dennis replied, 17 December 1941, that because of -

'...the supply of materials and conditions applicable on the job it is impossible to work the men more than 9 hours per day for six days per week. The men themselves consider that this is about their limit of endurance under the heat and conditions in which they are working...(re) working through the holiday period at Christmas, I have tried to persuade the men...but at a mass meeting of all my employees, they unanimously

MARYBOROUGH *(22/8/1943)*
This is the 'serviceability date' on Drawing 42-43/1219C. Notice that only two Bellman hangars are then indicated by the tarmac. [AA BP375 Item QL680]

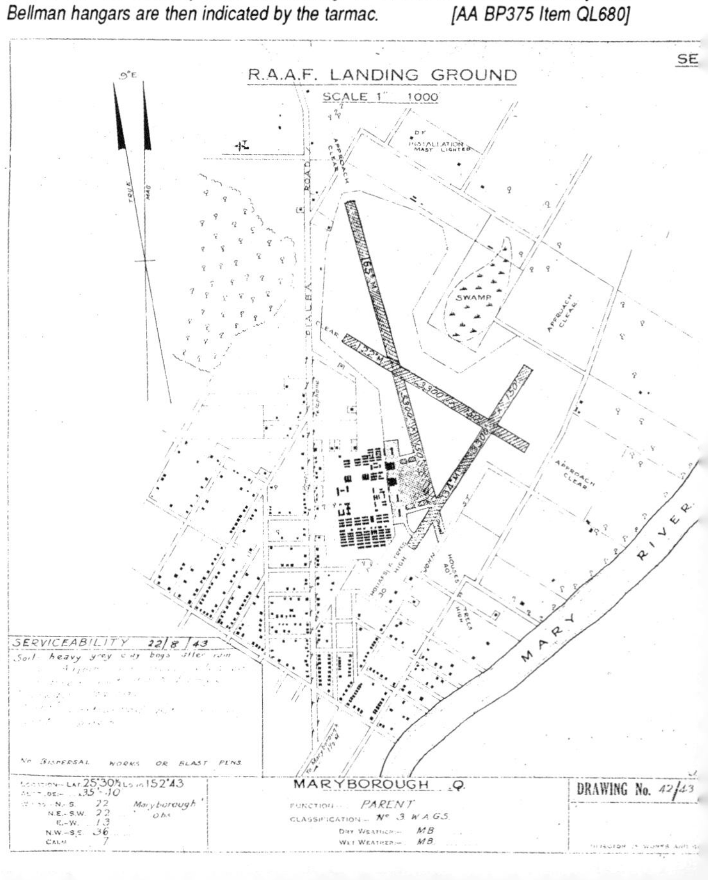

decided to take five days holiday... and to recommence on... 29 December.'

January 1942 was a period of preparation for the worst. To the construction authorities in respect of hospital and 'sick quarters' accommodation, the Director General wrote from Melbourne on 14 January 1942 -

'...The Director of Medical Services, Department of the Army has brought under notice the likelihood of service hospitals being required to accommodate battle casualties in the near future, and in view of the limited hospital accommodation available, requests that every degree of priority possible be given to the sick quarters requested or in course of erection at the following RAAF Stations Queensland: Amberley, Sandgate, Maryborough, Lowood, Kingaroy and Bundaberg.'

In July 1942 Fl Lt John Keays paid some attention to MARYBOROUGH as one of the aerodromes in the 'Brisbane Area'. As such he advised -

'...The sealing on runways is deteriorating rapidly. I do not think that the field is important enough to justify a further seal coat which is the only solution. It is obvious that unless full and complete sealing is applied pavements will go to pieces and sealing materials used will be wasted. There is little cover available for dispersal but good strips could be located in the area.'

Sam Hecker joined the Citizen Air Force Administration and Special Duties Section in 1942 and remained at RAAF MARYBOROUGH until 1945. In my quest to identify any associated ELGs, Sam Hecker spoke of the -

'...landing of a Wackett trainer on such a strip just off the Boonaroo Road southeast of Maryborough...'

From Sam's description, it was probably a straight section of a 'C2' class track at grid reference 940224 on the 1942 Military Survey mapping. Sam also spoke of a 'satellite' strip having been used on Fraser Island by Wackett Trainers but I've not been able to learn more. Presumably neither of these warranted any further evaluation by Keays or others seeking dispersal facilities.

The minutes of the AWC record 'aerodrome demolition' contingency works being carried out at MARYBOROUGH as with several other Queensland coastal airfields during 1942. Cancellation was ordered on 22 September 1942 and follow-up restoration work doesn't seem to have produced any 'mysterious' legacy, as at Rockhampton.

In April 1942 MARYBOROUGH was expanded to include Recruit Training. No 3 Recruit Depot (RD) functioned until July of 1943 and later from November 1944 until June 1945 No 6 RD continued this task. In all nearly 4 000 newly enlisted recruits received basic training here.

MARYBOROUGH was one of the Queensland RAAF Stations to take full advantage of the introduction of women to the Air Force. The Womens' Auxiliary Australian Air Force (WAAAF) was formed in April 1941 and while at first women were recruited mainly for clerical and domestic musterings this was steadily extended to a wide range of technical musterings.

Two Bellmans were already erected when a February 1943 report by DWO Brisbane focused on hangaring the Wackett trainers. With plywood and fabric covered wings they needed full weather protection -

'...it is only with much difficulty and loss of time that they can be put under cover. In the case of outbreak of fire, serious loss would occur owing to the close

MARYBOROUGH (-/7/1945)
The Royal Naval Air Service (RNAS) presence at RAAF Base MARYBOROUGH began in June 1945. Adjacent the four Bellman hangars are parked Sea Otter (bi-plane) aircraft, an RAAF Dakota and an Anson. The nearest building and the Bellman behind remain to this day. [AA CRSA 705/1 Item 171/89/113]

MARYBOROUGH (-/7/1945)
RNAS Corsairs dominate in this array of parked aircraft, when viewed from the tarmac in front of the buildings. [AA CRS A705/1 Item 171/89/113]

parking necessary. Some 36 Wackett trainers are now on the Station and 3 DH-84's are expected... the accommodation is insufficient for (this) number of aircraft...'

Eighteen aircraft of 37′ wingspan jig-saw parked in a hangar 96′ x 96′ looks impossible to me, let alone a daily ritual! The DWO, W Cdr G J Towers concluded -

'...Maryborough has now been classified as an Operational Base, class "BL", which may have some bearing on the question of hangar accommodation.'

I don't know what the 'BL' meant but two more Bellmans were eventually erected.

Although it may have been classified as an Operational Base, MARYBOROUGH certainly did not come in for significant attention by the Camouflage Authorities. A brief comment on their records dated 9 May 1943 reports an officer having visited the aerodrome where '(camouflage pattern) painting is in very bad condition'.

Sam Hecker was flying before the war but his war service duties kept him on the ground. Of the underpowered Australian designed Wackett, with its Warner Super Scarab 175 hp engine he recalled -

'a generous throttle opening was the order of the day... these motors gave much trouble... often overhauls were required at approximately 30 hour intervals.'

Sam recalled the Wacketts being flown to Fraser Island on training sorties and although this involved only a 15 mile water crossing, he maintained -

'...pilots and trainees carried revolvers to repel sharks should they have to come down in the water.'

MARYBOROUGH (20/4/1990)
The 'Wide Bay Aero Club' now known as Maryborough Aero Club Inc. still uses this WW2 period flight quarters building as its headquarters. The Bellman hangar in the background is the only one now remaining on the field. Post war, two of these four structures were declared surplus and sold to local interests where one is in service in the city with the Ford Dealership and the other at the Maryborough Sugar Mill. The current whereabouts of the fourth is not known to me. [Author]

Percy Yates was a RAAF pilot on Beauforts stationed at BUNDABERG. Percy recalled a distressed landing at MARYBOROUGH by an Anson, shortly after his own aircraft had landed there one day -

'...(we) were disembarking just as an aircraft came in bearing a court martial party... so... the place was littered with senior officers who were gathered on the tarmac... While waiting for the "brass" to move away, we heard a most peculiar noise - a cross between the rattling roar of a Walrus amphibian and a strange kind of whistling sound.... Then an Anson appeared... It shot straight over the fence and landed without making any kind of circuit. Naturally all eyes were on such a strange arrival, so its condition was viewed by all the "brass" when it taxied in. The fabric fairings were missing from both engines. There were rags of fabric flying from various places in the belly and both props were bent back for part their length into perfect "U" shapes. It seemed they had been indulging in some unauthorised low flying over the sea, had come too low and struck the water. By some miracle they had bounced off and with full throttles the pilot held his kite in the air at about 65 knots barely staying aloft all the way across the sea until reaching the nearest aerodrome. Very lucky! But Nemesis waited on the tarmac - Maryborough being a Training School, was renowned for its strict discipline and the pilot a Flight Sergeant, was an LAC before he even reached the Control Tower!

The thriving local timber industry and the long established metal foundries meant Maryborough boasted ready numbers of skilled wood craftsmen. An extract from a Defence Dept memo of 26 November 1943 to the Works Priority Sub Committee in Melbourne refers -

'...Because... demands made by the RAAF and USAAC(sic) for overhaul of airframes of operational aircraft have priority over demands for servicing of aircraft used in connection with the Empire Air Training Scheme (EATS), (relevant) facilities...in Queensland for (EATS) are inadequate... the other States are also fully taxed... The position has been complicated by the fact that a large erection programme of new Anson and Oxford aircraft for the RAAF has been undertaken by Servicing Contractors in NSW, while the transport aircraft overhaul programme commencing early in the new year will tax the facilities available to the utmost. Existing facilities in Brisbane for servicing of airframes from EATS aircraft have recently been surveyed by the Maintenance Division and the position may be summarised...'

Briefly, Aircrafts P/L already carried out airframe and engine overhauls at ARCHERFIELD and also in the Maryborough Naval Drill Hall. A second shift at ARCHERFIELD was out because of the lack of housing for any transferred tradesmen. The Dept's memo recommended in part -

'... alterations and additions to facilities already in use at Maryborough Naval Drill Hall (and) erection of two (more) Bellman Hangars at Maryborough aerodrome...'

The first happened quickly but the second took months. The General Index Cards of the Works and Services Branch of the Dept of the Interior, preserved in AA, indicate that various buildings at the Showgrounds in

MARYBOROUGH *(18/10/1945)*
This Dorland portable hangar is seen in process of erection on the northern fringe of the building area. The RNAS MONAB-6 Unit used several of these canvas covered steel framed hangars at Maryborough. *[John Oxley Library, Brisbane]*

MARYBOROUGH *(1945)*
One of the RNAS 1701 Squadron's Sea Otter aircraft in flight over the winding course of the Mary River near the airfield. *[P Groom via S Ryan]*

Maryborough were also pressed into service in this aircraft overhaul scene. The third and fourth Bellman hangars that were erected at MARYBOROUGH aerodrome appear to have been in place by 11 October 1944. Relevant documents show their suppliers as, typically, Australian Iron and Steel P/L Melbourne, Hume Steel Ltd Melbourne and Lysaghts P/L Newcastle.

A Minute dated 17 June 1944, on the Buildings and Services file (RAAF Maryborough) refers to a 3 WAGS request for -

'...a workshop annex to existing Bellman Hangar No 131. Proposed building is required to accommodate increased aircraft repair facilities necessitated by increase in aircraft establishment from 35 to 57.'

Many of the additional aircraft had flown to MARYBOROUGH from Central Flying School PARKES, NSW. However, a response to this Minute came -

'...As the future of the EATS Organisation was under consideration when this proposal was raised, it was held pending decision regarding the position of No 3 WAGS. At a conference held in CAS's office 15/8/44 it was decided to close 3 WAGS Maryborough and carry out all Wireless Operator Training at Ballarat.'

When 3 WAGS closed the facilities at the base were used by the RAAF's No 1 Radar School.

After victory in Europe early May 1945 the Royal Navy formed the British Pacific Fleet. A number of air bases in Australia were required for aircrew training and the assembly of aircraft. 'On shore' carrier aircraft training facilities were needed, including the simple marking out of a flight deck on a runway surface. The victory road to Tokyo was to be fast moving and if the Royal Navy was to earn a place in the front line with the USA, fully functioning mobility was the key. Mobile Naval Air Bases (MONAB) were formed. One such organisation commissioned in England under the name of HMS Nabstock, arrived by sea and proceeded to RAAF Base MARYBOROUGH on 1 June 1945. About nine Royal Naval Air Squadrons were based there for periods until late 1945. The men found service life at MARYBOROUGH much more relaxed for the obvious reasons.

Soon more than a dozen RNAS men were to return to Maryborough to settle. Peter Groom is one who returned. He was in the Royal Navy's, 1701 Sqn Air Sea Rescue (ASR) flying Sea Otter bi-plane amphibian aircraft -

'...Our "operations" as such were generally trips to the Commando Station on Fraser Is and patrols over carriers off Wide Bay when "flying on and off aircraft". The local Maryborough people were wonderful...'

After the airfield passed to the Department of Civil Aviation and eventually to its present day ownership by the Maryborough City Council. As Sam Hecker pointed out to me -

'...The present day airline passenger terminal was one of the war time RAAF huts and was used as a parachute packing centre. Benches ran the length of the hut and it was operated by WAAAFs who used long sticks to do the parachute folding.'

CHARLEVILLE

'...The aerodrome construction consisting of extensions, runways, taxiways etc. was originally commenced in accordance with the requirements of Civil Aviation requisition 1940/41-88 in which the Department of Air was particularly interested, requests (being received from the Director General of Civil Aviation on 30 October 1941... and 14 November 1941... stating that "the Department of Air have again contacted this Department and reminded us of the particular importance and urgency of this job." '

This is an extract from a reconciliation memo within the AWC/Dept of Interior 'system' dated 24 January 1945. The 'urgencies' had by then all passed but of course the job was not done until the paper work was complete.

This excerpt from a memo of 11 February 1942 under the signature of C A Hoy, Director General of Works, Department of Interior, serves to indicate the changes being forced on 'the system' by impending danger to this country.

'...With regard aerodrome works, the provision of two new runways and the widening of the third now being carried out under requisition CA41/42/88 is sufficient to meet requirements for the first stage of development, but plans indicating further requirement are now being prepared. You were advised on 6 February... that for all US/Australian Camps, such amenities as elaborate water reticulation, water borne sewerage systems and hot water services are not to be provided and the schedule now forwarded is to be disregarded in respect of those items. It will be noted also that the schedule calls for most buildings to be "Comet" type steel huts. It has been decided that in view of transport difficulties involved in supplying this type of building, huts constructed largely of rough timbers, Sisalkraft and galvanised iron may be substituted.'

Further in this same memo it is stressed the camp is -

'...for occupation at the earliest possible moment, and you are requested, before putting the work in hand in the normal way, to discuss... with the Queensland Co-ordinator General of Works to determine whether the work could be more expeditiously completed by means of the organisation of a State instrumentality.'

The 'State instrumentality' mooted became the AWC, an executive arm of the Commonwealth's Dept of the Interior. Its charge was to coordinate all building/construction work, both for the Australian Forces and the US Forces arriving in this country. History views the AWC's efforts in the face of great shortages with a mixture of genuinely deserved praise and condemnation. And somehow the sewerage system at the airfield did proceed!

During February 1942, the airfield saw the procession of most of the US 27 BG's luckless A-24s en route to Darwin and beyond. They had been assembled at AMBERLEY with all the smooth progress near-panic can produce. As James Rorrison notes in his ***Nor The Years Contemn*** -

'...One broke down at Charleville and was left there, much to the excitement of the townspeople who lined the airport fence to look at it.'

For a period later in 1942 CHARLEVILLE became the terminal for the Pacific Ferry Route over which heavy bomber aircraft were flown to the Southwest Pacific area. Not only was it safely located well in from the coast but it was virtually on a great circle route northward from the populated southeast corner of the continent. As a safe storage location for valuable aircraft its dry climate helped to minimise corrosive deterioration. Storage and maintenance were the intended dual roles for this base which eventually featured five significant hangars.

A 3 March 1942 newspaper 'tender notice' called for -

'...Erection of four heavy timber-framed buildings within five miles of Brisbane, (buildings are to be) 96' x 96' sheeted with galv' iron walls and roofs and with concrete floors.'

Seventeen Brisbane builders were separately alerted, including C P Hornick and T F Woollam. Both these names remain prominent in the industry today.

By 17 March a contract was let to T H Dennis and these buildings were to become hangars 3, 4, 8 and 9 at the Aircraft Erection Depot, Project No 2, EAGLE FARM. However, on 13 April 1942, the site for the work was changed to CHARLEVILLE and the time for completion shortened by a couple of weeks! -

Lt Walter H Johnson's daily journal of the US 19 BG's 435 BS first mentions CHARLEVILLE obliquely, but sufficient to show fresh B-17s were filtering through -

'...April 26 Sunday: Lt Johnson and supply staff - S/S Cuffel and M/S Patterson unload train of vehicles, but Headquarters squadron takes all except two tugs, one Cleatrac (sic), one Jeep and one staff car. Spieth and Beezley back, report Rabaul empty now, the cruisers and aircraft carrier being away. Surles goes to Charleville after a new plane...'

The SDCC included this aerodrome in its 'No 3 Division' centred on Dalby. Its records carry the following report dated 27 June 1942 when camoufleurs -

'...Oliver, Piper and Rowland arrived at Charleville on 23 June they contacted Col. Kruse CO of US Forces... and inspected the drome and a storage camp at the old wool scour buildings. The runways are nearly completed and the taxiways are in use... A dispersal scheme is badly needed. Mr Wyeth and P-O Keays proposed visiting Charleville shortly for the purpose of pegging the hideouts on the scheme laid out on the previous visit by Mr Johnson. Two large hangars are in course of erection and will shortly need toning down...'

The report continued on the need to paint the many -

'...tents dispersed throughout the timbered area...'

and concluded with the realisation that the -

'...Hangar of QANTAS Ltd which Mr Johnson had suggested should be removed will mostly be used by the Americans.'

CHARLEVILLE *((9/6/1943)*

This is the 'serviceability date' on Drawing 42-43/1327A. The short dotted runway part of the small internal triangle is indicative of the early 1940's development at CHARLEVILLE. Securely located far from the coast, this natural airfield location was quickly recognised for its value on the Inland Ferry Route and developed as a US aircraft storage and repair centre. *[AA BP375 Item QL 680]*

Australian Archives (Qld): Dept. of Services and Property; Registry Office, Qld. BP375 Correspondence Files, 1932-1973 ITEM: QL 680

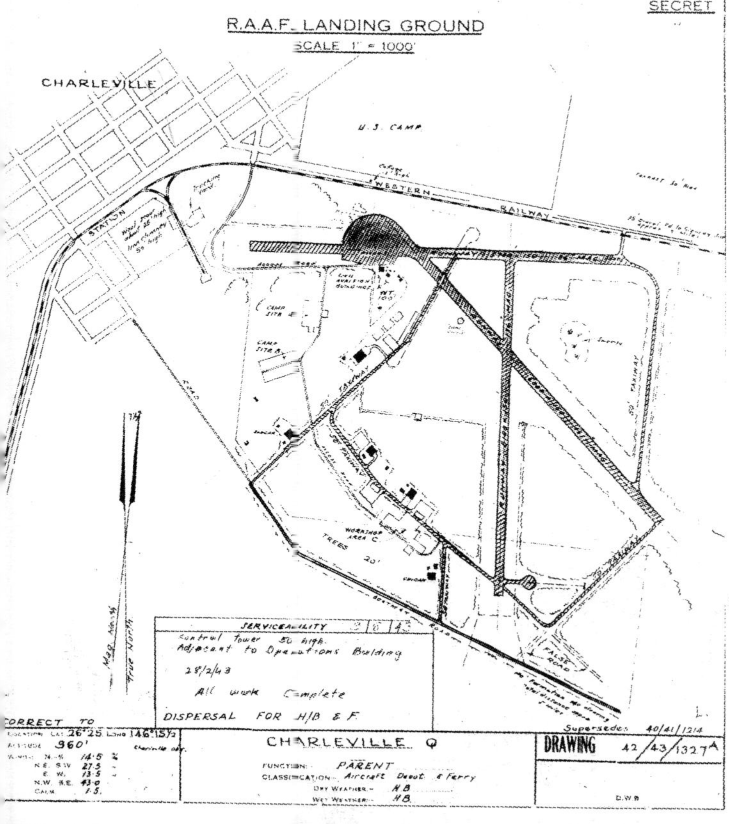

CHARLEVILLE *(26/6/1943)*
Frame PX 245, one of four companion obliques to the vertical runs in a systematic RAAF coverage. The hutted camps in the foreground on the fringe of the airfield were occupied by American servicemen during 1942. *[NLA (RAAF Official)]*

On 13 May 1942, a fifth and larger (184′ x 174′ with doors clearing 129′ x 31′8") hangar of similar design to the others, was added to Dennis' contract. This work was subject to all the difficulties of 'a remote location', and the urgency. The AWC wrote to Dennis 4 July 1942, in similar vein to their MARYBOROUGH approach of six months earlier. Hours per week were to be increased, with Industrial Court sanction, and the overtime would be paid. Whatever the reasons, the sorry result was no new hangars when they were most needed.

Fl Lt John Keays' report of 10 July 1942 stated -

'...The extensions to runways are complete including sealing. Approval has been given to widening of Boatman Road, road to gravel pits and cross road from railway to Morven Road all for use as dispersal taxiways. Schedule has been issued and corrected copy of drawing 41-42/3271 passed to Interior for submission to Air Board... A party of camoufleurs are in Charleville and it will be advisable... to take an early trip to hold them on the rails. In my opinion, dispersal only without camouflage, will be sufficient for present requirements and I only allowed £10 a hideout for clearing openings in scrub. Mr Jordan was advised of this and I suggested he withdraw all but one of the party who could stay and generally supervise the clearing and the small amount of painting that is necessary.'

Camouflage reports cite 10 July as the date on which control of the airfield was handed to USAFIA. The question of whether camoufleurs were further required was settled on 2 August 1942 when the American Authorities advised they would not order any work on establishments under their control beyond 100 miles of the coast.

Some interesting detail of the dispersal works can be gleaned from an MRC to Dept of Interior letter of 6 August 1942. It dealt with the need for stock-proof grids/openings in the airfield fencing -

'...No finality was reached before (Fl-Lt) Keays left for Townsville. It was agreed... a goat proof grid, to take a Flying Fortress from the drome out onto the Boatman Road Dispersal, was unnecessary. As guards would be on duty it was easier to arrange a collapsible fence from the drome. For the Fighter Dispersal north of the line, an open level railway crossing is necessary and details are now asked for regarding design. As fighter craft will have to cross the railway line, the width... must be sufficient.'

The US 45 Air Base Group had carried out servicing and repair duties at this 'reception depot' since March, and during July, B-17F aircraft had finally begun arriving to equip the 63 BS, 43 BG. A memo dated 8 August 1942 shows concern for effective liaison in the provision of buildings and facilities at the airfield. Major Willard Farrar CE HQ, US Base Section 3, advised the Deputy Director General of Allied Works Brisbane -

'...In order to better serve the needs of the 45th Air Base Group and the 174th General Hospital, now stationed at Charleville, we have arranged to have Lt John C Parker Jnr CE, together with a small staff, stationed at Charleville... (Please) instruct all members of the various Government agencies at Charleville that from time to time they may expect directions from Lt Parker in regard to the conduct of the work. Col. Kruse has had considerable difficulty in getting things built at Charleville from time to time, and it is believed that this method of handling the construction work will materially speed up the work. Mr Ross stated that he had heard nothing about the parachute drying tower, copies of plans for which were forwarded to you about a week ago. There is still no evidence of any construction whatsoever and the warehouse requested for the Glenroy Wool Scour Siding.'

It appears the timber framed fibro sheeted 'warehouse' was subsequently built by US troop labour. Incidentally, this John C Parker Jnr is likely the same officer who later served at IRON RANGE and in the Rockhampton district. Someone's careful collation of drawings and records of works carried out for and by the US Forces around Rockhampton endures to this day in the care of AA, Brisbane.

An inspection report dated 29 November 1942, found in Camouflage records presents a useful review of progress. It seems a visit was made in company with a Mr L S Steine who was to prepare an estimate for the painting of the hangars nearing completion -

'...A small number only, mainly US Air Force personnel, were in occupation. Dismantling of various portable buildings for transport elsewhere was in progress. The whole place presented a deserted appearance with the complete absence of any aircraft...'

There was a severe dust problem, where it frequently accumulated 'inches deep' on runway surfaces. Molasses and oils were tried but the vast barren areas surrounding the airfield ensured a fresh deposit appeared soon after treatment. Grass cover was slow because -

'...watering with bore water was dangerous to young growth owing to the high temperature of the water...'

CHARLEVILLE *(26/6/1943)*
One of the several verticals in a systematic RAAF coverage, this frame shows well the three intersecting runways. It also shows the five hangar structures, only one of which remains today, i.e. the second from the right in this photo. Some taxiways are paved while others have been left cleared and unpaved to serve as 'dummy roads'. *[CPE (RAAF Official)]*

CHARLEVILLE (9/10/1981)

This excellent oblique shows the close proximity of the airfield to the town of Charleville. Today the airfield is owned and operated by the Murweh Shire Council. The tree lined road entry to the current terminal buildings in the centre of the photo brings today's visitor to a point about 200 metres north of the 'Norden Bomb Sight Store Building' remaining from the wartime period. In the background just north of the railway line the local racetrack is visible within which emergency landing facilities existed for aircraft pre-war. [Department of Lands, Qld.]

The AWC's Queensland Construction Manager, Mr M C Cox is on record, 25 January 1943 describing the state then reached. Apparently the substantial number of buildings comprising the three Camps A, B and C were complete with water reticulation and sewerage to A and B. The sewerage pumps for 'C' were 'on site' but because Camp C served the Workshop hangars 103, 104 and 105, all not now likely to be used, no one seemed anxious to install them! The parachute building had been completed -

> *'...inclusive of electric wiring and installation of two blowers. The two blowers were subsequently removed by US Forces when they moved out.'*

Hangar 101, the largest of the five is reported as having been completed except for 'ripple iron on the doors which is now in transit'. Even the extensive educt fans, 30" in diameter were close to completion after having diverted some of the 25" fans from one of the smaller hangars. Seven prefabricated workshop buildings, each 80′ x 40′ were supplied by the US Forces; four were erected on the drome and the remaining three at the wool scour site as Stores. These had all been dismantled and removed by the departing US Forces. Also at the Glenroy Wool Scour -

> *'...three 25 000 gallon petrol tanks were supplied by the United States and installed on a US requisition. These tanks are more than half filled with fuel and some action should be taken to establish a guard to prevent pilfering or else the fuel should be removed.'*

CHARLEVILLE (-/7/1990)

VH-FDZ, a Beechcraft King Air of the Royal Flying Doctor Service on the tarmac entry to the surviving wartime hangar 104. This large structure served as the main shelter and relief headquarters building during the April 1990 severe flooding of Charleville when hundreds of local residents were forced to the higher ground of the airfield. [R Gill]

CHARLEVILLE (19/8/1944)

83 Squadron Boomerangs with an escorting 13 Squadron Ventura on the ground at CHARLEVILLE for a refuelling stop. In the right background under the wing of the Ventura can be seen hangar 101, one of the substantial hangars urgently erected in 1942. [D Oliver ex RAAF]

The report mentions various other headings but two stand out - one practical and one prophetic. Under 'Runways', some resealing was now urgent as -

> *'...at present four or five heavy machines are landing on the aerodrome daily...'*

Under 'Sirens', presumably of the Air Raid warning type -

> *'...six (6) sirens have been delivered... one for each hangar and one at Headquarters. These sirens have not been installed and directions as to their disposal are required.'*

A DWO Brisbane memo of 3 August 1943 to the Dept of Interior Brisbane relates to the transfer of certain hangars from CHARLEVILLE to the USA Air Service Command -

> *'...the large hangar 101 and small hangars 102, 103 and 105 as indicated on DWO drawing BRIS/U.2B... are (now) for removal to Eagle Farm. Hangar 104 has been retained by RAAF... at Charleville. It is understood USA 81st ADG would arrange for removal and re-erection at Eagle Farm of these hangars through AWC, but it appears this work is now being carried out by their own forces.'*

A US response dated 11 August 1943 confirmed -

> *'...these hangars are being dismantled by troop labour*

CHARLEVILLE (13/9/1989)

This small reinforced concrete building is located in the centre of the airfield complex about 300 metres north of hangar 104. During the author's inspection, the local ground staff were unable to explain the wartime significance of the building. Later archive study revealed it as a 'bomb sight store', clear evidence of the serious efforts by the USAAF to preserve secrecy of their Norden bomb sight. [Author]

and Mr Hanlon who constructed (them) originally has been hired to supervise their dismantling.'

AWC Minutes of 5 November 1943 refer to the laying of -

'...concrete floor 96' x 320' for the hangar to be removed from Charleville to Eagle Farm.'

As it turns out, no single floor of that dimension was constructed. Three of the nominal 96' x 96' hangars were transferred, one of them being termed Hangar 3A. Well after the war Hangar 3A was sold by disposal and relocated to the now defunct Chewing Gum Field Air Museum on the Gold Coast. It is not clear to me where the large Hangar 101 was relocated. Certainly the other two were used to augment EAGLE FARM Hangars 1 and 2, and in the case of Hangar 1 that structure remains in place to this day.

Strangely the RAAF Directorate of Works and Buildings' Schedule of Buildings 'document 112H' with the review date of 30 June 1944 shows Hangars 101, 102, 103 and 105 as being removed, yet the large Hangar 101 still appears in the background of a photograph dated November 1944 featuring a 13 Sqn Ventura and an 83 Sqn Boomerang refuelling at CHARLEVILLE. So when it was removed, and where to, are a mystery to me.

KINGAROY *(-/7/1943)*

This is the Department of Civil Aviation's Data Sheet for Landing Ground No 375. Only page one is reproduced with its locality plan showing aerodromes of WONDAI to the north and NANANGO to the south. *[Civil Aviation Authority]*

(Confidential Document)

DEPARTMENT OF CIVIL AVIATION **AUSTRALIA**

DATA SHEET FOR LANDING GROUND NO.375
KINGAROY, QLD.

ISSUE 3 - JULY, 1943. (2 pages) GOVERNMENT AERODROME "A" CLASS

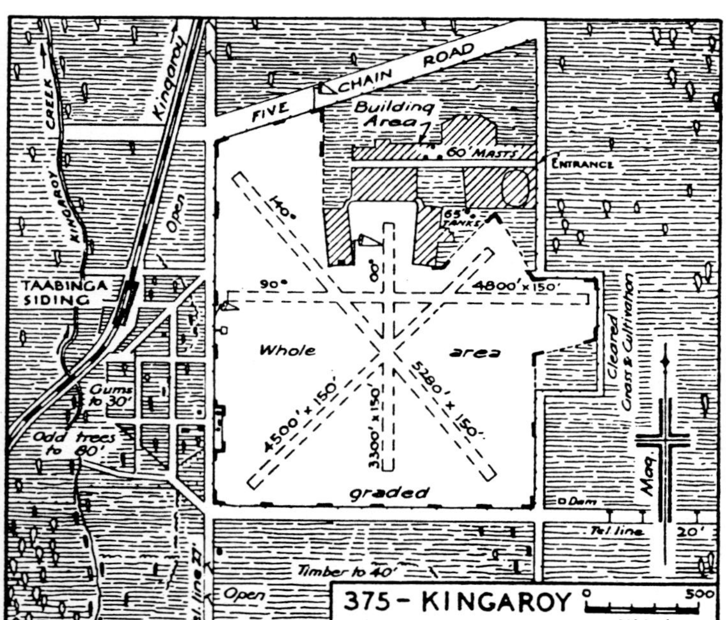

(For text see Page 2.)

D-1205 P.2815

KINGAROY

'...I was only there for about one month in August 1942 when No 3 Initial Training School transferred from Sandgate two thirds the way through 28 Course. I am sorry I have no recollection of either the place or time spent there after all these years, except that it was intensely cold after Sandgate... No, not quite true. I do remember there were no toilet seats - just the cold porcelain. If you washed out a pair of "goon skins" (overalls) at night they would be stiff with frost on the line next morning. Oh yes, and I do remember rolling fuel drums!'

This was Garth Parkinson struggling to recall KINGAROY. Garth went to England and the Middle East for further training prior to being posted to the RAF's 37 Squadron flying Liberators in Italy. The war was all over by the time many such Australians returned home.

The nucleus for KINGAROY was a small pre-war airfield known as the 'Taabinga Village site' with a few buildings grouped on the western boundary. It was identified in October 1940 as an appropriate location for the development of 3 SFTS. Eric Read, then in 23 Sqn, recalled various flights in a Wirraway aircraft out of ARCHERFIELD and AMBERLEY in search of airfield sites at that time. His log book entries align with archive records showing that a report was lodged by Fl Lt Read comprising reconnaissance notes, rough sketches and Plan X-7 of KINGAROY aerodrome. The same file recognised -

'...a licensed aerodrome controlled by a local council already exists on portion of the area... being considered for establishment of an Empire Air Training School...'

By March 1941, water supply from sub-artesian wells, was assessed as 'adequate', and the dark red sandy loam soil as having excellent carrying qualities after heavy rain. Importantly, there were two good relief landing grounds (RLG) within 15 miles. These proved to be HOME CREEK and MANNUEM.

Late April 1941, the RAAF's Air Marshall Richard Williams (then AMOE) outlined the case for KINGAROY in Air Board Agenda No 3276. Briefly, AMBERLEY had been chosen as the home for 23 Sqn pre-war. At the outbreak, AMBERLEY was not completed, so 23 Sqn was located at ARCHERFIELD. Subsequent demands on the Empire Air Training Scheme resulted in a proposal to locate 3 SFTS at AMBERLEY, but the difficulties of finding RLGs meant looking elsewhere. With the development of the site at KINGAROY for the transfer of 3 SFTS from AMBERLEY, 23 Sqn could then move there from ARCHERFIELD along with a second 'Service Sqn', 3 AD and 3 Air Observer School.

Doubts as to the adequacy of water supply persisted up until July of 1941 when a decision was made to proceed with the aerodrome project. It appears that the water problem didn't 'go away'. The ***History of the Queensland MRC*** states that organisation constructed a dam on the Stuart River to supply water for both the town of Kingaroy and the aerodrome.

By the outbreak of war in the Pacific the need for Bellman hangars at various locations in Queensland was certainly stretched. The Works Files for the Dept of Interior, Queensland show tenders had been invited -

'...returnable 18 December 1941 for construction, guttering and down piping to Bellman Hangars as follows: No 3 ANS & 3 WAGS Maryborough (looks like 2 out of establishment of 5 Hangars); 8 SFTS Bundaberg - 8 out of establishment 18 hangars; 12 EFTS Lowood - 4 out of establishment 4 hangars; 3 SFTS Kingaroy - 4 out of establishment 20 hangars.'

During January 1942, KINGAROY was one of the several Queensland Stations identified for expedition of

KINGAROY (23/3/1943)
This shows the 'all over field' nature of KINGAROY. The grass has been cut and windrowed on various runway strips. Four Bellman hangars are discernible in the building area with the adjoining barracks for the RAAF's No 3 ITS (Initial Training School). The prominent black shadow just southeast of the building group is cast by two elevated water tanks, the substantial timber stand for which remains to this day along with one of the tanks. The black 'creek-like' imperfection down the western half of the photo, is in fact a tear in the negative. [CPE (RAAF Official)]

'service hospital facilities'. The fear of imminent 'battle casualties' was growing. April 1942, 3 SFTS was disbanded in the national context of geographically rationalising flying training and the growing need to provide airfields for the gathering US forces. 3 ITS was to continue. KINGAROY also received the attention of the Camouflage Committee but there little progress was made. The SDCC report of 22 April 1942 mentions -

'...this drome has been taken over by the Americans, dull painting has not yet been commenced but Mr McColl (Department of Interior representative on the SDCC) reported that this painting would be commenced on Monday next.'

The next month's SDCC meeting recorded -

'...Mr Johntson suggested that to make any concealment of the drome effective the town of Kingaroy needed painting out as it stands as a pointer to the field, visible for a great distance... (However) Air Board (has) advised that Kingaroy is now to be regarded as a Training School, which may be used to accommodate fighter squadrons if required...'

KINGAROY (-/6/1942)
This could almost be a scene from some English airfield; pilots awaiting an order to 'scramble'. It is however at KINGAROY where these RAAF fighter pilots had assembled after their return to Australia in the dark days of early 1942. From left to right, Jeff Wilkinson, Pat Campbell, Lex Winten, Peter Jones, Charlie (?), Keith Gamble, Nat Gould, Frank Shelley, John Piper and ?. Some few weeks were spent here during reorganisation of the RAAF's 75 Sqn prior to its historic departure to New Guinea. [K Gamble ex RAAF]

KINGAROY (19/9/1989)
The sole remaining water tank on its elevated stand guards this airfield today. Just to its right and beyond can be seen the tree growth that has sprung from the earth mound of a 25 yard range adjacent to the reinforced concrete shell of the aircraft machine gun test butt. [Author]

Formal completion of work of erection of ninety six buildings at KINGAROY was achieved on 10 July 1942. A report of that date on 'aerodromes in the Brisbane area' by Fl Lt John Keays includes this reference -

'...field is in use but is not favoured by flying personnel.'

A reinforced concrete machine gun test butt structure remains to this day at KINGAROY. Archive documentation associated with that work shows that during its construction in September 1942 a 'steel deflector plate' was unobtainable so a design was issued for a reinforced concrete deflector wall instead. It was shown as applying to 'partially completed test butts at BUNDABERG and MARYBOROUGH' although someone picked up the error striking out MARYBOROUGH and writing in KINGAROY.

In October of 1942 KINGAROY came in for its allocation of construction of thirty six splinter proof pens for aircraft. It appears some such work did proceed, but certainly later events led to abandonment.

Duncan McConnel of Toogoolawah recalled his association with KINGAROY. He enlisted and -

'...went to ITS in Kingaroy in October 1942...there we slept in barracks on palliasses on the floor. When we had leave on some weekends I used to be able to travel down to Toogoolawah in the "service car" (a large touring car generally packed with passengers usually four in the front seat with the driver) for the weekend. At that time Colin Munro was in a 5 Squadron group camped at the airfield at Lower Cressbrook (Toogoolawah) and some of the members of the Squadron used to come up to home at Cressbrook. They had Wirraways then and one weekend one of the pilots flew me back to Kingaroy.'

A periodic airfield inspection late 30 December 1942 disclosed a lack of use of the airfield -

'...In the case of the parent station landing area, which is regularly used by Civil aircraft and may at any time again be in use by RAAF aircraft, it is assumed that regular maintenance is to be carried out... With regard to Home Creek and Mannuem direction is required... indicating whether or not it is intended to regularly maintain these landing grounds for use by aircraft.'

In the last quarter of 1942 WAAAF personnel began to appear 'on establishment' at various Queensland airfields. At KINGAROY, approximately six 'sleeping huts' were allocated by January 1943. when some concern for the adequacy of WAAAF accommodation was apparent.

Colin Munro recalled flying a reccé on 26 January 1943 from TOOGOOLAWAH to KINGAROY, where 5 Sqn was shortly due to move. The Squadron stayed for several months, training with Army formations. Colin spoke of artillery observation work 'south of Maryborough towards

KINGAROY (19/12/1990)
This recent vertical, from systematic mapping coverage, shows the now developed sealed runway 16/34 with alternate grass cross strip 05/23. The building area is now dominated by the sole remaining Bellman hangar and away to the southwest of it is the newly constructed Airport Terminal. The Kingaroy Airport is now named for Sir Joh Bjelke-Petersen KCMG, a long-serving former Premier of Queensland, whose residence and approximately east-west private airstrip adjoins this airfield diagonally to the southeast. [Department of Lands, Qld.]

STRATHPINE', where 5 Sqn had a detachment of two Wirraways based for a period.

Camouflage reports dated March 1943, show concern for the painting of most buildings on the Station -

'...(they) badly need repainting. They were done last June and some colours have faded badly, particularly "Darwin Stone" and the greens. There are also a number of new buildings and water tanks which have not been painted... (this) has been reported to DWO, who will probably get it done as maintenance.'

The constant need for maintenance of the airfield's grass surfaces and that of the satellite fields, periodically prompted review of such land holdings. The DASP recommended, November 1943, that -

'..As we have no requirement for RLG's in conjunction with RAAF Station Kingaroy, it is recommended these landing grounds be returned to their owners. It is also recommended that armament ranges be handed back...but any facilities provided there should be retained as RAAF property for the present.'

While 3 ITS activities continued up until war's end, 1945 saw the forming and/or reforming of several squadrons. Mosquito and Beaufighter aircraft were a common sight on the airfield then.

KINGAROY (19/9/1989)
RAAF veterans who spent time at KINGAROY may recognise the sick quarters (building 91) and the ambulance garage in the right foreground. Notice the lone faded red fire hydrant cabinet on the right. [Author]

CONDAMINE

'...There were never any Air Force personnel or other forces stationed there, or any buildings erected there.'

So answered the President of the Miles and District Historical Society in February 1991 when I circulated most such societies in Queensland on the matter of their 'local airfields WW2' - and she was right!

CONDAMINE was one of several fields commenced during 1942, charged initially to a 'Joint Aust-US account' and figuring somewhere in 'US (Project) 174'. I've found no veteran with 'landing experience' at CONDAMINE and while it is likely aircraft used it incidentally during WW2, I've found no record. It is also probable that CONDAMINE, being the most westerly of four 'Heavy Bomber (HB) fields urgently constructed as part of the *Brisbane Line*', proved to be the least attractive for 'reactivation' when the RAAF did finally commence formation of its Liberator squadrons.

It seems this site was one of several advised by the Wambo Shire Engineer as 'worth investigation' by the RAAF. P-O John Keays visited in company with F-Lt Higginson from AMBERLEY, on 21 March 1942. The 'bomber field' was conceived as part of the RAAF's proposed '73 Squadron (expansion) Plan freshly approved by the Australian War Cabinet, and RAAF Directorate of Works and Buildings (DWB) staff were then realising how the orderly peacetime procedures weren't working! Preparing full plans, specifications and estimates, and obtaining financial approval before starting construction, was not the way the US forces were advocating - even the Australian Army was not so constrained. A change was agreed to 'provided care and responsibility were retained', and so, dated 13 April 1942, the relevant Air Board Agenda was prepared for 'RAAF Station Condamine'.

By 20 April 1942 P-O Keays, reporting on the four sites of CONDAMINE, JONDARYAN, CECIL PLAINS and LEYBURN, said inspections of all sites showed -

'...progress to date is most unsatisfactory. Works had been commenced on site at Condamine and good progress was being made but no actual construction work had been performed on (the) other sites.'

CONDAMINE (1/6/1943)
This is the 'serviceability date' on Drawing 42-43/589A. This copy is from the 'hiring of site' (archive) file and was used in organising fencing of the airstrips for the return of most of the holding to the owners. [AA CRS A705 Item 171/93/5]

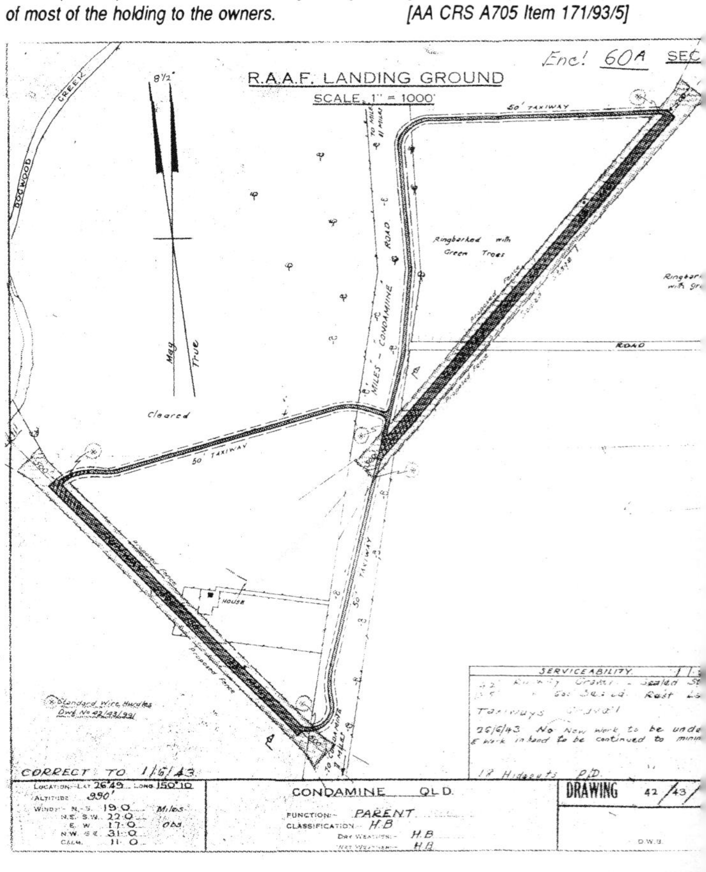

CONDAMINE *(27/9/1945)*
CONDAMINE is one of the airfields constructed to quite an advanced state during 1942-43 and not used. The taxiway within the north-south Miles to Condamine road reserve was an obvious inclusion in the eventual construction of that Main Road. For the pavements, in-situ soils were generally suitable with some small aggregate addition. However, the white 'blob' southwest of the 136° strip is believed not to be a quarry site, but simply a blemish on the negative. *[AUSLIG]*

Later in the report he described the N-E, S-W strip as -

'...40% cleared and it is anticipated that it will be available for aircraft within two weeks.'

Alan H Jordan was recruited in Victoria in April as a Camoufleur and promptly transferred to Queensland. By June 1942, he was in charge of 'Division 3 - Dalby' for the Department of Home Security. The SDCC report for 19 June 1942 touches on concerns apart from 'the work' -

'...Mr Jordan has arranged to meet Mr Wyeth and P-O Keays at Condamine drome next week...(to peg the) hideouts... Housing camoufleurs on these dromes is a problem. Huts can be provided by Main Roads but bedding must be found by the camoufleur.'

While planned as the 'conceal only' type, such hideouts did not proceed.

By this time the search was on for an Aircraft Depot site and while CONDAMINE was considered, it was discarded as being too far from the Railway. Attention turned towards Dalby and eventually, OAKEY was chosen.

Identification of dispersal strips was part of the full development of these HB airfields. A Minute of 25 August 1942 shows the CAS had approved work on -

'...Condamine and Leyburn for use by Operational Squadrons and consequently for dispersals at Miles and Goonalah for the strip at Condamine and Sandy Creek as a (dispersal) strip for Leyburn.'

Meanwhile up north, crisis point and the 'wet season' were approaching. Resources were so scarce and stretched that some rationalisation had to occur. The AWC direction of 29 August 1942 to its Queensland office included -

'...Department of Air advise that they require the work on the parent aerodromes at Jondaryan, Cecil Plains, Condamine and Leyburn continued and completed as early as practicable, but they agree to the work on all dispersal strips being held in abeyance... work on the dispersal strips for Condamine and Leyburn may be proceeded with when plant and men are available from the North Queensland area.'

By the end of 1942 it was clear to the RAAF that its own Heavy Bomber aircraft were a long way off. Understandably, proceeding with the 'squadron camp' building at CONDAMINE came into doubt. On 6 December 1942 the Camouflage Committee reports indicated that 'the Condamine camp is to be a replica of the other Downs drome camps'. AWC minutes of early February 1943 show this job being assigned to Contractor Strohfeldt who was doing similar work at LEYBURN. However, the death knell for further development came 26 June 1943, with the signal from DASP to DWB -

'...In view of - (a) the improved strategical situation and (b) the availability of suitable HB dromes at Jondaryan, Cecil Plains and Leyburn, further works at Condamine should not be undertaken and work now in hand should be completed to the stage of minimum serviceability.'

Here was another airfield deemed surplus to RAAF requirements. At least, the return of property use to its owner(s) was an obvious move. On 2 May 1944 a review on record showed the present stage of completion as -

'...two gravel runways (sealed at ends), and perimeter taxiways partly sealed. No dispersal.'

There were 'nil' buildings completed although water supply was shown as 'complete'. Arrangements were in place through the AWC for 'care and maintenance of landing ground and other services by local authority.' The decision by the CAS was 'to offer the site to CA and if not required, to retain for 6 months and bring up for review.'

Because of its distance from Miles this airfield was not much used by the Department of Civil Aviation, post war. Only the 42° runway was retained and its license lapsed after some years. However, Murilla Shire Council has recently revived the field and the license as the current aerodrome for Miles.

CONDAMINE *(1972)*
The 1:100 000 topographic mapping shows this airfield 'thirty years after'. *[AUSLIG]*

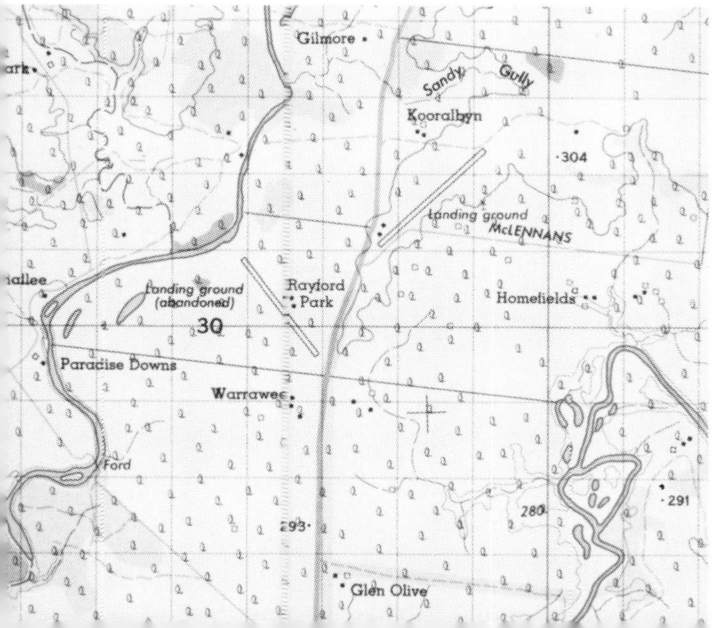

TOOGOOLAWAH *(25/12/1942)*
Ten officers of RAAF 5 (Army Cooperation) Squadron sit down to Christmas dinner in the Officers' Mess TOOGOOLAWAH. From left to right F-O's Ray Goldsworthy, John Hayward, Don Hutton, Bob Staley and Clive Freez, Capt Bill Hansom, F-O's Ron Forsyth, Fred Watchom, P-O Allan Saunders and Aub Grady. The bottle of beer nearest the camera bears the popular label of the day, Bulimba Pilsener Beer.
[I Olorenshaw ex RAAF]

TOOGOOLAWAH

'...Vintage, Warbirds, sport and recreational aviation have been the only real growth areas in aviation in recent years. Watts Bridge, will meet the leisure aviation demands of SE Queensland - the fastest growing area in the Nation - long into the future.'

This is the closing statement in an information leaflet produced recently by Watts Bridge Memorial Airfield Inc with a postal address, Box 698 Archerfield Q 4108. Watts Bridge, as an airfield name, stems from the nearby bridge over the Brisbane River washed away in 1974. The airfield was known simply as TOOGOOLAWAH, during WW2.

The Empire Air Training Scheme, instituted after the outbreak of war, prompted Local Authorities and keen minded individuals to promote local aerodrome potential. Aircrafts P/L, an airline linking Kingaroy and other country centres with Brisbane, tried unsuccessfully to interest the RAAF in development of the 1930s landing ground about 1 mile north of Toogoolawah.

When Japan entered the war the Australian Army became vitally concerned in the operation of what were termed Army Cooperation Squadrons. The RAAF's 5 Sqn was then operating from TOOWOOMBA which was considered an unsatisfactory site. The Army's need seemed to be better served by a site in the Toogoolawah area. Colin Munro was then a Flight Leader in 5 Squadron. His log books show a reccé flight over Toogoolawah on 18 May 1942. On 20 May 1942, in company with the CO, W Cdr A D Charlton, Colin made the first landing at the new site letting down from the west. He recalled -

'...only realising the presence of the N-S telephone lines crossing that strip as we rolled underneath them.'

The following extract from a report by the RAAF's P-O J Keays on files now held in Australian Archives states -

'...As requested, this site was inspected Monday 15 June 1942 with Messrs Calder and Lowe of MRC, Mr Hill of Esk Shire, Mr Wyeth of Home Security and Lt Chester of 1st Australian Army. The site originally selected was condemned as it is subject to flooding during the wet season... An alternative site selected by Mr Hill was inspected and although it is not entirely satisfactory, it appears the best site available.'

TOOGOOLAWAH *(19/7/1942)*
This oblique (P 773) taken by 5 Squadron shows the western approach on runway 121°. The furrowed appearance on the near section of the strip is cut grass. Silverleaves Road crossed runway 121° and in fact telephone poles are barely visible along this road with one appearing almost on the airstrip centre line.
[RAAF Official courtesy I Olorenshaw]

This was the site duly developed. Keays continued -

'...Col. Bleachmore was advised that works could be commenced next Monday and Mr Calder is making arrangements accordingly. Action taken on this matter has been entirely to assist the Army as I was advised in Melbourne that it was not an Air Board matter. I was to give all assistance possible in locating site and arranging layout.'

A month later John Keays reported the field under construction by MRC for the Army.

This demarcation in responsibility for airfields was reinforced by a RAAF file Minute of 22 August 1942. DWB sought clarification from the CAS saying it was -

'...undesirable that Army should instruct Interior to construct aerodromes without reference to this Department, particularly in view of the critical position regarding man power and plant. It is also desired to know whether the previous assumption is correct, that the Army would provide, through their own Engineer Units, any landing grounds required for operations of the Squadron for Army purposes, with the exception of the base aerodrome, which the RAAF is to provide.'

While John Keays was occupied elsewhere, P-O J Campbell was detailed to check progress on the drome -

'...As instructed an inspection was made on 16 September 1942 as to the suitability of the aerodrome site as a base for Army Cooperation Squadron. The site is 5 miles east of Toogoolawah township... Two runways have been constructed along high ground with good longitudinal and cross grades...'

He noted the disadvantage of only scattered trees in which to effect dispersals and some constraints on possible strip extensions. Under a heading of 'Obstructions', he stated -

'...strips will not be available for use until both telephone and power lines are removed. At present the telephone trunk line is being placed underground...'

Campbell described the graded 118° strip as well -

'...consolidated (with) new grass taking very well.'

While tree cover was scarce, he described the field as -

'...difficult to locate from the air, owing to the excellent growth of new grass on runways which run between existing farm houses and cross gravelled roads'.

He recommended it as the Base for the Army Co-op Sqn.

During November of 1942 TOOGOOLAWAH features in the regular reports to the State Defence Camouflage Committee (SDCC) as one of the various airfields in Southeast Queensland requiring the construction of aircraft hideouts. It seems no such work was actually carried out at TOOGOOLAWAH although it was at that time set down for twenty-four such hideouts.

Techniques for improving the load carrying capability of in-situ soils generally by the addition, mixing and curing of cement, received urgent attention during 1942. TOOGOOLAWAH was, for a period, approved to have a 'soil cement pavement' constructed on runway 121°. Though cement modified gravel pavements later became common, this early move was eventually scrapped in April 1943. There was a lack of confidence in the economical

TOOGOOLAWAH *(8/12/1991)*
Taken from the windy open seat of a Piper Cub, this is 'Watts Bridge Airfield' today. The airfield is now used for recreational flying by the Queensland Vintage Aeroplane Group Inc., and others. Some members have recently erected the first hangar, (right background). The hill profile of 'P 773' has changed little in fifty years. [Author]

TOOGOOLAWAH *(16/8/1942)*
This Frame V3125 is one of several verticals taken of the airfield about the time it was in use by RAAF No 5 Squadron. The scattered trees between the field and the Brisbane River to the north offered little cover for dispersed aircraft, fuel dumps and tented camp establishment, yet that is where such activities were sited. Note the local road deviation skirting the airfield to the north west to a bridge over the Brisbane River known then as Watts Bridge. Note also the gravel scrape in the southern quarter and the short haul road to the 121° strip. [CPE (RAAF Official)]

TOOGOOLAWAH *(-/9/1944)*
This is the publication date of the 1:63 360 (1 inch = 1 mile) topographic map from which this extract has been taken. The grid squares each measure 1 000 yards, which roughly converts to 1 kilometre. [Department of Defence]

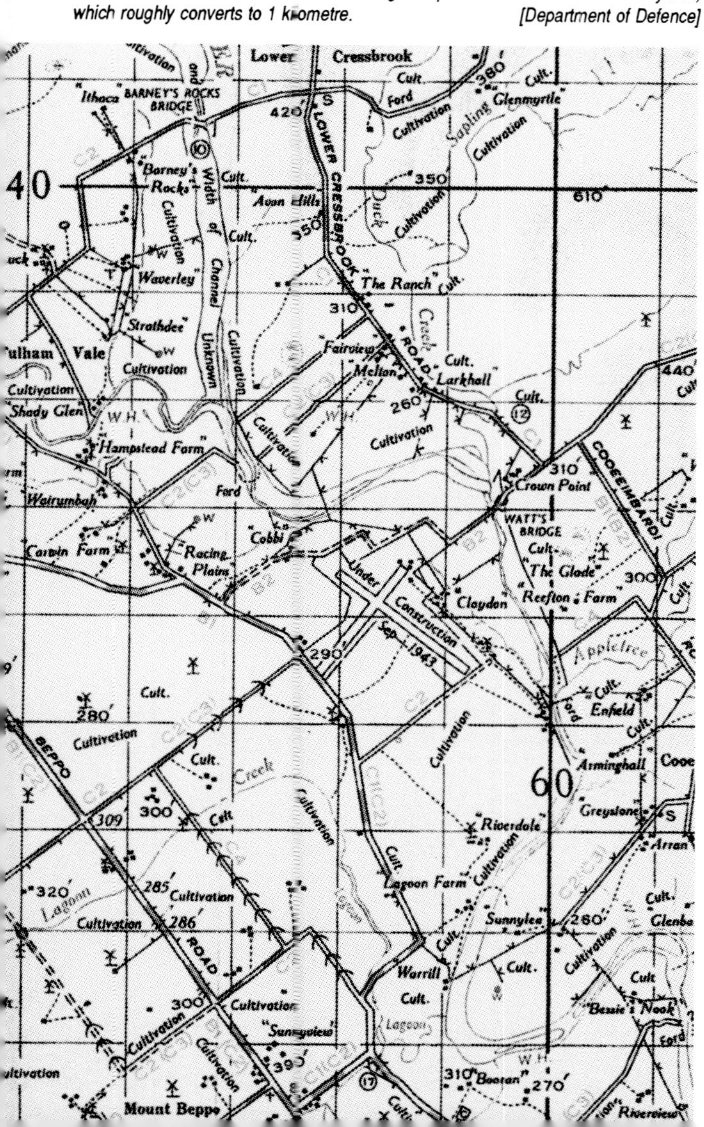

conduct of such experimental work and of course the war situation had by then, much improved.

Airfield related 'buildings' were scarce to non-existent at TOOGOOLAWAH. The camouflage reports around December 1942 mention the airfield being used for -

'...Army Cooperation work and regarded as purely temporary. Unit is mobile and no structures are required.'

Colin Munro's period at TOOGOOLAWAH extended only into January 1943. He remembered basically 'tents'! He added his stay there was -

'...pleasant with comfortable access to the local pub.'

Drawing 42-43/1929A (undated) shows 'explosives' and 'petrol storage' against two building-like rectangles in the northern quadrant. The 'B' amendment, dated 10 August 1943, shows various black rectangles all unspecified. On the Schedule of Buildings (sheet 225H) from the inspection of this date, several buildings are listed, but it is likely they relate to the AWC construction camp, marked for take-over by the RAAF. While 'nil hangars' are listed on these report sheets, one 'workshop' appears. I have heard it described as a tent-like structure, and as such I can only conclude it was similar to other(s) at GOOLMAN.

While W Cdr G J Towers, DWO Brisbane, consulted with 'Army' and concluded the 'extreme mobility' philosophy for the Army Co-op Sqns in December 1942, the SDCC report for 10 January 1943 mentions -

'...only ablutions, latrines and an explosive store are to be built. A 25 yard range is also to be installed and other camp buildings are to be mobile type and supplied by Army'.

Gravelled runways were signalled in lieu of 'soil-cement', around 6 April 1943. Near the end of May, DWO confirmed to AWC that work on these runways was to be -

'...temporarily suspended (so) that construction forces may be made available for works of greater priority at Horn Island and Jacky Jacky. This is on the understanding that this cessation of work is only temporary and that recommencement will be made as soon as additional men, already requested, arrive from the south.'

Around 13 July 1943, W Cdr Towers queried DWB -

'...Work on these runways has now been resumed and gravelling is about to commence. As the Army Co-operation Squadron, which was to have been based on this drome has now moved North, it appears there is now no unit waiting to occupy this drome on completion. Under the circumstances, it is recommended (bitumen) sealing of these runways be deferred...'

The deferment of sealing led quickly to reappraisal of any further work. The Camouflage organisation was winding down its field role when its 14 August 1943 report said -

'...At DWO, F-O Dallimore advised that Toogoolawah is to be used as an RSU instead of being scrapped.'

An amount of planning was done for this RSU in lieu of GOOLMAN, previously chosen, but it seems to have been a very half-hearted move. On 18 September 1943, DWO Brisbane submitted the 'scheme' documents to DWB Melbourne. The cover letter said -

'...Gp Capt Douglas CO Amberley, acting on behalf of HQ, No 5 Maintenance Group, concurs in principle on location proposed workshop and camp sites.

The enclosed drawings and schedules included for -

'...strengthening and sealing the 121° runway on which work was terminated... (on) 23 July 1943.'

On 'termination', certain rounding off work was instructed. However, this 18 September statement reported a -

'...state of minimum serviceability, viz
a) the 5 000' x 121° runway is gravelled to consolidated thickness of 4".
b) the proposed 3 600' x 32° runway was only partly consolidated and after the "rounding off" work, will revert to a natural surface strip.
c) relocation of the overhead H/T line on approach was not carried out.'

AWC did 'round off' both strips and the drainage. On

12 January 1944 DWO reported the NE-SW strip as -

'...Gravelling of the 32° runway has been carried out approx 4" consolidated thickness for a length of about 2 200' from the NE end.'

This statement by W Cdr Towers concluded with -

'...AWC buildings have been removed from site...The 32° runway may now be regarded as having reverted to natural surface, and it is not thought that any additional drainage work should be carried out on this runway to maintain work already effected.'

Dated 12 February 1944 a report apparently lodged by a Civil Aviation inspector Mr B G White, placed a different light on the 32° strip, saying this NE-SW -

'...runway is dangerous being overgrown with grass and it has never been finished off. What makes it extra dangerous is the fact that the old main road runs parallel with the runway and across the NW-SE runway. This old road is also the same colour as the NW-SE runway. To bring this point out, a plane hit a tree on the western end and crashed on the road, both men in the plane being killed: this happened 10 weeks ago.'

Mr White suggested to the Defence Dept that various items be tidied up, two of which were ploughing up the old road to make it 'less like a runway' and some formal appointment of the local telephone exchange operator, Mr F Cannell who is -

'...the Volunteer Air Observer Officer in this district. Mr Cannell more or less keeps an eye on the drome!'

Records show an Australian built De Havilland Dragon A34-19 operated by the RAAF's 2 Aerial Ambulance Unit based at KINGAROY crashed here on 17 September 1943. The pilot, F-O Thorley is remembered today by the naming of the entry road to Watts Bridge Memorial Airfield.

The ***MRC History*** published in 1949 mentions, in part -

'...Toogoolawah was provided with an aerodrome in 1943...Two runways, one of 5 000' and one of 3 600' were formed and about 2 000' of longitudinal collection drain pipes provided with transverse pipe drains under the runway. The value of this undertaking was demonstrated after the first runway was gravelled when a large US bomber which had run out of petrol in bad weather was able to make a forced landing...'

Other sources indicate that the bomber was a B-24 Liberator, and I infer that the date was after August 1943.

After the war, TOOGOOLAWAH was licensed by DCA for a short while. That lapsed early 1947. It may have taken the RAAF all of fifty years to revive an interest in the airfield. Recently, 38 Sqn with its Caribous, now based at AMBERLEY, 'rediscovered' the field as suited to their practicing night flying operations.

TOOGOOLAWAH *(8/12/1991)*
A colourful group of aircraft attending a 'fly-in' to Watts Bridge. Even a 'guard dog' is in attendance. *[Author]*

BRYMAROO

'...Yes, we use it from time to time as an alternate.'

So answered one of the Army Aviation officers at OAKEY when I asked the question in November 1990. BRYMAROO is only 25 kilometres north west of OAKEY which is an operational Army Aviation centre today.

BRYMAROO was a flurry of activity in part of 1942, being commonly perceived as one of the dromes forming the *Brisbane Line*. Archive records indicate a 19 March 1942 inspection by P-O John Keays and F-Lt Higginson, a General Duties officer from AMBERLEY. Keays wrote -

'Jondaryan...considered by Fl Lt Higgensen as excellent site...'

Water quality appeared to vary but Keays saw no trouble -

'...(being) experienced in obtaining... 10 000 gallons/day from several bores located around the area.'

Here was a classic case of dual names for an airstrip development! Perhaps it didn't matter - probably those people and authorities who were directly involved were adequately briefed and 'knew the site'. Perhaps it didn't hurt to have such confusion. If enemy intelligence read such 'reports' as two airfields, so much the better!

Although this project was urgent, it took some time to get going. P-O Keays' report to 20 April 1942, notes -

'...Inspections of all (aerodromes - West of Brisbane) sites have been carried out during last week and progress to date is most unsatisfactory... The Jondaryan site was inspected on the 14th with MRC District Engineer Mr C H Patterson and an assurance was given that plant and equipment had left Brisbane and would be on the site within the next few days. I have since been advised... works commenced on 17 instant.'

In May 1942 the RAAF's Works Director Air Services wrote to the AWC to tidy up confusion in some airfield naming. To avoid ambiguity, the Coordinator General of Public Works for Queensland was requested -

'...to refer to dromes near Brisbane by the names used by this Department as follows...(in the case of) Brymaroo, (please adopt) Jondaryan.'

The RAAF wanted 'JONDARYAN' yet its own 23 Sqn continued to produce photographs titled BRYMAROO drome up to at least January of 1943!

June 1942 was 'teething time' for effective liaison between the RAAF and various camouflage authorities. A memo from P-O Keays to Air Board Melbourne dated 16 June 1942 sets out in part, his concerns -

'...Camouflage features of aerodromes at Leyburn, Cecil Plains, Jondaryan, Condamine, were planned with Mr F Johnston of the Dept Home Security. I find on

BRYMAROO *(16/1/1943)*
Oblique 'PX42' is one of four such prints accompanying several verticals, all taken by 23 Squadron. This view is looking east with the turnouts to some of the several aircraft hideouts evident in the foreground taxiway leading to the 54° runway. *[NLA (RAAF Official)]*

my return to Brisbane Mr Johnston's services have been dispensed with and that I will be required to go over the complete schemes with his successor... A change of personnel in the middle of such a job will seriously affect the time of completion of the designs and I would be glad of any action that can be taken to prevent similar occurrences. I found Mr Johnston had a considerable knowledge of the subject and was quite prepared to adjust his ideas to suit engineering requirements. I found him very easy to collaborate with and I am satisfied that the schemes mutually agreed upon are good ones and are in marked contrast to those already completed at Archerfield and Amberley....'

Keays went on to discuss input on taxiway lengths from a 'Mr Hughes, USA Camouflage Officer' who -

'...suggests a maximum limit of 3 miles...'

The RAAF had instructed Keays that taxiways were to -

'...be kept as short as possible and that 1-1½ miles was the limiting distance... In view of the fact that these fields will in all probability be occupied by USA forces, I have maintained a close liaison with these officers and would appreciate instructions on these matters...'

Keays appealed for guidance in this interaction, camouflage with dispersals, right down to asking -

'...who is the referee in case of dispute... I understand that the Dept of Home Security has considerably increased its strength in Queensland and that they will now be taking an active interest in all works under construction. It is most desirable that the whole procedure be clearly set forth and that the Dept of Interior and MRC be made familiar with the procedure.'

There has always been professional rivalry between engineers and architects. John Keays was an engineer and many of the Dept of Home Security's camouflage people were trained in architecture or related graphic arts. Keays appreciated input from professionals other than engineers and it is not surprising that he was able to forge that productive link with the camoufleur. Such team work didn't always develop. Rod Voller who was senior camoufleur in charge of the DHS operations at Townsville, has on several occasions explained to me that often 'the RAAF just didn't want to know the camoufleur'.

Geof Collings and Frank H Johnston knew each other in London before the war and Geof tells me his services -

'...were "dispensed with" too, at one stage, in New South Wales. It was some disagreement with Authority, yet two weeks later they reappointed me to Charters Towers'.

By design, a Parent airfield required at least two dispersal strips. Not surprisingly many potential dispersal sites were identified and even documented without ever proceeding. For JONDARYAN, one can see a touch of humour in the following excerpt from a report by P-O J W Campbell to Air Board dated 3 August 1942 -

'...Quinalow - (Dispersal Strip No 1) Location: The possibilities of this site were first recorded when Sqn Ldr Quinan visited Irvingdale site on 27 July 1942. On his drawing attention to closeness of the site to the parent aerodrome, it was jokingly suggested that another special site be found and be named after him. A search was made further up the adjacent road. A narrow road through almost virgin bush disclosed possibilities in Portion 148V, Parish of Rosalie. It is 2 miles SW of a small village called Quinalow. The Sqn Ldr was satisfied with the possibilities of the site including its proposed name...'

Campbell detailed the other satellite site 'Moola Road, 5 miles west by air'. Like 'Quinalow', no work proceeded. The need shifted and resources were directed elsewhere.

Around 16 August 1942, an advance party of men from the Dept of Interior had arrived at JONDARYAN to commence work on an arch type hideout. The SDCC report of 23 August 1942 mentions -

'...Mr Jordan and Mr McColl of Dept of Interior were told (by the Camouflage Committee) that no work could be commenced on arch type hideouts without approval of the Technical Director and that no camouflage scheme had yet been approved by Dept of Air.'

The Technical Director of Camouflage, Professor W Dakin, was headquartered in Canberra. At this time the 'arch type hideout' was just coming into vogue. It was probably attractive to many camoufleurs and indeed to many sites, but the Technical Director rightly saw the need for a responsible channelling of such effort. Some experiments had been carried out and clearly the Technical Director was in the best position to approve where further experimentation should, or should not proceed.

By early September 1942, Mr Jordan, the camoufleur at JONDARYAN had other things to worry about. He discovered that the contractors on the job had suddenly erected their camp on what had been chosen as the site for the RAAF's squadron camp. As feared, the site was badly damaged by undue clearing. Mr Jordan reported -

'...There is a lack of coordination between Departments, or perhaps between branches of the Dept of Interior... Plans arrive for jobs which are not ready to start, while men arrive and make a start on other works without any advance knowledge or plans being available to me. The things that we are here to watch and control start over night and are usually discovered by chance - after damage has been done. Although camoufleurs are living on the job, it is quite impossible to police areas several miles across.'

During October 1942 the AWC minutes indicate a few weeks of indecision on the construction of a 'temporary hutted camp'. Work did resume on 22 October and the SDCC weekly reports of that time carry into December with repeated statements like that of 6 December 1942 -

'...Good progress is being made with camp buildings and it is expected the construction of hideouts will commence next week.'

Jondaryan certainly did receive its 'eighteen conceal-only hideouts'. A remarkable aerial photo record of their location and nature survives in AA. Erection of these 'slung type' hideouts finally got under way by 10 January 1943. They were completed in May.

By this time both runways had also been sealed. A file Minute dated 18 June 1943 mentions the following 'operational requirements Brisbane area as submitted to this HQ by Allied Air Forces' -

'...Interceptor fighter - Archerfield and Eagle Farm; Long Range Fighter - Amberley; Reconnaissance - Loganlea; Dive Bomber - Jondaryan; Flying Boat - Brisbane River; Strategic Air Route - Leyburn; Army Cooperation - Toogoolawah...Loganlea is partly constructed (and) Strathpine/Petrie is the current fighter aerodrome.'

JONDARYAN now entered a period of great inactivity! There was no occupation. In fact subsequent reports show an increasing concern with the growth of weeds and the grass itself on the flight strips adjacent the runways. Indeed, attention was being turned to the use of Bermuda Couch or Queensland Blue Couch and Kikuyu as better grasses to sow on these strips in that they grew a mat rather than tall grass. Repairs and minor maintenance for JONDARYAN was then placed in the hands of 14 OBU located at LOWOOD. A report dated 11 January 1944 brings up to date the state of JONDARYAN -

'...There are two strips bearing 131° and 54°. Both are allegedly sealed but the sealing is particularly thin and almost non existent. Grass and weeds have grown through... on both runways, and are apparently rapidly spreading. These will have to be poisoned. Sealing in its present state, even apart from the weeds, would not stand up to light traffic. A B-24 landed a few days before my visit and had broken through the seal when it turned. The surface below the sealing appears to be good, but its recommended that the strip be sealed properly before occupation by a squadron... the taxiway is not sealed and was very wet... It nearly bogged a Jeep and certainly would not stand a heavy aircraft.'

The ***MRC History*** lumps JONDARYAN in at the end of a brief description of CECIL PLAINS and goes on to note -

'...repair and maintenance work continued until the end of hostilities...'

Phil Dynes, now the official historian for the RAAF's 200 Flight, told me that his records show RAAF's 99 Sqn formed at LEYBURN on 1 February 1945 and became a 'lodger' unit with 23 Sqn. When 200 Flight followed a couple of weeks later, messing facilities were most inadequate for the total 1300 odd personnel involved. 99 Sqn was instructed to move to JONDARYAN, which was achieved one week later.

John Burley was with 99 Sqn at JONDARYAN. He recalled the 'bare strip' nature of the place and the inadequacy of the hardstandings even after 'repairs'. He described the camp about 1 mile south of the strip -

'..Bare buildings served as the kitchen and mess for Officers, Sergeants and OR's. There was a water tower and an open-air shower block. We had to dig deep pit latrines and grease traps and put up tents and lay out duck boards...'

John was the Unit's Warrant Officer Disciplinary (WOD) -

'...I left with the advance party to Darwin about April. We had just received our aircraft then. For the last two or three months of the war the squadron flew operations from Jondaryan, fuelling up and being armed at Darwin on the way. Life was very spartan at the camp but there was a Pub a couple of miles down the road...and Toowoomba at weekends.'

JONDARYAN or BRYMAROO is one of the airfields nearer to Brisbane that I have visited more recently. I found part of it owned by members of the family of Alex and Mavis Campbell who won the right to their holding in the WW2 Service Veterans' Ballot scheme early after the war. Alex was in the RAAF and appreciated the airfield's relevance to the point of being instrumental in the mounting of a post-WW2 Canberra jet bomber near the Brymaroo Hotel.

In recent months the Defence Dept acquired essentially the southern half of the 131° strip. Freshly fenced and with cultivated areas returned to open grass, the old airfield is again being used 'from time to time', albeit by rotary winged aircraft - certainly not B-24s.

BRYMAROO *(30/6/1945)*
This extract from systematic map coverage upon enlargement, shows twelve Liberators of the RAAF's 99 (HB) Squadron in occupation at the time. None were parked in dispersal hideouts so late in the war. The squadron camp is visible at 'centre bottom'. Today, the 1:100 000 topo map shows a property named 'Tarko' centred on the old camp. *[AUSLIG]*

PETRIE-STRATHPINE

'...Verbal advice has now been received from Senior Air Staff Officer RAAF Command, that an arrangement has been reached with the Director of Operations 5th AF, for the 3 strips A1-Petrie, A2 and A3-Strathpine to be made available for RAAF fighter operation.'

This brief Minute appears on file dated 9 February 1943 in respect of 'Aerodromes-East Area'. It represented the solution to the RAAF's search for a fighter airfield for the continued defence of Brisbane.

As early as November 1940, RAAF files record the search for what was termed a 'Service' aerodrome near Brisbane. This is believed to mean a fully developed drome for one or more 'Service' squadrons as distinct from 'Training' units. W Cdr J Waters, CO RAAF Station Archerfield, in a letter to Air Board Melbourne, dated 4 January 1941 reported on reconnaissance efforts -

'...suitable sites have been discovered at... Fort Lytton, Meeandah (near Eagle Farm) and Strathpine. No further places have yet been located in the vicinity of Brisbane (except) Bundamba Racecourse and Eagle Farm aerodrome, both of which are, at present, being used as satellite aerodromes for 2 EFTS. It will be noted that of the three areas...Fort Lytton is considered to be a suitable site for a full EFTS, (and) Strathpine... for a half EFTS. Meeandah could be prepared as an ELG for the Service Squadron presently stationed at Archerfield.'

Aerial photos followed this letter. Dated 8 January 1941, those of STRATHPINE showed the area west of the railway line and down to the South Pine River flats where later A3 strip was sited. Apparently, little more happened until Japan entered the War. American interest in developing an aircraft erection facility close to the wharves quickly focussed on the EAGLE FARM site. Slow construction progress there only sharpened the need for associated dispersal strips.

The Minutes of the AWC, preserved in AA Brisbane, commence with 'a meeting held in Brisbane, 23 May 1942'. These 'meetings' carried on almost daily. Around July the first mention of STRATHPINE appears - and then it relates

PETRIE-STRATHPINE *(11/3/1943)*
PETRIE A1 was a grass strip and at the time of this photograph, so was STRATHPINE A2, partly visible, bottom of this photograph. Camouflaged arch hideout structures were introduced to the SWPA by the US Forces in this location. The prominent tree-like black objects just north and immediately south of the western end of PETRIE strip (A1), were arch hideouts. *[CPE (RAAF Official)]*

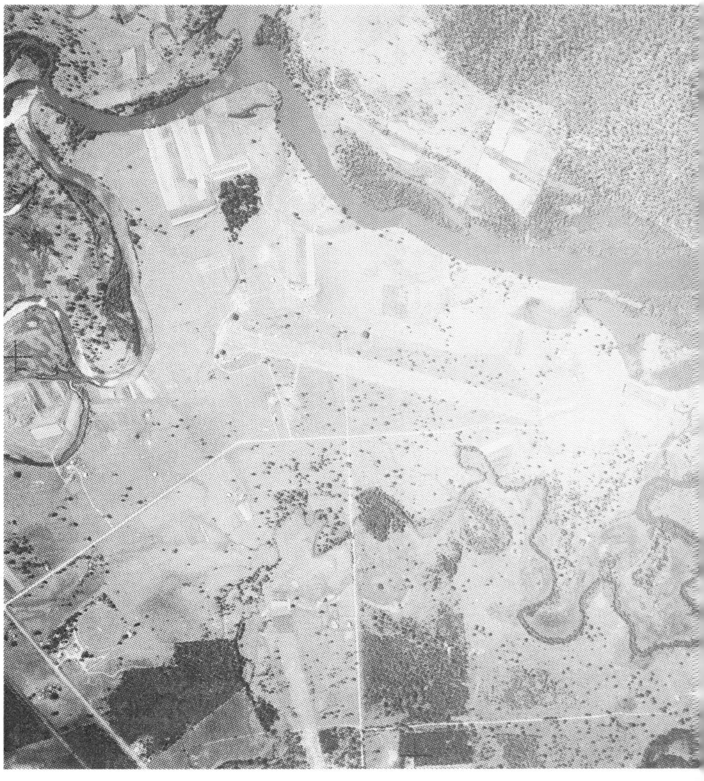

to 'camouflage work'. The ***MRC History*** only mentions this airfield site in terms of 'May 1943' activity and goes on to be quite misleading and inaccurate. Accepting that these initial months of 1942 were 'early days' in the US forces' build-up, and they didn't yet have the masses of construction equipment later to appear, we are left asking who and with what equipment were the early STRATHPINE strips 'built'?

The General Index Cards of the Works and Services Branch of the Dept of Interior, Queensland, record 21 May 1942 as the 'date of communication' from 'Base Section Engineer USA, Brisbane', for 'Survey of Landing Strips... dispersal areas, camps and Gun Pits etc' for what came to be known as Strips A1, A2 and A3. The mere listing of 'Gun Pits' suggests to me, that these works had already substantially commenced and that maybe the 'Survey' was needed only to tidy up matters of whose land was affected! Indeed, US records show the 80 FS with its P-39 Airacobras, stationed at PETRIC (sic) from 10 May 1942, after moving from LOWOOD. It seems most likely that MRC, or some other civilian works organisation was commissioned directly by the US Army or its civilian consulting engineering agent, Sverdrup & Parcel to clear and prepare these strips, around April 1942.

Rod Voller was CHS' senior Camouflage Officer for the North East Area centred on Townsville and his duties brought him on a few occasions, back to Brisbane. His diary notes reveal that he first met -

'...Hughes and Bresse (sic) in Brisbane 5 June 1942. There was also a Mr Murphy present - not sure what part he played. They were in Brisbane to "sell" their arched structures for hangars and camouflage covers. We went to Newstead to see a sample under construction, then to Strathpine strip to see their camouflage hangars.'

It appears this was the earliest work on the 'igloo' structures undertaken in Australia. While American official historical publications credit the development of the 'camouflaged arched aircraft hideout' personally to Brigadier Hugh J Casey, they give some credit to a French engineer, Brizay by name. Karl C Dod in his work ***The Corps of Engineers:The War Against Japan***, mentions a -

'...most widely used... igloo warehouse, modelled on a hangar designed by Mr Brizay, a French engineer who had come to work for the Base Section Engineers in Brisbane after fleeing from Malaya...'

PETRIE-STRATHPINE *(1/6/1943?)*

The 'serviceability' date on this drawing is hardly legible. A number of existing hideout structures were ultimately repositioned when this field was taken over for RAAF use early in 1943. *[AA BP375 Item QL 680]*

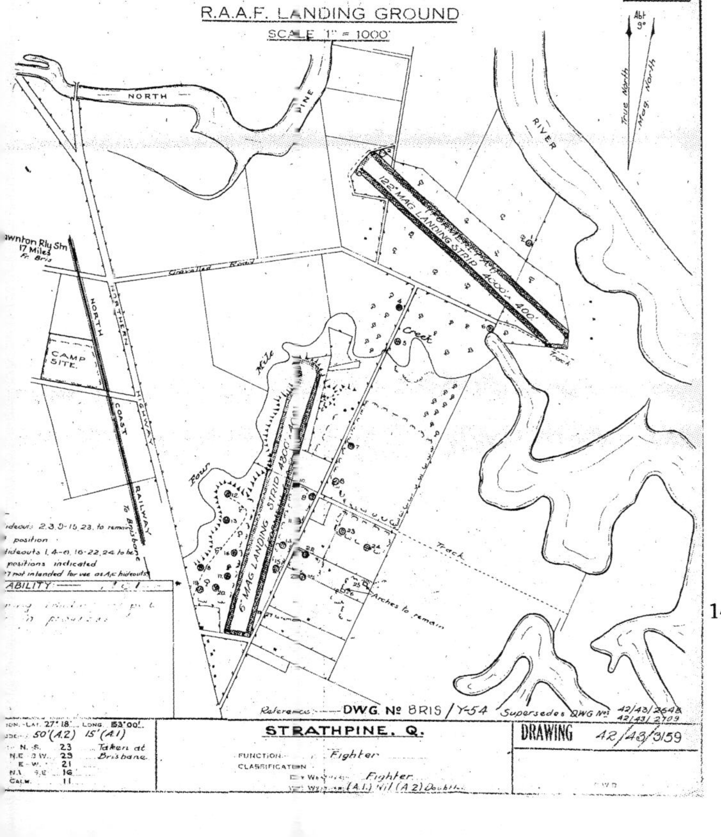

Ms Waveney Browne was for many years, secretary to the late M R (later Sir Manuel) Hornibrook. Her 1974 biography of MRH, titled ***A Man of Achievement***, refers -

'...The now famous igloo type building was designed in the Hornibrook office by Emil Brizay, a French Engineer, who came to Brisbane from Singapore when it fell to the Japanese. The first of these 200' x 104' buildings put up by M R Hornibrook (Pty) Ltd at Breakfast Creek, was a real nine days' wonder and caused much interest...'

This structure served the rest of the war as the Allison aero engine workshop of the EAGLE FARM complex. It was the same 'Newstead' job Rod Voller saw on 5 June 1942.

In my search for the origin of this igloo design, I note in AA, several detailed drawings (tracings) preserved from that period. One is numbered *T* (for Townsville as its point of origin or T - Tropical?), *T412*, titled *Igloo Type Warehouse No 2 Base Section Engineers USASOS SWPA - Dept of Interior, Works Branch Townsville Q*. Alongside the undated plan is *'designed by E Brissez (sic)'*. While the tracing is undated, its number sequence suggests between 20 and 30 July 1942. The correct spelling of the designer's name seems equally in doubt!

F-Lt John Keays inspected STRATHPINE aerodrome and with many others, in his role then as Air Board's field Liaison Officer. His report dated 10 July 1942 includes -

'...Strathpine - single strip complete with dispersal and hideouts selected designed and constructed by USA. Is a particularly good job but suitable for fighters only.'

The Airacobras of the US 80 FS departed around 20 July 1942. Their formation and training stay had seen the loss of at least four pilots in crashes nearby. Early in 1943, the RAAF's DWO reviewed 'Aircraft Landing Strips - Brisbane Area'. W Cdr Towers' memo of 8 February to HQ Eastern Area said, in part -

'...the three strips near Petrie, i.e. A1-Petrie, A2 and A3-Strathpine still remain under the nominal control of USA though at the moment, a detachment of 23 Sqn is located at strip A2. A formal recommendation for the retention of certain strips and the abandonment of others has now been made to RAAF HQ, including your recommendation, that A7, A10, A11 and A12 be kept properly cleared as "dummy" aerodromes.'

In a detailed report two days later, Towers expanded on how the existing A2 strip could be improved. He outlined the idea of a new 65° strip with its western end intersecting A2 near its southern end. The philosophy was still that the 'three strips/legs of the triangle' were needed. Of the PETRIE A1 strip, Towers commented it -

'...could be widened on its present location and extended to a total of 4 600' which would suitably accommodate a hard-surfaced runway 4 200' long. This strip is reputed to be occasionally flooded but as the last flood occurred in 1931, would appear to be a fair risk.'

In a letter signed by W Cdr Gordon Grant for Air Vice Marshal Bostock, AOC RAAF Command AAF, to Air Board Melbourne dated 11 February 1943, the firm request for a hard surfaced runway was made -

'...In view of the fact that it is probable that a Fighter Unit will normally always be located at Strathpine (Petrie) in future, it is requested that direction be given for a strip to be completed with a hard surfaced runway at an early date.'

None had previously existed here, and then only A2 was so developed, although the ***MRC History*** mentions two!

The camouflage people certainly came into this planning exercise. Their report for 21 February 1943 included -

'...visited area in connection with camp site and selected about 8 acres W of railway and just S of Lawnton - situated on a well drained ridge with good cover.'

Again on 28 February 1943, the comment was made -

'...Strathpine - with F-O Dallimore, checked up on positions of US arch-type hideouts and re-arrangement of dispersal to suit 24 medium bombers. This will preclude the use of existing US fighter hideouts.'

PETRIE-STRATHPINE (-/7/1943)
One of twenty-two P-39 Bell Airacobras taken on charge by the RAAF and used during 1942-43, this aircraft (A53-7) is photographed facing west at the southern end of STRATHPINE A2. The 83 Squadron pilot posing by the aircraft in khaki shirt, shorts and flying boots is Nigel Pugh who later lost his life in the Arafura Sea north of Arnhem Land. [D Oliver ex RAAF]

Henry (Harry) R Parker was a Fitter 2A posted from Townsville to 83 Sqn, STRATHPINE, about March 1943 -

'...we lived in tents. Trenches were dug for toilets with two saplings to sit on, surrounded by hessian for privacy - which was to a certain extent transparent. We had no aircraft or landing strip. A strip well down below the camp was too wet and dangerous to land aircraft. A team of bulldozers and other machinery turned up and made a landing strip from which the aircraft had to take off and land over the main road. We received Boomerang aircraft some of which on arrival ground looped on the strip. I used to come home to Brisbane every night in a bulk milk truck...'

Harry explained the truck took him to where he could catch a tram home to his wife, at Hawthorne. Also, Harry's hessian screens 'caught fire' like others I've heard of!

As the main Fighter drome for Brisbane, this airfield was slated for 'splinter proof pens'. It was argued dispersal areas were either not readily available or parking close to the strip meant quicker 'scramble' times. In any case, the RAAF Chief of Air Staff personally took a hand in this matter around April 1943. He advised Air Board -

'...As a result of my personal observations it is now intended to provide only minor improvements to the two existing strips and minimum Buildings and Services for the camp. It is intended that these two strips be used for normal operations of a squadron during dry weather which is about eight months of the year. Under adverse conditions, the squadron would move to another aerodrome as a lodger on the present tenants. A scheme has now been prepared for widening the existing strips at Petrie A1 and Strathpine A2... No runways, taxiways or hardstandings will be provided... Camouflage is provided in the form of removal, re-erection and overhaul of (existing US) Fighter hideouts.'

It didn't take long for the CO of 83 Sqn, Sqn Ldr C W Lindeman to report adversely on STRATHPINE A2. Dated 4 June 1943, he wrote, after two months there -

'...The Strathpine (A2) strip was made the parent aerodrome and all aircraft and maintenance facilities were moved to that location on 4 April 1943. The strip has not been used very much as the squadron is well below establishment with aircraft. This report is submitted now as it is anticipated that the aircraft strength will be greatly increased very soon and, therefore, the amount of traffic on the strip will increase considerably. The surface of the strip is natural soil, about 60% of which is covered with grass. The soil does not appear to be of a good binding nature.'

Lindeman described the bumpy, tufty surface left after the soil was blown away by aircraft use. Rain showed up the lack of drainage and -

'...when the site was cleared originally, all stumps were not grubbed and some were cut down only to the level of the ground...These stumps are now protruding above the surface and have to be constantly cut down.'

A1 deserved and received less attention from the CO -

'...In view of the fact that a considerable sum is being expended on a permanent camp site... it is necessary to take immediate steps to prevent the deterioration of the A2 strip, and to improve both strips so that they will not become unserviceable after rain.'

SDCC reports during July 1943 deal with some re-siting of the arched hideouts and their overall renewal. By then, DWO Brisbane had advised one only strip (A2) was to be gravelled and the road to the east (Bells Pocket Road) was to be utilised as a perimeter taxiway.

Ron McDonell was a pilot with 83 Sqn. He recalled -

'...arriving there in July 1943. The squadron was equipped with Airacobras which were replaced later with Boomerangs...our camp was to the SE of the strip... in a timbered area adjacent the road. The strip was about 3/4 of a mile from the Strathpine Railway Station. There was a Pub and a few shops and small farms covered the main area. The Strathpine Pub was burnt to the ground while we were there - probably the result of a "blue" with the management about grog. Beer was very short in those days and closing hours were always good for an argument...

I well remember a farm house near the main gate where we could buy home made pies. We reckoned they were the best pies in Queensland... Our squadron was occupied identifying aircraft inbound from America which had to be checked prior to reaching the North tip of Stradbroke Is. We were always full with ammunition, 20 mm and .303. We also checked unidentified radar "blips" in the area, from Tweed to Noosa.'

Ron recalled nasty gutters on the edge of the strip, after rain. He said the Boomerang was tricky to handle -

'...What with a very short wheel base and poor brakes, if a swing developed on landing one was lucky to avoid a prang. I was involved in an accident ending up in the trees after a brake failure - didn't get hurt really but the aircraft was badly bent...'

PETRIE-STRATHPINE (-/7/1943)
Sitting on the wing of one of their 83 Sqn Boomerang fighters, are (from left) Paul Bash, Geoff Cole, Roy Ayre, Eric Dawson, Gordon Crossing and Clive Winnett. Bob Mutch is standing by the wing mounted Hispano cannon. [D Oliver ex RAAF]

PETRIE-STRATHPINE (-/9/1943?)
83 Squadron (RAAF) operated light aircraft for its general communication duties. This DH 94 Moth Minor A21-8 served as 'Squadron Hack' and is photographed in front of one of the camouflaged arch hideouts here given a roof, but still unmistakable with its nailed wooden arch frames. Ron McDonell felt that it may be Bob Moody in the cockpit of the aircraft. [J D McLean courtesy F Smith]

Two other incidents came to Ron's mind -

'...one occasion whilst flying No 2 to a character, we flew down the Pine River at high speed. The other aircraft, about 20' in front of me and half a wing span on my left, flew straight through high tension lines somewhere near Petrie, cutting all three cables carrying power north from Brisbane. I saw a blinding flash which caused us to pull up with No 1 heading for Strathpine. On landing it was found the windscreen (approx 3" thick and bullet-proof glass) was cracked straight across. The plane had lost about 18" off the rudder and stabiliser and all three prop blades were scored as they bent back over the cowling. I forget what resulted - perhaps 14 days CB (i.e. confined to barracks). In those days we did some silly things. I now criticise young fellows for doing mad things - yet the exuberance of youth is the greatest contributing factor in winning wars, provided it is guided by discipline... Two of us flew under the Story Bridge at high speed from the city side towards New Farm, completely ignoring the presence of the cliffs - real smart! We were caught by some observant type who took our "ident letters" and we did 14 days Orderly Sgt duty.'

Fred F Martin was a Liaison pilot with the US 61 Field Artillery Bn, part of the 1st Cavalry Division. Fred flew Aeronca L3, Piper L4 and Stinson L5 'lighties', for several months here in 1943. He wrote of making friends with a -

'...number of Australian Boomerang pilots. I have flown into some of their fighter strips and traded flight time... they flew my L3 and I flew a "Moth" trainer of theirs.'

Fred's log entry for 4 September 1943 reads -

'...From: PETRIE STRIP to: Australian Ftr Stp (from where he) flew their Moth Minor low wing 90/90/90.'

PETRIE-STRATHPINE (1991)

The three strips are here superimposed on the current-day UBD map. A2 strip was effectively 'right down the middle of a suburban street'. As a boy I first visited STRATHPINE for a motor racing carnival. Forty years later I only found the site again by noticing the name 'Spitfire Ave' on this map. [Universal Press P/L]

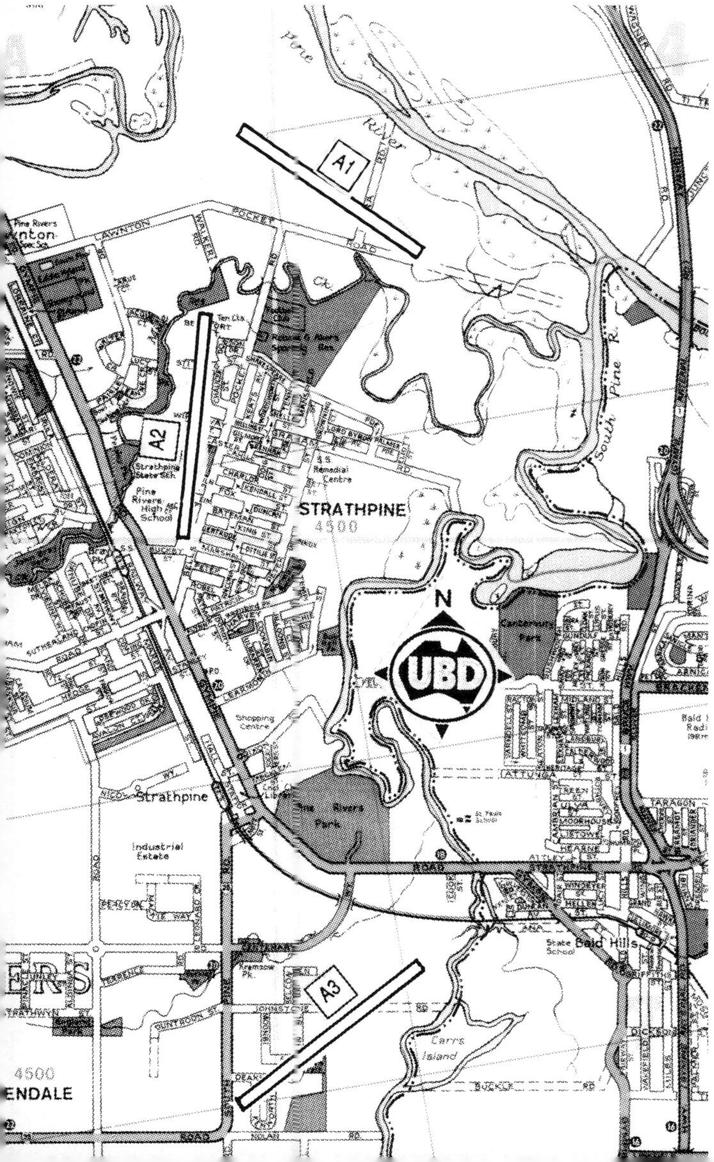

When I queried this jargon, Fred replied -

'...it took off at 90 mph, cruised at 90 and landed at 90!'

Fred's principal base/strip for the months from July to December 1943 is logged as 'PETRIE', apparently the A1 strip. His work sessions often took him to 'Toorbul Point' where any suitable clearing or length of open road served as the strip. During November 1943 Fred flew consistently from 'Flinders Range', again a suitable open grassed area, which I gather was somewhere northwest of Beaudesert.

There doesn't seem to have been any developed landing grounds in the Pumicestone Passage to Bribie Island area. Although Army veterans of nearby coastal artillery installations tell me of seeing planes 'descending below the trees', this was probably only to drop off target drogues.

On 9 September 1943 the AWC instructed MRC Brisbane on 'further works at STRATHPINE', this time the construction of 'a sealed gravel runway 3 900' x 150' on 6° bearing'. The job recognised the rehabilitation of some twenty-four aircraft hideouts, the other natural surface runway, and extended to fencing the aerodrome to make it secure as stock were straying onto the runways.

83 Sqn moved to A1 strip when gravelling got under way on A2. Ron McDonell feels this was near their time of departure north. While at A1 torrential rain fell and -

'...we had to evacuate in a hurry. The strip started to flood so we flew out to Amberley, and later north to Gove. Half flaps were used for take off using full throttle on the brakes, and as the tail began to lift, brakes off and as the run commenced, down with half flap and we were off in about 350'. According to the book you never use more than 15° of flap in take off...'

An October 1942 marked up cadastral map extract on file shows the disposition of various airfield buildings. Amongst the 'hangars, latrines, HQs, gunpits and ammo dump' markings, there is shown, south east of A2, a 'wireless mast'. Ron McDonell remarked -

'...something that fascinated us was the radio mast practically in the circuit area - about 700' high. It wasn't amazing to us at the time, but can you imagine the trauma if that condition existed today!'

On 28 February 1944 the US 5 Air Force asked RAAF Command Brisbane about the need to retain Landing Strip A3 STRATHPINE. They were considering abandoning the strip and handing the ground back to the owner.

Lawnton and Strathpine figure as places where the two RAF Spitfire Sqns, 548 and 549, were formed at the end of 1943. Manned as they were by RAF pilots, the units were intended to join with 54 Sqn and complete the Wing the British Government promised in the aerial defence of Australia. Hundreds of RAAF ground staff made up the majority of these two Units. History tells of the low morale period of the early months of 1944 spent awaiting delivery and commissioning of the MKVIII aircraft, crated from England. A Fitter 2A involved in 548's aircraft preparation, spoke of some RAF pilots having to content themselves flying Tiger Moths or a few Wirraways, until the Spitfires arrived. Wishing to be nameless, he told of his only trip in a Tiger Moth, flown by an RAF Sgt pilot -

'...(he) asked me and I jumped at the chance. We stooged out over the Pine River and the Hornibrook Highway. We put a bloke out of his boat down there with our daring - we dived down on him in the first instance and he didn't move so we dived down again and this time he fell overboard...I believe there was a bit of a stir over it - everybody was told to be a bit more careful.'

Doubtless such 'buzzing' was treated on merit. However I am reminded of an RAAF pilot being cashiered by the RAF for buzzing some beach crowd in the Middle East!

After the war, STRATHPINE along with various other strips in Queensland, was intended by the RAAF to be kept as one of its airfields of responsibility. However by April of 1946 it was decided to return A1 PETRIE to its owners and to retain the hard surfaced A2 strip for emergency landing purposes only.

COOMINYA

'...The time when all the aircraft suddenly came over to Lowood, the Douglases and the Airacobras - from Archerfield I think - all our Tigers' were sent to Coominya. Three days of wet weather were spent on guard duty at either Coominya or the Lowood drome. The reason for this panic was an unidentified Aircraft Carrier which had arrived near the tip of Moreton Island!'

Vic Waddell, ex RAAF Survey Flight, recalled this of COOMINYA. He couldn't date the event, but it's clear the site would have been that of the old RLG southeast of Coominya township. Other RAAF veterans have recalled it as 'just an open grassed field with maybe a windsock'.

During WW2 there were two aerodromes 'at Coominya'. Unfortunately, the ***MRC History*** is misleading in its fusing together of these -

'...A landing ground at Coominya as a Relief to the Lowood drome was authorised late in 1941 and work began early in 1942. It included clearing, grubbing, surface formation and preparation of a landing strip with taxiway, dispersal strips and hideouts in heavy timber. Also fencing, draining with reinforced concrete pipes and the construction of a half mile access road...'

It is true that an RLG was required for LOWOOD under its EATS role. This was satisfied by the relatively small 'all over field nature' of the RLG southeast of Coominya township. Elsewhere in the ***MRC History*** it is stated -

'...The Americans used Lowood for a few months in the middle of 1942 but the surface did not please their pilots and a further landing strip was provided for them west of Coominya.'

The General Index Card file of the Works and Services Branch, Department of Interior, Queensland held now in AA, carries a card for 'COOMINYA-RAAF'. Under date 21 May 1942, a request appears for 'survey of landing strip, dispersal area, etc. for US Army (Portions 143-148)'. AWC Minutes of 13 July 1942 refer to 'camouflage work having been done at STRATHPINE, KINGSTON, LOGANLEA, WACOL, COOMINYA, Darra and AMBERLEY'. It

COOMINYA *(25/4/1944)*
This is the 'Left Oblique' Frame (Run C107 L) of some experimental Trimet photography conducted over the Crows Nest map area by the RAAF. The photo-aircraft looks like a twin-engined Beaufort. The COOMINYA No 2 airstrip is clearly visible in the centre with dispersal roads leading northward to the bottom right and back in towards the railway in the foreground. This site was initially constructed only to grass strip state for the US Forces in 1942. It is about 8 km north of LOWOOD, the open area of which can be seen in the middle distance. On the extreme left hand edge under the engine nacelle and east of the railway line, lies the all over field referred to as COOMINYA No 1. *[AUSLIG]*

was referred to as 'dummy hangars', 182 of them, generally 62′ long with one large hangar 170′ x 98′ span. The Minute goes on to record that Hornibrooks were to be paid for the work done, though 'no requisition was made'. These 'dummy hangars' at COOMINYA were the thirteen odd nailed wooden truss camouflaged hideouts built for the Americans when they initiated COOMINYA No 2.

F-Lt John Keays, then Air Board's field Liaison officer for Queensland, made this brief reference in a 10 July 1942 report on 'Aerodromes - Brisbane Area' -

'...Coominya - the USA Engineer section have constructed a strip but I have not inspected it.'

Like PETRIE-STRATHPINE, it seems more likely the MRC 'constructed' this strip under US direction.

I asked Colin Munro, ex RAAF, if he had any knowledge of COOMINYA strips. He was in charge of a 'Flight' of 5 Sqn Wirraways at TOOGOOLAWAH during 1942. After consulting his 'log book' Colin told of sending his fellows on a search exercise in which they were to locate an airfield just constructed at COOMINYA -

'...they had great difficulty finding it.'

This may be explained in part by the fact that the strip itself was parallel and close to a section of local road.

It is not clear to me how much use the Americans actually made of COOMINYA No 2, which they also referred to as 'A-6', or 'LOWOOD No 2'. After they moved on in the second half of 1942, the RAAF's resumption of responsibility for LOWOOD focused on providing dispersal facilities including some twenty seven aircraft hideouts. LOWOOD had nowhere near adequate natural tree cover to provide this, so attention turned to COOMINYA No 2. Work proceeded early in 1943 on the up grading of the strip there, construction of taxiways and the development of nine 'conceal only' hideouts. None of the earlier arch type structures were then considered worth incorporating into the finalised system.

I interviewed William Lubke, then in retirement, in Lowood. As a 'Ganger' with MRC/CCC he recalled being involved in the 'gravelling of the strip and taxiways at COOMINYA'. The RAAF even considered scrapping the work already done at LOWOOD and shifting that whole operation to COOMINYA, such was the comparative superiority of the terrain. Remember, it was the undulating nature of LOWOOD that put the Americans off it. In the final analysis, so much expenditure had been committed at LOWOOD that no really strong case existed for shifting. Anyway, by then the urgency had waned.

SDCC reports throughout the first half of 1943, show much effort was put into 'hideout' siting and planning. 'Approval' was still not through by 30 May 1943 and an October 1944 reference suggests it was never given. A DWO Memo to RAAF HQ, Eastern Area reported -

'...This landing strip...has two hurdles of 100′ span to provide access to the strip from public roads and intended originally to give egress for aircraft taxiing to dispersal points. As the proposal for dispersal of aircraft at Coominya was abandoned, and as these wide hurdles are sagging...(and) are not stockproof...'

On the other hand the Airfield Data Sheet, by handwritten overwriting of typed detail suggests -

'...Gravelled taxiways have been constructed to dispersal areas with nine (9) conceal only type, hardstanding surfaced hideouts, medium bomber size.'

The RAAF's Airfield Data Sheet for COOMINYA No 1 under a review dated 17 May 1943, states in part -

'...no dispersal areas or hideouts available on this field. Six dummy aircraft are on the field.'

No doubt these are the ones the SDCC report of 18 April 1943, noted as 'dummy fighter planes...in bad state of repair'. I've not found any photos which definitely relate to these 'dummy aircraft', but some do exist of others erected in NSW. It appears fairly sophisticated 'dummy' P-40 Kittyhawks were brought in by the US forces in 'kit' form, and taken as far north as GARBUTT.

The RAAF's 11 RSU spent several weeks in camp at

COOMINYA around December 1943. The location served only as a field toughener for this approx five hundred strong Repair and Salvage (later termed Servicing) Unit. Unlike their forthcoming duties in New Guinea, they didn't 'mend planes' at COOMINYA. A shipping delay meant this bush site helped with final training and morale.

A recent visit to COOMINYA No 2 revealed that while the airstrip has truly reverted to rural use, a small section is currently devoted to the flying of model aircraft. It could be said that aviation is still alive and well on this strip!

EAGLE FARM

'...No provision has been made for watering which should not be required if work is done in October or November.'

This comment appears on a handwritten estimate dated 29 August 1941 for the preparation of a Relief Landing Ground at EAGLE FARM for 2 EFTS, ARCHERFIELD. Since the latter became 'Brisbane Airport' early in the 1930s, hosting some gliding activity was about the closest EAGLE FARM had reverted towards the day it welcomed the 'Southern Cross' in June 1928. This estimate by Mr V G Crawford, an engineer for the Dept of Interior, referred to a drawing marking out the land north of and adjoining the straight six furlong (1320 yards) extension of Doomben race track. Crawford's estimate totalled £1 727 and the drawing was finalised 11 December 1941. Considering this was only days after Pearl Harbor, and the planning then quite unpretentious, it can have been nothing short of staggering for people such as Mr Crawford to see EAGLE FARM, 1942.

The US Convoy en route to the Philippines when Pearl Harbor was attacked, was logically diverted to Brisbane, arriving late December 1941. The prospect of unloading crated aircraft, assembling and flying them on to their Clark Field destination seemed simple. However, the realities quickly validated the early January 1942 decision to develop EAGLE FARM as a major aircraft erection depot/drome from which the planes could proceed.

The ***MRC History*** published in 1949 addresses this work in terms of its total cost over some three years to the end of 1944 as £559 687 and involving huge equipment utilisation extending to 'one hundred trucks' -

'...a week after the work of construction had started (February 1942) heavy rain fell and continued for nearly three weeks, converting the ground to a quagmire. Thousands of feet of ashes and coke breeze were carted from the gasworks and deposited in order to keep trucks moving with borrow material from Nudgee. On several occasions two fire brigade pumps were used to get the water from the boxing...'

The US Archive sourced ***History of the 5th Air Force (and its predecessors)*** mentions not this rain delay -

'..By 18 February one hangar was started and one runway (the SW-NE) nearing completion. However a week later the situation was unsatisfactory, for although airplanes were already on the field, the runways had not been making satisfactory progress. Col. (Alexander) Johnson (US Commander, Base Section 3) complained...'

The ***MRC History*** continued -

'..About this time the American authorities advised that the runway was urgently needed for fighter plane protection of Brisbane. With the clearing of the weather, two ten hour shifts were set to work, using 100 trucks which hauled 33 000 cu yds of borrow material over a 5 mile lead. 12 000 cu yds of gravel over a 6 mile lead

EAGLE FARM (-/8/1944 and 4/7/1944)
These four photos relate to comparative trials undertaken by ATAIU. Charles Whistler (USAAF) and Ron MacDonald (RAAF) both participated - Charles sent the top two and Ron the bottom two photos. XJ004 was the US serial given this Japanese 'Oscar' MkII. Whistler is preparing to taxi, while Lt A V Jackson looks on. Igloo Hangar 7 is the backdrop and the white painted tail on the right is 42-27622, a P-47 Thunderbolt also involved in the trials. Ken James (RAAF) flew this MkVIII Spitfire for the trials. The Curtiss Kittyhawk P-40N, A29-803 was flown by Ron MacDonald, though most of the participating pilots 'had a go in each other's aircraft'. At bottom is one of at least two 'fly-by' sequences, showing XJ004 in a low pass. Snapped from the Control Tower atop Hangar 3, Doomben Racecourse forms the middle ground. Below the Oscar is what looks like a US Navy PB4Y-1 with an Erco nose turret. [C F Whistler ex USAAF] [R S Macdonald ex RAAF]

EAGLE FARM (4/7/1944)
Two overlapping frames of a low level pass by XJ004 clearly show in panorama the hardstand fronting the hangars, with Hangars 7 and 8 in the background. Note the pools of water on the ground. [C F Whistler ex USAAF]

and 1 250 cu yds of screenings for bitumen on leads of up to 55 miles. This section was completed and a squadron of P-39 (Air cobras) (sic) landed on Sunday afternoon 29 March 1942.'

The difference in scale emerges when one refers back to V G Crawford's RAAF RLG estimate -

'...fill holes with approved gravel - estimate 10 cu yds per acre - i.e. 750 cu yds at 6/- per cu yd = £225.'

Col. Johnson's complaint was symptomatic of the gigantic construction and supply tasks fast appearing. The ***5 AF History*** covered the evolving solution -

'...Recognising the need for a coordinating agency for all Australian construction of a military nature, an Allied Works Council (AWC) was formed on 25 February 1942... The principal agency for carrying out the (work) was the Civil Construction Corps (CCC), into which labor and manpower could be drafted...'

The 'straight six' section of the Brisbane Turf Club's track was promptly swallowed up by this new EAGLE FARM. It meant ready-made apron areas to the hangars soon to come. Though it would be years before the matter of compensation was settled, never again would real horse power thunder along this section.

EAGLE FARM and its environs got their first mentions in the SDCC Camouflage Reports of January 1942 -

'...Emergency Military Camps - the toning down of these (Ascot and Doomben Race Courses are two) has been completed and is very effective from the air, the tents blending remarkably well with their surroundings.'

EAGLE FARM (20/12/1941)
EAGLE FARM was the Australian land fall for the epic trans Pacific flight by Charles (later Sir Charles) Kingsford-Smith and Charles Ulm, landing here in 1928. The field on which they landed, is the area just east of Doomben racecourse (centre of photo). When this was taken as part of systematic mapping (KD 1039) US Forces had commenced their encampment in the grounds of this racecourse. The 'straight six' part of the track is clearly seen as the east-west tangential extension. The little white square on the far eastern side of the field, is the bare floor of an early hangar, the structure of which had, in the early 1930s been relocated to ARCHERFIELD. [NLA (RAAF Official)]

Suitable spray guns were scarce. It looked like tent flys would have to be laid flat and painted with brooms!

Hangars 1 and 2 were probably built by Dept of Interior 'day labour' out of sheer urgency. Their comparative readiness was a direct result of General Brereton's November 1941 'secret' visit to Australia. However, 'Tenders' were called early March for Hangars 3, 4, 8 and 9, of similar 'heavy timber framed' design. Won by Contractor T H Dennis, these were the ones quickly 'redirected' to CHARLEVILLE. Eventually, three of these were returned to become the doubling in length of Hangars 1 and 2 and the then smallest of all, Hangar 3A, just west of 3. Interestingly, Hangar 3A, 96′ x 96′ was well after the war relocated again to where it survives to this day on the Gold Coast. It served for several years as the nucleus of the then Chewing Gum Aviation Museum.

Some US published sources place the advent of Anti Aircraft Artillery units to Brisbane during March and April 1942. One US Army GHQ 1946 history, lists -

'...The 94 CA(AA) with the 101 and 104 CA Battalions (AA) and several Australian units were disposed to protect the vitally important wharf and warehouse areas and the Archer and Eagle Farm and Lowood aerodromes in and near Brisbane.'

By 22 April 1942, the Camouflage people reported painting had commenced at the drome -

'...in accordance with a scheme that has been prepared. Models of all the hangars were made. The treatment of the field, on which tremendous earth works are being carried out, will present a difficult problem.'

This was 'early days' when serious thought was applied towards 'hiding' large airfields! Hangars 1 and 2 would have been addressed but surely the Camoufleurs would have been aware of the 13 April direction to Contractor Dennis to pack up work on 3, 4, 8 and 9 and build them at CHARLEVILLE.

The SDCC report of 14 May 1942, touches on AA Batteries around Brisbane -

'...The concealment of both the Australian and US Army batteries is being taken very seriously by those responsible. Mr (R P) Cummings has visited all the sites and has evolved several methods of concealing the guns and the associated instruments. The problem of hiding the American guns is not so difficult as they are smaller and semi-mobile. Fortunately the diameter of the Australian gun pits has been reduced. Arrangements were made to have an umbrella constructed from the specifications forwarded by the TDC. Drawings prepared by Mr Cummings for sliding nets, or alternatively, roofs, for American batteries were handed to the TDC and are now being constructed.'

Ideas abounded. These were civilians helping -

'...A "rat trap" cover for A-A machine guns was constructed in model by Mr (Athol) Bretnall of the Dept of the Interior, and one has been erected of lawyer vine and garnished netting. It is very successful and Mr Dorey (camoufleur) said that it is similar to a cover

used by him with success at Port Moresby.'

Harry Kyle was in the Australian Army April 1942 and soon after at Lytton for basic training. Very shortly after he went to EAGLE FARM with 'No 56 A-A Battalion (Searchlight Unit)' until about September 1942, when he 'transferred' to the RAAF. Harry described the setting up of searchlight pit(s) just east off Schneider Road -

'...about 200 yards north of the (Allison) engine test stand and about 200 yards south of Chris Gill's Model Dairy...the pit was about 30' in diameter. There were long arms on the base of the light to manually rotate it.'

He explained a problem with some 'sound locator' equipment they were expecting. I gather this was an ex WW1 device like the old 'His Master's Voice' trademark - designed to pick the direction by maximising the sound of approaching aircraft -

'...anyway, we were told we couldn't use them 'cause the nearby A-A guns' noise could burst our ear drums!'

It was to be some three months before building work started on Hangar 3 and still more before the four igloo hangars, 4, 5, 6 and 7 were commissioned. While Emil Brizay designed his nailed wooden arch aircraft hideouts in Hornibrook's Brisbane office around May 1942, other engineers were busy adapting the idea towards the major warehouse and hangar/workshop role. Indeed, such was Hornibrook's rapport with the Americans that his men were up at Charters Towers on 'hideout' construction about the time his organisation had completed the first 'hangar/workshop' in Brisbane, on the left bank of Breakfast Creek. It is quite likely M R Hornibrook's men had at least a guiding hand in the initial two warehouse structures completed by mid August 1942, in the US Engineers' Depot near GARBUTT railway siding. The 'igloo' hangar was then ready to take off. Problems with strength and supply of nails and the very hardness of the Australian timbers were overcome. The AWC minutes of 7 October 1942 read as though that organisation didn't yet have ideal overall control -

'...noted US Army have commissioned M R Hornibrook to erect four large stores at Eagle Farm aerodrome...AWC to check US Base Section on details.'

These very likely were Hangars 4, 5, 6 and 7. There is some chance the minute related to the parallel construction of 'four' of the five propped lattice truss warehouses immediately south of the railway line. Both projects involved fresh railway work. Brig Gen Hugh J Casey's ***Engineers of the Southwest Pacific 1941-1945 - Vol VI Airfields and Base Development*** states -

'...By the last quarter of 1942 the depot for the assembly of aircraft was complete. It consisted of five Igloo hangars and five warehouses with two loading platforms along a new railroad siding.'

Interestingly, three weeks later the same AWC minutes mention some mysterious disappearance of a large stock of 'turncoated' iron, apparently ear marked for hangars at

EAGLE FARM (20/7/1942)

The cross strips are operating and in the southwestern sector, close to the railway line, Hangars 1 and 2 have appeared. Taxiways lead from them to the nearest point of the airstrips. At this stage no construction has proceeded along the edge of the 'straight six'. [Ponting Family]

EAGLE FARM (27/9/1942)

This photograph carries identification indicating its origin with the US 81st Air Depot Group and is one of several prints donated to the MRC Engineer Mr Ponting sometime after the wartime construction was successfully completed for the American Air Force. Both Hangars 1 and 2 are clear in the foreground with the structural frame (timber with bolted steel gussets) of Hangar 3 fronting the 'straight six'. Further similar hangar construction had been urgently contracted but for reasons of preserving strategic security for aircraft repair services, the contractor had been ordered to dismantle some work and continue the erection at CHARLEVILLE. Hangar '1', nearest the camera, and '3' both remain to this day. A P-38 Lockheed Lightning is blowing dust on to Doomben racecourse[Ponting Family]

EAGLE FARM. The record shows that it was agreed -

'...fresh supplies would be released from stock for the Eagle Farm project and that investigations would be conducted to establish what had happened to the missing material.'

Waveney Browne in her biography of M R Hornibrook, notes the igloos were first roofed with galvanised iron -

'...when (it) was restricted, terne-coated iron was used.'

It's strange how a discrepancy in spelling arouses the curiosity! I would have bet the more exotic 'terne' coating, long associated with a lead-tin alloy skinning of steel for fuel tank manufacture, was here misapplied. Surely this was used roofing iron simply 'turned over' to extend service life. However, the manufacturer's archives show zinc for galvanising was then in short supply and hence, Waveney was right!

As a lad of seventeen, Tom Whyte recalled his experience 'employed by the Americans, or whoever operated their Officers' Club/Canteen'. Tom spent about six months at AMBERLEY before coming down to the 81 ADG at EAGLE FARM, 'when General Kenney came on the scene'. Tom's duties included ensuring (?) all 'visitors signed the visitor's book'. Until Tom received his RAAF call-up around March 1943, he shared an unusual friendship with 'a lot of the American pilots testing newly assembled aircraft, such as Farrior, May, Handley, R Selman and C P Gies' - all members of what the pilots themselves referred to as the 'Fair Dinkum Flying Circus'!

Charles F Whistler arrived in Brisbane with most of the pilots of the US 348 FG, in the *Henry Gibbons*, 14 June 1943. The Group's P-47 fighters arrived separately and were off loaded for assembly at EAGLE FARM by the men themselves. In a further four weeks Charlie and most of the 342 FS had arrived in New Guinea -

'...Eagle Farm in July of 1943 was a beehive of activity. My memory is that P-38s were also being assembled there for the formation of the 475 FG and others in New Guinea. I put the slow time on the engine of my P-47 between 4 and 8 July on five separate trips averaging the order of 2 hours each. On 8 July my aircraft, was the first to get its 10 hours. I made a low pass at about

EAGLE FARM ca(-/12/1942)

Numbered from left Hangars 6,5,4 and 3 are shown in this aerial oblique probably taken by Capt Paul C May, a US test pilot member of the 'Fair Dinkum Flying Circus'. During 1942, May along with Lts J Handley and R Selman and Capt W Farrior and others, was responsible for flight testing newly assembled aircraft. The only aircraft visible in this photo is a lone Tiger Moth (centre bottom). It was probably assigned to the US 81 Air Depot Group. ***There Will Always be an 81st****, published in Brisbane around May 1943, mentions opening of the Control Tower atop Hangar 3, as 1 January 1943. The 'tower' has its wind sock up and flying in this photo. From the left can be seen Warehouses 22 (part), 21 and 20, with its roof as yet unpainted. Hangars 4 and 5 served many years post war, as domestic airline terminals for TAA (now QANTAS) and Ansett, respectively. [T Whyte ex RAAF]*

50' over Eagle Farm at about 500 mph for which I was reprimanded and put in charge of moving 342 FS to New Guinea. However it was worth it for the morale boost it gave those assembling the P-47 at Eagle Farm. Our crew chiefs had taken lots of ribbing from P-40 and P-38 crew chiefs because the P-47's big radial engine needed this initial "slow time".'

Prominent in various photos of EAGLE FARM is a perimeter fence intended not only to keep unauthorised people out but also to stop them seeing the activities within. Security was probably at its peak in respect of the work in Hangar 7. Early in 1943 captured and salvaged enemy aircraft were shipped here from New Guinea, in very much a 'knocked down state'. To learn the relative strengths and weaknesses of the enemy's equipment, particularly its aircraft, was a critical facet of wartime intelligence effort. Besides airmen gaining that vital edge in combat, the close examination of enemy materiel sometimes gave valuable clues to the state of enemy

EAGLE FARM (23/5/1943)

From left to right Hangar 2 (half of its eventual length) and Hangar 3, both to an Australianised RAF Type 'A' (1917 style) design, phased out in England around 1935. The next three structures fronting the 'straight six' paved area are of 'igloo' construction and are numbered 4, 5 and 6. Hangar 7 has also been constructed near the bottom right, rather apart from the others. This degree of 'isolation' suited the coming role of Hangar 7. South of the railway line, five large warehouses have been erected, four of which remain to this day. Numbered from the left, W24 is nearly completed. The prominent building in bottom right corner is the Brisbane Headquarters of the Ford Motor Company. Aircraft of several types have begun to appear on this very busy airfield. A relatively uncommon aircraft stands apart from the others. It is a US B-18 snub nosed twin engine bomber of mid 1930s origin. [Neg A29900 USAF Photo Collection (NASM) Smithsonian Institution]

EAGLE FARM (1942-43)

Somewhere over Moreton Bay, this DH82 has been photographed from a high wing 'lightie'. Probably painted in foliage green with the US white star and a curvaceous blonde on the cowl, the aircraft was one assigned to the 81 ADG[T Whyte ex RAAF]

factory production. After December 1941, US Intelligence realised it had a lot to catch up.

Much of the story of the formation and work of the Allied Technical Air Intelligence Unit (ATAIU) here at EAGLE FARM has been told by Robert C Mikesh in his ***Made in Japan...Tested in America***, June and July 1982 alternate issues of the US Magazines *WINGS* and *AIRPOWER*. Mikesh interviewed Frank T McCoy Jr and Francis M Williams, key members of this select unit which centred on Hangar 7. Up against not just the language and cultural differences, the group had first to collate field information, importantly 'descriptions' of new planes that aircrew could only glimpse in battle. Mikesh relates how confusion induced by the increasing number of Japanese types 'caused Fran Williams to suggest to McCoy in casual conversation that they assign code names to these aircraft to simplify matters'. The 'system' caught on immediately.

An apparently new version of the premier Japanese fighter, the 'Zero' was causing great concern. Observed with clipped wings and engine differences, examples were recovered from battered aerodromes in the Buna-Gona area of New Guinea. Field inspection by TAIU men produced some answers but rebuilding and flying it in comparative trials with Allied fighters was the clinical way to go. Field or maybe better termed jungle/swamp recovery of such 'wrecks' was a saga of its own. The indigenous New Guineans star there, for without their physical superiority over their natural terrain, it might not have been done.

To Hangar 7 these wrecks were transported. Painstaking work over several weeks brought the day of reckoning. Mikesh wrote, in caption to one of his photos -

'...Captain William O Farrior became the first to fly this earliest of rebuilt "Hamp" at Eagle Farms, doing so on very short notice. The intended pilot was killed in an

EAGLE FARM ca (-/7/1943)

Inside Hangar 7 several air frames are seen in varying stages of either being dismantled or rebuilt. Most are Zero and Oscar fighter elements with the notable exception of a tail assembly of a 'Betty' Bomber at the far end. [J Selwood]

EAGLE FARM (-/7/1943)
The Australian War Memorial's caption to this photograph reads, in part - **'A Mitsubishi A6M3, Naval Type O Model 32 fighter aircraft outside Hangar 7. This aircraft is probably the first 'Hamp' rebuilt by the Technical Air Intelligence Unit (TAIU) in Hangar 7 at Eagle Farm.'** The US 5th Air Force had recognised an urgent need to technically evaluate the many enemy aircraft appearing in the southwest Pacific theatre. [AWM Neg P1097/35/05]

EAGLE FARM (12/8/1943)
The first square wing tipped A6M3 'Hamp' fighter on the hard stand being readied for in-flight movie filming. This aerial gunnery training footage was shot from 'Sally' as another of its varied tasks. Standing by the 'Hamp' is Staff Sergeant Jim Helton the first engineer crew member of 'Sally', the B-17E in the background. This Flying Fortress was by then dedicated to providing transport requirements for US 5th Air Force commanding General George Kenney. In the interest of historical accuracy. Dick Graf, 'Sally's' Radio operator, took the photo, at EAGLE FARM, with other crew members gathered by 'Sally's' starboard side rear entry door. [G R Graf ex USAAF]

AT-6 crash the same day, when he declined to fly this "Hamp", due to faulty brakes. Farrior, "borrowed" from a nearby fighter pilot replacement pool, was to become well known for flying practically all aircraft rebuilt at Eagle Farms. He later moved to Anacostia with TAIU and remained with intelligence his entire AF career.'

My curiosity was aroused by the mention of an AT-6 (Texan/Harvard)! I'd not heard of any US AT-6s here during WW2. With Robert Mikesh's help I was able to reach Bill Farrior. He wrote back, in 1989 -

'...you are correct. We had no AT-6 aircraft. The intended pilot was Paul May; a Captain, US Army Air Corps. He and one other were killed in an A-20 accident on a flight returning from Amberley Field to Eagle Farm. We had delivered three fighters, P-38s I think, to Amberley and my aircraft had mechanical problems so I flew it back to Eagle Farm for repair. May picked up the other two, Lt Handley and Lt Wallace for the return flight. Handley was killed with May. Wallace was badly injured but survived...'

Bill's letter was handwritten and easily read, but somewhere in Bob Mikesh's manuscript or possibly a handwritten source document 'A20' had been misread as 'AT6' -

'...At that time I was the Engineering Officer and a "Test" pilot for newly assembled aircraft at Eagle Farm. I was asked by Frank McCoy to test hop the Japanese fighter being put together at Hangar 7. Then with the help of a captured Japanese pilot and an interpreter I got a cockpit check and later (few days) we took a 30 minute flight. That was 20 July 1943.'

In recent time Tom Whyte asked me, for any information I might have on the circumstances of the death of Lt Handley. Tom explained that his interest sprang from his friendship with Gabby Kennard, the first Australian woman to fly solo around the world in 1989. In her book ***Solo Woman***, she tells of her strange relief when, aged eighteen she learned from her troubled and angry mother, of her real father - relief because she felt it accounted a little for her love of flying. It transpired that Gabby, well aware that Lt James J Handley was her real father, had sought whatever further information might become available about the circumstances of his death.

How could I help? Might Bill Farrior know if 'Wallace' was still alive? Might Ralph Dietz, another 81 ADG man know any further? Ralph, who felt the third man in the crash was 'Harrison', was able to tell a little more -

'...see Handley was riding up on that shelf in the back and Harrison was in the back compartment, and May was flying it. They were good pilots these two young men - but it was a careless act. They let that A-20 sit there and idle, which you can't do with a radial engine. When it sits there idle, them bottom cylinders don't keep throwin' that oil up - it builds up down there, the rings can't handle it and the plugs foul up... they said he just got in and opened the throttle up and tried to take off. The way I heard it, he never got off the ground - just ploughed out through the trees... oh, and Harrison survived - he was hurt pretty good but he survived.'

I know of two books produced by the 81 ADG at the time. I borrowed a rare copy of the later one entitled ***81st Air Depot Group 1943-1944.*** Sure enough, page 35 told of the dangers and sorrow that the 81 Depot Repair Sqn faced -

'...deepest was the loss on 17 July 1943, of Major Donald Simpson, Sqn CO, and Capt Paul May and Lt James Handley, test pilots. In the course of "routine" work they had flown three P-38s to a nearby field and were returning in an A-20 when the fatal accident occurred. They were given full military honours in the presence of the entire squadron under a sky in which flew a formation of P-38s with three empty positions.'

Simpson apparently died on 19 July. Ralph Dietz had remembered the '...son' in his 'Harrison', and there was a Lt Luther E Wallace in the test pilot group. For a while I reasoned Bill Farrior's memory was mistakenly connecting 'Wallace' with 'Simpson' as in the topical romance between American divorcee Mrs Wallis Simpson and King Edward VIII of England! However, a closer study of the 81 ADG's 1943-44 book, under a heading of 44 Depot Repair Sqn, revealed Lt Wallace was also on board! He -

'...was the only one to recover,and is now back on duty.'

So Bill Farrior was right!

Archive records cover the matter of dismantling and returning the various hangars from CHARLEVILLE to

EAGLE FARM (14/11/1989)
Looking somewhat dilapidated today this is the eastern end of Hangar 7 with those wartime warehouses 'W 23' in the background across the railway line. The igloo structure has been modified along its northern side but is otherwise essentially unchanged. Hangar 7 is today listed by the National Trust as a building worthy of historical preservation. [Author]

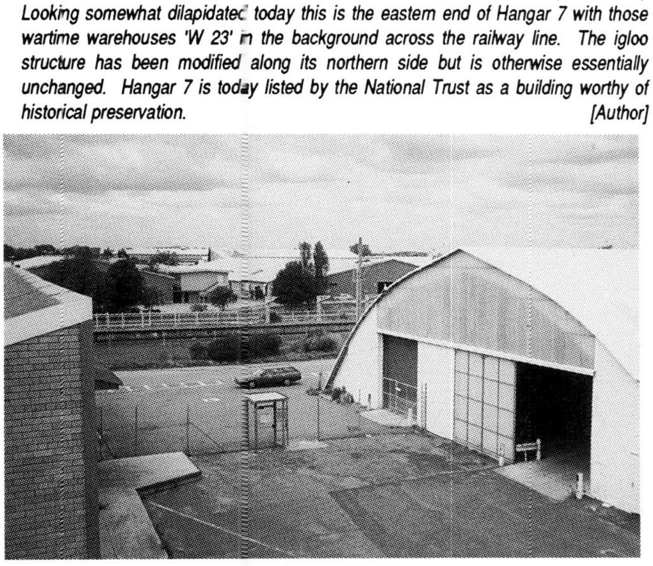

EAGLE FARM, August 1943, onwards. AWC were involved only in providing the concrete floors. 81 ADG provided their own 'troop labour' for the take-down and re-erect. The AWC was apparently quite concerned that the US troops might omit some necessary upgrading. Lt Col Edward D Rosendahl, CE, replied in mid November 1943 -

'...our Structural Engineers are checking the matter of providing additional bracing as suggested and your office will be advised regarding the action taken in due course.'

Arlo Gilpatrick came to Queensland by ship with the ground echelon of the 22 BG. Arlo became Engineering Officer of the 19 BS on Walter Maiersperger's transfer to 'squadron operations' early after the Group's B-26s reached Townsville late March 1942. He had spent the first few weeks at AMBERLEY 'uncrating and assembling some P-40 fighters which were intended for the Dutch'. Arlo stayed with the 19 BS through its IRON RANGE and WOODSTOCK periods leaving it November 1943 to become 2 I/C of the 5 AF Air Service Command's Experimental Section in Hangar 2 at EAGLE FARM. Walter Maiersperger was his CO. The 'section' had been headed by 'Pappy' Gunn, and certainly part of Arlo's work load continued in 'armament improvement'. With a voice and smile which would both pass for John Wayne, Arlo described some other 'projects' -

'...one thing I was most proud of, was the smoke screen tank installation... and they gave me credit for the whole thing. But I really don't deserve it... General MacArthur wanted a capability of laying a smoke screen 9 miles long and a certain depth and so forth... laid by aircraft to conceal landing areas in the islands. And the Navy had this function... when I got there they were playing around with some of the Navy equipment, but they weren't getting anywhere... and they had to have it in three weeks as I recall... So I personally tackled the thing myself and analysed what was going on... We settled on an airplane, the B-25... We had a tank in the bomb bay, loaded with this chemical...'

Arlo went into some detail. It seems the main problem was conventional valves and pipework quickly 'leaked' -

'...We had a civilian engineer working there, and I have his name somewhere... he and I worked on it and he came up with this idea...'

The suggested material was very inert and had good natural lubricating qualities but its name eluded Arlo -

'...we machined the valve from a solid block... anyway within a week we had this thing operating and we'd flown it, and everyone was just so delighted. We strung a cloud over 11 miles long and several hundred feet in height with the thing, which exceeded what MacArthur had specified...(they) later on adapted it for use for "defoliation". I remember Owi Island just off Biak was the first island that they defoliated with it...'

I guess like most people these days, to me 'defoliation' is synonymous with Vietnam and 'Agent Orange'. When I prompted Arlo he explained 'it was done for pest control'. Walter added that the spray was DDT -

'...which doesn't "defoliate"...it got rid of the lice and the Typhus. I suspect the vegetation on Owi was burned off.'

Arlo followed up with the name of the Aussie 'boffin type' as Walter Maiersperger remembered him. Bless me if it wasn't George H Birkbeck, my lecturer emeritus, in Soil Mechanics/Civil Engineering, University of Queensland! While George remembered the work on 'bolstering the structural shell of the B-25 to withstand the huge muzzle blast pressures of the .50 cal machine gun blister-pack', he couldn't recall Arlo's 'smoke generator'. George had slipped into Aeronautical engineering from Civil and was in England working with Vickers Supermarine at the outbreak of war. On return to Australia, he was welcomed into the Government's Department of Aircraft Production.

Mikesh's account has the TAIU ordered back to America June 1944 but the 'shift' seems to have taken several months. Also around June 1944, the Allied capture of airfields near Hollandia on the northern coast of Dutch New Guinea, produced several restorable Oscar MkIIs. While Mikesh reports XJ005 and another (XJ006?) being repaired at Hollandia and flown 'locally', the AWM believed, in July 1992, that 'one Oscar II was flown from Hollandia to Eagle Farm and another three or four, shipped there in pieces'. The diary of the RAAF's 27 OBU at COOKTOWN, carries an entry for 5 July 1944 -

'...rebuilt Oscar escorted by B-25 passed overhead enroute Hollandia to Cairns.'

Charles Whistler was at Wakde Is 'with 186 combat missions up' when he was pleasantly surprised. He was directed south to fly the P-47 in trials with the MkII Oscar.

'...When I arrived there I was told that the Oscar MkII would soon be ready to fly in mock combat against the P-47, P-38, P-40N and Spitfire. No pilot was available for it so Lt Jackson, Lt Strand and I flipped coins to see who would fly it first and I won. I flew the Oscar on various dates 4 through to 11 August 1944. I think it was one of the most fun aircraft I have ever flown; with all that power and lightweight it was a 'hot rod'. For sheer joy of flying, it was a dream, but I would not wish to fly it in combat against the Allied aircraft that we had. Its lack of fire power and inadequate self sealing fuel tanks made it a death trap... It did have excellent armour plate behind the pilot, but hits in the fuel tank were disastrous for it.'

Somehow, XJ004 was snapped on the tarmac at GARBUTT as evidenced by ex USAAF photographer Arch Fraley's photo currently displayed at the Townsville International Airport. Charlie Whistler was on his way back to Port Moresby after his involvement 'in the trials' -

'...While I was at Townsville (RAAF Station) Lt Strand came through there on his way to Nadzab New Guinea in the Oscar II... I heard it went "US" there in a landing mishap'

Ken E James and Ron S MacDonald both had logged considerable Spitfire time against the Japanese in the Northern Territory. They were both detailed to participate in ATAIU trials at EAGLE FARM during the second half of 1944. Ken James wrote -

EAGLE FARM (1942-43)
The legendary Paul 'Pappy' Gunn, snapped in front of a B-25 with Doomben racecourse in the background. 'Margaret' was probably being used to test the 'blister pack' .50 cal guns on the side below the cockpit. [T Whyte ex RAAF]

EAGLE FARM (9/1/1944)
This low level oblique looking to the southeast shows much activity in the erection of fresh aircraft. Most visible in this photograph are A-20 Boston twin engine light bombers. Hangar 3A (half length of its neighbour Hangar 3) has by this time been returned from Charleville and erected here at EAGLE FARM. Behind Hangar 3A it is possible to see a doubling in length of the initial Hangar 2, this additional structure again being part of the buildings returned from CHARLEVILLE. [Ponting Family]

EAGLE FARM (17/3/1944)
Nakajima Ki-43 'Oscar' MkI aircraft is seen here after being rebuilt by the TAIU. This extremely manoeuvrable lightweight aircraft was the work horse of the Japanese Army whereas the Zero aircraft served with Japanese Naval units. Bill Farrior's Flight Record (Form 5) shows he flew this at EAGLE FARM on 17-18 March 1944, for a total of 50 mins only. [Mrs Ralph Chan]

'...Ron and I both flew the Spitfire, P-40, P-47 and P-38. He also flew the Oscar... I hadn't flown a P-38 or P-47 before. The P-47 was a big burly aircraft - it needed a lot of air space to manoeuvre. The P-38 was fast and smooth and a real delight to fly. I remember the clear forward vision particularly. Capt Sam Hepford was one of the USAAF Officers involved...'

Charlie Whistler rightly saw EAGLE FARM as a 'beehive' of activity in 1943. With Supreme HQ SWPA located in Brisbane, the airfield was a real hub through which most important visitors transited. It was also from here that some strange 'proving flights' originated.

Most people with only a passing interest in the history of flight will know Charles A Lindbergh as the first solo conqueror of the Atlantic, from New York to Paris in 1927. The highly emotive kidnap and murder of his baby son and his support for America's isolationist policy of the late 1930s kept him in the public eye. Queenslanders who've not read ***The Wartime Journals of Charles A Lindbergh*** published in 1970, may not recognise -

'...Rest of morning spent walking alone along the streets and through the stores of Brisbane... nothing in the Australian papers about my being here; consequently, no one recognises me, and I am able to move about with complete freedom...'

EAGLE FARM (ca -/5/1944)
Used as a utility 'hack' aircraft this Curtiss Falcon is pictured at the northern end of hangar 7 with two notable buildings in the background. On the left is part of the prefabricated Butler Hangar 3, approximately half of which remains to this day. On the extreme right the gable profile of the western end of the Ford Motor building is seen. This Curtiss Falcon is believed to have been one of several originally destined for the Netherlands East Indies Air Force and side tracked for US service through developments in the Far East. Ron MacDonald's log book shows he flew '777' on 14 August 1944 for 'familiarisation and aerobatics'. [J Selwood]

EAGLE FARM (22/10/1991)
In a special hangar at Scone Airfield, NSW, this was the Nakajima Ki-43 Hayabusa 'Oscar' MkI c/n 750 belonging then to Mr Col Pay, well known 'warbird' aircraft restorer/operator. Believed to have been shipped to Australia from Bougainville sometime after the war it is similar to the first 'Oscar' rebuilt and flown at EAGLE FARM. Col intended restoring this aircraft to flying condition but has since sold it to Sir Tim Wallis' Alpine Fighter Collection, New Zealand. Other MkII sections, some thought to have passed through ATAIU's EAGLE FARM operation, were held until recently by the AWM. [Author]

The civilian master of maximum range flying had just spent four weeks 'travelling in New Guinea on Navy orders'. US Army HQ in Brisbane now very tartly asked -

'...what right the Navy had to send anyone into Army territory without first getting permission...'

Lindbergh chose to help the US war effort in his beloved aviation advancement. Opportunities to improve airplane design and performance impelled him to tackle the job at the front, actually flying in combat! So it was that he was summoned by MacArthur, early July 1944, to explain. Lindbergh wrote 12 July 1944, that with pleasantries over MacArthur turned -

'...immediately to the subject of combat radius for fighters, saying that Gen Sutherland had told him that I said the P-38's were capable of 700 miles...'

Maximum range comes with flying at the maximum economical speed, for the changing weight of the plane -

'...MacArthur said it would be a gift from heaven if that could be done and asked me if I were in a position to go back up to New Guinea to (so) instruct the squadrons...'

Somebody in New Guinea had slipped up in not sending the relevant copy of Lindbergh's 'orders' on to Brisbane when he'd first arrived. Some informal radio communication with Kenney seemed to have righted this, but 'the old Army-Navy friction' had persisted and he'd finally been summoned. Now it was cleared up, Kenney -

'...greeted me with a smile... told me he was going north in the morning and asked if I would like to go along. I thanked him and said I would like to hold over in Brisbane for one day. I asked if there might not be a plane I could ferry north the following day. Kenney said he could let me have a P-47 any time I wanted it.'

Gen Kenney tried to persuade Lindbergh to be -

'...satisfied to fly around in a bomber instead of a fighter after you get there.'

But Kenney knew he wasn't getting anywhere! The irony of Lindbergh's departure from EAGLE FARM came with his lodgement of his flight plan -

'...Army operations did not want to clear me non-stop to Horn Island, the captain in charge saying that a P-47 was not capable of making the flight from Brisbane to Horn Island without refuelling, and that he knew because he had gone up there several times in P-47's. I asked him what his fuel consumption had been. "About eighty gallons an hour." I told him that the R-2800

engine in the P-47 should be capable of cruising in the vicinity of fifty gallons an hour. He said he could hardly believe that and that he had never been able to approach such a low figure. However, he was very pleasant and considerate and cleared me direct to Horn Island as I requested...'

Unusual tourist that he was, his diary notes -

'..Dropped down to 200' above the ground over the valleys of northeastern Australia in the hope of seeing some wild kangaroos or wallabies, but sighted only cattle and an occasional small ranch building... arrived over Horn Island with enough fuel to reach Nadzab with almost a two-hour reserve. Tried to contact Horn Island by radio to extend my flight plan to Nadzab, but without success. Landed at Horn Island at 1511.'

Two weeks later, flying a P-38 with the 475 FG, on a fighter sweep to Ceram Is, Lindbergh shot down an enemy fighter in combat. August 13, 1944 he wrote -

'...Gen Kenney landed at Biak yesterday and left orders that I am to do no more combat flying! Well, I am about to start home anyway...'

Lindbergh reported back to Brisbane on 21 August. Kenney was firm on the 'Hullabaloo' back home if word of his doings got about, but he and MacArthur were very appreciative of the benefits gained. MacArthur asked me -

'...how many Japanese planes I had shot down. I told him, "One." "Where was it?" "Off the south coast of Ceram." "Good, I'm glad you got one."...'

It is natural that repair or modification to an aircraft should be followed by its flight testing. Walter Maiersperger and Arlo Gilpatrick worked closely on many projects. Walter's log shows countless proving flights like the P-47 'buzzing of the Toowoomba Boys' Grammar School, where a friend's son boarded.' It also shows probably the only visit by two 'Yanks' in one P-38 to CUNNAMULLA. The way he tells it, his friend Arlo 'Gil' Gilpatrick had -

'...targeted a beautiful red head in Brisbane whom I eventually met with her mother. The father was running a sheep station near Cunnamulla. The girl, Helen, invited Gil and me to attend a sheep shearing festival there - racing, banquet and dance. I got permission to use a P-38 we had for the purpose of installing an additional fuel cell behind the pilot's head, under the canopy, by relocating the radio elsewhere...'

EAGLE FARM *(4/7/1944)*

*This is XJ003, the TAIU's restored 'Tony' aircraft, in flight. The camera plane is unmistakably a Curtiss Falcon CW 22B such as an 81 ADG 'hack' at EAGLE FARM. Tony was developed to use the licence-built Daimler Benz 601 inverted 'Vee' 12 liquid cooled engine of the German Messerschmitt 109. The engine proved unreliable with the Japanese Armed Forces and not much better in the hands of the TAIU. Ex RAAF pilot Ron MacDonald's log book shows several hours flying P-40N '803' and Spitfire '424' in **'comparative tests against Jap Tony'** between 29 September and 7 October 1944, at EAGLE FARM. Charlie Whistler sent this photo, given him by another USAAF pilot Dick Kent,in July/August 1944 at EAGLE FARM. Australian historic aviation researcher Len Lindorff has this same photo in a series, two of which are uncropped and still show edge identity windows reading **'Test Flight Tony 4 July 1944'**. Some of the on-ground photos are clearly in and around Hangar 7. While Robert Mikesh's article and Bill Farrior's 'Form 5' both show a Tony, very likely XJ003 being test flown in USA during 1945, there is strong evidence that XJ003 is here pictured over EAGLE FARM. [C F Whistler ex USAAF]*

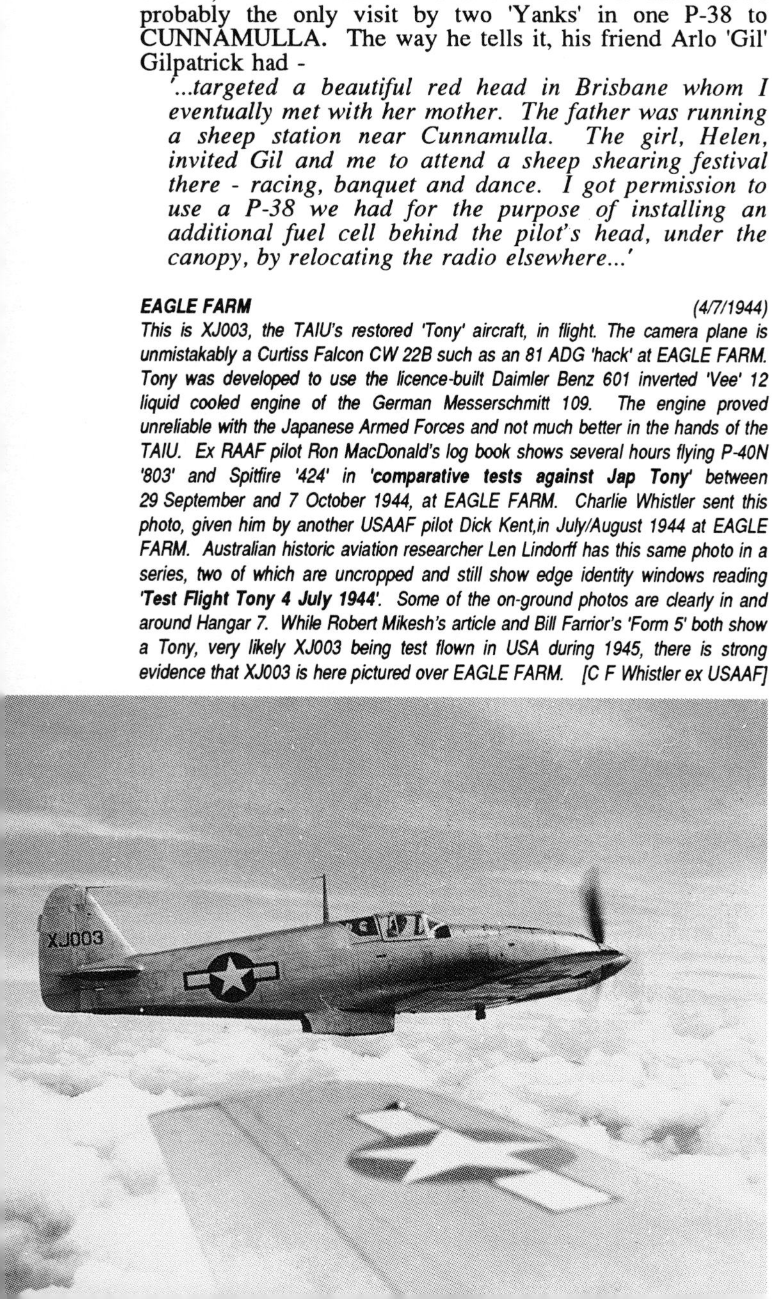

Walter described how the fighter was temporarily adapted to carry a passenger 'sans parachute'! The mission was most carefully planned; he calculated the part-fuelled and two-up configuration and -

'...made a few practise take offs and landings so I could approach at the minimum speed at Cunnamulla, as the (Civil) field had three runways only 2 400' long...'

He was conscious of the sparsely populated, relatively featureless terrain he'd be over -

'...the thing to do was stay on course, or north of it, until one hit the railroad. We topped off the plane and headed for Cunnamulla on 26 May 1944 (my 28th birthday). We hit the field on the nose just short of the railroad and town but noticed it full of sheep. Still on course, I passed over, swung around to make another pass over it at about 20' to scatter the sheep, swung around again for the landing. A Home Guard agreed to guard the plane until the next day. A car picked us up and took us to the race track where everyone for miles around were in attendance, just over the railroad tracks on the other side of town. On the way there, Gil, whose position faced aft in the plane, asked if I had noticed that on the downwind I had passed over the horse race starting line and scattered the horses each time I passed over. I had also passed directly over the stand, upwind. 'No I hadn't' I replied, I was looking ahead to make sure I didn't crack up the plane in the outback 450 miles from its home base.

We were united with Helen and her mother who had even provided a date for me. That night her mother came over and said some "gentlemen" wanted to talk to me. Two bruisers approached and asked me if I had noticed how I had scattered the horses every time I passed over. I assured them that I hadn't - my eyes were forward and I had only been concerned to make a safe landing. Their next question - "Major, do you know how much money was bet on that race?" Of course I hadn't, but was glad they didn't beat me up. The visit was a delight. Of course the Aussies believed we two were just cutting up for the fun of it - "trust the Yanks"...

In May '92 Gil and I were house guests with Helen and her husband Jim Shaw in Buderim, Queensland. His uncle was the McKay...(who) was a major contributor in starting the (Royal) Flying Doctor Service (with which I became acquainted by reading the novels by Neville Shute Norway). There was to be a benefit luncheon of the RFDS in Toowoomba which we attended as guests of the Shaws. They told the story of our earlier visit to Cunnamulla to the President of the Chapter. When the guest speaker had concluded, we were introduced and our story repeated to the membership. It brought the house down.'

Not to take any of the shine off Walter's account, I found some stabilising background in a closer reading of his log. This temporary carrying of a passenger in 'P-38 H-1-LO 42-665491' should not be construed as too 'experimental'. It is the first P-38 to appear in Walter's log, October 1943 at Mascot, Sydney. Several names appear over the next twelve months, prefixed by an '(X)', denoting carriage as a 'passenger'. I noticed a like entry for 26 May 1944, at Cunnamulla - '(V) Miss Helen Pierce' but Walter explained -

'...(V) meant "visit". Helen got no ride!'

In January 1946 Dept of Air advised its post war requirements of EAGLE FARM to DCA. This was only that a Communications Flight would be established there. Certainly the prime role that EAGLE FARM played post war was to resume service as Brisbane Airport, later becoming a fully recognised International gateway. Today the wartime airfield has completely lapsed with the commissioning in 1988, of the new complex 5 km northeast. Buildings of WW2 period relevance remaining

EAGLE FARM *(14/6/1991)*
This recent aerial shows the Gateway Arterial motorway passing through where Hangar 2 once stood and between Doomben racecourse and the old Eagle Farm airfield. The current Brisbane International Terminal is visible at centre right, where taxiways lead north eastward to the new Airport. *[Department of Lands, Qld.]*

at the old EAGLE FARM airfield today are Hangars 1,3,7 and 8. Hangar 8, the extreme eastern one was, at war's end, represented only as a US Butler prefabricated steel frame gabled structure, in this case metal clad throughout with canvas curtain entry. Some time in 1949 it was foreshortened and added to by the transfer of the 'saw-tooth roof' steel trussed hangar of large proportions which disappeared from ARCHERFIELD in 1948. In this form Hangar 8 remains to this day serving a somewhat strange role as covered storage for various bulk commodities.

In 1992, Hangar 7 enjoyed some recognition in being cited on the register of the National Estate as a building worthy of preservation. However that situation falls short of being awarded 'a preserve at all costs' status. Hopefully it may become viable to incorporate this historic structure into some local museum scene worthy of its WW2 heritage.

In recent time, where truly historic aircraft once lived and breathed, a ten minute ride in a diminutive 'go-kart' is the closest one can approach such exhilaration.

OAKEY *(4/11/1990)*
The airfield is now the home of the Australian Army Aviation Corps. A small museum sponsored by this arm of the Defence Force exists on the base and is open to the public at prescribed times. One aircraft on display is this twin engine Beaufort bomber looking quite the worst for wear. It's believed to have been recovered from Tadje, New Guinea and is the subject of long term restoration. *[Author]*

OAKEY

'...As Army Airfield Oakey is categorised as an Operational Base, the Unit has been instructed not to release current imagery to you for the purpose of reproduction in your proposed book.'

So answered HQ 1 Military District in December 1991 when I sought appropriate current photography of OAKEY. I hasten to make clear that what I have presented here has been approved by the Unit involved.

Although this site was identified and developed as the base for No 6 Aircraft Depot RAAF during the WW2 years, it was relinquished by the RAAF a few years after the war. It was utilised as a Civil Airfield until July of 1969 when the Defence Department did acquire the site. Since then OAKEY has become the home of the Australian Army Aviation Corps. Civil air traffic uses the airfield under a joint-user agreement.

As early as April 1942 the search began for a suitable site for the RAAF's 6 AD. It was needed to take some of the load developing at AMBERLEY. Sites in the Chinchilla district were considered and found wanting. Interest shifted to CONDAMINE for a while, then Dalby and elsewhere. P-O J W Campbell in a memo to Air Board Melbourne reported -

'...In passing through the town of Oakey on 1 August 1942 a good strip was located near Wilhorn (sic) Railway Siding. The site appeared attractive for a maintenance echelon. As the Bell Road country was

OAKEY *(1943)*
This extract from the 1:63 360 (1 inch = 1 mile) Military Survey map of the time shows by dotted outline the runway works projected at that time. Note also the small circle/airfield legend immediately south of the intersection of those runways. The proximity to disused coal mines such as that figured at grid ref. 789951, has probably fuelled stories of war surplus equipment being hidden/buried locally.
[Department of Defence]

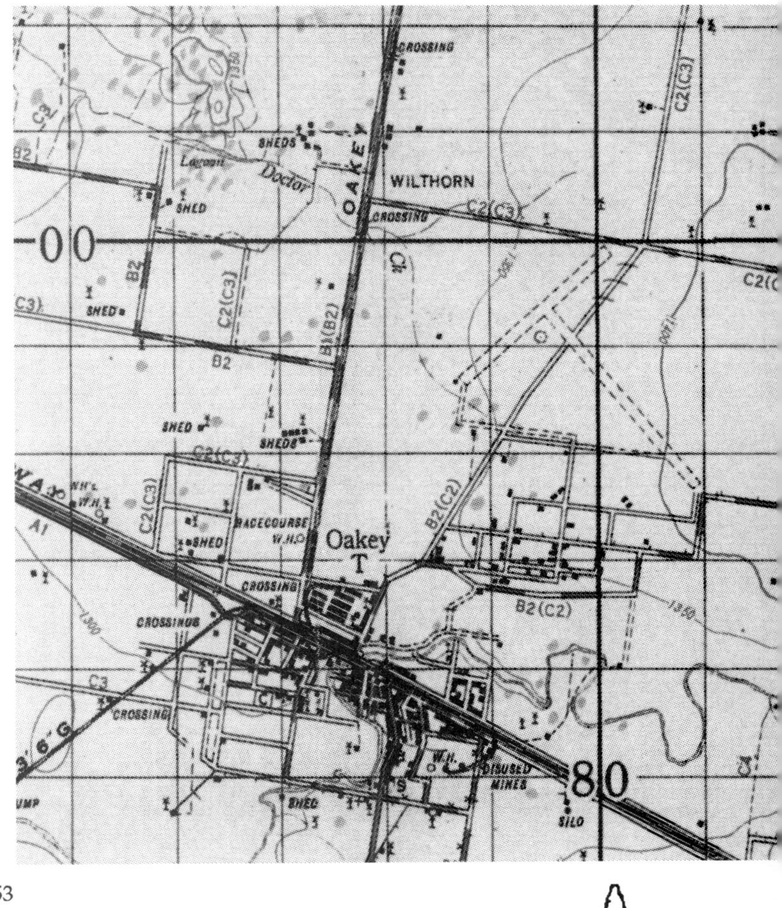

found to be unsuitable Air Board was signalled on 3 August 1942 as to the suitability of Oakey from operational aspect. The site was further investigated on 4 August 1942 in company with Mr Wyeth of DHS and was found to be more extensive than at first thought. Air Board was signalled stating site was good enough for a main HB aerodrome and would definitely be wasted as a maintenance strip only.'

With two thirds of 1942 gone, and the wet season up north calling for the diversion of construction resources, the RAAF settled down to carefully plan OAKEY. A single project, estimated to cost around £500 000 deserved that!

AWC Minutes early in February 1943 show the MRC were then ready to work on OAKEY. However, planning was in no way complete! The periodic reports of the SDCC support this. Dated 14 February 1943 it was noted -

'...Some workshop buildings have been re-sited and a new plan prepared. DWB suggests standard huts dispersed instead of the suggested regular placements of buildings as extension to township. Wyeth advised against this as cost of camouflaging would cost more than the savings made by changing to standard huts.'

By 14 March 1943 these Camouflage reports added -

'...no dispersal of planes is to be done here at present. Buildings are to take precedence over drome works (runways etc) and latter are to be commenced only when buildings somewhat advanced as it is required that buildings and runways are ready together...'

The lesson on completing all construction simultaneously had been learned.

Building commenced in March 1943. Contractors H W Thompson & Co and F T Eastment & Sons were to erect thirty-seven and forty-four buildings respectively. The site was large and the natural tree cover such that none could be wasted. Although the National Security provisions of the day 'regulated' tree felling etc., someone had to police them. Some humour appeared in reports -

OAKEY *(2/10/1945)*
An extract from systematic mapping of the 'Jondaryan' map area some few months after the cessation of hostilities, shows OAKEY as it then was. Aircraft surplus to the RAAF's needs were being parked immediately south of the southeastern end of the 153° runway. The taxiway leading southward from the airfield into the building area, laid out as an 'extension of the town', passed the duty pilot's tower just to its east. South of the duty pilot tower can be seen an easterly seeking darker pavement leading to machine gun test butts. Just southwest of this junction the first Bellman hangar shown is the most significant wartime building remaining on the airfield today. *[AUSLIG]*

OAKEY *(1944)*
This somewhat hazy photo shows an impressive line of RAAF 79 Squadron MkVIII Spitfires off the north south taxiway just near the duty pilot tower. *[F Smith]*

'...Mr C F Oliver, Camoufleur at Oakey has received his push bicycle direct from the PMG Stores. He asks whether he is expected to deliver letters. The bicycle is pillar-box red with a PMG number plate. He has been told he's only expected to deliver "the goods".'

The SDCC report of 2 May 1943 noted some -

'...64 buildings in various stages of construction and 18 not yet started. There are now sufficient buildings ready to make a start on the dull painting.'

The NW-SE runway was commenced in June 1943. Bitumen sealing of both runways and various taxiways proceeded during September 1943. Hence it appears that things went as planned, in as much as the airfield construction works and the building work of the depot came to completion as the unit took up its station in October 1943. It was an impressive result, complete with full camp facilities including water-borne sewerage.

The Museum of Australian Army Flying is located on the now Army Aviation Base of OAKEY. The Museum's current information leaflet records 6 AD as having, in its time, erected and test flown Beaufort, Boomerang, Kittyhawk, Mustang, Norseman, Spitfire and Wirraway aircraft in addition to the marine version of the Spitfire (Seafire) which was erected for the Royal Navy. It also mentions work on 'Lancaster' which presumably relates to the two Avro Lancaster aircraft which came to Australia, one in June of 1943 and the other in November 1944. The first one, A66-1 did not see any war service but was utilised successfully in the campaign to secure War Bonds in this country. The other aircraft was presented to the AWM and still holds pride of place in Canberra.

Mr W D (Don) Parker was a member of the RAAF's 7 RSU working 'up in the Northern Territory'. Don tells me he spent some six weeks or so early in 1945 at OAKEY where he ran into an ex CO of his, W Cdr 'Bugle' Campbell. One of his two recollections of OAKEY were that for some 'unexplained reason', W Cdr Campbell set him to work on the restoration of a lone Beaufort Bomber there at the time. It apparently wasn't in anywhere near as bad condition as that presently in OAKEY's Museum of Australian Army Flying. The other vivid recollection Don has of OAKEY, is that it was the 'coldest place he can remember'.

LOWOOD *(28/9/1991)*
As though it were landing at the wartime field, this modern day crop duster is seen during a low level demonstration. A small part of the original runway is used from time to time by small aircraft. Mt Tarampa is clear on the skyline but for the rest it is simply a case of back to 'rural civilian activity'. *[Author]*

LOWOOD

'...The construction of an aerodrome at Lowood in the Upper Brisbane Valley commenced in September 1941... and work of maintenance and repairs went on until 1945... The Americans used the field for a few months in the middle of 1942 but the surface did not please their pilots and a further landing strip was provided for them west of Coominya. The RAAF again took over the Lowood drome in September 1942 with heavier machines'.

Apart from a few further details on the overall cost of work associated with the Main Roads Commission, this is about all that is said of LOWOOD in the ***MRC History*** published in 1949. The Empire Air Training Scheme embraced LOWOOD as the site for establishment of No 12 Elementary Flying Training School (EFTS) in July 1941. Two Relief Landing Grounds were approved; WIVENHOE in a pocket of the Brisbane River and COOMINYA (No 1), just southeast of the township.

Buildings at LOWOOD were quite concentrated. Barracks were close to the four Bellman hangars and the whole was arranged with no regard for the likelihood of it coming under attack. Vic Waddell recalled his introduction to LOWOOD in January 1942 -

'...with the Japs moving south, it was time for our unit, 12 EFTS, to move (from Bundaberg) to Mt Tarampa, Lowood. A few weeks after we arrived we were struck by heavy storms over three consecutive nights - the first being the worst. Hangars were built but doors had not been fitted. Many aircraft were out in the open. We lost somewhere in the vicinity of twenty. They were Tiger Moths, and during the height of the storm one was carried backwards across the drome from the top of the barracks area and dropped on its tail on the lower road. By the time we had restored some order, with most of our tents gone, it was late at night. Our very thoughtful sergeant arranged for an urn of coffee and rum - the most enjoyable I ever tasted.'

Colin Cutler as a young RAAF trainee pilot on 20 Course at 12 EFTS recalled a 'rather quaint regular occurrence'. To avoid interruption to the daylight operation of aircraft the all-over grass field was mowed at night! Colin also told of a period of 'field punishment' whereby for some unremembered misdoing a group of his course was given the option of several miles of pack laden running or some equally arduous period on 'picks and shovels' extending the development of machine gun pits and slit trenches on the airfield.

While Vic has not been able to date this other event, it does appear from other sources that about 5 February 1942

LOWOOD *(17/7/1942)*
Looking over the natural observation high point of Mt Tarampa in the foreground, to the building development and the all-over field nature of the airfield at this time. Several bomb store buildings were constructed in the timbered western slope of Mt Tarampa. A semi-underground Operations Building 35 m x 7 m was constructed on the north slope overlooking the field. [CPE (RAAF Official)]

LOWOOD had its introduction to American aircraft -

'...We were indoors receiving gas training lectures and going through the gas chamber. We could hear much aerial activity outside - and not Gipsy engines either! Imagine our shock when we returned to the hangars, to find all our aircraft gone, and the drome covered with Airacobras and Douglas Dauntless A-24s. Our aircraft had been flown to Coominya. 'What happened was that an unidentified aircraft carrier was sighted off Moreton Island. It later turned out to be American. Then followed two or three wet days either at Lowood or Coominya - everybody did their share of guard duty.'

This was the arrival of the carrier USS *Langley*, an historic vessel in that it was originally a collier to which had been added a high mounted flight deck. It thus became the first US aircraft carrier. Confusion as to its identity prompted a precautionary dispersal of the very valuable aircraft then being assembled at ARCHERFIELD and AMBERLEY. Being several miles inland from Brisbane and the coast made LOWOOD a safer haven.

Anti aircraft elements of the US Forces then arriving in Australia were detailed to join with some Australian units, in the near Brisbane area to defend -

'...the Archer, Eagle Farm and Lowood airdromes.'

The General Index Cards of the Dept of Interior Works and Services Branch show the advent of WAAAFs to LOWOOD. Dated 25 February 1942, the card notes -

'...WAAAF accommodation - alterations to Huts 71, 73, 74, and omit lining Recreation Room.'

The US 8 Pursuit (Fighter) Group faced two major hurdles on arrival in Australia early 1942. Their P-39 Airacobra aircraft had first to be assembled and commissioned, and many of the young pilots had then to learn to fly them. This fighter proved to be 'well ahead of its time', but it demanded an experienced pilot, as events in combat with the lighter, more manoeuvrable Japanese aircraft soon showed. One RAAF veteran I have spoken to told of the tentative way in which some young pilots taxied their Airacobras around LOWOOD in this training phase. He described seeing an undercarriage suddenly collapse while taxiing, believing this was evidence of the nervousness attaching to such a compressed training programme.

Early April, it became untenable for 12 EFTS to remain at LOWOOD. It was dispersed into other similar units down south. The SDCC report dated 20 May 1942 mentions the drome as having by then -

'...been taken over by Americans'.

Remembering that LOWOOD was an 'all-over field' lacking paved runways, it is not surprising to hear veterans speak disparagingly of it. Geof Gillmore, a pilot in the RAAF's 71 (Reserve) Sqn flying Avro Ansons wrote -

'...The lack of paved runways created a problem for pilots after heavy rain as our planes were liable to sink into the soft earth and bog. We did overcome this by taxiing at fairly high speed and carrying out all cockpit drills while on the move to the take-off point and not stopping before commencing our take-off run.'

Perhaps this explains the P-39 undercarriage collapse! In any case the Americans moved to have a strip constructed urgently just west of Coominya township. Dissatisfaction with LOWOOD is reinforced by a report on RAAF files by F-Lt John Keays dated 10 July 1942 -

'...Lowood Aerodrome is now being used by 23 Squadron. It is a very poor field for operational purposes. It is visible from a considerable distance and there is next to no cover available for dispersal. The camoufleurs have been on the site, but to date I have taken no action to initiate a dispersal scheme.'

The viability of LOWOOD was placed in doubt. The process of searching for, selecting and planning for aircraft dispersal there was fairly unrewarding. The only timber cover of any significance was adjacent the creek on the northern boundary and near the isolated knoll of Mt Tarampa. Quite serious thought was given to scrapping

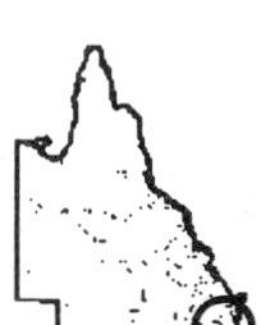

LOWOOD *(25/4/1944)*
This 'right oblique' frame from experimental Trimet photography by the RAAF, shows that by this time a formal runway had been developed. The dark sealed apron area surrounding the four camouflage painted Bellman hangars is prominent towards the centre of the photograph. Mt Tarampa is south of the station in the centre of the photograph. The black areas within the airfield circuit are most likely burned off grass. On the airfield side of the slopes of Mt Tarampa a reinforced concrete Operations Building has been constructed. As with most structures of this kind it was positioned and camouflaged to represent nothing other than a nearby farm house. The structure remains to this day. *[AUSLIG]*

LOWOOD *(25/4/1944)*
One shot removed in the Trimet photography, this vertical Frame '97', shows the northern half of the runway which by now had necessitated in a deviation of local roads. The rather sparse tree cover fronting the creek to the north of the runway does provide for some nine aircraft hideouts and taxiways leading thereto. The lack of adequate natural cover was one of the factors against further development of this site. The checkered patch, bottom left, is the effluent dispersal area from the station's sewerage treatment plant. The two 'white bits' just east of this disposal area are not known to the author. *[AUSLIG]*

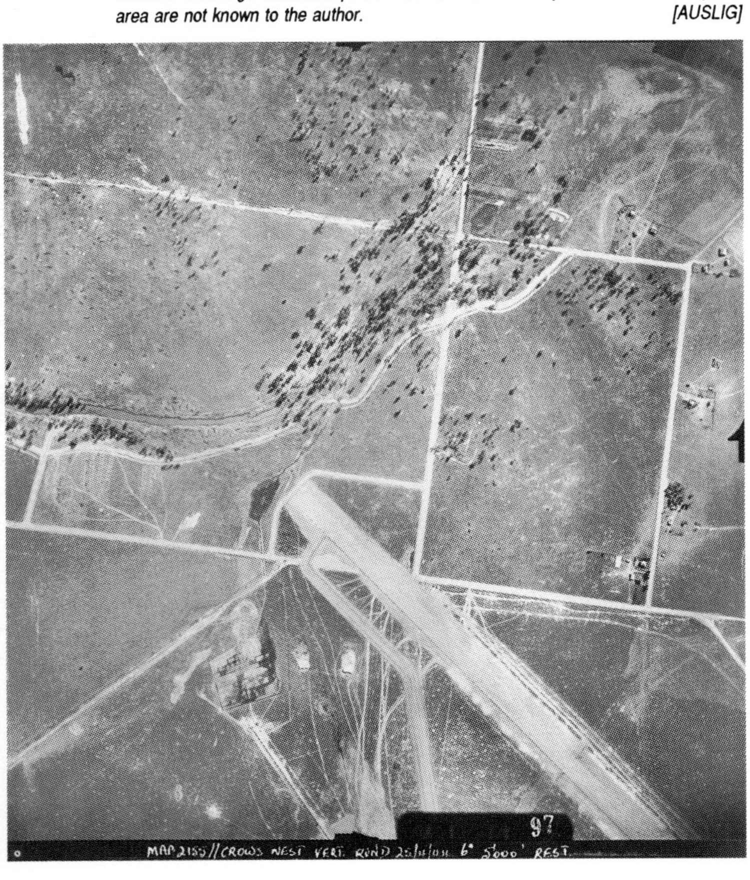

the whole establishment in favour of redevelopment at COOMINYA No 2. As the strategic situation improved, the risk of attack diminished. February 1943 saw a settling down to develop LOWOOD as an Operational Base with COOMINYA No 2 as an ancillary landing ground.

Percy Yates was a navigator with 71 Sqn. He recalled a pre-dawn take off at LOWOOD typifying the hazardous conditions that it presented to the heavily laden twin engine Ansons. The airfield had a -

'...rather disconcerting downhill slope terminating in a field of tree stumps which had a multi-line telephone or electric wire system crossing it...the Anson just cleared the fence coming to rest in a very decrepit condition amid the tree stumps. The body of an aluminium sea marker burst and the three crewman, emerged from the wreckage covered with fine aluminium dust looking rather like space men from some distant planet. The pilot told me that when he looked up from his inverted position one of the two 250 lb bombs was poised over his head...arming vanes on the nose had come off and the nose cap was staring him in the face just waiting to be struck so it could set the bomb off! It was some time before he stopped shaking.'

Geoff Gillmore answered my question about Air Traffic Control at LOWOOD. He described a Duty Pilot's Tower -

'... which was in telephone contact with Brisbane. It was the Duty Pilot's duty to notify Brisbane of all aircraft movements. We had only W/T communication hence we had no voice contact with the Duty Pilot and all communication was by means of our signal light on the aircraft. For landing we had a large white 'T' on the ground in front of the tower to indicate the wind direction for our landing. The Duty Pilot would indicate that it was safe to land by giving us a green light on his Aldis lamp. I might add that we had little option but to land as soon as possible as we were operating on our maximum fuel endurance so that we could remain on escort duty with convoys as long as possible.'

Percy Yates commented on 71 Sqn's involvement in the largest loss of life in Australian waters due to a sinking by Japanese submarines, 14 May 1943 -

*'...The big news headline while at Lowood was the sinking of the hospital ship **Centaur** just off Cape Moreton. We were not allowed to escort a hospital ship, but one of our kites found the survivors and guided rescue vessels to the scene. The rescuers were given great praise, but the Air Force was criticised by all and sundry for not escorting **Centaur**.'*

Rod Voller spent virtually all his time in north Queensland on duty with the DHS's Camouflage Services. While he remembered little of his visit to LOWOOD, the SDCC report by Mr R P Cummings, shows that gentleman visited with the Technical Director of Camouflage and -

'... Mr Voller on 23 May 1943. The MRC workmen have not yet completed the draining and loaming of taxiways to hideouts in the dispersal area.'

The relative cost and merits of some 'slung type' hideouts were touched on and it was reported that -

'...A little work has still to be done on the treatment of A-A guns and also on the covering of petrol drums.'

Professor W J Dakin had just toured projects out Dalby way and this visit had embraced both COOMINYA aerodrome (No 1), the No 2 strip and AMBERLEY. Further building at LOWOOD was mentioned in the SDCC report of 30 May 1943 showing work had started -

'...on the new W/T building near the drome...(and) the new (semi-underground) Ops Bldg... Treatment of drum stacks on petrol dump adjacent to dispersal area is continuing, also...treatment of machine gun positions.'

Mid August 1943, this Operations Building was close to a stage of back filling, so maybe it was 'completed' by the end of 1943. Effective commissioning seems to have been delayed - the General Index Card for 'Lowood RAAF', features work of 'electrical reticulation' and other 'fitting out' of the 'Operations Building' and the 'OBU 14

Transmitter Site', in the first quarter of 1945. Work of construction of a gravel (later sealed) runway was about 60% complete by March 1944 and the 'Lowood RAAF' card early 1945, shows 'cabling etc' for Electric Flare path!

The RAAF's 21 Sqn was reformed at Gawler SA on 13 September 1943, after an eighteen month quiescence. Equipped with Vultee Vengeance aircraft and forming part of the RAAF's new 10 Operational Group, the squadron relocated to LOWOOD in November 1943, preparatory to embarkation from Brisbane. This American dive-bomber, ordered by the RAAF when little else was available, proved difficult to handle and quickly outclassed by the USAAF's own growing armoury. As the Squadron's published History ***Demon to Vampire*** tells of LOWOOD -

'...confusion in shipping arrangements resulted in difficulties for No 21 Squadron. There was no suitable camp site at LOWOOD and the whole of its equipment was dumped in a creek bed. After a few days heavy rain fell, the creek overflowed and washed out the squadron's camp. Most of the equipment was ruined and had to be replaced. In these chaotic conditions the Vultee Vengeances could not be kept serviceable for long, with the result that the crews obtained very little practise before flying to New Guinea.'

Both 21 and 24 Sqns, flew Vengeances with 10 Operational Group in New Guinea. On their withdrawal, 24 Sqn relocated to LOWOOD, where it became the first RAAF unit to re-equip with Liberators. It departed to the Northern Territory during June 1944.

A few months after the end of the war, LOWOOD was categorised by the RAAF as an airfield to be 'acquired only', the runways of which were 'not to be maintained'. The 'Lowood RAAF' card records 8 October 1946, the registration of a communication from the 'Estimating Officer' concerning the removal of the Duty Pilot's Tower and its re-erection at EAGLE FARM. I wonder if it happened?

The single runway and taxiway circuit early in the 1950s was considered almost 'ready made' as a racing car venue, by the Queensland Motor Sporting Club.

LOWOOD *(14/6/1991)*
Find the old airfield *in this extract from systematic mapping photography! The original runway and circuit/taxiway which was utilised in the 1950's for motor car racing, is just discernible amongst the acreage development of the area. A significant length of the runway now provides the foundation for a sub divisional road. The concrete slab bases of the four Bellman hangars can still be seen, some now serving as residence foundations. Where the extension of the sealed 128° runway now meets the southeastern border of the old field, which is itself a local road, a memorial stone has been erected to commemorate this site as an important RAAF Airfield.* *[Department of Lands, Qld.]*

CECIL PLAINS

'...Being the Senior Fitter Armourer, with the exalted rank of LAC, I was assigned to the CO's kite. I never got to speak to him, let alone get to know him. He had served in Britain and always wore an RAF battle jacket with a whistle hanging from the left lapel. He had to report to Melbourne regularly so his crew got their hours up that way. As a consequence a number of us had to endure a fair amount of passive time - reading under trees out of general observation.'

After twelve months or so training in the operation and maintenance of Liberators both in the Northern Territory and down south, this is how Goff Bauer recalled the period with 102 Squadron late in the war at CECIL PLAINS. 102 Sqn formed there around the end of May 1945. It was the second unit to use the airfield following on from the RAAF's No 12 Squadron.

This airfield site did not get a glowing appraisal on early evaluation. P-O John Keays reported 22 March 1942, noting the abundance of dead trees up to 100′ high all over it. He saw the scarcity of clearing and construction equipment as the major problem, at JONDARYAN, CECIL PLAINS, LEYBURN and CONDAMINE, all Heavy Bombardment fields requisitioned by the Americans as part of US Project No 45. All formed part of the *Brisbane Line* whereby some defence strategists reasoned that the more closely populated

CECIL PLAINS *(3/10/1945)*
Nine of 102 Squadron's Liberators were 'at home' on the field when this systematic mapping photography was flown. On enlargement, one can be seen taxiing eastward on the southern connector straight taxiway, while the others are strung out along the 45° runway. Note the classic cross runway layout and the orderly tent lines amongst the scattered tree growth on the extreme right. *[AUSLIG]*

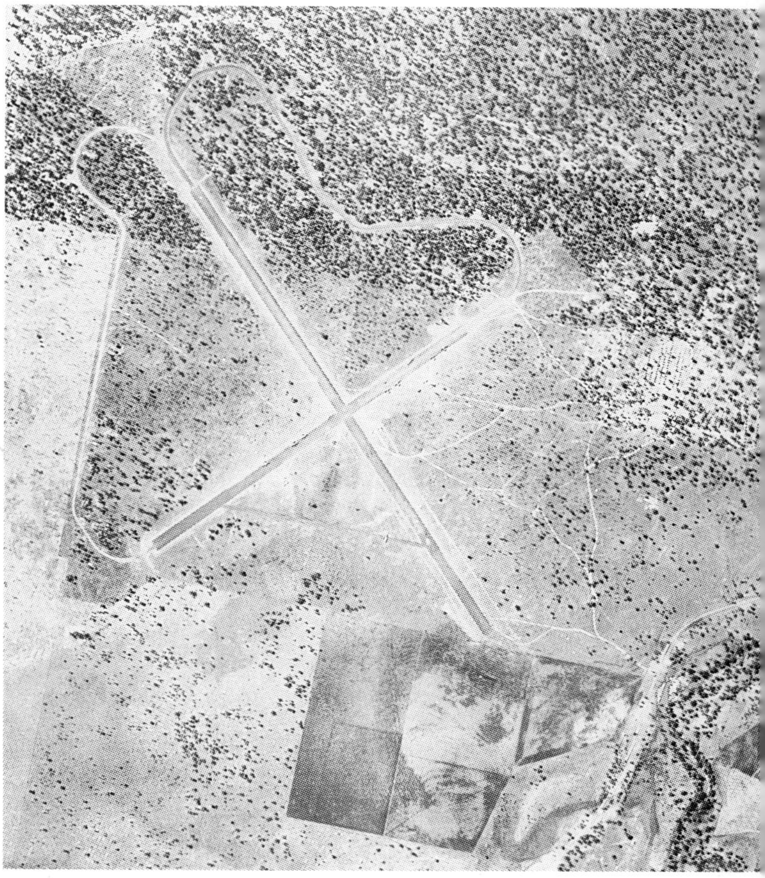

and industrialised southeast portion of the continent may have to be more heavily defended if invasion came.

Provision for the dispersal and camouflage of aircraft on these fields was of great importance in 1942. The State Defence Camouflage Committee (SDCC) had been set up and while its collaboration with the RAAF staff was often less than desired, John Keays was one to recognise this need. He criticised the disruption caused by the change of key camoufleur personnel 'in the middle of such a job'. He later reported again after inspection with -

'...Mr Campbell Brown (MRC) and on 28 June with architects and camoufleurs and on 30 June with Messrs Jordan and Wyeth. Good progress has been made with comparatively small unit and gravelling completed for 6 000' on NW-SE runway...'

Selection of hideout locations had been completed and Keays noted that the Dept of Interior were to construct -

'...splinter proofing and camouflage'.

The SDCC reports over July and August 1942, covered the usual items of project status. There was also the occasional social comment -

'...arrangements have been made for camoufleur C D Wood to reside at this drome. He has been provided with camp gear and accommodated in an MRC hut...wire netting is being pre-garnished by voluntary labour in Dalby... Large supplies of steel wool garnished netting have arrived and it is proving very suitable for dummy trees...'

The Camouflage Report for 16 August 1942 is particularly of interest. It describes more fully, some of the work -

'...Logs for splinter proof pens have arrived and a start is expected to be made early next week on No 3 hideout. Experiments in the construction of larger dummy trees have been made. They are made of steel wool or scrimming and stand about 15' above the ceiling of the hideout and cover an area of about 15' in diameter. The results are satisfactory...'

Steel wool later proved too prone to deterioration by rusting.Under a heading 'General', this report carries on -

'...it is disappointing that it has not yet been possible to obtain aerial photos or to make an air reconnaissance. Many organisations and groups, including Boy Scouts, Red Cross, VDC, school children and business girls, have been organised to garnish wire netting in their spare time. Mr Jordan and Mr Steine have been actively assisting with this work during evenings and weekends. Mr Jordan has now made contact with the CO of the Squadron at Toowoomba, whom he finds he knew at Laverton and has been able to arrange a reconnaissance flight.'

CECIL PLAINS (1/9/1942)
This low level oblique titled 'Tipton Dispersal Point 1942' is taken looking to the northeast where the taxiway loops round to join the northern tip of runway 45° at centre right. Some 'dark covering' is in progress on the turn-in to hideout No 13, at centre bottom. Aircraft hideouts here were a 'slung type' involving steel wire cables supporting the camouflage net on top. To some, the airfield was named TIPTON after the property and small school a few miles north. [NLA (RAAF Official)]

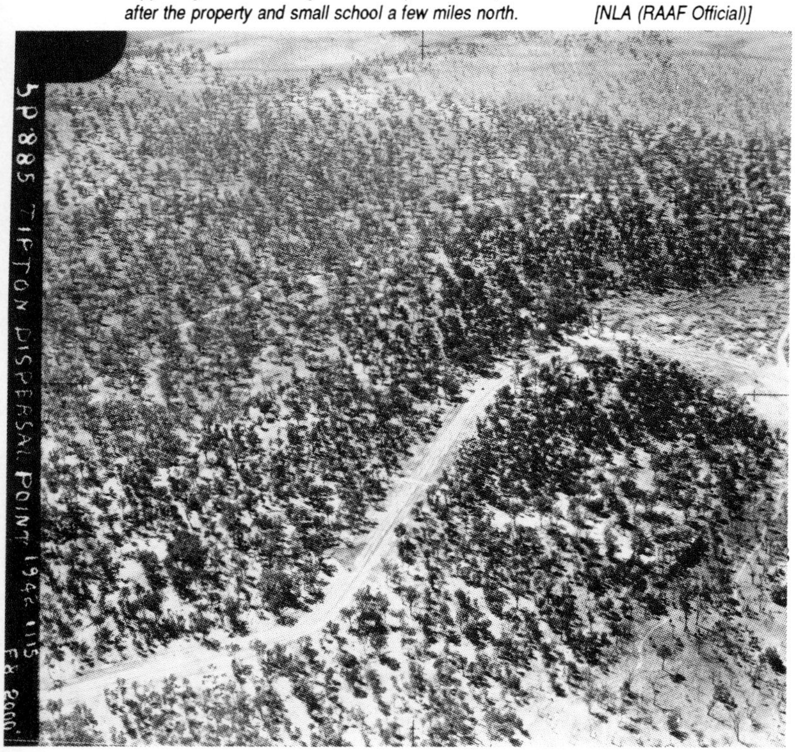

CECIL PLAINS ca(-/8/1945)
These are typical of the 'non permanent' buildings at many airfields, especially towards the end of WW2. This view shows part of the workshop and stores buildings. Corrugated iron roofs, bush timber frames and in many cases hessian walls were the order of the day. [A Bailey ex RAAF]

P-O J W Campbell took over from John Keays when he was posted elsewhere. Campbell reported to Air Board on 26 August 1942 after a visit with Sq Ldr Dawkins, Eastern Area and Mr Wyeth, DHS. Cross drainage and pavement preparation for bitumen sealing were addressed. On the camouflage side he noted -

'...good progress being made on construction of hideouts. Attached is copy of plan showing final location of hideouts.'

By the end of August 1942 there were very pressing requirements in northern Queensland to ensure that operational aerodromes were sealed by the start of the wet season. The Allied Works Council was busy finding additional resources to send north. Cutbacks in Victoria, New South Wales and southern Queensland did occur but as far as CECIL PLAINS and other 'western dromes' were concerned, Dept of Interior, Melbourne relayed to the AWC, Brisbane -

'...Dept Air advise that - they require work on the parent aerodromes at Jondaryan, Cecil Plains, Condamine and Leyburn continued and completed as early as practicable, but they agree to the work on all dispersal strips being held in abeyance.'

Dispersal strip sites had been located but none did proceed.

There also appears to have been some hesitation creeping into the 'hideout' work. The SDCC report of 30 August 1942 includes for CECIL PLAINS aerodrome -

'...men working on camouflage are now making garnishing and dummy trees, pending decision on hideouts. Work commenced on lead-in to hide-out No 13. Fl Lt Munro to take aerial photos including finished hideouts. Mr Jordan is to prepare scheme and estimate for camouflage of RAAF camp site.'

Photo 5P 885 is likely one of the photos Colin Munro's observer/photographer took on that occasion. Colin was with the RAAF's 5 Sqn and he vaguely recalled the sortie from Toowoomba. This hesitation extended to the buildings, as a report dated 4 October 1942, noted -

'...scrimming of three hideouts completed, and two others almost completed this week. F-O Harrison considers completed hideouts very satisfactory. The builders have started work on the mess building on the approved new site...On 2 October, Wyeth with F-O Harrison and K Patterson, engineer, also O C C White, assistant to Mr Cox Dept of Interior, proceeded to Condamine parent drome where general inspection was made. Here and at all other Downs dromes, all work on taxiways is suspended pending instruction from Dept Air. They visited Cecil Plains where inspection was made...On 3 October, O C C White received signal from Air re possible stoppage of erection of hutted camps at Leyburn, Cecil Plains and Jondaryan.'

While the AWC minutes of 12 October confirmed this halt, ten days later they recorded a change of mind -

'...temporary hutted camp at Cecil Plains (and others) - building is now to proceed.'

It seems the hesitancy on taxiways had to do with disguising wide gravelled pavements to make them appear as something else. The building decision was connected with the RAAF's plans for its own Heavy Bombers.

Concerning taxiways, hideout entrances and the runways, the Camouflage report of 15 November 1942, clarified -

'..All entrances to Nos 4 to 16 have been graded and gravelled. No 13 has had 2" of top soil spread over the gravel to promote the growth of grass... an experimental 2" dressing of topsoil has been applied over the edges to a distance of approx 150 yards of taxiway to reduce the width and give the appearance of a country track. This was fused into entrance of hideout No 13... Nothing further has been done with the runways at Cecil Plains, where construction of drains is still in progress but nearing completion. Grass has grown freely up to the edges of the gravel helping to eliminate the straight edges of the runways.'

On 4 December a request appeared for a further two hideouts to be constructed at CECIL PLAINS. A positive move towards securing the water supply was also made. The 27 December SDCC report mentioned this in passing -

'...the buildings at camp site are now well advanced and some of them should soon be ready for toning down... The weir across the Condamine is in elementary stages of construction... during the week the 3 Downs dromes were visited by Messrs E Lindsay Thompson, D Annand and Max Dupain accompanied by the Deputy Director, and Mr Jordan. Particular attention was given to hide-outs and numerous photos were taken by Mr Dupain at all projects.'

This was the Max Dupain who 50 years later so helpfully put me in touch with other surviving camoufleurs. He became a legend in the photographic world and until his death recently remained active in his profession.

The first quarter of 1943 saw painting of buildings still in progress and details like road access to the camp site receiving attention. All new hideout construction ceased as from 3 March 1943, but 'repairs' were then being considered. The 27 June 1943 report said -

'...Garnishing has been lifted by means of straining the supporting wires to the specified height of 19'. Steel wool which has badly rusted has been removed especially in cases where it would be directly over the engine of an aircraft and has been replaced by scrim.'

On 16 July 1943 CECIL PLAINS figured in a review of unoccupied aerodromes in Australia as one of several in Queensland. It was classified as a Heavy Bomber field with buildings and services for 450 men and shown to be 100% complete. December of 1943, the record shows No 14 OBU at LOWOOD was currently charged with undertaking inspections and maintenance repairs of a minor nature at various RAAF stations including CECIL PLAINS. Note that the distance between LOWOOD and CECIL PLAINS is every bit of 160 km (100 miles)!

The very real likelihood of loss through neglect of maintenance and unauthorised entry and pilfering were well appreciated by officers of the Department of the Treasury. A memo direct to Department of Air requested 'consideration and comment' upon a 14 March 1944 inspection of the Downs dromes by a responsible officer -

CECIL PLAINS ca(-/7/1945)
This airfield was not used early in its lifetime, and only came to life with the RAAF developing its Liberator force. Here a B-24J aircraft of 102 Squadron, possibly A72-344, is being serviced on the edge of the 45° strip. Work on engines is in progress, and with jacks under the starboard wing and by the cockpit, it is likely the main landing gear is being checked. Note the camp in the right background. [A Bailey ex RAAF]

CECIL PLAINS ca(-/8/1945)
Entertainment on the base was usually the visit by the mobile cinema unit of the Australian Comforts Fund. Here the vehicle is in place with its on-board projection equipment sited towards the fabric screen up on poles. [A Bailey ex RAAF]

'...Close to the building group at Cecil Plains are four structures stated to have been left by AWC, all of which should be valuable elsewhere. One hut approx 45' x 20' roofed with corrugated fibro cement and sheeted with rough 6" x 1" cyprus pine is packed with sections of portable 3-ply huts, two 3-ply huts approx 18' x 14' are partly filled with ply hut parts, floors, camp fitments, oil drums, etc. and the fourth is a temporary latrine - fibro roof and hessian sides. These items will probably be a complete loss if they are not soon removed and reconditioned... There is lush growth of grass surrounding all these aerodrome building groups and (soon) fire risk will be considerable.'

By 5 September 1944 the place seemed a bit closer to some utilisation as an airfield. A file minute reads -

'...Cecil Plains is likely to be used by Heavy Bomber Units in the future. In the first place, for the formation of RAAF Squadrons prior to movement to northern areas and secondly, as a probable location of Heavy Bomber squadrons arriving from overseas. It is noted however that the 135° runway conforms with the specifications for Heavy Bomber, i.e. 7 000'. In these circumstances additional expenditure on the other runway is not considered justified.'

Extension of the 45° runway was being contemplated but it didn't happen. This mention of 'Heavy Bomber squadrons arriving from overseas' possibly refers to contingency plans for the location, if only for training purposes, of US B-29 Superfortress units in Australia.

The RAAF's No 12 Squadron, which had converted from Vultee Vengeance dive bomber aircraft to the B-24 Liberator first occupied CECIL PLAINS around 14 December 1944. The squadron assembled in early 1945, its CO W Cdr N G Hemsworth arrived on 3 February 1945 and the first Liberator aircraft arrived two days later.

CECIL PLAINS ca(-/8/1945)
Hygiene was always important, but doubtless 'Skippy' has found something in the tent lines worth closer investigation. [D Carroll ex RAAF]

With all the care and effort put into the eighteen aircraft hideouts at CECIL PLAINS, those most responsible fortunately probably never saw a RAAF inspection report of 13 March 1945, nor the 12 Sqn ORB. Apart from the recent wet weather conditions, and some pavement failures generally over culverts draining under the runways, the Inspectors warned -

'...Taxiways are little used and in good order except that hideouts are generally overgrown, but it is not considered necessary to maintain these in view of present changed "war" position. It is obvious that this field is not built to B-24 standard but it will stand up to use by same, provided aircraft are grounded after a fall of say 2" of rain for at least 2 days.'

An initial summary in the ORB described the hideouts as -

'...scarcely wide enough... more suitable for fighters'.

12 Sqn soon found Dalby better placed as its postal and supply centre. The ORB 'Review for February 1945' mentions sporting and recreational activities, the splendid hospitality extended by citizens of the district, and -

'...A fifty percent increase in the Dalby beer ration was negotiated and arrangements made with the local Hotel proprietors to open between the hours of 1700 and 2000 on Friday nights to enable the men to take advantage of this increase.'

12 Sqn moved out at the beginning of May and 102 Squadron commenced to form at CECIL PLAINS at the end of May, with its twelve B-24 Liberator aircraft arriving over the period 4 July 1945 through to 1 August 1945. They were also allotted a twin-engined Anson aircraft and a Tiger Moth. Alf Bailey was a Flight Mechanic and like Goff Bauer arrived at CECIL PLAINS while 102 Squadron was forming during June of 1945.

The squadron commenced its training but never saw active duty. It did however have the honour of flying a nine aircraft mass formation over Brisbane on 16 August 1945 to celebrate the end of the war.

After the war the RAAF did acquire CECIL PLAINS but on a basis of 'acquire only and not maintain'. The Department of Civil Aviation saw fit to use it for a while but the airfield is not used as such today. I understand Mr W H Cook of nearby 'Kantara' property may still own this site. That name appears in records as the WW2 period owner and also as 'caretaker' around 1946.

Where, for a few months Liberators did finally fly, cattle now graze over the cross runways of CECIL PLAINS.

TOOWOOMBA

'...The aerodrome at Toowoomba was not required by the RAAF and there has not been any RAAF occupation there known to this Department.'

In January 1944 the RAAF's Directorate of Works and Building (DWB) responded thus in writing to the section concerned with 'hiring of properties' in respect of property which had been utilised by the RAAF in 1942/43. Local residents of Toowoomba and users of the airport will know that even now 'it's difficult to see one end of the runway from the other because of the hump in the middle'! So it's not surprising that overall, history has tended to lose sight of the fact that TOOWOOMBA did play its part, in WW2.

Early 1942 while AMBERLEY was seeking to satisfy its training field requirement for 3 SFTS, TOOWOOMBA was considered for that role. Patriotic fervour knew few bounds during that year. Mr William Rankin who owned land adjoining the southern boundary of the Civil aerodrome, wrote 12 February 1942, to Gp Capt Lachal CO of 3 SFTS AMBERLEY -

'...Sir, In reference to recent telephonic conversation with your Assistant, in which I expressed my desire to assist the Defence Authorities by placing my Toowoomba property at the disposal of the Air Force without charge for the duration, and to a call today by an Officer of your Department requesting me to place the matter in writing, it gives me pleasure therefore to confirm my offer as follows: a) To give my property at Toowoomba, known as Werrington Park, to the Defence Department for use of the Air Force unconditionally and absolutely without charge for the duration of the war and for six months thereafter...'

Various fences were legitimately removed to allow the dispersal and better operation of 5 Sqn's Wirraways. Inevitably, cattle made it onto the aerodrome, at times causing the obvious hazard. By June 1942, 3 SFTS was disbanded and DWB was advised by the Director of Training, of no further requirement for TOOWOOMBA. 5 Sqn was another matter, which was similarly resolved in a few more months. So, with no RAAF operation, the cattle took over. When Mr Rankin approached the 'powers that be' to have various fences re-erected, he was confronted with outright denials that his land had even been used by the RAAF! However persistence won out and satisfaction was finally achieved.

Colin Munro recalled how he came to be here -

'...I was back at Laverton in Victoria during the latter part of '41 for the formation of No 5 Army Co-op Squadron. With the rank of Flight Lieutenant I took over A Flight and moved to Toowoomba in Queensland on 12 May 1942. The Squadron operated Wirraway aircraft and commenced training with the Australian 1st Army Corps. principally the 7th and 9th Divisions camped and dispersed generally east of the Dividing Range between Beaudesert and Gympie. The essence of the Squadron's work was to provide tactical reconnaissance often involving aerial photography by the observer in the rear seat of the aircraft.'

Photographs of 5 Sqn's tenancy are rare. Colin Munro was able to recall the disposition of at least two Flights during their period at TOOWOOMBA. Flight A was quartered and its aircraft were dispersed in trees across the Western Highway some 200-300m north of the Wilsonton

← **TOOWOOMBA** (-/9/1943)

This is the compilation date of the 1:63 360 (1 inch = 1 mile) Military Survey Map showing the strips in outline only. This map extract reaches southward just far enough to show the RAAF No 7 Stores Depot developed at Drayton, particularly the large rectangular buildings hard on the eastern edge of the railway line.

[Department of Defence]

TOOWOOMBA *(1942)*
RAAF No 5 AC (Army Co-operation) Squadron members relaxing close to their dispersal area. From left to right Ian Olorenshaw (on the swing), P-O John Unkles, P-0 Bob Wilkinson and F-Lt Colin Munro. In the background it is possible to make out hessian drapes which may well be screens around latrines. The squadron was flying Wirraways from TOOWOOMBA during this period. [I Olorenshaw ex RAAF]

Post Office. Colin recalled the other Flight as having an area generally west of the airfield, again in scattered timber. Their operation in support of Army formations, relied heavily on mobility. The aircraft had to be able to move to and work from rudimentary strips, but there was a background need for a 'home base'. TOOWOOMBA served that purpose for a while until the Sqn 'looked at TOOGOOLAWAH' and moved on to KINGAROY and northward.

During the rest of WW2, most mention of the RAAF at Toowoomba is of the large No 7 Stores Depot that was developed at Drayton to the south. That centre had no direct communion with the airfield.

Today the TOOWOOMBA airfield serves quite adequately the small regional commuter airline activity and private charter requirements. Interestingly, it is a developing hub for recreational flying and the private restoration of antique and classic aircraft. While active under his Eastland Air Charter banner, Jim Czerwinski persevered for years to restore a Mk16 Spitfire in his 'off' time. The aircraft still requires work to see it back into the air. Well known 'warbird' owner/operator Guido Zuccoli has also chosen this airfield as 'base' for his impressive and growing collection of airworthy machines.

TOOWOOMBA *(5/10/1945)*
This extract is from systematic mapping photography and shows the four graded strips, criss crossing the airfield with the white circle marker in the centre. [AUSLIG]

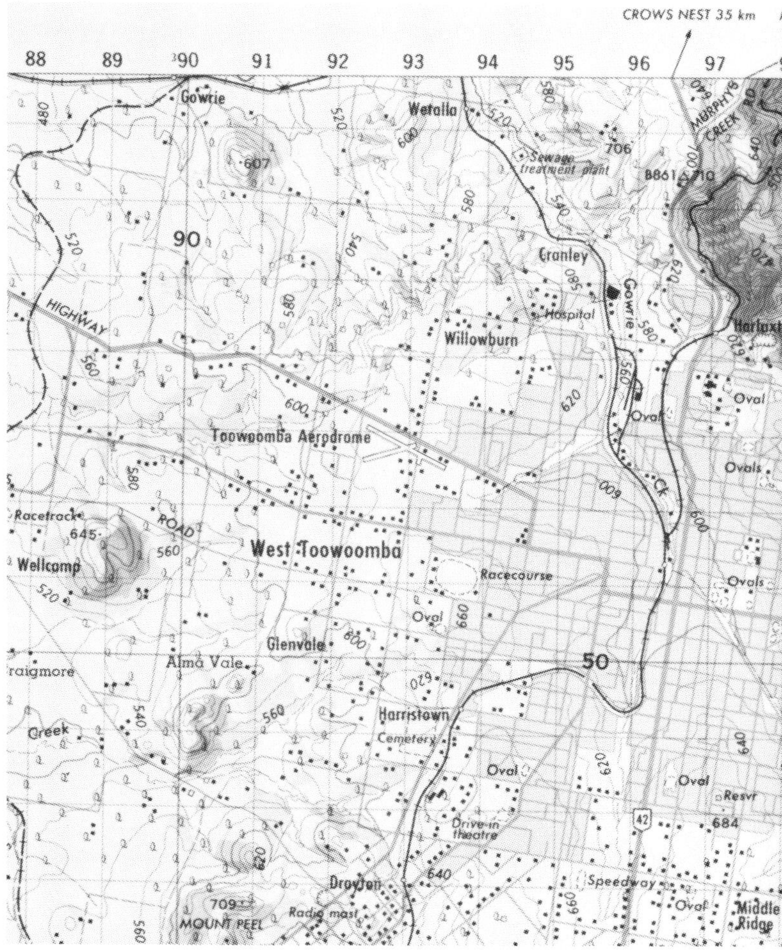

TOOWOOMBA *(1974)*
More recent 1:100 000 topographic mapping shows the current day strip in outline. The RAAF's WW2 period 7 Stores Depot location is grid 933482. [AUSLIG]

ARCHERFIELD *(-/1/1988)*
Pictured here with some interesting background is an airworthy Wirraway owned by Jeff Milne of Albury. '652' was visiting to take part in an aerial display to mark the opening of Sanctuary Cove marine residential development. Largely obscured by '652' is Col Pay's MkVIII Spitfire. The gabled hangar now serving Royal Queensland Aero Club is of historical interest in that it originally served as a hangar at the 1930s period EAGLE FARM airport for Brisbane. The building was relocated to ARCHERFIELD in 1931. [M Denning]

ARCHERFIELD

'...The base was commanded by an RAF Officer (on loan) W Cdr W R Hartwright, the EFTS by Squadron Leader G C "Skipper" Matthews (a WW1 AFC pilot) and I commanded 23 Squadron. On 7 December the Japanese attacked Pearl Harbor and we were immediately put on "war alert". The aircraft were dispersed around the boundary of the airfield to minimise damage in case of attack and duty crews, for both Wirraways and Hudsons, were on immediate standby at all times. Seaward patrols looking for enemy shipping and subs were instituted by the Hudsons and the Wirraways were ready to protect Brisbane... Narry an enemy ship nor aircraft was seen... We were very disappointed that we were so far away from the war.'

This was Deryck Kingwell's opening to me when I asked of his time at ARCHERFIELD. He need not have been too worried about getting closer to the action because after some few weeks he found himself commanding the RAAF's 32 Sqn and up at Port Moresby.

The RAAF's own Airfield Data Sheet of 22 July 1943, just after extension to the W and NE, described an -

'...all over field, no runways or strips with runs up to 6 000'... useable at all times except after exceptionally heavy rain. Aircraft do not bog (but), have tendency to slide on greasy surface during and after heavy rain'.

Twelve months before, the best this could have read was about 5 000', with the wind right! ARCHERFIELD had become Brisbane's airport with the discarding of the EAGLE FARM site. It was home to the civil airline operators, among them, QANTAS and AOA/ANA, both well established there. While AMBERLEY was being built, ARCHERFIELD hosted 23 Sqn and 3 Flying Training School (FTS), soon renamed 2 EFTS.

ARCHERFIELD *(27/11/1941)*
Put up towards a mirror the top edge title of this photograph reads 'Decoy aircraft Archerfield'. It's not clear which are the decoy and which are the real aircraft, so possibly some benefit would have come from the deployment of 'decoy' aircraft! Note the non-camouflage painted state of the buildings. Like many other prints held in the collection at the National Library of Australia, it has only been possible to date this photograph by reference to the RAAF Survey Negative Register now held by AUSLIG in Canberra. The road running northward along this eastern side of the airfield is Beatty Road and the nearest residence to the camera on its right side is believed to have belonged to Mrs Grenier. The palm tree in front of this house appears in another photo here presented. *[NLA (RAAF Official)]*

ARCHERFIELD *20-25/11/1941*
Indisputably this is ARCHERFIELD, looking south to 23 Sqn's buildings, by Mortimer Road. Trevan Jackson gave me the photo ***'taken from an open window, upstairs on the unfinished Administration Building'****, which survives to this day. Comparison with other photos 'dates' this as certainly before March 1942 and likely before 27 November 1941. The aircraft is likely to be a US B-17D Serial 40-3067, probably then under the command of Lt Mel McKenzie, in process of transporting Maj Gen Lewis H Brereton on his brief but momentous visit to Australia, from the Philippines.* *[T V Jackson]*

Cyril Broome was Chief Flying Instructor with Royal Queensland Aero Club at ARCHERFIELD at the time WW2 broke out in Europe. He was instructing cadets who had undergone initial training at Point Cook. He recalled -

'...The following Monday I carried on flying instruction as "3 FTS" with my eight cadets at the time.'

Training aircraft from ARCHERFIELD used EAGLE FARM as it was then, for landing practice and a licensed ELG at Runcorn, near the railway station.

Eric Read was a young pilot in 23 Sqn here in 1940. He recalled the period when, as soon as construction at AMBERLEY produced an adequate landing surface -

'...we used to fly their mail over to them in a DH-94 Moth Minor aircraft. We also flew Wirraways in and out of Eagle Farm at that time. I recall a programme of testing parachutes using a 200 lb wooden dummy - known to us all as "LAC Wood".'

Eric flew sorties out of ARCHERFIELD with Sqn Ldr J F Jackson, September 1940, in search of bomber airfield sites. This was also a time of seeking auxiliary strips close to Brisbane. Sites near KINGSTON and west of AMBERLEY at Rosewood were discounted while one at STRATHPINE was viewed more favourably.

On 21 November 1940, RAAF Station HQ ARCHERFIELD became aware of a 'strange ship 75 miles east of Noosa Heads steaming south'. Japanese shipping still plied this coast a full year before Pearl Harbor. The Operations Record Book (ORB) entry read, in part -

'...Ship was intercepted at 1315 hrs, 115° Point Lookout, 59 miles. Identified as ***Hitati Maru****. Photographs of ship sent to HQ by Air Mail.'*

A RAAF file Minute dated 21 March 1941, demonstrated the day to day development then of ARCHERFIELD. Gp Capt Knox, for DWB, queried the AMOE -

'...In view of the fact that it will be some time before service squadrons are transferred from Archerfield to Amberley, should we proceed with the proposal for one squadron bomb dump at Archerfield please? '

Air Marshall Richard (later Sir Richard) Williams answered promptly that expenditure on bomb dumps -

'...at Archerfield is not justified at present.'

Delays in the completion of AMBERLEY had caused the temporary basing of 23 Sqn, a 'service' squadron, at ARCHERFIELD. Air Board agenda No 3276 dated 29 April 1941, under the signature of Air Marshall Williams, shows some of the chess-like moves made -

'...the demands of the EATS (here) resulted in a proposal to locate No 3 SFTS at Amberley, it being the

surest way of having that school functioning in the time laid down in the programme. 23 Sqn was left at Archerfield and an EFTS was also developed there. Subsequently "Z' scheme called for another Service Sqn in the vicinity of Brisbane. This could not be located at the Civil Airport at Archerfield and, as that airport was becoming congested, it was decided to look for a site near Brisbane which would house two Service Sqns and take 23 Sqn from Archerfield...'

The chosen solution was to develop KINGAROY for the SFTS role, use AMBERLEY for two Service Squadrons, and leave 2 EFTS at ARCHERFIELD.

General Index Cards of the Works and Services Branch of the Dept of the Interior, now held in Australian Archives, indicate a move to construct 'A-A machine gun emplacements' at ARCHERFIELD in July of 1941. Dated 7 August 1941, those same cards indicate the issuance of instructions for preparation of EAGLE FARM aerodrome as an RLG for No 2 EFTS. That same Department's Works Files, on 27 August 1941 referred to -

'...requisition DWB 2640/41-42 for improvements to the surface of the RLG at Eagle Farm. The site was recently inspected by F-Lt Towers who reports that the surface can be made suitable by grading off the minor undulations and rolling... As this area is in constant use by 2 EFTS Archerfield, arrangements should be made with... (them), that works in hand are so arranged as not to interfere with training operations.'

1941 saw the construction of the new administration building at ARCHERFIELD. Architecturally it mirrored the state of the art in such buildings with forerunners at Mascot and elsewhere in the country. On its commissioning late 1941, the 'Control' function passed from the previous tower structure just north of the new building, to the glass cubicle on the new roof.

By now events in the East Asia region and America's diplomatic responses were pointing to a conflict with Japan. 23 Sqn's ORB for September of 1941, indicates aircraft involved on at least six occasions on 'blackout checks' over southeast Queensland. The community was being educated to a wartime need for such precautions. Immediately after Pearl Harbor, excavation of slit trenches and air-raid shelter construction came swiftly.

ARCHERFIELD (23/3/1942)
This is one frame from a mosaic of the whole airfield. With the outbreak of war in the Pacific came the urgent camouflage painting of building roofs, etc. and the worm-like 'excavations' which are actually air raid personnel protective trenches. The several rows of tents appearing on the southwestern quarter of the road intersection (Beatty and Mortimer Roads) comprise the embryo American Forces camp which developed over in the SE quadrant into 'Camp Muckley'. The buildings in the southeastern corner of the airfield are essentially those erected for the RAAF's 23 Sqn. At the northern edge of this photograph the building on extreme left is the airfield administration and control centre with tower facilities on the roof. In the open paddock area east of the field the large igloo structures of the Kerry Road complex are yet to be commenced. The 1930s quarry (4 o'clock-centre) is filled with water and a rectangular benching appears on its northern edge. [NLA (RAAF Official)]

ARCHERFIELD (1942)
Believed to be Lt Paul C May, this US airman is photographed atop a Douglas A-20 twin-engined medium bomber in front of ARCHERFIELD Administration Building. Note the bomb splinter-proof high wall erected around February 1942, in front of the building's ground floor. As the risk of enemy air raids receded, this protective wall was removed, though it's not clear when this was done. (T Whyte ex RAAF]

A US convoy of eight ships docked in Brisbane on 23 December 1941. This was the *Pensacola* Convoy, so named for its escort the cruiser USS *Pensacola*, enroute for the Philippines and diverted to Brisbane. While the US Archive sourced ***History of the Fifth AF (and its predecessors)*** credits the US Military Attaché in Australia Col. Van S Merle-Smith with 'securing of suitable erection facilities...(at) Amberley and Archer Fields', it appears most, if not all of this convoy's seventy crated A-24s and P-40s, were trucked to AMBERLEY for erection. ARCHERFIELD later shared in the erection of both fully crated and deck cargo delivered aircraft.

The General Index Cards show 10 February 1942 as the 'date of communication' associated with -

'...job L-138: splinter proof protection to control building by landscaping around ground floor and taping all windows.'

Such 'landscaping' appears to have included earth filled timber sheeted traverse walls.

February 1942 was a busy month, in which Deryck Kingwell handed over command of 23 Sqn to F-Lt M V Lewis. On 17 February initial parties of Dutch airmen arrived aboard three Lockheed Lodestars - two days before the Japanese bombed Darwin. The day the Dutch arrived six B-17s of the 88 Recon Sqn, 7 BG, landed. A journal attributed to Lt Walter H Johnson described their island hopping journey, their stopover on the politically unsettled French island of New Caledonia and their arrival over Moreton Bay to find rain over the mainland -

'...While five ships circled in the bay, Maj (Richard H) Carmichael flew in under the inland rains to search for the airport. It was 45 minutes before he came back to lead the flight inland through the rough air. He had found Archer airport and there on its grass, we set down... Townsville was being hit by winds of hurricane proportions... So we stayed in the tents at the American

ARCHERFIELD (-/9/1942)
This photo was offered with caption ***'B-17F late '42 at Townsville?'*** *Clearly the scene is ARCHERFIELD with the characteristic administration and control building in left background and also without any evidence of ANA's large steel frame saw-tooth roofed hangar built in 1943 just south of this administration building. Discussion with Bruce Hoy indicates that this B-17F 124548 flew to Australia in September 1942 and was assigned initially to the 403 Bomb Sqn. Some time later in its career it was possibly named 'War Horse'.* *[R Williams via D Vincent]*

camp, tents with grassy rain soaked floors. During the early evening an Australian Airliner ploughed into (Harold N) Chaffin's ship, damaging a wing, the tail and part of the fuselage. This ship was parked, no one was in it. The liner also knocked out of the war a Dutch bomber parked alongside Chaffin's ship.'

The Station HQ ORB entry that day read, in part -

'...AOA aircraft DC3 VH-ACB, pilot Keith Virtue, whilst taxying collided with USA B-17E No 12430 and Dutch Lockheed Lodestar No LT922. Fuselage of the latter completely wrecked. Starboard wing, elevator, fuselage and tail plane of B-17E damaged.'

No injuries were recorded in the DC3. Joan Priest's ***Virtue in Flying***, published in 1975, quotes Keith Virtue -

'...There was three or four inches of slush and water sitting on top of the hard ground. We could see all the water and decided not to use the flaps, as I might break them. So I landed towards the east, and we didn't know it but three Fortresses had arrived from the American Air Force, and Dutch aircraft had parked just around where I'd be finishing my run. I saw them in time but the brakes were no good on that greasy stuff, so I tried to ground loop to dodge them but I couldn't We went through a Lockheed Lodestar belonging to the Dutch Air Force (and my old acquaintance, Van Messel, was very unhappy), damaged a wing on a Fortress and poked the wing of the DC3 into the side of it. The Americans weren't happy either but I hadn't been warned they were there.'

ARCHERFIELD ca(-/3/1943)
Here General Douglas MacArthur has just completed the field award of the Distinguished Service Medal to Lt General George C Kenney. Kenney in mid salute, is standing in front of the crew of 'Sally', a B-17E assigned to duty as his personal transport. The members of 'Sally's' crew (from nearest the camera) are pilot Wilbur B Beezley, navigator (2nd) Walter R Friederich, co pilot (2nd) Ken Rasmussen, radio George R Graf (only shoes showing), engineer (2nd) George Powell, tail gunner (2nd) Shelby Griffin and side gunner Donald Aderman. Behind General MacArthur are the characteristic arch frames of construction work on the initial pair of igloo hangars for the Department of Aircraft Production, Kerry Road. Under 'Sally's' nose is Mrs Grenier's palm tree. *[G R Graf ex USAAF]*

The repair of 41-2430 must itself have been an interesting if not impressive saga in its own right. Lt Johnson's journal has it arriving at Townsville to rejoin the body of the original 12, on 27 February 1942. Some four more days ground work were needed to make it combat worthy again and ready for 'dispersal to CLONCURRY' with the others. Joan Priest's account mentioned it was 'flying again in two days'. After Virtue's incident at ARCHERFIELD there were five B-17s flyable and Johnson says all soon transferred to AMBERLEY 'which had paved runways'. 41-2430 is reported at AMBERLEY on 22 February and if that is correct, just how it got there is not clear to me. I fancy I heard or read of an outer wing panel being airlifted from a derelict B-17D up Darwin way, to replace the damaged one. Such damage repair stories are legion and often the start of 'legends'.

Greg McGregor was a Fitter/Armourer posted to 23 Sqn direct from Armament School Point Cook. He wrote -

'...I arrived at Archerfield on 25 February 1942. 23 Sqn was equipped with Wirraways fitted with 2 Vickers Mark V 303 machine guns firing through the propeller with synchronising gear. The observer had either one or two Vickers gas operated machine guns. Bomb load at that time was 2 x 250 lb bombs or 8 practice bombs.'

Greg's diary notes are short and typify the daily work -

'...February 27: belting .303 ammo for Vickers machine guns and March 4: worked on bomb dump filling practice bombs. Mar 5 and Mar 6: bomb dump. Mar 7: dentist, tooth ache and also Mar 8. Mar 9: 3 North American B-25's and 1 Boeing B-17E Fortress arrived. Mar 10: 1 Douglas Dauntless, 1 B-17E, and several Bell Aero Cobras arrived. Small twin engined Beechcraft tested and flown to Philippines to pick up officers, wives and families. Mar 11: leave-in Brisbane. Mar 12: leave - trip around Moreton Bay.'

The B-25s were possibly the earliest delivery of those purchased by the Dutch, probably flown via West Africa and 'taken over' by the US for immediate use by the 3 BG.

Recorded as being formed at AMBERLEY on 16 March 1942, the RAAF's 76 Sqn is shown in ORBs for ARCHERFIELD as arriving there on 14 March. On 16 March Greg McGregor was posted to this squadron being equipped with some of the RAAF's first Kittyhawk (P-40) fighters. His diary entry says 'posted to 76 Sqn, saw Kittyhawks close up for first time on guard duty'.

Dated 22 March 1942 and under the signature of F-Lt S L Dawkins for the AOC NEA, the Dept of Interior was instructed on aircraft dispersal works -

'...Mr Crawford (MRC) knows the schemes and in conjunction with the COs, could formulate the final details necessary. Air Board has been requested to provide funds £15 000 for Amberley and £16 000 for Archerfield but owing to the extreme urgency of the case it is requested that work be commenced immediately and that all available plant and material be utilised. This is a matter of real urgency and it is requested that work be commenced without delay.'

Greg McGregor's diary notes show March continuing with a rising background of activity -

'...Mar 18: worked on Wirraway A20-317 - took off electro magnetic bomb releases, overhauled synchronising gear. Mar 19: spent morning working on Kittyhawks and learning armament - damn near got shot when Browning .5 went off. I wonder where it landed. Got paid. Mar 20: working on P-40's. Mar 21: took guns out of P-40 - on guard duty - then another turn of guard duty because a bloke got sick. Mar 22: leave in Brisbane, swimming at the Valley Baths. Mar 23: busy day - put guns back in P-40, 4 Martin B-26 Marauders arrived and a nice looking aircraft they are. Mar 24: busy day on P-40s. Mar 25: very busy - rained heavily, 4 - B-26s arrived over Archerfield and 2 tried to land on the wet grass field. The first one is a write off, landed, then braked and skidded on wet grass, hit the tail of a DC3 and then went through the fence

across the road and second fence and ended up with nose inside of a house. The second B-26 also skidded on wet grass and smashed into a B-25 Mitchell. Fortunately no fires. Then a B-25 Mitchell stalled and pranged on take off - interesting day.'

It appears that while Greg McGregor was swimming at the Valley Baths on 22 March, 'A' flight, down to two B-26s of 19 BS, 22 BG, arrived. Led by Walter Maiersperger, these were very likely the first B-26s to arrive in Australia having winged their way across the Pacific from Hawaii. Walter explained he arrived at ARCHERFIELD on 22 March, transferred to AMBERLEY on 26 'for more room' and then on to GARBUTT on 28 March with his trans ocean wingmen, Chris Herron (No 2) and Robert Hatch (No 3). Herron's aircraft had blown a tyre on landing in New Caledonia. It was only as a result of Maiersperger's round trip back there on 24 March, with 'jacks and a spare wheel', that Herron was able to rejoin the trio. RAAF ARCHERFIELD records show four B-26s arriving on 22 March, four on 24 March and another three on 25 March. Although the dates are not likely to be wrong, it's quite possible Maiersperger's aircraft was double counted when he returned on 24 March. In any case it appears two other B-26s arrived on 22 March, unbeknown to the 'A' flight pair. Though a competitive claim appears on behalf of Franklin S Allen in the 1980 publication entitled ***The Martin Marauder and the Franklin Allens: A Wartime Love Story***, the Individual Flight Record (US Form 5) for Capt Frank Allen clearly dates his arrival at ARCHERFIELD as 22 March 1942. Walter is confident his flight was the first of the 22 BG aircraft to make the long ocean crossing, proving and adjusting the route for the others to follow.

Walter well remembered being surprised by the short grass field at ARCHERFIELD but the CO of the 33 Sqn was the only one to overshoot. In his 1992 biographical note on Maiersperger, titled ***The Americans Arrive at Archerfield***, Ted Wixted added -

'...33 Sqn CO 1st Lt (William A) Garnett (and) Co-pilot Larsen... continued through the fence and across the (Beatty) road to make the famed intrusion into the local Barber shop.'

Trevan Jackson, a welder with AOA/ANA, described the result in his unpublished memoirs -

'...A skidding Marauder shot through the fence across Beatty Road, spun around and put the nose through Butch Freeny's kitchen and crushed his water tank with a wing leading edge...'

Freeny's barber shop was on the corner of Beatty and Boundary Roads with the residence behind. After this incident and another involving a Lockheed Ventura with Gough Whitlam, later Prime Minister of Australia, in its crew, and 'several trailing aerial strikes on the roof', Trevan told me the house was relocated!

McGregor's diary tends to confirm B-26 numbers -

'...Mar 26: very busy - we now have all the Kittyhawks serviceable. Had pep talk from CO - appears we are getting brand new aircraft tomorrow and will go north shortly. 8 B-26's took off probably for Amberley.'

A Dept of Air memo to Dept of Interior, dated 30 March 1942, suggests the Dept knew the airfield was too small. It enclosed requisition number -

'...DWB 1025/41-42 for extensions to the landing area at Archerfield aerodrome, in accordance with drawing 41-42/2182 at an estimated cost of £10 000.'

ARCHERFIELD (15/1/1990)
Rebuilt by Ross Smith of Rolleston, Queensland, this classic Boeing Stearman stands roughly where General MacArthur decorated General Kenney back in 1943. Seen through the port wings are the still serving Kerry Road igloos. [Author]

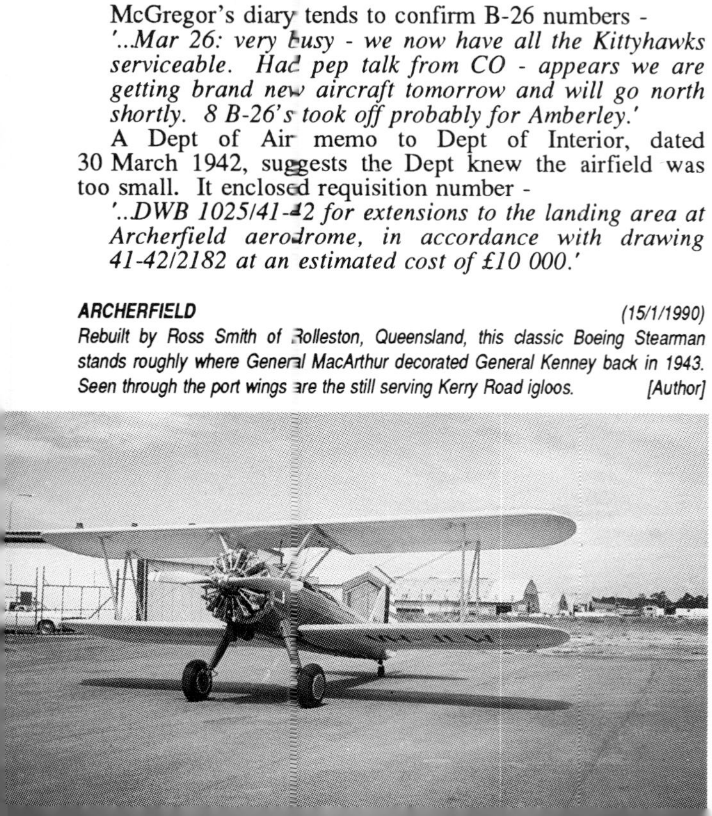

ARCHERFIELD ca(-/1/1944)
This complete photograph is presented partly to show the value of identification inserts in its top border. 'Sally' is the name of the B-17E assigned as personal transport for General Kenney. The aircraft serial 12633 features in the right hand slot. The roll number here shown as Roll V, has helped in dating this photograph. While it may not be possible to discern much detail of the Kerry Road igloo development, it is clear the middle two hangars are complete, that the far left (eastern) hangar is only half built, and that the far western one has not yet been erected. Richard Graf, the photographer, explained this 'event' related to experimental air dropping of crated supplies without a parachute. Unfortunately it's not possible to see the scattered and dismal result on the ground. [G R Graf ex USAAF]

The extension on the northeastern corner, involved the severance of Boundary Road westward from its intersection with Beatty Road. The gravel pavement was interestingly preserved, although graded over, so that from the air there was some appearance that Boundary Road still existed and suggesting that the airfield did not extend beyond it. The old Aero Club buildings on Boundary Road were shifted.

March 1942 also saw strengthening if not the very establishment of some anti-aircraft defences in the Brisbane area. A 1946 US Army publication ***Anti-Aircraft Artillery Activities in the Pacific War***, lists various units assembling here from March 8, most -

'...armed with .50 calibre A-A machine guns only. The Commanding General 40 CA Brigade (A-A) was designated as Commander of the North Eastern Command which unified the operational control of Australian and American Anti Aircraft troops from Brisbane north. The 94 CA (A-A) with the 101 and 104 CA Battalions (A-A) and several Australian units was disposed to protect the vitally important wharf and warehouse areas and the Archer and Eagle Farm and Lowood aerodromes in and near Brisbane.'

Two WW2 period Bellman Hangars remain to this day. I had hoped they were the two in the 30 June 1943 photos from AA. The following extract from a 1 April 1942 letter to the Dept of Interior suggests otherwise. From the Aircraft Production Commission (later DAP), it appealed for the priority allocation of the services of a -

'...local building firm for the fabrication and erection on Archerfield Aerodrome of a Bellman Hangar... The hangar space at present available to AOA at Archerfield... for the repair of RAAF aircraft has for some time been totally inadequate for the purpose, and aircraft urgently needed by the Air Force have consequently taken much longer to repair and return to service than they should. AOA Ltd are the only civilian contractors in Queensland approved by the Dept of Aircraft Production and the Air Board for the repair and overhaul of Douglas DC-2 and DC-3 transport

ARCHERFIELD *(30/6/1943)*
Inside the Bellman being operated by ANA as an Airframe Repair shop. Both aircraft in the photo are B-25 Mitchell medium bombers. Through the open door in the rear wall it is possible to see south to the entry into a further Bellman which was operated by ANA, as a Sheet Metal Repair shop. *[AA BP34 Item 2012]*

aircraft. As the RAAF have only a few of this type of aircraft,they can ill afford to have them laid up for long periods owing to congestion in the hangar of AOA Ltd.'
The General Index Cards show a 6 April 1942 communication from the 'Air Product' Com' for two new hangars. Trevan Jackson confirmed the location of these two new Bellmans as 'over near Mortimer Road'. A couple of months after, the General Index Cards show registration of Job S7 'Erection of Hangar and Workshops for Australian National Airways Ltd'. This saw-tooth roof structure, the largest free span hangar then on the airfield, was to take just over twelve months to complete.

Greg McGregor's diary covers 76 Sqn's departure -

'..Apr 13: very busy, pilots doing gun firing practice and we had a hard job refilling and rearming the aircraft all day. Too tired to go out. Apr 14: busy day, half the squadron went on leave, packing gear. Apr 15: local leave. Apr 16: busy day getting clearances from Archerfield, left at 1800 hours and were given (an) hours leave in town.'

A week after, 2 EFTS was disbanded.

The State Defence Camouflage Committee had commenced its function before the outbreak of war in the Pacific. The minutes of its twenty seventh meeting dated 22 April 1942 are of interest in their listing of attendees and some mention of this airfield. Present were -

'..R A McInnis (Chairman), R P Cummings (Civilian Member), J D McColl (Dept Interior), Lt Cdr V W Bowden (Navy), Lt A T Longland (Army), A B Leven (State) and Dr W H Bryan (Specialist Advisor), W Cdr Hartwright was unable to attend.'

From Army camps to woollen mills, and sugar mills to wharves; all came in for attention. On ARCHERFIELD, the aerodrome was being extended -

'...across a bitumen road and a gravel road and Mr McColl informed the committee that the existing roads are to be retained. W Cdr Hartwright (had) informed the chairman that he has a photo of the drome but could not allow a copy to be taken away. There is no photo of Loganlea satellite and no planes available to take a photo...On 13 April Messrs (O J) Besly and (R F) Gordon put their cars in commission again to assist them in inspecting dispersal areas and hideouts but this work was held up owing to shortage of petrol... Later the Chairman and Mr Cummings inspected the dispersal area and found that the taxiway leads direct from the drome to a small area thickly studded with hideouts, no effort has been made to plan it for concealment.'

This first dispersal area was close to Oxley Creek just south of the now popular Archerfield Speedway.

Works files of the time show some of the rising intensity of construction work. Contractor H Trelour had earlier undertaken 'erection of 49 buildings' here for the RAAF. He had left to commence urgent RAAF work at LOWOOD. Arrangements he'd made with another contractor to finish off a then 'temporary Headquarters', failed. Technically, Mr Trelour was responsible but the Dept of Interior's architectural supervisors finally had the outstanding work done by others. Dated 26 May 1942, the Inspector's report duly recommended Trelour not be charged the £7 because he'd not been formally cautioned. A week later Japanese midget submarines raided Sydney Harbour, also without any caution! Such stunning developments had little effect on other important matters such as 'compensation re Wotley's well'. Allied Works instructed MRC, 13 June 1942 -

'...Wotley's well approx 42' deep is included in the acquisition of areas involved in extensions of Archerfield aerodrome. This is their only water supply - a suggestion of town supply being connected inflicts on them an unnecessary expense of water rates which would necessitate compensation being paid by this Department to the extent of £400. It is deemed advisable to construct another well and you are instructed herewith to carry out this work immediately at the site indicated by the water diviner and as shown your foreman by Mr Mowbray of Civil Aviation.'

This all shows that nothing much changes with time.

A little of the camouflage activity can be gleaned from this extract from R A McInnis' report, 20 June 1942 -

'...Archerfield Aerodrome - SE dispersal; only 4 hideouts remain to be treated; work is proceeding. Mr J Goldburg camoufleur, took over from Mr O J Besly... on 19 June 1942... P-O Keays also had a discussion with Mr Cummings regarding Archerfield and Amberley. They waited the rest of the week to contact Mr Hughes of the US Camouflage Section who was to come from the north and inspect Archerfield and its satellites with them. He did not arrive. Mr Annand who was to come down and initiate Mr Wyeth into this work has not come. I have not wired him to do so as the work he is doing in the north is more urgent and Mr Wyeth is able to handle the work here until Annand can get away.'

On 1 July 1942 the RAAF closed down its Station and ARCHERFIELD was then given over entirely to US operations, the ongoing civil airlines' needs and importantly the established repair and maintenance facilities. John Keays' report of 10 July 1942 mentioned -

'...Landing area has been extended but new area is not available for use because of continued wet weather. Station has been taken over by USA who are reconstructing dispersal scheme...'

On 19 July 1942 the regular Camouflage report added -

'...photos of SW and SE dispersal areas received from Maj Taylor US Army on 17 July 1942... On 13 July Mr Wyeth inspected Loganlea and Waterford which were (taken over by the Americans)... as dispersal strips for Archerfield, with Mr Cummings.'

Of the eventual three Bellman hangars erected in the SE corner of the airfield, near Mortimer Road, one was complete around December 1941. Mid 1942 ANA occupied it as a DAP Sheet Metal workshop. The Camouflage report of 26 July 1942 says -

'...A new hangar and some smaller buildings have been painted to conform to general toning down scheme.'

It appears this too was given over to ANA for air frame repair work for the DAP.

ARCHERFIELD *(30/6/1943)*
*Australian Archives caption shows this as **'Exterior of Bellman hangar, Archerfield, operated by ANA as Airframe Repair Shop'**. It appears to have been 'building 70', of which only the floor slab is clear near Mortimer Road, in the 23 March 1942 photo. The aircraft is an American Liberator.* *[AA BP34 Item 2001]*

ARCHERFIELD *(27/4/1944)*
Extracted from the systematic mapping photography of the 'Beenleigh' sheet, this photograph shows the continuing 'all over field' nature of the landing area. The north easterly extension can be seen with its angled western boundary and the purposeful retention of the graded out section of Boundary Road crossing that part of the field and tending to perpetuate its function as a road - all part of the camouflage effort of the day. The Kerry Road complex stands out with its extensive reflective hardstand surrounding the four building sites. Work on the western hangar is underway. In the extreme southeast corner of the photograph, the aircraft dispersal taxiways and hideout locations can be seen and a similar loop taxiway shows in the bottom left near Oxley Creek. The taxiway to the south eastern dispersals passes close to the old quarry, with the earlier (23 March 1942) benching now not so evident. Allied Works Council (AWC) Minutes for 8 January 1943 show that, although the project had not been properly 'requisitioned' a bomb proof building for US Air Corps Operations/HQ had by then been built! The nature and precise location is not known, but it may have been built in this quarry bench, as certainly that was the case at AMBERLEY. *[AUSLIG]*

ARCHERFIELD *(19/8/1943)*
The second of ultimately five igloos in the Kerry Road complex is here under construction and QANTAS are already at work. While by this time EAGLE FARM was strongly operating as an aircraft erection depot, the significant maintenance and airframe repair resources of civilian organisations such as ANA and QANTAS were much in demand at ARCHERFIELD. The building is now owned by Hochtief Ltd a major construction company in Germany, and affiliate of Thiess Contractors P/L. In the right background can be seen the characteristically curved roof of the Piper hangar over at the original ARCHERFIELD hangar complex. *[AA BP34/1 Box 4]*

In September of 1942 ARCHERFIELD was host to the RAAF's 22 Sqn freshly equipped with Douglas A-20 Boston aircraft. They had arrived from Richmond on 4 September and departed about three weeks later. At that time F-Lt W (Bill) Newton, later to be awarded a posthumous Victoria Cross (VC), was on the strength of this famous Australian unit.

Interesting background to the imminent development of the Kerry Road complex appears in the Camouflage report for 13 December 1942. A Mr Meldrum of DAP -

'...interviewed W Cdr Towers, F-O Harrison and Mr Wyeth, also the Deputy Director and Mr Cummings re area for Dept of Aircraft Production Hangars at Archerfield. The proposed site was close to, or actually in area set aside for aircraft dispersal. Other areas were suggested by F-O Harrison and inspected. Messrs Meldrum and Wyeth and F-O Harrison made reconnaissance and discovered that Civil Aviation contemplated extension of runways east of Beatty Road, and action is being taken by them to acquire a strip of land 15 chains deep parallel with Kerry road... (It) is to become the main access road to the drome and at some future date it is intended to move some of the existing hangars at Archerfield and place along this road...'

Contractor William Hughes & Co P/L commenced the first of the igloo works around February 1943. There were eventually five large igloos and many ancillary buildings. The igloos are a testament to the then Commonwealth Government's philosophy of dual purpose planning, of high cost projects for wartime and post-war. While they and their surrounds have been 'reworked' over the years, minimal dry rot in some original heavy plywood gussets and break-up of some concrete floors are about all that one owner, the present Caterpillar equipment distributor, Hastings Deering (Aust) P/L, can point to.

John S Alcorn in his 1981 publication ***The Jolly Rogers*** describes 'the nose turret mod' to the B-24D. This adding of a powered twin .50 cal tail turret to the nose became somewhat obvious to Col Arthur H Rogers around August 1942 as the 90 BG was working up in the US. While the idea was slow to advance in semi-official circles, Rogers revived it here early 1943. By then he had first-hand experience to add to conviction in a theatre where the 5 AF Commanding General was already 'on side' to extra

ARCHERFIELD *(1945)*
Precise dating has been difficult on this low level oblique. It shows the completed Kerry Road complex and its taxiway connection to the airfield. Notice on near top left, the large saw-tooth roofed ANA hangar erected in 1943 and removed to EAGLE FARM by 1948. The prominent 'hole in the ground' in centre of photograph is, in fact, the residue of a railway ballast quarry which operated in the 1930s and here the presence of a rectangular building is seen sited in the 23 March 1942 benching. Recently the water level was pumped down in an effort to check for dumped wartime equipment. *[A Bailey ex RAAF]*

ARCHERFIELD *ca(1945)*
Viewed from the north on Kerry Road, this shows the large stormwater drain laid under this last constructed extreme western igloo in the complex. The building currently serves Thiess Contractors P/L as a repair/paint shop. *[T Kayser]*

frontal armament in his medium bombers. He found an enthusiastic ex CO, Brig-Gen Carl Connell in charge of the 'local (Brisbane) Air Depot'. Alcorn records 41-23765 as the first SWPA B-24 so modified and freshly named 'Connell's Special'. It was back at Port Moresby by 3 March 1943. Trevan Jackson recalled some personal involvement in this 'initial fitting' up in the 'Bellman by Mortimer Road' with Col. Fox and in two recent books on QANTAS history, several photos show that Company's Beatty Road hangars busily involved from April 1943 onward. 41-23765 was reported 'materially' as distinct from 'slightly' damaged at IRON RANGE on 16-17 November 1942, and while I can't be sure, it seems likely it was the plane Jack Atkinson 'repaired' so it could be flown 'some thousand miles south to the Depot'.

The RAAF was looking to locate a fighter squadron in the Brisbane area early 1943. This led to the redevelopment of STRATHPINE A2. A relevant DWB letter dated 25 January 1943 provides an interesting review of some local strips, many of which were not then in use -

'...Most of these strips were constructed under requisitions issued by USA Forces. The 5th Air Force requires strips A1 - Petrie, A2- Strathpine and A3 - Strathpine for Fighter operation, but the other strips are not required for any operations that are known at present. With the exception of strip A6 - Coominya No 2, which has already been allocated as a dispersal for Lowood, it would appear that strips A4 onwards are available for use by RAAF if required. Some of these strips are badly constructed and are useless for aircraft operation, but others may have a certain value and be worth retaining for present or future use.'

While the views of RAAF HQs were sought, this DWB letter went on to suggest that -

'...strips A4 - Coominya, A7 - Bremer River, A10, A11, A12, and Rosewood be abandoned and that A8 - Warrill Creek, A9 - Gailes, Loganlea and Waterford be retained... Strip A8 is suggested for retention as it maybe required if the Fighter squadron on the Development programme ever is established at Amberley, and Loganlea and Waterford for the same reason if a Fighter squadron is located at Archerfield. Though it is a bad strip, A9 would appear to have some value owing to its proximity to the Gailes Ranges and its retention is suggested.'

Security procedures operated to prevent precise location of installations such as A-A guns appearing on most paperwork. Indeed I've only found such map references in RAAF photo negative registers without, so far, finding the photos! However the Camouflage report of 7 March 1943 addressed the siting of a CCC camp for the 'new hangar work' on Kerry Road. It noted -

'...Mr Bradhurst of AWC suggested a site for CCC camp north of Kerry Road and discussed same with F-Os Dallimore and Cathie White. Later... decided to move site for camp to Beatty Road to keep clear of A-A guns.'

Work on taxiway clearing and gravelling for extension of the SE dispersal area down to what is now Amherst Street, finally got under way around April 1943. Extracts from the last of the regular Camouflage reports of 14 August 1943 suggest little progress on what clearly had received a low priority. The same report added -

'...Saw-tooth hangar on drome being dull painted. No painting yet to 2 igloos for DAP on Kerry Road...'

Under a separate heading, some progress was noted on a pre-fabricated Medium Bomber Hideout, a long running idea the DHS had been pushing. Mr Athol Bretnall, then with the Dept of Interior, was involved and it was decided to erect it at ARCHERFIELD. Like lots of such endeavours I have no idea if it was ever built. However, the same report ended on a 'General' note -

'...The bicycle that was purchased lately from PMG's Dept for use at Oakey, has been brought down to Brisbane for use at Archerfield, where the dispersal covers a considerable area.'

During November of 1943 matters came to a head in respect of the then very taxed facilities available for the overhaul and servicing of airframes of aircraft involved in the Empire Air Training Scheme. These Avro Ansons and single engine Wackett trainers were until this time being overhauled by Aircrafts P/L, a civil contractor operating at ARCHERFIELD, with a subsidiary location at the Maryborough drill hall. The result was the expansion of facilities at MARYBOROUGH. Both the skilled craftsmen and their housing were more readily found there.

The last two igloos to be erected featured their abutment pins raised a couple of metres above the ground on sturdy timber frames, resulting in much greater clearance. Their 'registration' on the General Index Cards is dated 21 September 1943. The drawings are dated October 1943 and copies are secured in a strong room to this day by one present owner. As usual, the most difficult site was left until last. The General Index cards show a 15 December 1943 registration titled 'drainage of western igloo not yet built'. The January 1944 AWC minutes showed some awareness of the risk -

'...Culvert under hangar - Mr Kemp referred to the culvert which is to be constructed under a hangar at Kerry Road. He suggested that it would be wise to increase the (drainage) capacity considerably beyond that which calculation indicates as necessary.'

Wise advice then and it remains so today.

The Kerry Road igloos were shared initially by ANA

ARCHERFIELD *(28/6/1990)*
Recent mapping photography by the Queensland Government shows the suburban and industrial surrounds to the airfield. Sealed runways now command what was, during WW2, an 'all over' grassed field. *[Department of Lands, Qld]*

ARCHERFIELD 9/12/1992
Guido Zuccoli's Boomerang A46-206 'Millingimbi Ghost' appears in front of the 'Control' building in preparation for a ceremony commemorating the WW2 presence here of the US 5 AF. Behind the 'Ghost' is Barry Hempel's Antonov AN-2, recently imported from Europe. [Author]

and QANTAS, in the repair and maintenance work for the DAP. When US forces reached the Philippines, their Kerry Road work ceased. From February 1945 the middle two hangars were given over to the UK's Royal Naval Air Service. Their Transportable Aircraft Maintenance Yard No 1, referred to as TAMY 1, assembled, repaired and flight tested Corsairs, Seafires, some Grumman Avengers and Fairey Barracudas, all for the Fleet Carrier force sent to the Pacific after Victory in Europe. Colourful are the recurring accounts in aviation interest literature of the large scale dumping that was soon to occur of much of this 'Lease-Lend' equipment from the decks of carriers, fifty plus miles off the east coast of Australia.

Today, ARCHERFIELD is one of the few airports nationwide that is still Government owned. Managed by the Federal Airports Corporation, it is under siege from the encroachment of surrounding residential and industrial development. A vigorous private and recreational aviation community will be needed to secure the airfield into the 21st Century.

GAILES (27/4/1944)
This extract from mapping photography shows the location of the 'crash strip' southeast of the Wacol Railway Station and the extensive munitions dump area on the right. The wavy clearing north south in the centre is part of a fire break. In top right of the photo the section of fire break running south easterly from 'creek crossing to creek crossing' is today the alignment of Sycamore Street, Inala. [AUSLIG]

GAILES

'...Struth!, here's another one.'

I recall this as my unspoken reaction to my careful examination of the old WW2 period '1 inch = 1 mile' topographic mapping of the area around Brisbane. The strip appeared on the 1951 map, in dotted outline, with the words, 'ELG (overgrown)'.

GAILES or WACOL or SERVICETON or 'A-9' or the 'crash strip' as it was sometimes known, was a site initiated by the Americans in 1942. I have found brief reference to it in AWC Minutes. The Brisbane based contractor M R Hornibrook carried out construction of 'hangar and associated camouflage work' at this, and other sites near Brisbane directly for the American Forces, around May 1942.

Other Archive documents include reference to periodic inspections of GAILES. These describe it as a poor site with somewhat unacceptable grades and relegate it to an Emergency Landing Field status and the other colourful role of 'dummy airfield'. I have not been able to establish precisely where such 'dummy hangars' were built, but it seems most likely they were of the igloo arched hideout style built at STRATHPINE-PETRIE and COOMINYA No 2.

In 1943 the nearby bombing range appears to have been used by 'dive bombers' from LOWOOD. Certainly its value as an ELG close to that activity, saw responsibility for 'general maintenance' assigned to that station's 14 OBU.

Today the suburb of Carole Park has developed and no trace of the WW2 strip remains on the ground.

AMBERLEY (12/3/1988)
This 'fish-eye lens' aerial from 5 500' altitude has been kindly provided by the RAAF. In the centre of the photograph can be seen the covered readiness bays generally for the RAAF's F-111 Strike Force. The twin concrete bases appearing angled just to the south, are in fact the original floors of the Trans Pacific Ferry Terminal igloos. In top centre of the frame stands the huge Maintenance hangar of the recently formed Logistics Unit, 501 Wing, incorporating both 3 AD and 482 Maintenance Sqn. Immediately left of the 501 Wing Maintenance hangar, the concrete floor slab from the southernmost wartime igloo hangar is visible. The WW2 main hangar close to the intersection of runways is still in existence, though it has been significantly modified. For several years now it has served an important role as a jet engine repair facility. [CPE (RAAF Official)]

AMBERLEY

'...Throughout the long ocean trip, we knew that we would arrive somewhere where the P-40s on board would have to be put together for flight. There were 25 pilots, 25 crew chiefs and 25 armourers. The senior officer on board had us go through the Manuals and at least get in the mood to be prepared to do this job. I was the only "Engineer" so was assigned that position. I had studied Aeronautical Engineering at school. On arrival at Amberley much to our delight we found the RAAF already had a production line going. So we just fitted in, until we moved to Eagle Farm in about August 1942.'

William O (Bill) Farrior was fresh out of Flight School and into the US 35 Pursuit Group when his Unit was alerted to embark 'for a place code-named "Plum" '. What to Bill was shaping up as a peaceful cruise changed on 7 December 1941. His unit was diverted to the freighter, *President Polk*, arriving in Brisbane 13 January, 1942.

The RAAF's 1938 decision to base one 'Citizen Force Squadron' in Queensland resulted in the construction of AMBERLEY. Site selection was not a quick process. Brisbane's Bicentenary Sports and Entertainment Complex at Boondall now stands close to the site first chosen but later discarded because of flood risk.

Construction was underway when war broke out in Europe. The ***Telegraph*** newspaper reported 13 December 1939, on progress of '23 General Purpose Squadron's defence aerodrome', mentioning in part the Main Hangar -

'...the whole of the steelwork having been welded into position without the use of a single bolt. The attractive entrance to the hangar is like a large hotel.'

Modelled on the RAF's 'C' type of the post WW1 period, this hangar was similar to another at GARBUTT. Several other red brick buildings foretold the 'permanency of the aerodrome'.

Contractors Chesterfield and Jenkins (Qld) P/L are

AMBERLEY *(3/3/1941)*
Viewed from the north this oblique shows the diamond road layout enclosing the parade ground of the RAAF Station. Ends of the runways are visible and in the centre left are two large buildings. The nearest is a saw-tooth roof hangar with a small control tower developed on its nearest left hand corner. The other building is a workshop. These were standard designs used by the RAAF at this time having been used also at GARBUTT in Queensland, and in respect of the workshop, at BUNDABERG. The large 'C' shape apron with several rectangular flight/quarters huts will soon be developed fully with their Bellman steel frame metal clad hangars. *[NLA (RAAF Official)]*

AMBERLEY *ca(-/5/1942)*
Thirteen Bellman hangars have been in place since mid 1941. There is no sign yet of the initial lengthening of both runways. Taken very likely from a B-25(A?), the date has to be after mid March 1942. *[T Whyte ex RAAF]*

recorded as carrying out initial clearing and site development. In 1982 the RAAF's own in-house magazine ***Strike***, reviewed the history of the base -

'...It was during this initial construction period that Mr Chesterfield had a small strip cleared and levelled towards the southern end of the main "paddock" and flew the first aircraft into Amberley. The aircraft was a DH-87A Hornet Moth registration VH-UUW, and was thereafter used regularly by the contractors.'

The General Index Cards of the Works and Services branch of the Dept of Interior, show 6 July 1940 as 'date of communication' associated with Job K5 - Relief Landing Ground, ROSEWOOD. This was a doubtful site in black soil country just 2 km from that township. Photos and records show minimal grading was done and equally minimal use made by training aircraft. By September 1940 this 'K' series of consecutive RAAF Job/file numbers advanced to 'K39' with the entry -

'..Delivery of 17 Bellman Hangars'.

At the outbreak of war in Europe, the Australian Government offered Britain up to six trained squadrons for equipping with aircraft as they appeared from the UK factories. The RAF and/or the British Government quickly took a longer view, steering Australia toward the Empire Air Training Scheme (EATS) through which eventually some 26 000 trained aircrew left this country from stations such as AMBERLEY. They were diluted throughout the full spectrum of RAF Commands, mostly to serve in bombers over Europe. Australia's emphasis on recruiting and training aircrew for such dispersed overseas operational experience backfired when Japan entered the war.

Britain's contribution to this offshore recruitment included part provision of the aircraft and instructors. Hugh J F 'Speed' LeGood was an RAF flying instructor who came to AMBERLEY October 1940.

In January 1941 while site preparation of the AMBERLEY runways was proceeding, the matter of an adequate number of RLGs surfaced. The GOOLMAN site was agreed as an obvious one. DWB hoped one RLG would suffice and that ROSEWOOD would not be needed. Most of this problem disappeared mid 1941, with decisions to proceed with KINGAROY for the location of 3 SFTS. A 17 May 1941 file Minute deals with a DWB submission to AMOE. It concerns the adequacy of water supply at KINGAROY but moves on to note the accommodation for both personnel and aircraft at AMBERLEY -

'...Existing accommodation at Archerfield for 23 Sqn consists of sectional steel huts. These huts will be dismantled and stored at No 3 Stores Depot for erection as required at RLGs and Advanced Operational Bases. Accommodation exists at Amberley for 1800 personnel and additional accommodation for approx 500 will be required. Amberley has been planned to accommodate 108 aircraft, but the Bellman hangars have not yet been erected. Action has been taken to cancel the erection of 3 Bellman Hangars (there are 14 such structures there to this day) leaving sufficient hangar accommodation for the 50 aircraft of the AOS (Air Observer School) and 2 Service Sqns plus the allocation of the permanent hangar to the Aircraft Depot.'

Abe (Joe) Howsan first arrived at AMBERLEY in December 1940, on posting to 3 SFTS. He stayed until early 1942, being promoted to 'Technical Corporal' in July 1941. While he went on to serve at BUNDABERG and elsewhere, he was to return later to 3 AD AMBERLEY. Joe told me of the practice whereby fitters responsible for fresh work done on the aircraft were 'invited' to ride in it during the proving flight. 'Speed' LeGood's log shows Cpl Howsan as 'passenger' with a Sqn Ldr Millett on a round trip to KINGAROY, 19 December 1941.

The General Index Cards show 24 July 1941 as the 'initiation date' of K310-Construction A-A Machine Gun Emplacements. One may well wonder about the resulting progress when one reviews the matter of Squadron Bomb Dumps. A Departmental letter of 11 October 1941, advised this work had been -

'...re-advertised in accordance with drawings prepared by you and lowest tender received was £4,778.'

It detailed how the price could be reduced by -

'...deletion of sapling reinforcing to mounds and embankments, deletion of all plastering, hardwood mullions, sills, heads and hardwood frames to shutter openings and substitution of one only 9" x 6" concrete pier between each two roof trusses and between each truss and dividing partition or gable... and substitution of timber cleats to roof in lieu of angle iron cleats.'

This was reckoned to drop 7½% off the tender price. Two days after Pearl Harbor, 'K553' on the Index Cards signalled excavation of 'shelter trenches'.

Trainee pilot, LAC Walter B (later Sir Walter and now ex-Governor of Queensland) Campbell was given his 'Wings Test' by Chief Flying Instructor Sqn Ldr LeGood in Avro Anson W2095, on 7 December 1941. While LAC Campbell and the Ansons promptly became part of 67 Reserve Sqn, patrolling the east coast, AMBERLEY was preparing its 'aircraft erection squadron'.

The RAAF's ***Strike*** magazine of 1982 records the advent of USAAC personnel and the change in pace -

'...on 15 December 1941, following the attack on Pearl Harbor, personnel of the United States Air Corps commenced arriving on the Base and were accommodated at No 3 SFTS. The total strength of the USAC on 31 December 1941 was 998 personnel, for whom accommodation facilities were provided at No 3 Recruit Depot's quarters and in tents erected by the United States Army. Throughout December 1941 the American personnel with the assistance of members of No 3 SFTS assembled P-40E Kittyhawk and A-24 Douglas Dauntless aircraft. This event foreshadowed the change in role of RAAF Amberley to a major centre for the assembly, maintenance and salvage of combat aircraft for the Pacific theatre.'

AMBERLEY (17/7/1942)
Some extension of runways has proceeded. In the left hand middle distance, south of the airfield buildings and across Warrill Creek can be seen the suggestion of crossed runways. These were developed for the two-fold purpose of decoy deception (dummy airfield) and the practical need for emergency landing strips. Note the 'U' shaped objects like aircraft dispersal revetments on the extreme eastern side, near to camera. These are probably 'revetments' hastily built, soon condemned and possibly some of them relocated to the dummy airfield. [NLA (RAAF Official)]

AMBERLEY (1942)
With '31' just discernible on its right nose, this B-26 fuselage is pictured arriving at the Main Hangar '76'. While nothing is yet available to me on its background, I feel it is probably one of a dozen odd which were shipped out to rejoin the US 22 BG, as deck cargo after being damaged in some way. [T Whyte ex RAAF]

General Brereton's November 1941 visit to Australia concentrated on the Pacific Ferry Route for US aircraft to the Philippines, particularly involving Darwin. By mid December 1941, it was clear that AMBERLEY could be needed for the erection of the seventy crated aircraft on the Pensacola Convoy. Training of the pilots accompanying these aircraft was another challenge.

Hugh LeGood, by his brisk and purposeful yet friendly nature, had earned the nickname 'Speed'. US Archive sourced minutes of an urgent 29 December 1941 meeting at AMBERLEY acknowledged his continuing contribution. Titled 'Conference on Further Training to be given Pilots of US aircraft' and listing attendees as, Air Cmdre F M Bladin and Sqn Ldr LeGood, RAAF and Majs E S Nichols and J H Davies, Capt F Pell and Lts H Lowery and G M Keenan, USAC, it reads, in part -

'...Instructors: The responsible US officer is to detail one instructor for each type (aircraft). They can be helped by selected pilots from the RAAF at Archerfield and Amberley, provided that the Operational Sqns are not rendered immobile thereby. Gp Capt Lachal has detailed Sqn Ldr LeGood to assist Maj Davies and has requested that the services of F-O Jackson, ex Tomahawk Pilot of 3 Sqn, be made available. Maj Davies intimated that Lts Lowery and Keenan, USAC, would be available for instruction on A-24s and P-40s respectively. Sqn Ldr LeGood will assist Lt Lowery at Archerfield in the organisation and training of the pilots on A-24s. F-O Jackson, if and when available, will be detailed to assist Lt Keenan on P-40 training.

AMBERLEY (1986)
The 1940 Medical Section building located near the northern tip of the 'diamond' enclosing the parade ground, was still used as such at least up until 1986. [R Alford]

Syllabus: Maj Davies will give Gp Capt Lachal the desired syllabus for each type of aircraft to include air to ground firing (front gun), night flying and dive bombing, for pilots. In this regard it is the considered opinion of Air Corps and RAAF officers that a minimum of 300 rounds per pilot should be allotted for front gun firing and six bombs per USA pilot for dive bombing. It is to be noted that none of these USA pilots has previously fired front guns... It is estimated that the average flying per pupil on each type (aircraft) will be 15 hours...'

LeGood's log book shows 31 December 1941 as his 'initial flight on type' for A-24, 31 January 1942, for Kittyhawk and 31 March 1942 for Airacobra. By 1943, after eleven years in the Air Forces, 'Speed' LeGood's logs show in excess of 4200 hours spread over sixty-five different aircraft types. During the second half of February 1944, LeGood was to add to this. As second pilot, he contrived to fly AMBERLEY to Dobodura in a Dakota and a Lodestar. From there he piloted a Beaufort and a Beaufighter on strikes on Rabaul Harbour and troop concentrations, before returning to Brisbane as second pilot in a Martin Mariner flying boat. The war was changing for LeGood. He returned to Britain to fly Lancaster bombers on two tours of Pathfinder operations. Sir Walter Campbell quickly remembered 'Speed' LeGood as a no-nonsense flying instructor who once told him to -

'...do me a forced landing in that spinney meadow.'

The crated shipment of 'complete knocked down' (CKD) machinery by sea had advanced with technology over the years. US Aircraft factories were industry leaders, yet somehow early January 1942, serious deficiencies appeared. Under the heading 'Erection of Aircraft', the ***History of the Fifth AF (and its predecessors)*** records -

'...The P-40s had no Prestone, (a vital anti-freeze liquid coolant component) and none was found on the convoy. All the Prestone in Australia was rounded up and proved to be just enough for the 18 P-40s in the convoy. The problem of completing the A-24s took longer, because these aircraft lacked solenoids, trigger motors, gun sights, and self-sealing tanks. Again and again the cargo of the convoy was searched but the shipment of these vital parts had been omitted. Washington was informed on January 3, 1942 of the shortages, and air shipment requested for the gun sights, solenoids and trigger motors. This was arranged and the parts were started on their way via the African ferry route.'

AMBERLEY (7/10/1942)
RAAF men uncrate a P-39 Airacobra just north of the large saw-tooth roof hangar '76'. Assembly was carried out in this hangar. Of interest just behind the unopened crates is the Station Headquarters building with observation deck evident on the peak of the camouflage painted roof. [Neg SC165926:US National Archives]

The problems of non-appearance of such vital combat items became a sore point in history. James Rorrison in his ***Nor the Years Contemn*** published in 1992, exposed some of the controversy. It has been commonly held that such small items were accidentally left nailed to the inside of timber crates which were burnt as rubbish. However, not only is there strong evidence pointing to the resourceful use of such boxes, it also appears the A-24s were 'second hand'. Rorrison explained, that while the Army Air Force ordered the SBD-3A model, with armour plate protection to the two-man crew and self-sealing fuel tanks, somehow a mix-up occurred. Uncrated, they were found to be an earlier Navy version, SBD-2, with -

'...neither self-sealing tanks, nor armour plating. Incredibly, there were other deficiencies. The main undercarriage tyres were worn bald, and mud still caked undercarriage legs and open wheel wells from use in America. Their Wright radials already had excessive engine time on them...'

It will probably remain a mystery as to where the missing gun parts were. At least this background shows the consignment was no ordinary crating of factory fresh aircraft by 'state-of-the-art' packers. The end results were primarily that combat worthiness of the A-24s was delayed and the difficulties of mustering and/or adequately training pilots came to the fore.

Lawrence Moore was a 'technical specialist', with the 11 BS, 7 BG, arriving in Brisbane in the *Pensacola* Convoy. Larry wrote, dated 7 December 1991 -

'...I surely do remember this day 50 years ago - was on a troop ship on the way to the Philippines, had just crossed the Equator prior to the 7th. Never before had they done this going there - think "someone" knew something but has never admitted it!'

Larry's role was in the uncrating of the aircraft. He wrote -

'...Some of the RAAF personnel that hauled away the tops, sides and ends said they would make use of this lumber to build buildings as it was soft wood (pine) and easy to work with as most of their timber was quite hard... I cannot imagine them burning the bases as there was much good heavy lumber in them... These solenoids must have been found or quickly replaced as these A-24s left for the "combat zone" either late Jan or early Feb. They surely would not have done so with the guns inoperative. I almost had to go with them either to the Philippines or New Guinea as they wanted radio operators for calling ahead for weather reports. There were about 8 of us that were to go. We had our bags packed and were standing by the planes ready to take off when our CO asked the Officer in Charge of the planes if he could guarantee 100% that we would get back to our Sqns. He stated that there was no guarantee but that he would try to get us back. Our CO said that was not good enough and he could not afford to lose that many trained radio operators and mechanics. So we did not go with the planes.'

Many factors operated to hinder effective mobilisation of the US 'provisional' P-40 squadrons. While it took some weeks before the Americans were to divert some P-40s to allow formation of the RAAF's 75 and 76 Sqns, their young US pilots needed 'transition trainer' aircraft. 23 Sqn's ORB records February 1942 -

'... - F-Lts M V Lewis, L D Jackson and K W Lloyd gave dual instruction on Wirraway aircraft to twenty officers of the USAAC.'

In return these RAAF pilots successfully -

'...completed conversion course on Kittyhawk P-40.'

Jim Steenson was an RAAF member of the 'erection squadron' at AMBERLEY for about 18 months from the end of March 1942. 3 AD formed 16 March 1942 and presumably Jim's group was part of it. Jim wrote -

'...our prime objective was to assemble the Kittyhawks (P-40) and 'Cobras (P-39) alongside the American

AMBERLEY *(1943)*
Some of the airmen's sleeping quarters of the day. Some veterans with time at AMBERLEY in early 1942 speak of the much prized softwood packing crates used to freight aircraft to Australia. Such timber crates finished up being used for 'huts' on at least the adjacent satellite field, of GOOLMAN. [A Howsan ex RAAF]

engineering staff in an effort to get fighter aircraft into action as soon as possible. We worked around the clock in two twelve hour shifts. I was on the 6 p.m. to 6 a.m. shift, seven days a week for a period of a few months. We were given a meal at midnight (on the job) of sandwiches and tea... The aircraft were stored in crates in the bush just outside the station, guarded by American soldiers. They were placed on the apron outside the hangar, with mobile cranes pulling the crates apart. Then on low wheeled platforms they were taken inside - four assembly lines - two Australian and two American. From memory, the best assembly rate was eighteen in a 24 hour day!'

The RAAF's ***Strike*** magazine, in 1982 said of 3 AD, that -

'...By the end of April 1942... 123 P-39 Airacobra fighter aircraft had been assembled and delivered to preparational squadrons.'

Dispersal strips were needed urgently. As the SDCC Minutes of 22 April 1942 show the Committee visited -

'...Warrill Creek with F-Lt Starr where considerable clearing is being done by US coloured troops.'

They also minuted some enthusiasm for a solution to -

'...the problem of disguising the field (camouflaging Amberley itself!) .. from a study of an excellent mosaic held by F-Lt Starr, it was decided that in conjunction with a certain amount of treatment of the real field, another dummy one could be constructed very easily.'

The dummy field site started out just north of AMBERLEY, but actually proceeded on the south side and never really gained RAAF HQ support.

The SDCC also inspected the dispersal taxiway and aircraft hideout work. While they acknowledged better siting than those at ARCHERFIELD, the criticism was -

'...the hideouts for fighters are sited further from the drome than those for bombers, but it was pointed out that the main bitumen road, with passes through the

AMBERLEY *(10/3/1943)*
Left of centre is the large saw-tooth roof RAAF hangar '76' in and around which the urgent work of erecting fighter aircraft freighted to Australia in large timber packing cases was carried out from late 1941. [Neg SC239567:US National Archives]

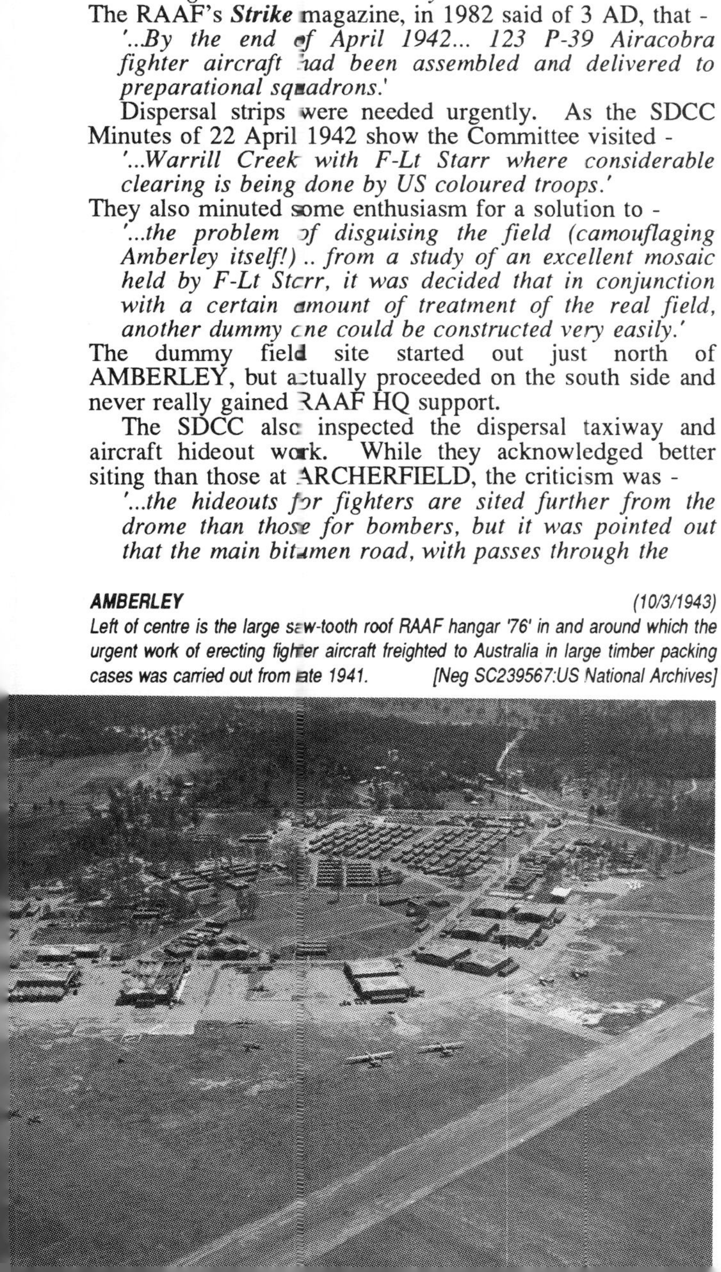

AMBERLEY *(1943)*
These are some sleeping quarters assigned to WAAAFs at AMBERLEY. They appear to be prefabricated, ply sheeted with waterproof bitumenised building paper roofing. While veterans have spoken of 'huts being built from aircraft packing crates' and a 5 January 1943 review of WAAAF accommodation does list several sleeping huts here of aircraft packing case construction, these are not some of them. It appears 'Mickey Mouse' was alive and well in hut 5. [A Howsan ex RAAF]

dispersal area is to be prepared as a runway for them to take off. Provided this is successful, the scheme may be a good one.'

The camoufleur involved, Mr R F Gordon was -

'...engaged upon netting the air raid trenches which have been constructed all over the station, and are very prominent from the air. A-A machine gun posts have been constructed round the drome. So far there has been no attempt to conceal them.'

Camouflage and deception are synonymous. It was noted -

'...Good dummy A-A guns had been very badly placed (by the Americans we understand). They are simply standing in the open with no attempt at simulated protection or occupation. This fact has been pointed out to the US officers in charge of the A-A guns in Brisbane.'

Apparently some other prompt camouflage efforts at AMBERLEY were not endorsed. On 27 April -

'...Mr McColl informed (the) Chairman that all splinter proof pens on this drome were to be either removed or camouflaged within two days. For this purpose a camoufleur and a large gang of men were taken from the munitions work. It became necessary to decide whether and where a dummy landing field was to be made as some of the pens would be moved to it. It was decided to make it on the south side of the drome across the creek in preference to the site previously suggested to the north of the drome. Some trees were to be removed, grass cut to simulate runways and eight of the pens were to be transferred to it... Splinter proof pens are now to be used in special cases only, where hideouts are close to dangerous areas, and they will then be of the log-palisade type, and not the massive A-frame type that had been constructed at Amberley. Those of the old type that remain on the drome will be used to house decoys only.'

Early May 1942, the layout of accommodation huts was addressed by the SDCC. A Mr G T Dyson of the -

'...Victorian Committee was sent up to arrange a scheme for the dispersal of personnel. He reported...it was proposed to remove 38 huts, every second one, and place them nearer Ipswich as a suburban extension.'

Emergency Landing Strips (ELS) received continuing attention. The 14 May SDCC Minutes noted -

'...Fl Lt Starr told Mr McColl and Mr Cummings that it had been decided not to gravel the flight strip on the Rosewood road to the west of the drome and that it

AMBERLEY (26/7/1943)
Although deteriorated with age like many other prints surviving in the collection of the National Library of Australia, this print is presented as one of the few showing in the foreground, the winding dispersal taxiways to the west of AMBERLEY RAAF Station. In the centre left of the photograph are the two large igloo hangars serving as the Trans Pacific Ferry Terminal. With Ipswich in the left background, the road stretching westerly towards the camera, actually crosses the 135° runway and carries on westward to Rosewood. The Amberley to Warwick road turns off near centre of photograph heading southward. Just in the southwestern quarter of that junction is the Amberley school building which served during the war years as an administration centre. It survives to this day. Southward along the road to Warwick and towards the right edge of this photo is the site of the well camouflaged Operations/Communications Centre. [NLA (RAAF Official)]

would only be used in dry weather as an ELS... None of the US Army coloured troops have been employed clearing flight strips since the last meeting. The work is now being done by the Dept of Interior.'

AMBERLEY (29/11/1943)
This is the 'serviceability' date shown on Drawing 42-43/580C. Two igloo type hangars built by M R Hornibrook have appeared on the extreme southern edge of the built-up (camp) area. These have both since disappeared to make way for the huge maintenance hangar of what is now structured as 501 Wing incorporating the previous long-standing 3AD and other RAAF Units. Notice the dispersal taxiways to the west of the station. [AA BP375 Item QL680]

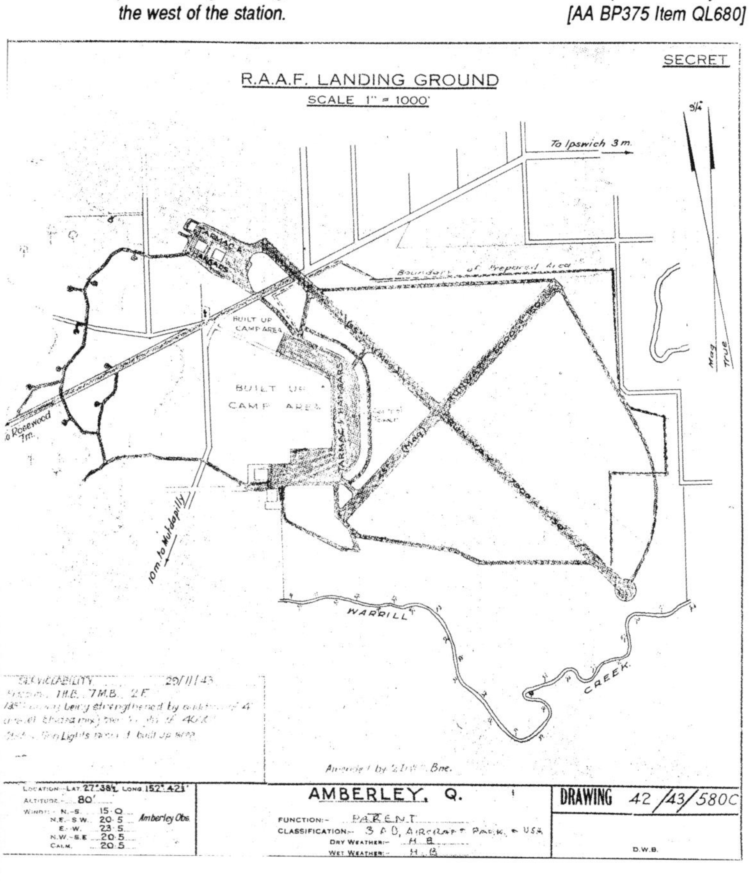

F-O John Keays is on record having reviewed these dispersal strip sites mid May 1942, in company with Camoufleur F H Johnston. Of BREMER RIVER, he said -

'..Dispersal is fair and timbered areas to SE and W will take 10-12 aircraft. Construction has been carried out by Barracks Officer, Amberley at practically no cost. Strip and dispersal areas affect 5 farms but only improvements affected are fences. Strip has been used.'

The AMBERLEY Barracks Officer then was F-Lt J Starr and he had been very energetic in this urgent development of dispersals. Keays also checked on WARRILL CREEK, which he called WARWICK ROAD -

'..Longitudinal grades appear excessive and the country rises from the south end at a grade of at least 1 in 30. Arrangements have been made for Dept of Interior to survey the strip, prepare longitudinal section and check grade on approach... All strips are to be correctly located and resumption details obtained. Timber cover is good and satisfactory dispersal obtainable. Drainage will be necessary in places... (at) Goolman - construction (is) in progress.'

The Minutes of the 28th Meeting of the State Defence Camouflage Committee dated 14 May 1942 mentioned some lesser known aspects of the art. Items such as -

*'..**Decoy planes** of an improved type are now being manufactured in numbers in NSW... (and) **Dummy A-A Guns** of an approved design are being made by Wunderlich, but the best means of creating a flash from them is still undecided. Dummy planes should contain inflammable material that will cause a decent fire when they are hit.'*

A 16 June 1942 report by P-O Keays to the Deputy Director General of Allied Works shows what official support the dummy airfield got. Keays wrote that -

'...Air Board has no knowledge of the dummy aerodrome being constructed. I advised... that I did not favour the scheme and informed them that I had advised Mr Lowe of MRC that no plant should be used there if it was required for runway extensions. I informed Mr McColl by telephone that I would not agree to demolished pens being re-erected on the dummy site. During an inspection today, W Cdr Hartwright informed me that he was opposed to the construction of the dummy field and had not given any instructions in regard to performance of works thereon. I noted that pens had been erected on the dummy field. I would suggest that the following matters be investigated:

a) who authorised construction of the dummy field?
b) who authorised erection of pens on dummy field?
c) what funds are being used to carry out the works?

Mr Cox has received complaints from the owner of the property and I understand he is lodging a considerable claim for compensation. It is recommended that all fences be replaced and repaired immediately.'

A couple of days later John Keays expanded on this. The RAAF Station CO, Gp Capt L V Lachal had bought into the idea saying the swampy northern site was too close to the dispersal area and a site near the southern boundary was better because it could serve as an ELG too.

July 4, 1942, the regular Camouflage reports noted -

'..Amberley Aerodrome - camoufleur E D Read - a visit was made with Mr Bretnall on 1 July to inspect hideouts and the dispersal. Mr Brizay, US Engineers, is building an experimental arch-type hangar for a B-17 in the dispersal area... Two dummy Wirraways are being built by Dept of Interior, and US Fighter types are being planned.'

Apparently the B-17 hideout was completed two weeks later, and work on remaining medium bomber hideouts, then 60% complete. The report of 19 July mentions -

'..Arched type for medium bombers being built, US Fortress hideout complete. Awaiting vertical photo of flying field as at present before making final suggestions on ground treatment... Amberley Strips -

AMBERLEY (1986)
Viewed from the east the permanent nature of the guard house as part of the original grand design for AMBERLEY can be appreciated. That original design did not envisage WAAAF telephonists who came to staff many needs from 1943 onwards. Indeed the patched on nature of the toilet facilities appearing through the roof line were added in mid-1943 for just that purpose. The function of this building has now changed in as much as, in very recent times, the public road has skirted further west of this guard room location resulting in a modern entrance to the Base being located further south. [R Alford]

dispersed hideout positions on Warwick Road and Rosewood strips selected with Mr Wyeth and Mr Annand and camoufleur Read on 14 July 1942, positions to be approved by P-O Campbell. Photos of all strips to be taken by Major Taylor, US Army.'

John Keays' report of 10 July 1942 is another of those items which aids dating of an aerial photograph here presented. It touches on -

'...gravelling of runway extensions and new taxiways has been completed and sealing is in progress. Gravelling by MRC, sealing by Messrs Chesterfield & Jenkins. Works inspected on 9 July with Mr Chesterfield... Dispersal taxiway has been partially primed with crude tar. This has been for camouflage purposes.'

Keays reviewed design progress on dispersal strips. He'd refrained from finalsing some details because -

'...we wished to cooperate with Mr Hughes of USA Camouflage Section. Amberley has changed hands a number of times and it is desirable that any works carried out there should be set out in collaboration with the USA officials. Station is at present operating as No 3 AD pending completion of the works at Chinchilla or alternative site.'

AMBERLEY (29/9/1948)
Extracted from systematic mapping photography, this view shows the RAAF Base some three years after cessation of hostilities. RAAF AMBERLEY is a prominent serving RAAF Station to this day and understandably the Defence Department restricts publication of 'current day imagery'. Retired from service Liberator aircraft can be seen parked in the open and on the eastern perimeter taxiway. Other smaller aircraft are parked at extreme north and south. [AUSLIG]

The phrase Trans Pacific Ferry related importantly to fresh bomber aircraft. This extract from the US Archive sourced ***History of the 43 BG*** covered the arrival of four 63 BS B-17s at AMBERLEY. These were recorded as serials 124355, 57, 81 and 84 piloted by Sogaard, Barnett, Scott and Capt McCullar. Three were named: 'Dinahmite', 'Pluto' and 'Momma Maxie' but specifics aren't given -

'...Wednesday August 6th, at 11:45 a.m., the four bombers, still intact touched their wheels on Amberley Field, Brisbane. Australian soil at last. Immediately upon landing the crews reported to public operations. As each man walked in, a thermometer was popped into his mouth and thereupon followed a hasty physical check-up with favourable results. The flight was informed that it would proceed to Charleville next morning where final checks would be made on all aircraft before allowing them to continue into combat zones. The four crews were given lodging in a small hotel in Ipswich, a village near the airdrome. When all were somewhat refreshed, they were taken to Brisbane, thirty miles distant. Here an emergency pay was drawn from the US Finance office that issued Australian currency. Wednesday evening was well spent looking over Brisbane and tasting some of its social life, although lights were limited due to blackout regulations...'

On 6 September 1942, the SDCC report included a passing mention of some presence of dummy fighters.

'...Amberley - 6 arch type medium bomber hideouts close to completion. Some cannot yet be used as Main Roads people have not constructed cross-overs from roads. No further word on proposed extension of the dispersal scheme has been received. A prompt decision will allow already depleted staff to carry on with this work. The new vertical water tank is being dull painted. Three P-40 type fighter dummies have been completed and look very well indeed.'

In its 1982 issues, the RAAF's ***Strike*** magazine touched on the USATC. It referred to the advance party of -

'...USAFIA Ferry Division, Air Transport Command Williamtown, commenced operating from the Station on 10 September 1942. This party, which comprised initially of 2 officers and 7 airmen were required at the terminus of the Trans Pacific Ferry Service to supervise off loading and subsequent delivery, and the outward loading of aircraft proceeding between Australia and the USA. At about the same time the US Army Airways Communications Service arrived on the base, and then in November 1942, the 22nd Service Group of the 5th Air Force also settled in on the Station. Both these units were to remain till 1945.'

Facilities for this freight and ferry service were set to be enhanced. It was in the SDCC regular report of 13 September that an indication came -

'...On a recent visit to the drome, the CO indicated that a considerable extension to one runway is contemplated. This will involve diverting the main road outside the dispersal area. Developments are being closely watched but nothing has so far been finalised.'

As usual, the camouflage activity reports have proven a valuable diary-like source of development progress and comment. Dated 4 October 1942, various items figured -

'...apart from a little touching up, six arch-type hideouts and one wire-type hideout are complete. All are for medium bombers. Mr (camoufleur C W) Atherton flew over dispersal on 2 October and reports favourably on Dept Interior hideouts... Dummy planes have been built at Amberley at the request of the CO but on 2 October the Dept of Interior was instructed to stop this work by Dept of Air. Three P-40 dummies are complete and another two of this type are nearing completion. The hessian covering on the latter two is being coated with plaster before painting. Three Wirraway dummies have

AMBERLEY (26/2/1943)

The Military Survey (1 inch = 1 mile) map of the 'Ipswich' area was reproduced June of 1945 with the annotation 'Information correct to January of 1924'! It shows a quarry west of the Amberley to Warwick road at grid reference 831691. It does not show any of the AMBERLEY air base development. The quarry was an obvious choice of site for a reinforced concrete Operations/Communications centre built in 1942. Here the centre is photographed in its camouflaged state. The flat roof of the structure has been built level with the adjacent top edge of the old quarry. Camouflage net and other stretched and suspended fabrics bridge from the roof to that undisturbed top edge of surrounding ground. The entry to the 'underground' structure is via the small residual 'hole' which presumably the camouflage experts reasoned could be read by enemy aerial photographic interpreters as the original quarry. With the camouflage material totally stripped away, the substantial building remains to this day. [NLA (RAAF Official)]

also been completed. Two of the P-40 dummies are in use at Lowood drome. Experiments are being made in breaking the shadows of aircraft parked in the open... US Air Force CO on station Col Gilchrist, has asked for estimate of cost to efficiently garnish large fortress arch-type hideout... A-A Batteries - the treatment of the battery at Amberley is expected to be complete in about ten days. A plan has not yet been received from Army Camouflage officer...'

AMBERLEY (16/3/1992)

The WW2 Operations/Communications building has long since been stripped of camouflage which melded it with the surrounding brink of the old quarry. It now serves as the Defence Department's Tri-Service Fire Training Centre. Fire Service personnel for Army, Navy and Air Force attend here, at some stage. This photo is taken looking towards the east and while not entirely clear, the building on the hill, partly visible through the trees, is an 'old Queenslander' farmhouse which survives from the war-time period. [Author]

A scheme for the camouflage of the new Operations building was approved by the TDC by 18 October, 1942 and Air Board approval was awaited. The Station CO Gp Capt E Douglas, was independently seeking to hasten Air Board approval. This was still outstanding by 8 November 1942 yet several airfields were advised as due to have upwards of eighteen fresh hideouts each added. In AMBERLEY's case the number was twenty-four.

Air transport was to prove absolutely vital to the Allies' conduct of the Pacific War. The ***History of the Fifth Air Force (and its predecessors)*** noted a start, January 1942 -

> *'...an Air Transport Command was created under USAFIA. Capt Paul I Gunn was assigned as CO with HQ... at Amberley Field, Brisbane. Aircraft assigned to the ATC by 6 February 1942 included the following: 6 C-53s; 3 Beechcraft; 3 B-24s; 3 B-18s; 1 B-17.'*

The Australian Government impressed a wide assortment of civil aircraft, from both airlines and private owners, most suitable only for communications and training duties. The major airlines chipped in, professionally and heroically. General MacArthur advanced his HQ, SWPA from Melbourne to Brisbane, 20 July 1942. Shortly after, the Directorate of Air Transport, (DAT), Allied Air Forces, SWPA was formed in August 1942, under the command of a noted Australian pioneer airman, Gp Capt Harold Gatty, then a Regional Director of Pan American Airways. DAT operated in Australia and offshore to the near north, but by the end of 1942 the Pacific Air Wing of the giant USATC was showing the way to trans ocean freight and personnel transfer. AMBERLEY was the SWPA terminus until even that moved to Townsville in 1943 and northward beyond. Casey's ***Engineers of the South West Pacific 1941-1945*** recorded the effect on AMBERLEY -

> *'...Considerable maintenance became necessary on these runways at Amberley which had not been designed to withstand the exceptionally heavy loads landed in ATC transports. Assembly operations carried out at the airfield by the (US) Consolidated Aircraft corporation under contract to ATC also immediately necessitated the provision of two large hangars. Construction of these was initiated by the Allied Works Council in November 1942 and in accordance with instructions of the Chief Engineer GHQ SWPA, their design was based on a modified igloo type hangar consisting of three point bearing arches which would support a travelling chain hoist.'*

Casey's HQ staff paid direct attention to this critical project, but timber priority favoured front-line New Guinea works. Completion dates were missed, and -

> *'...In any event, delayed delivery of electrical transformers from the US would have prevented utilisation of these hangars prior to the end of May and by that time finishing touches were being applied to the structures.'*

The two large igloo structures which emerged on the northern fringe of the base were to serve till the 1970s redevelopment. Bare concrete floor slabs are still evident from the air, incorporated into the asphalt hardstanding.

Work started on the camouflage to the new 'Ops building' by 10 January 1943, with -

> *'...garnished netting... being prepared by female labour at Dept of Interior, Brisbane.'*

AMBERLEY (1944)

Looking south to the 3AD hangars with several RAAF Liberators in the servicing area. [A Howsan ex RAAF]

Such field installation of camouflage to the Operations building was reported 90% complete by 31 January 1943.

The other two igloo hangars to be erected at AMBERLEY were heralded with the issue of DWB requisition 976/42-43 dated 18 May 1943. With a 'K' number of 1849, these were part of an extension to 3 AD's facilities. With the prospect of the RAAF 'soon' to receive its own B-24 aircraft, maybe the A2 (instead of A1) priority listing was prophetic! The two 170′ span structures appeared under job USM-218, a contract awarded to M R Hornibrook P/L. They were located on the southern fringe of the building area and became well known to RAAF crews ferrying B-24s across the Pacific, and those of the RAAF Liberator squadrons picking up their assigned machines in 1944 and 1945. These hangars served until their demolition to make way for the huge maintenance facility now operated by 501 Wing.

Strike magazine's 1982 review addressed the expansion of 3 AD noting particularly the WAAAF numbers, which in June 1943, totalled 1856. Over these war years, many RAAF Sqns actually 'formed' here, including an AD unit -

AMBERLEY *(-/1/1946)*
Discovered amongst a set of photographs of unknown origin in the Courier Mail newspaper Photo Library, this frame was catalogued as 'Eagle Farm'. Unmistakably it is AMBERLEY viewed from the north west. The skyline with Mt Lindsay on the far right confirms this. The Trans Pacific Ferry Terminal igloos are in the centre. In the immediate right foreground is the camouflaged top surface of the northernmost arched aircraft dispersal hideout, one of several shown on Drawing 4243/580C. A 30 August 1942 SDCC report described two such structures to be 'waterproof covered, camouflaged and used for stores dispersal'.
['RAAF or US Official' via Courier Mail]

AMBERLEY *(1951)*
It took some years after WW2 for the Military Survey Mapping programme to show AMBERLEY, although its development as far as this map is concerned had changed little. This extract is framed to show the relationship to the parent drome of WARRILL CREEK ELG on extreme left centre and GOOLMAN ELG in bottom right corner.
[Department of Defence]

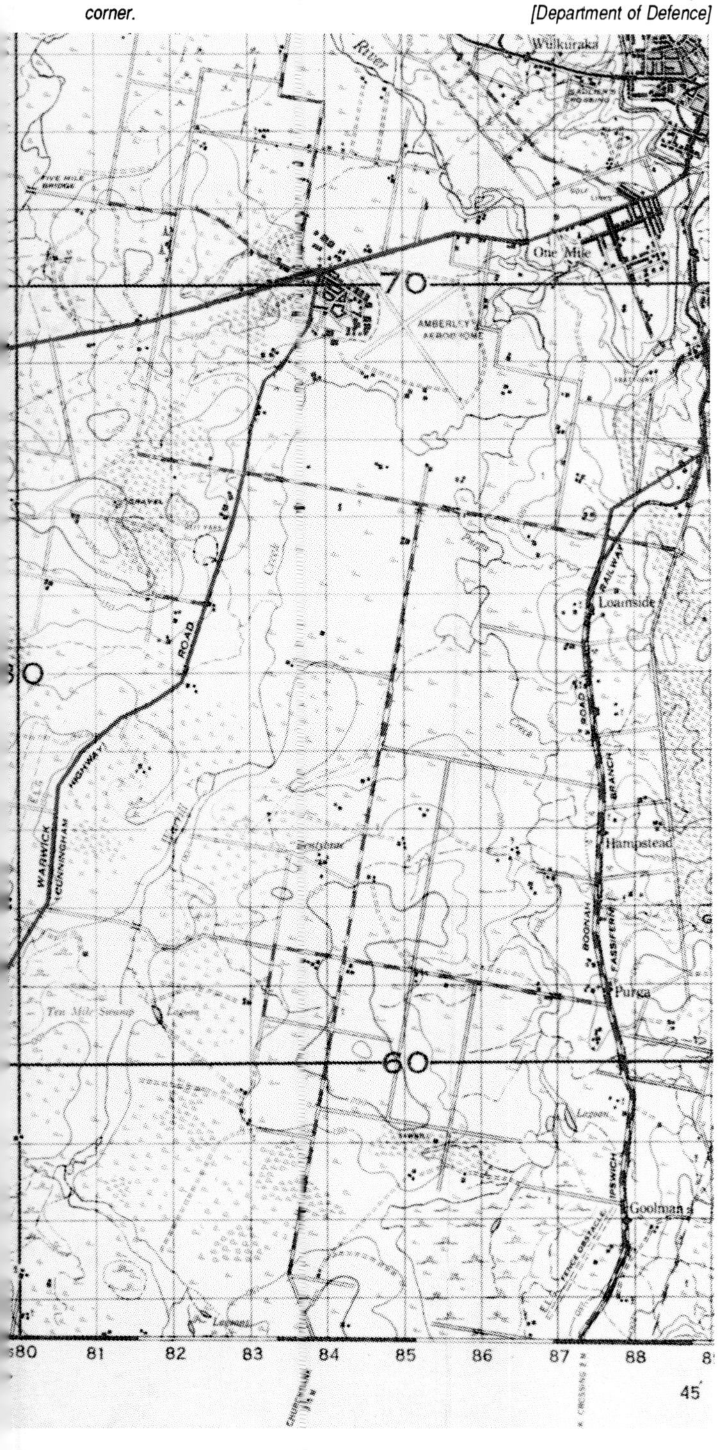

'...On 14 October 1943, 6 Aircraft Depot (6 AD) was formed at Amberley and departed soon after (for Oakey).'

Dudley Curwen enlisted in the RAAF 5 November 1943 with 'stars in his eyes'. After 'rookies' at Sandgate he was posted to AMBERLEY; mustering - cook's assistant. Most of 1944 he spent at the Sergeant's mess -

'...24 hours on, 24 hours off... The main road from Ipswich to Rosewood went straight across the main runway... they had guard huts on each side. When traffic approached, they would open the boom gate to let them across. This went on day and night.'

Wally (later Sir Wallace) Rae was one of the RAAF ferry pilots flying one of the first half dozen B-24Js to go to 24 Sqn, the first RAAF HB Sqn to form. Landing A72-35 (or 38?) at AMBERLEY on 1 June 1944, Wally went on to serve in and ultimately command 'Test Flight' there early 1945. Reminded of the road over the runway, Wally recalled he -

'...caused some excitement when with a circus crossing the runway, my take-off in a Lib' resulted in two days spent recapturing elephants and other animals... it certainly speeded up re-routing of the road!'

The veteran Lancaster W4783 'G-George' which participated in ninety sorties for RAF Bomber Command with the RAAF's 460 Sqn arrived from Tontouta, 8 November 1944 captained by F-Lt E A Hudson, DFC (and bar). 'George' was overhauled at 3 AD and departed for Rockhampton on War Bond business, 10 November. This 'Lanc' now occupies pride of place in the Australian War Memorial's Aeroplane Hall, in Canberra.

Unlike the majority of airfields for which little if any post-war future existed, AMBERLEY has prospered. Its multi-million dollar presence in the Royal Australian Air Force of today was very aptly heralded in that ***Telegraph*** newspaper article in December 1939.

WOODRIDGE

'...It is understood that constructional work is now being carried out by US Engineers in this vicinity, but this is not with the authority of this Department.'

This was the 19 August 1942 answer given by the Dept of Air to the Dept of Interior, Canberra, when residents 'asked' about airfield construction on their properties.

During September of 1940, while the Battle of Britain was at its peak, the RAAF began looking for an Operational Base (OB) in the KINGSTON and WOODRIDGE areas. Surveys revealed 'disadvantages' with sites in the area from what is now Wembley Road south to Browns Plains road. In 1941, the RAAF turned its attention northward to the Strathpine area, but when Japan came into the war, American Forces urgently sought airfields in the Brisbane area for pilot training and the dispersal of fighter aircraft. They were quick to review sites in which the RAAF had previously expressed interest.

WOODRIDGE *(27/4/1944)*
This extract from Frame 314 of Run 5 of Mapping Photography, is an opportunity to show the proximity of KINGSTON A-11 (top) and A-10 to the south. Both proved a dismal choice of sites on the part of their US initiators, but doubtless served well in the overall role of 'dummy airfields '. Browns Plains Road now skirts westward from the southern end of A-10 where it joins Chambers Flat Road. *[AUSLIG]*

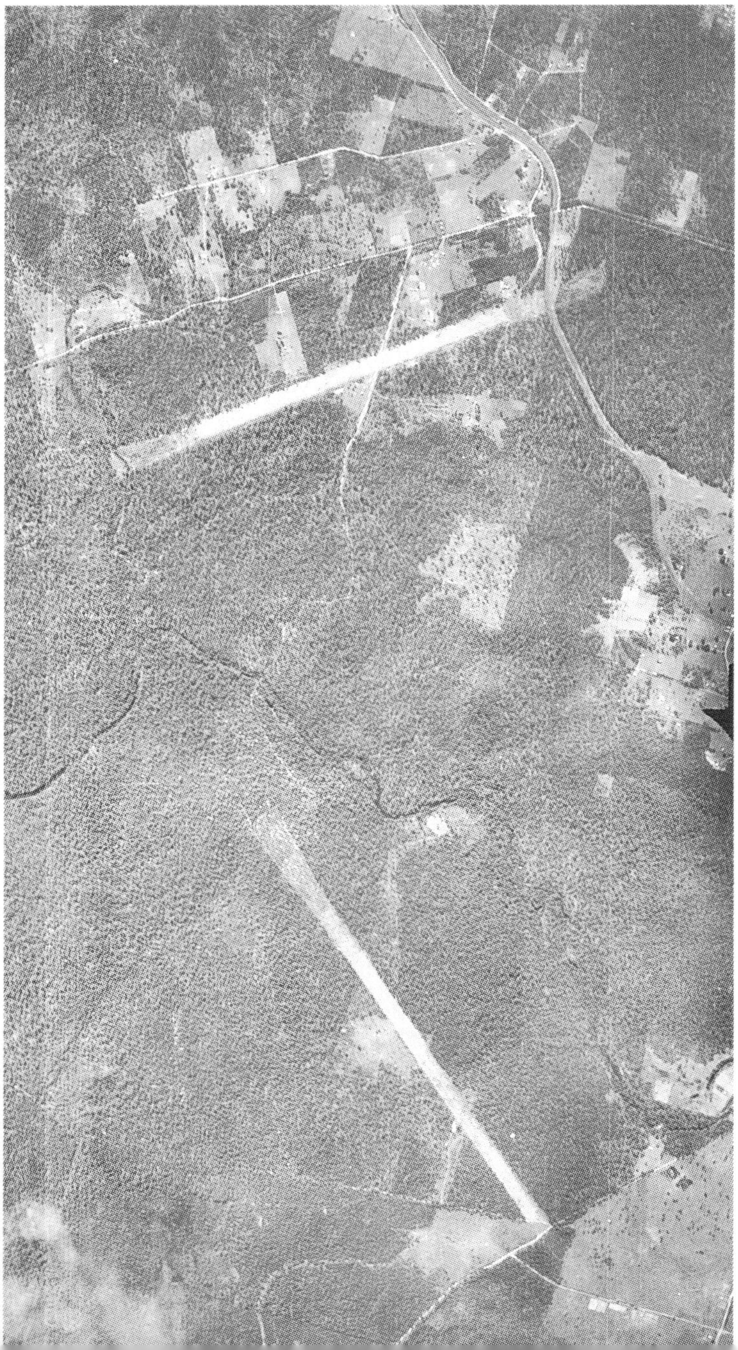

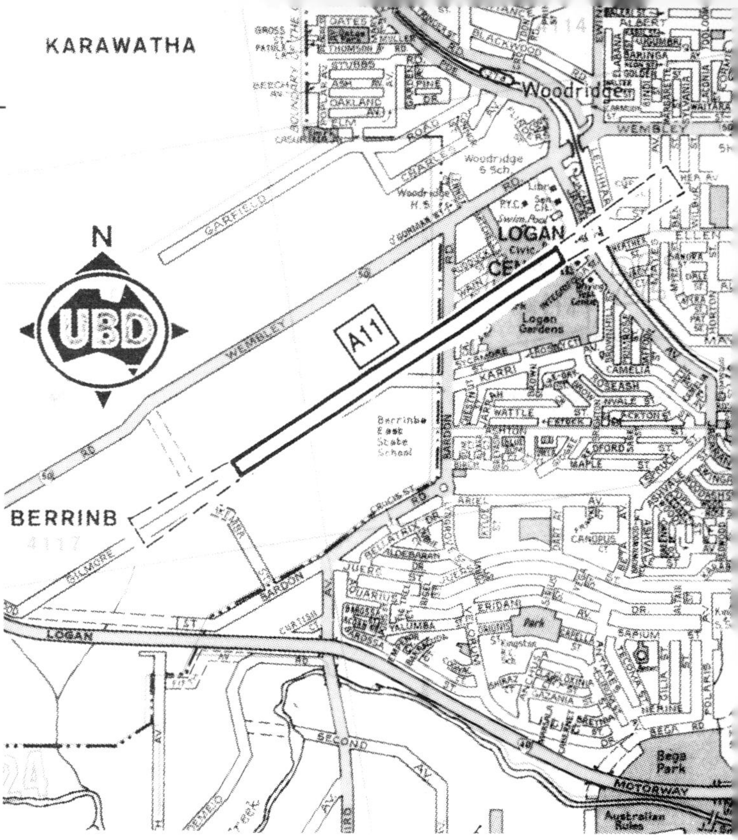

WOODRIDGE *(1991)*
It's probably not widely known that the NE end of this strip (A-11) was in the 'backyard' of today's Logan Civic Centre, in Wembley Road. *[Universal Press P/L]*

KINGSTON/WOODRIDGE A-11, KINGSTON A-10 and A-12 further south, resulted from this search. In January of 1943, when the US Forces had essentially lost interest in this area, the RAAF recorded assessments of these Kingston fields as follows -

> *'...A-11 - severe slopes, unsuitable for operational use, may have some value for emergency landings. A-10 (Woodridge) and A-12 - steep slopes - useless.'*

By February 1943 all three A-10, A-11 and A-12 were on the list of strips 'to be abandoned'. The RAAF's 1" : 1 000′ plans of these strips, show them all as 'dummy' airfields, which is probably the only realistic use they did serve.

LOGANLEA *(17/7/1942)*
Photographed as one of the several 'KINGSTON Landing Strip(s)' this is another of the sites urgently developed by the MRC at the direction of the US Forces. Comparison with current day mapping shows that the Logan Motorway runs east-west through almost the intersection of the airstrips. Loganlea Road runs north through the centre of the photograph with its prominent by-pass kink to cross the railway line at the south. At least one arch type hideout was developed at this site, probably at the edge of the cleared area between the two strips just west of Loganlea Road. *[NLA (RAAF Official)]*

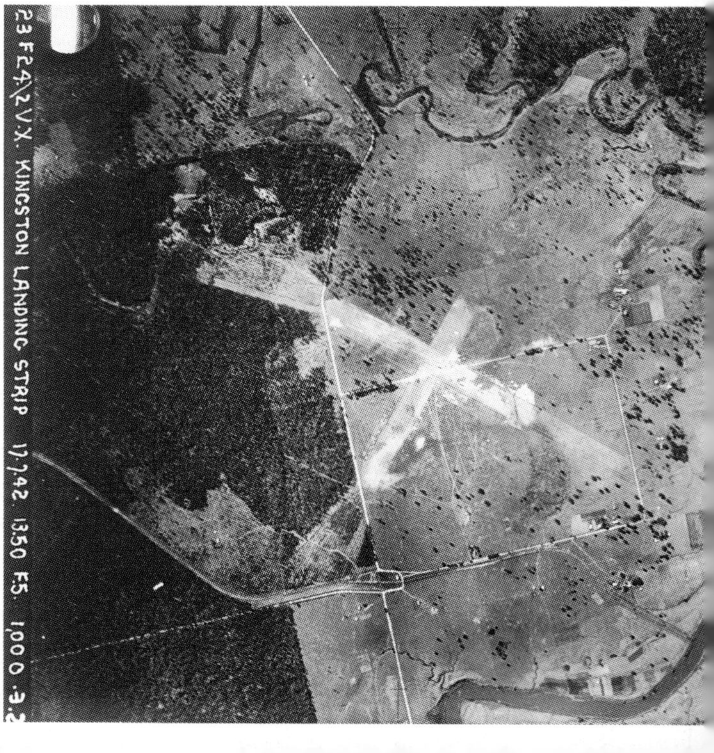

LOGANLEA

'...In most cases the hideouts are well hidden.'

They certainly must have been! The archive records of the SDCC carry a note dated around July 1942 -

'...Loganlea and Waterford strips - these were visited on 2 July with Mr Wyeth. Well dispersed US arch type fighter hideouts have been built but access to most of them has not yet been cleared. In most cases the hideouts are well hidden...'

Brisbane contractor M R Hornibrook P/L is recorded in AWC minutes as directly engaged by the Americans, in 'camouflage work' on these airstrips. The engagement was on an urgent and personal basis, so given the turmoil of the period, it's not surprising the payment took many months to come through.

On 10 July 1942 F-Lt Keays had reported -

'...this field was commenced as an RLG to Archerfield but has been changed to a satellite strip. Construction... was under RAAF supervision but extension of area has been under USA control. Soil and drainage conditions are not the best and in my opinion runway should be gravelled if it is required for use in wet weather. Dispersal scheme is under design by USA Camouflage Section and construction of taxiways and hideouts is in progress.'

To this day I've not been able to establish where such hideouts were constructed, neither at LOGANLEA nor WATERFORD. Some hint is given on the RAAF 'Landing Ground' plan for LOGANLEA by the extent of clearing on the western edge and it seems likely that at least one such hideout was located in that north western corner.

LOGANLEA *(1991)*
This extract from Map 72 of the current UBD Brisbane Street Directory, shows the present day scene. Notice that the newly developed Logan City Community Hospital is located in the southwest corner of the old airfield. The isolated strip in the SW corner of this map was WATERFORD. *[Universal Press P/L]*

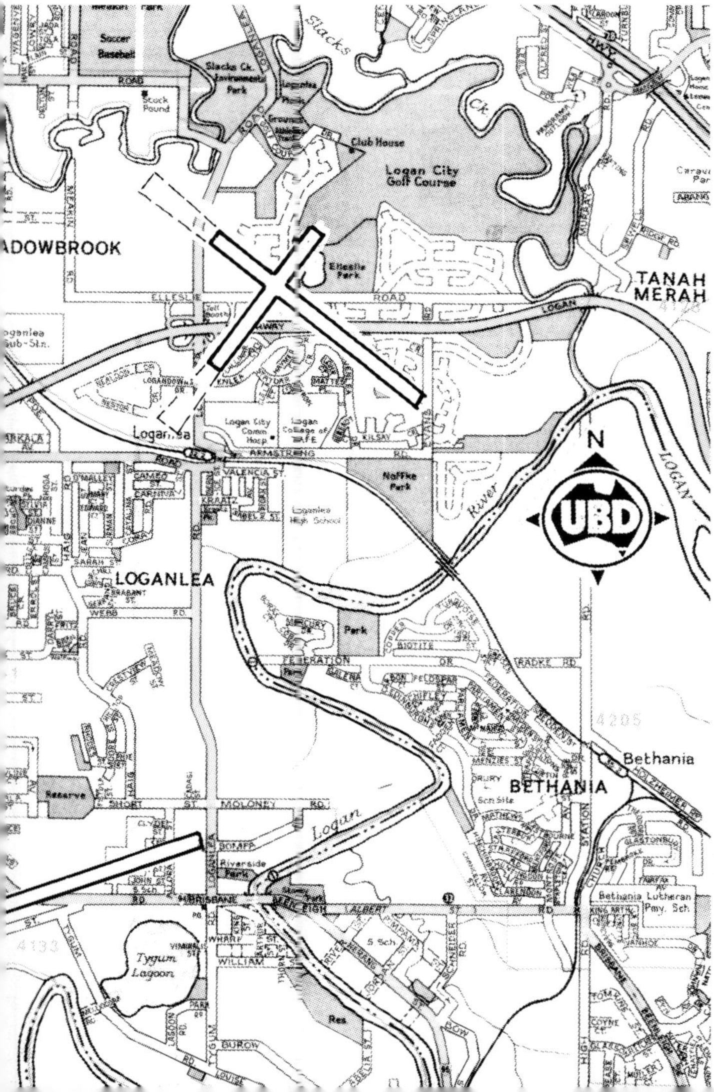

When the US requirement for LOGANLEA expired, mid 1943, the RAAF reviewed its needs. A Directorate of Air Service Planning report of 18 June 1943, to the RAAF's Chief of Air Staff had reviewed operational requirements for the Brisbane area as follows:

*'...**Interceptor fighter**: Archerfield & Eagle Farm; **Long Range fighter**: Amberley; **Reconnaissance**: Loganlea; **Divebomber**: Jondarayan; **Flying Boat:** Brisbane River; **Strategic Air Route**: Leyburn; **Army Cooperation**: Toogoolawah... Loganlea is partly constructed, Strathpine/Petrie is the current fighter aerodrome.'*

It appears such 'plans' did not proceed. The airfield lapsed and the fate of these hideout structures is not known.

WATERFORD *(27/1/1991)*
Standing near what would have been the western end of the airstrip and looking generally along its east-north-easterly bearing, it is difficult to believe this site once served as it did. The RAAF Airfield Data Sheet for WATERFORD records 'that (at least) one arch type camouflaged hideout suitable for fighter aircraft only has been constructed'. My efforts to establish where such hideout(s) was located have so far been unsuccessful. The Data Sheet goes on to mention that the strip was made available for practice landings and take off by US fighters by July of 1943. [Author]

WATERFORD *(17/7/1942)*
Looking west along Kingston Road where it crosses a right angle bend in the Logan River, the WATERFORD airstrip is seen in the centre of this photograph. The result of the siting of this strip was the need to deviate the road southwest near the far end of the strip bringing into being that part of Beutel Street which exists today. The through-going part of Kingston Road has long since resumed its function and the old river crossing has given way to a new bridge on the direct projection of Kingston Road. In the right background, can be seen the clearing of KINGSTON A-10 and on the extreme left hand edge in the background the south westerly aligned clearing of KINGSTON A-12 can just be made out. [NLA (RAAF Official)]

WATERFORD

'...Igloo hangars, offer of purchase.'

This is the initial entry on the old Works and Services branch property card for WATERFORD, RAAF. The entry is dated May of 1945 and appears to represent an offer from the IWC Brisbane, which may have been the Irrigation and Water Supply Commission.

An archive report by then F-Lt John Keays dated 10 July 1942 in respect of WATERFORD strip, briefly states -

'...This strip is adjacent to Loganlea strip was commenced under RAAF supervision but went over to USA with Archerfield. Dispersal scheme is under construction.'

With a 'serviceability date' of 8 June 1943 the following notes appear on a 1" : 1 000′ scale RAAF Landing Ground drawing No. 43-44/139 now in archive records -

'...Graded strip which was compacted and allowed to come back into grass, used by US trainer aircraft, soil will not carry aircraft in summer rainfall period. One fighter arch type hideout located to south across road.'

Presumably that is the 'igloo hangar' that is scored on the property card in reference to the offer to purchase. I've not been able to find anyone with intimate knowledge of that airstrip and while of a low key nature, it remains one of the interesting wartime strips located nearer to Brisbane.

GOOLMAN

'...We could find no record of... an airfield south of Amberley named Purga. Was the runway sealed? Perhaps it was only a small emergency/satellite drome for Amberley?'

Such was the reply by the RAAF's Historical Officer in February 1987 when information was sought by the Watts Bridge (TOOGOOLAWAH) Memorial Airfield Association. Doubtless the Historical Officer's records would have shown GOOLMAN, but without reference to maps which show that Goolman and Purga were adjoining sidings on the now defunct Fassifern branch rail line south of Ipswich, the connection would not have been made.

GOOLMAN certainly was a Relief Landing Ground (RLG) for AMBERLEY. Its initial need was established when the Service Flying Training School (SFTS) was planned for AMBERLEY. GOOLMAN appears first on the Property cards of the Works and Services Branch of the Dept of Interior, Queensland, on 28 February 1941: 'Preparation of Site'. The rest of 1941 passed without more than further 'consideration' being given. Early 1942 contour surveys etc. were hastily organised, and by 10 July 1942 RAAF HQ field liaison officer F-Lt Keays reported -

'...formation has been completed and gravelling should be completed in a week or so. Site was inspected with Mr Wyeth with a view to location of hideouts. Suggest 1 and 2 to the west of south end, 3 to 8 to be located in timber on either side of centre of runway. I marked a plan for Mr Payne showing boundary of area required. Dept of Interior have carried out extensive boring tests for water but with little success. I do not anticipate that any camp will be built at the site...'

In August of 1942 the Camouflage Committee reported the CO at AMBERLEY, Gp Capt Douglas as not liking -

'...strips at Warwick Road, Rosewood and Bremer. They are too short. New strip at Goolman is very satisfactory and dispersal scheme will be fixed with Mr Wyeth on his return from the west.'

Often when property owners affected by airfield works of this nature came to realise that certain lands were 'earmarked' for further work, out came the axe! The camouflage report of 6 September 1942 mentions -

'...Goolman: since the dispersal points were selected the owner of that property has been cutting timber and two or three hideout positions will have to be changed. The cutting is being stopped.'

A subsequent camouflage report of 10 October 1942, shows some work on 'buildings' -

← ***GOOLMAN*** *(14/9/1943)*
A detachment of 11 Repair and Salvage (late Servicing) Unit (RSU) was deployed here for about the second half of 1943. Here Boomerang aircraft are seen undergoing service in rather primitive workshop conditions. The nature of the canvas(?) covered 'hangars' appears very much improvised. There are on record various mentions of 'scrap aircraft packing crates' being used to provide camp buildings at GOOLMAN. This puts me in mind of the classic comment by Michael Caine acting the part of a Squadron Leader in the film 'The Battle of Britain' when he was told to take his squadron to some airfield where the only building was that of a small Flying School - he remarked that he certainly knew the place - it had 'a nasty little shack full of dead flies!' [F Smith courtesy R Hourigan]

← ***GOOLMAN*** *(17/7/1942)*
This site was identified during 1940 as one of the few reasonable locations for satellites to AMBERLEY aerodrome. Development commenced in March 1942 being pressed ahead through the outbreak of the Pacific war. This photograph is taken looking northward, and in the distance left of centre it may be possible to make out the open area of AMBERLEY airfield. [NLA (RAAF Official)]

GOOLMAN (14/9/1943)
Boomerang A46-15 is standing by improvised workshop facilities operated by 11 RSU (Detachment). *[F Smith courtesy R Hourigan]*

'...Mr Wyeth went to Amberley and then with Barracks Officer (F-Lt J) Appleby proceeded to Goolman re camp, buildings of which have been constructed from scrap plane cases. They are under fairly good tree cover; but need dull painting. They also inspected sites being used for storage of aircraft and crated engines. These are well sited under trees.'

This tends to confirm recollections by Jim Steenson, a RAAF member of the 'erection squadron', then at work assembling American fighter planes at AMBERLEY. Jim spoke of the crated aircraft being stored 'in the bush off the Station'.

Records show a faltering in the advancement of the airstrip towards full bitumen sealing. A 200 yard length (about 180m), the high impact area at each end of the strip was set to be sealed by late 1942, but this was deferred. It was given 'A2' priority in late March 1943, and apparently completed by August 1943. By September, the idea of actually basing a full RSU here finally lapsed in favour of TOOGOOLAWAH. The fact that such work didn't proceed there either simply demonstrates the fluidity of airfield development decisions at the time.

11 RSU spent a long yet productive fourteen months in its formation, training and preparation for active service in New Guinea. C R (Rex) Taylor commanded this unit for a little over two years of its war-time service and recently published its history titled ***I Sustain the Wings***. Of the field training phase in Australia, Rex wrote -

'...The reason for our Detachment at Goolman was to gain some experience of climatic conditions somewhat like those which lay ahead in New Guinea and further north west. Also, as there was considerable operational flying practice in the Queensland area at that time, the crash rate had increased requiring salvage and retrieval of all types of aircraft exceeding the capacity of 3 AD to handle. In this regard, the air ambulance aircraft, A34-19 (one Rex mentioned in his book), was typical of this kind of work and the wreckage was so badly damaged that it was returned to 3 AD for write-off.'

Rex was referring to the DH-84 Dragon which crashed killing its pilot F-O Thorley, for whom the Watts Bridge Memorial Airfield Association has now named its entry road to that airfield.

The 1" = 1 mile Military Survey Map reprinted in June 1945 was marked 'information correct to January 1924'. On the GOOLMAN site it shows 'emergency landing ground' and a straight piece of 'A1' road (by legend, 'concrete or sealed - sufficient width for two traffic streams'). The only other piece of 'A1' class road shown for miles around, is the section of Ipswich to Warwick road through the unmarked AMBERLEY airfield!

Mrs Jean Haughton-James (nee Philp) lived as a young girl on her parent's property 'Wyaralong' a few miles south of GOOLMAN. She left home around February 1942 to join the WAAAF. Jean told of a landing area there, on which the early Queensland pioneer airman, Jack Treacy, often landed in his Avro 504 to visit her parents in the mid 1920s, 'probably flying the silent movie films to various towns'. This landing area was also frequented by Royal Queensland Aero Club planes doing their 'cross-country flights' and 'on occasions the boys conveniently gave some of us a lift to the City'. No doubt the ELG by Goolman was a similar unofficial pick-up spot.

'Wyaralong' appears in the RAAF's 23 Sqn ORB as a bombing and gunnery range, and while landings there were not authorised, Jean recalled pleasant swim and picnic occasions, after a Wirraway or other aircraft would fly over, dropping a message streamer -

'...saying they'll be expecting engine trouble!'

Jean even remembered the Sqn's DH-94 Moth Minor which she said the pilots referred to as their 'Minorschmitt'.

On 1 April 1945 Archive records indicate that post-war, the Dept of Civil Aviation -

'...requires the following dromes which are now in use - Goolman and Loganlea, Queensland. These cannot be handed over until no longer required by RAAF.'

Such interest by DCA lapsed and around the next 'April Fools' Day', 1946 the same file shows GOOLMAN as one of many not further required for Civil Aviation purposes.

It would have been nice to have located an RAAF veteran who had spent time at GOOLMAN in those 'packing case days'. However, to date that has not occurred.

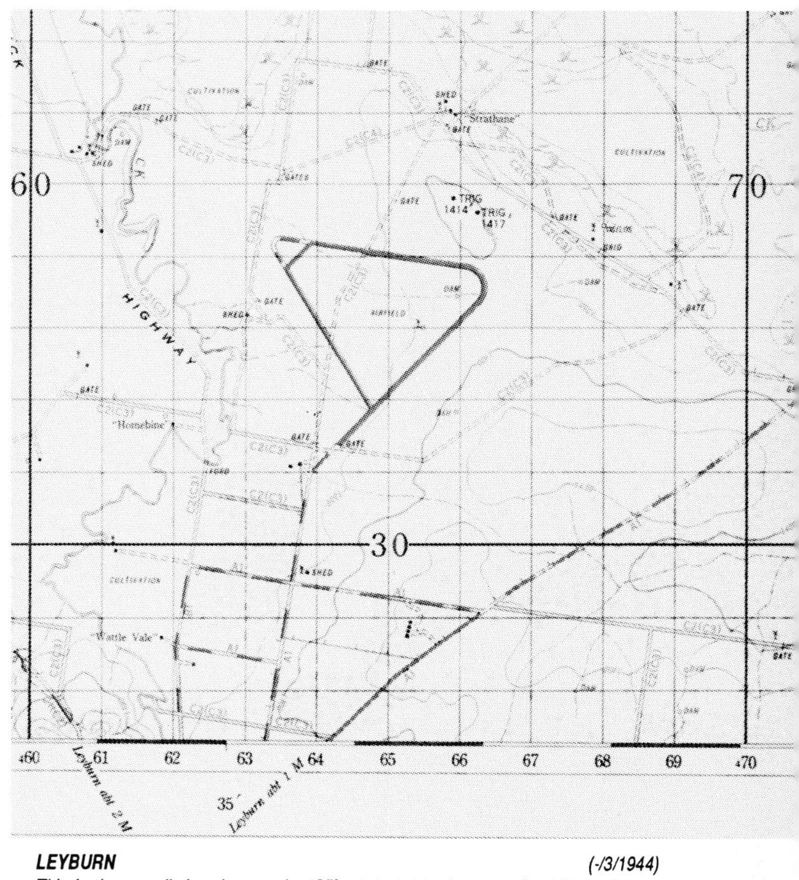

LEYBURN (-/3/1944)
This is the compilation date on the 'Clifton' 1:63 360 (1" = 1 mile) Military Survey Map of the time. Map Grid reference 653288 is roughly the centre of the camp facilities for LEYBURN. While dispersals and hideouts were planned for this field the need passed without their being completed. *[Department of Defence]*

LEYBURN

***'...Location: 120 miles west of Southport, 35 miles southwest of Toowoomba, 6 miles north of Layhuon.'**(???)*

Many of us can lay claim to having the 'world's worst handwriting' - I know I can. However, I also remember John Keays having highly competitive handwriting! So it's not surprising to see the typist at work on P-O J Keays' report of 7 March 1942, typing 'Layhoun' for Leyburn. This report was on a proposed airstrip site on a property named 'Strathane', close to that finally developed.

While John Keays was involved in the location and assessment of new airfield sites during this urgent part of 1942, the credit for finding this LEYBURN site probably falls to Eric Read, while a member of 23 Sqn. Eric quoted his log entry as -

'...September 23 1940, A20-100 (Wirraway aircraft) Jackson and myself - Warwick, Strathane.'

Eric explained that he and then F-O J F Jackson were stationed at ARCHERFIELD and suitably instructed to conduct an aerial reconnaissance along with ground checks, searching for suitable sites 'west of Toowoomba'. Little did Eric expect that some five years later he would be stationed at this airfield in command of a Special Duties Flight.

LEYBURN was one of four fields provided for operational flexibility of the USAAF's HB force in the event of a Japanese landing 'near Brisbane'. Keays noted -

'...Sites at Jondaryan, Cecil Plains, Leybuon (sic) and Condamine were inspected with F-Lt Higginson. General Duties officer from Amberley on 19, 20, 21 March... comments are... Leyburn - F-Lt Higginson was very impressed with this site except in so far as it is 20 miles or so from the rail head.'

Keays' 28 March 1942 signal to Air Board referred to preparation and design works being needed on some of these, but at LEYBURN he anticipated rapid progress by the MRC. At both here and CONDAMINE he said -

'...Clearing on these sites (is) very light and sites could be available for reception (of) planes in 1-2 weeks.'

While Keays predicted fast construction progress, the constraints of the pre-war Departmental procedures quickly threatened. Maybe it took the pivotal Coral Sea Battle to swing the 8 May 1942 approval by War Cabinet for urgent works to 'proceed on the basis of plans estimates and schedules to be submitted when available'. In fact, P-O Keays' report of 5 May 1942, recorded -

'...Works inspected on Sunday 3 May. The NW-SE strip had been cleared and graded for a length of 7 000' and was fit for use by all types of aircraft in dry weather. Camp was erected, tracks graded to gravel pits and construction of "chinamen" commenced. Gravelling of NE-SW runway was to commence on 5 May. One strip will be available continuously from now on. Water is available in Canal creek and could be carted to camp site. MRC will assist in this matter if so required.'

Also around this time the RAAF's Works Director Air Services sought clear adoption of the name LEYBURN for this field so as to remove confusion. Some authorities were calling it 'Strathane'.

During June of 1942, these four airfields 'west of Brisbane', received close attention to layout and design of dispersals and associated taxiways and their camouflage consideration. In this scenario, John Keays worked well with the DHS field staff. However, by 16 June he found it necessary to write, pleading for retention of particular proven personnel in this close liaison between the Camoufleur and responsible RAAF staff. As described more fully in the context of BRYMAROO, both parties had contributions to make. Both had to recognise that ultimate responsibility lay with the RAAF.

During July, the search was on for requisite dispersal strips for these fields on the Darling Downs. Near LEYBURN, several sites were considered, but none proceeded, principally because the need had passed.

The report for 9 August 1942 from the 'Division 3 - Dalby' component of the SDCC, indicates the non-cooperation encountered by some camouflage personnel. Mr C F Oliver had been transferred from the DHS' Victorian organisation to LEYBURN -

'...As there were no facilities for living at the drome and blankets ordered have not arrived, Mr Jordan stayed for a few days at Clifton with Mr Oliver and drove him to the drome. He was thoroughly instructed in the procedure of the work. Arrangements have now been made for him to live at the drome, but he will be required to "find" (provide his own food) and cook for himself.'

On 20 September 1942, the regular Camouflage Report noted some of the camp buildings -

'...will be ready to start dull painting in a few days. The work in progress on lead-ins (from taxiways to individual hideout), etc. is being closely watched together with the lopping of trees...'

By October 3, informal instructions reached the field about a possible halt to erection of 'hutted camps' at LEYBURN, CECIL PLAINS and JONDARYAN. Confirmation came on 7 October, yet the Allied Works Council minutes of 22 October 1942, record fresh instructions to recommence work on the 'temporary hutted camp' at LEYBURN. Such 'hiccups' in the progress of this sort of work were not uncommon at that time. The danger of invasion had passed and the question of the future role of this airfield had to be addressed. Australia was aiming for its own Heavy Bomber Squadrons though more than a year was to pass before that materialised.

Provision of aircraft dispersals 'hotted up' late in 1942. In the 'pre-planning' part of the 8 November Camouflage report, the then RAAF liaison officer, P-O Harrison -

'...advised Wyeth that Cecil Plains, Jondaryan, Leyburn, Condamine and Lowood require 18 hideouts each and Archerfield, Amberley and Toogoolawah... 24 each.'

By November, detail of such hideouts was clarified in that the eighteen at LEYBURN were to be of the 'conceal only' type, with no provision being made for splinter proofing. From subsequent Camouflage reports the work on these hideouts was, to say the least, slow and maybe half-hearted. Possibly 'someone' knew the impending outcome. On 7 March 1943 it is recorded -

'...Leyburn aerodrome (DHS Project No.) 145 - inspected by Mr Piper with Mr McInnis on 2 March. The lead-ins to hideouts are not yet constructed. When work was stopped, poles and wires had been partly erected on 3 hideouts, and poles are lying on the ground at most of the others. The walls of the camp buildings have been satisfactorily painted a grey-green.

LEYBURN *(5/3/1943)*
This oblique shows the freshly completed airfield, with catch drains neatly graded to protect the pavements from saturation by heavy run off. Runways are sealed, ready for occupation, which did not come until mid 1944. [CPE (RAAF Official)]

The fibro roofs are only primed.'

Early in 1942, particularly in North Queensland, airstrips were often 'completed' and useable long before adequate camp facilities appeared. At least at LEYBURN, that mistake was less evident. Both facilities were completed around April 1943. The airfield then settled down to inactivity! It joined a list of some seventeen unoccupied aerodromes in Queensland during July, 1943. It was quoted as a 'heavy bomber field with buildings and services for 450 personnel (no sleeping accommodation)' and 100% complete. By 18 June 1943, it appeared in a DASP review of operational aerodrome requirements in the Brisbane area. Its assignment to **Strategic Air Route** foretold the important clandestine role it was later to play. The rest of 1943 was to pass before the RAAF won an allocation of B-24 Liberators for its HB Sqns.

A glimpse of forthcoming activity for LEYBURN is seen in an archive report dated 11 January 1944, wherein the CO of Eastern Area addressed No 2 DWO Brisbane. He pointed to the need for a 'stores building' because of the 'heavy rain known to fall during the wet season'. Quite what function(s) such building served I'm not sure, but it spelled forthcoming habitation -

'...for your information these strips together with several others, will be used for the formation of Heavy Bomber Squadrons commencing approx July 1944.'

March 1944 brought an abrupt and disappointing end to the RAAF's Vultee Vengeance active service in front-line New Guinea. All three squadrons, 21, 23 and 24, were withdrawn to the mainland. Personnel from 21 Sqn were largely dispersed to other postings when it was decided to rebuild and re-equip it with Liberators. Phil Dynes, as the Official Historian for the RAAF's 200 Flight, commented on 21 Sqn's tenancy at LEYBURN when he wrote, probably drawing from relevant ORBs -

'...21 Sqn - on 6 July 1944, instructions from HQ Eastern Area that 21 Sqn was to move from Camden to Leyburn Qld. Advance party, 1 officer and 29 ORs (other ranks) proceeded... equipped with 2 Chev trucks, 1 Dodge Weapon carrier, blankets, mess gear and 7 days rations. Arrived Leyburn 9 July 1944.'

Horrie Norling was one of the few 21 Sqn men to remain with it after its return to the mainland. His story continued up to the point where the Sqn departed LEYBURN for the Darwin area. Horrie was -

'...in charge of all of the ground crew who had formed up at Camden NSW... When we arrived at Leyburn the camp consisted of only one big building - a mess hall and kitchen. I had the job of supervising the digging of latrines for officers, sergeants and airmen, then erecting tents for all the crews when they arrived. The Liberators didn't arrive for some time after the camp was ready. The landing ground had no buildings on it at that time. The camp was situated about 3 miles away from the drome hidden in a pine and she-oak forest on the Leyburn to Clifton Road... The Senior NCOs of all musterings had the task of ordering all the spare parts, tools and equipment needed to service and maintain the Liberators - quite some job! When the pilots were trained in handling the aircraft they were then flown to Fenton in NT. The ground staff then went to Brisbane by tender (...the term used for trucks) the equipment went by boat to Darwin and all the ground crew went by rail from Roma Street Station to Mt Isa via Townsville then across country by tender to Fenton.'

LEYBURN ca(-/12/1944)

'In a tent at LEYBURN - just after supper on steak and toast - from left to right, Joe, Alan, Reg and I'. *WOE (Warrant Officer - Engineering) Horrie Norling was one of an advance party for the RAAF's 21 Squadron arriving at LEYBURN in July 1944. Horrie was not able to date this photograph but knowing the supper time temperatures around LEYBURN on the Darling Downs during winter, I'd say these fellows aren't rugged up enough for winter! Notice the wartime issue sandshoes, the plywood butter box on the folding table, the bakelite radio case and the photographs of wives or sweethearts.* *[H Norling ex RAAF]*

23 Sqn joined 21 at LEYBURN. Phil Dynes explained that its nucleus moved there from Menangle, NSW on 19 October 1944. It too reformed as a HB Sqn with W Cdr R A Dunne as CO. LEYBURN was set to become over-crowded! Phil Dynes' account had 99 Sqn, the first of the new B-24 units to form, arriving there 1 February 1945 -

'...to become a lodger unit with 23 Squadron. In accordance with HQ Eastern Area direction every assistance was (then) given to 200 Flight in its formation... Messing facilities are most inadequate since the formation of 200 Flight as there is now between 1 000 and 1 300 personnel on this station.'

200 Special Duties Flight was born out of the desire of the Allied Intelligence Bureau (AIB), to have its own secure capacity to insert trained personnel into enemy held areas, by air. This applied principally to the Netherland East Indies (NEI) when Japanese forces had essentially been by-passed by MacArthur's campaign towards the Philippines and beyond. Supplying and maintaining contact with such operatives were obvious supplementary aerial tasks. Modified Liberators were the obvious aircraft for the job. LEYBURN was chosen partly for its isolation, and the Flight formed on 15 February 1945.

The overcrowding was reduced by ordering 99 Sqn to move to JONDARYAN/BRYMAROO, but not before that fledgling unit, with less priority than 200 Flight, was robbed of trained B-24 ground crew. The shift took place over the last week in March 1945. John Burley was a member of 99 Sqn during its brief period at LEYBURN -

'...We lived at Clifton some distance from the strip which was out of bounds to us as the secret 200 Flight was operating from there... 99 Squadron did not have aircraft then... I was the Acting WOD (Warrant Officer - Discipline) and took 30 chaps over to Jondaryan to prepare for the full squadron, about 1 000 men...'

During November of 1944 the early experience with Liberators at both LEYBURN and JONDARYAN revealed deficiencies in the bearing capacity of the runways at both fields, particularly the junction of LEYBURN's two runways and the flanks of its '139°'. By 12 March 1945, a RAAF memo referred to a -

'...conversation between W Cdr Yeaman (Forward Echelon RAAF HQ) and this office (DWB) relative to the procuring of Marsden Mat from US Army for urgent use on aerodromes in Brisbane area. It has now been ascertained that a total of 180 tons (American) i.e. 275 bundles, are required for the provision of hard-standings at Amberley and Leyburn for use by Liberator aircraft. It is requested that steps be taken to obtain the above mentioned material and it would be appreciated if this matter could be treated as urgent.'

A detailed report dated 13 March 1945 generally assessed the 37° runway as in 'excellent order', the 139° as doubtful on the shoulders due to aircraft being parked on one edge leaving little room for manoeuvring on the remaining paved surface. The same report criticised the loop taxiway 'which is badly wheel marked in places'. A separate gravel hardstand was then provided within the southern corner of the triangular airfield layout. Whether it was ever overlaid with PSP is not clear.

Bob Long, one of the 'Z' Special Unit party in the initial foray, code named 'Semut 1', referred in his 1989 publication ***Operation Semut 1***, to the build-up problems 200 Flight experienced -

'...There were also no reserves of Liberators and very quickly the six held were all required for operations; the intensive operational training being flown in the same aircraft which quickly consumed the valuable air frame and engine hours between overhauls. Five aircraft were on strength by February 27 and everyone was working hard to get the unit operational. The sixth aircraft arrived on 10 March and five days later three were dispatched with the "Semut" party...'

Apparently 'Semut' means 'ant' in the language of the Borneo region in which these men were to operate. It was on the way home from this drop that the aircraft of the CO, Sqn Ldr Graham Pockley, disappeared. After that, W Cdr Eric Read took over command.

Mark Carr was a pilot in 200 Flight. He wrote -

'...we started with 6 full crews and 3 supernumerary pilots of which I was one. The supernumeraries were trained as pilot, navigator and in radio, radar and gunnery. However, by the time the replacement for Sqn Ldr Pockley's aircraft (which has never been found) came, one supernumery had been shot in the leg, another had been upgraded to 2nd pilot and the third one, (me!), was sent on the "milk run" to all the aerodromes where we might require a pilot in a hurry.'

Eric Read remained with the Flight until deactivation after the war. He described LEYBURN, starting with -

'...My bed in our sleeping quarters, had a thick layer of newspapers followed by two blankets, two sheets, and seven blankets. It was too awkward to sleep on one's side so "on the back" it was. We were in tents - cooks and "bottle washers" were in properly constructed buildings. All maintenance work on the aircraft was carried out on the hardstanding located on the cross strip which we never used for landings or take-offs... The maintenance of radio and our "special" equipment was carried out in tents... we did not have a "control tower" as such - we provided our own "separation", when carrying out local flying, mostly the dropping of paratroops, in training for their missions.'

Eric covered the typical off-duty recreational and social options. There was -

'...the odd visiting entertainment group but due to the remoteness of the base, that was mainly provided by locals, supplemented by some of the musically minded troops. Films were screened outdoors...'

Eric's stay embraced the local winter and spring, with -

'...clear frosty nights followed by sunny days...'

I had asked about the 'locals', both wild life and human. Eric recalled 'no problem' with either. He volunteered -

'...The "Royal Hotel" (in Leyburn) was "out of bounds" but as we also had an Army contingent on the base there was, through "devious" means, a supply of beer and we were able to turn on free beer to the troops at intervals.'

Eric explained their B-24M Liberators had their camera hatches fitted with a retractable slide for those jumping to exit the aircraft. Parachute rip cords were attached to a static line in the aircraft and more than one 'operator' jumping was known as a 'stick'. It was possible in a fast 'stick' for five operators to clear the aircraft under ten seconds, which was necessary so that when they reached the ground they were in close proximity. These aircraft -

'...carried the "top secret" "S" phone. This was fitted to the operators before leaving the aircraft, and it was possible thereby to communicate during descent and on reaching the DZ so we were able to report, on return to home base, the success of that particular drop. It also provided a "homing" facility. When re-supplying a party we would arrive in the area and establish communication. With a "left/right" indicator on the pilot's panel it was possible to "home-in" and the party would tell us our distance. We would then identify their position often by white ground strips, which had to be positively identified in case of a capture. Also a smoke signal was usually provided. Unfortunately some parties were a bit over cautious and would not use the equipment. We fixed that by saying "no signal, no supplies" and were determined to carry out the threat. The message got around.'

Within the space of two months, 200 Flight was to loose two more aircraft, their full crews and members of the Army unit they were transporting. Crews led by F-Os A McL Clark and K Emmett were lost over Timor and North Borneo, on 17 and 21 May 1945 respectively.

Keith Gamble, (ex 75 Sqn) told of his one only visit to LEYBURN in a Wirraway, as a Flight Commander in Ferry Flight. Why did he recall LEYBURN? He answered -

'...I remember getting arrested!'

Keith's brother was in F-O Archibald Clark's crew, which Keith believed may have been 'shot down by rifle fire'!-

'...he went missing and I couldn't find anything out about him so, I said to the CO one day - I was stationed down at Parkes (NSW) at the time - I wanted to fly up to Leyburn and find out just what had happened... It was a "top secret" squadron - everything had to be "burnt before you could read it"! And even though I knew people after 4 or 5 years in the Air Force - high up - I couldn't find out anything. It was a very secret mission because if the Japs knew anything about these blokes we were dropping, well... I realised it had to be secret... (but) you know, the family was getting worried and he'd been missing for a couple of months. So I flew up there. I landed and a couple of guys came over (I thought) to say "hello" and they said - "You're under arrest Sir... this is an unauthorised airfield - you can't land here!" I said, "Well I've landed!... I'm here and I want to find out about my brother"... So they took me up to the CO...'

Keith was able to collect some of his brother's belongings, but not much more by way of information. Getting 'arrested' was one thing - getting the plane back to Parkes, on time, without Air Board learning of the unauthorised flight, was another story.

Archive documents show that early August 1945, preparations were in hand for the construction of a cantilever workshop hangar at LEYBURN along with other up-grading works to sections of hardstanding and associated water supply improvement and its connection to this hangar. It certainly seems the construction of no such building ever proceeded at LEYBURN, although at least one photograph has appeared in a publication showing Vultee Vengeance aircraft being serviced close to such buildings, allegedly at LEYBURN.

200 Flight disbanded in December 1945 and the airfield fell quiet. Its triangle of runway and taxiway surfaces came back briefly into prominence with the 1949 staging of motor racing's Australian Grand Prix. After a few years that too lapsed. Today, the Roll of Honour fixed on the Memorial stone near Leyburn town reads, in part -

'...we remember the following RAAF aircrews and the undercover ground operators of "Z" Special Unit who departed Leyburn and gave their lives in Borneo and Timor.'

LEYBURN *ca(-/4/1945)*
In the southern corner of the triangular airfield these non-permanent buildings were assembled to provide some workshop and storage facilities for the Liberator Squadrons using the base. Note the rudimentary duty pilot/observation tower and note also the mobile fuel tanker on the right. *[P Dynes ex RAAF]*

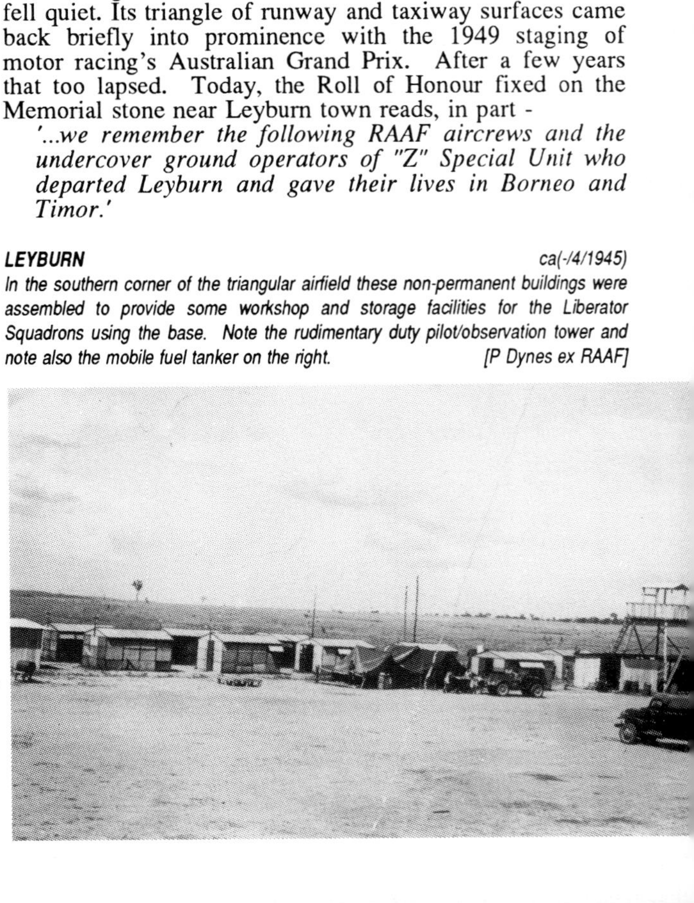

APPENDIX 1

LOCATIONS

Absolutely crucial to the historical research process for an airfield is its LOCATION.

WW2 period development took place directly on and around some pre-war airfield sites such as **GARBUTT**, Townsville.

With others, it embraced a pre-war airfield but extended significantly off to one side - for example **MARYBOROUGH**.

Many airfields were born in the flurry and urgency of early 1942 when the war threat to our continent became real. Several were abandoned as the need subsided.

Air power was rightly seen as the medium by which we would initially be attacked and as the same medium by which we, with the help of the USA, could only hope to counter that threat.

In Queensland alone, upwards of forty new airstrip sites were hastily planned. Most urgent were those around Townsville and inland along the railway as a logistical life line. These were to bolster **GARBUTT** in its key role as the only all-weather airfield 'up near the action' with three hard surfaced runways and some industrial and transportation infrastructure to back it.

Ferry routes, to actually get fresh aircraft to the north and north-west around Darwin, took on added importance. Airstrips sprang up on these routes, both east-west through **CLONCURRY**, and southward through **WINTON** and **CHARLEVILLE**, the latter being viewed as safely inland from the coast where invasion would likely deny ferry traffic the usual coast route.

Cape York was part of this vital ferry route. To fly much needed fighter aircraft to Port Moresby, you had to fly first to the jumping off point on **HORN ISLAND** near the tip of Cape York.

BRYMAROO *(1946)*
Often referred to as JONDAFYAN airfield, it is here shown and named 'BRYMAROO AERODROME' on the 1946 issue of the 1:63 360 (1 inch = 1 mile) military topographic map. Clearly the name has been drawn from the nearby farm centre to the east. *[Department of Defence]*

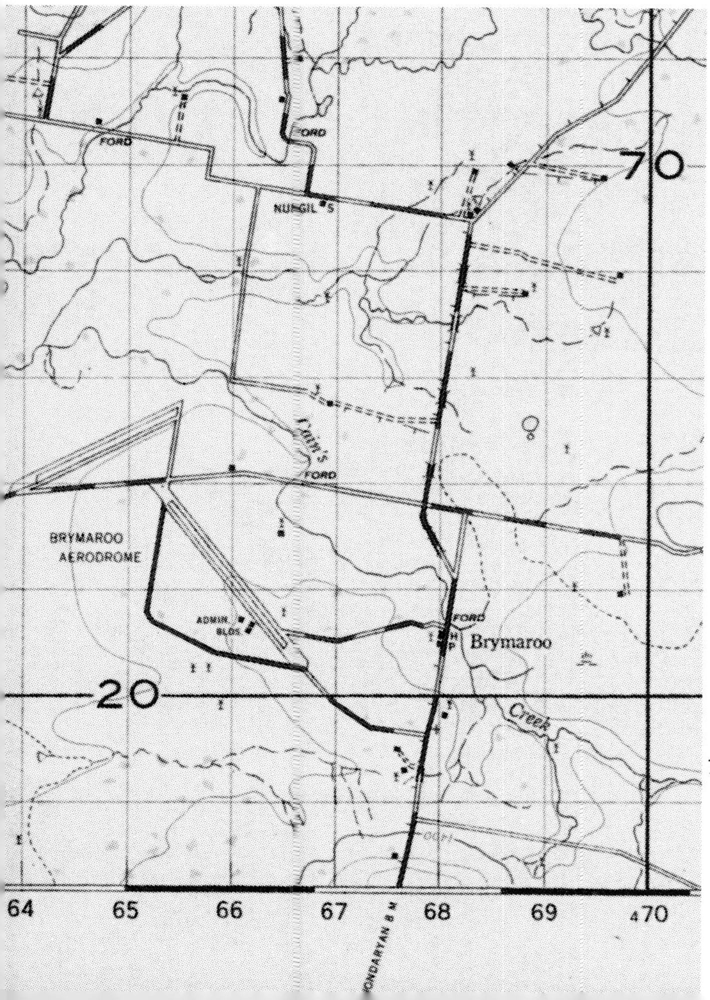

There was of course the airfield activity inland from Brisbane, commenced as part of the controversial *Brisbane Line*.

Australia was correctly viewed by the US as the safe land bastion of the South West Pacific. While Australia looked to its own north coast in review of its aerial defences, as early as 1938, the US very sensibly visited and took stock of the airfield scene, up to just before Pearl Harbor.

With a combination of shipping and trans-Pacific air ferrying, the US could see Australia as the outpost from which to counter Japan's military expansion towards the oil fields and the rubber plantations of the Netherlands East Indies and Malaya.

Obviously if the US had to meet a hostile threat from Japan, conflict anywhere but on its own home soil was preferable. If the Philippine Islands could not be held, there was the Australian continent, so long as its populous south eastern areas could be held.

So various airstrips were planned. Construction by Australian civilian instrumentalities started and languished in the procedural red tape of the early 'forties. When the American forces arrived in significant strength, they were quick to force the pace. They brought equipment - whilst it was 'never enough', it was generally better and more than we could have applied to the task. So more airstrip sites were started or taken over and advanced by the Americans. At an early stage in 1942, when their equipment supply lagged behind the arrival of large numbers of their engineer troops, the men worked with little more than axes, picks and shovels on urgent airfield works.

Simplistically, it could be said that airstrips were built, and thus aircraft were able to operate, leading in turn to improvement in the military position. Invasion by the Japanese forces did not occur, and equally simplistically the need for many planned or part-developed airfields vanished.

Fifty years later, both the military aviation history enthusiast (MAHE ?) and the mother of school age children may ask or be asked 'Where did that happen?' 'Where was that airfield during WW2?' 'What was it like then?'

While the MAHE, if he's lucky, may have a few fast dwindling WW2 veterans to ask in furtherance of his research, who can the Mother turn to for such answers, and relief? I want to believe, that if something deserves to be recorded, someone will make the effort to do so. Then following generations will have better than no starting point at library level.

Appendix 1 has been compiled as a basic level of reference. I started by listing all Queensland Airfields mentioned in the two volumes of ***Australia in the War of 1939-1945*, *Series 3 (Air)*** viz. ***Royal Australian Air Force 1939-1942*** by Douglas Gillison and ***Air War against Japan 1943-1945*** by George Odgers.

Archive study and scrutiny of Aerial Photo Flight Diagrams of the 1940s provided names and general location of further airstrips.

The National Library of Australia (NLA) in Canberra, advised that they held an extensive collection of aerial photography of the Australian Continent, including various airfields, dating from about 1935 to 1960.

In addition NLA reported holding -

'...some photocopy pages of Airfields - World War II from RAAF Museum records, and the Byrne Goodrick Special Collection of lists of wartime airfields together with strip maps of routes and landing ground cards with diagrams and notes'.

Though I believe others sharing my interest in WW2 airfields had already visited NLA and obtained various copies of these cards and diagrams, when I asked I was told in May 1989 that the Byrne Goodrick Collection required permission of Civil Aviation Authority (CAA) before copying. On visiting NLA in June 1989, such 'permission' was still not available from CAA and the RAAF Museum document copy could not be found.

Some weeks later I received a telephone call to say the RAAF Museum document had resurfaced at NLA but it 'now had to be cleared' before relevant copies could be mailed to me. This, remember, was apparently a display item at the RAAF Museum Point Cook, Victoria, but it was decided the Department of Defence would require to 'clear it' before I could gain copies of pages relating to Queensland. Fortunately for me, an airline pilot friend loaned me a mint condition copy of the full document he'd been given by a veteran. This is the RAAF Publication No 694 ***RAAF Aerodrome Guide*** - see Note 1.

I have found difficulty in establishing the precise nature of forerunners to Publication No 694 of January 1946. I have found reference to RAAF Publication 184C - Queensland which appears to have taken the form of Civil Aviation landing ground diagram/note sheets of the time, reviewed and issued to RAAF personnel as required. There also appears to have been some form of 'Notice to Airmen' issued to RAAF pilots, first instituted week ending 18 November 1941.

Flying Boat Bases and Alighting Areas - Australia is the title of RAAF's publication ACD 157 dated June 1943, catering obviously to Flying Boats.

Clearly the US Forces, both the Army Air Corps and the Navy had publications of this Airfield Directory nature. Note 2 refers to one which has been incorporated in Appendix 1. Others are:

- ***Aircraft Facility Directory North East Australia (Queensland)*** produced by United States Navy Department Hydrographic Office, Washington, DC 20, July 1943.
- ***Airdromes Guide South West Pacific Area.*** Published by authority of the Commanding General, Army Air Forces, Office of the Assistant Chief of Air Staff, A-3, HQ, Far East Air Forces November 1944- (First Edition??), and Second Edition, 1 July 1945.

REID RIVER *(1974)*

This extract from the National Mapping 1:100 000 sheet 'Mingela' reaches southward to embrace an area known as Cardington on the Haughton River. Lux Foote of Charters Towers lived there as a boy and recalls various moves by survey parties and others to locate and construct an airfield in that area also. The post-war landing ground shown is in fact quite close to one which was 'pegged out at that time'.

[AUSLIG]

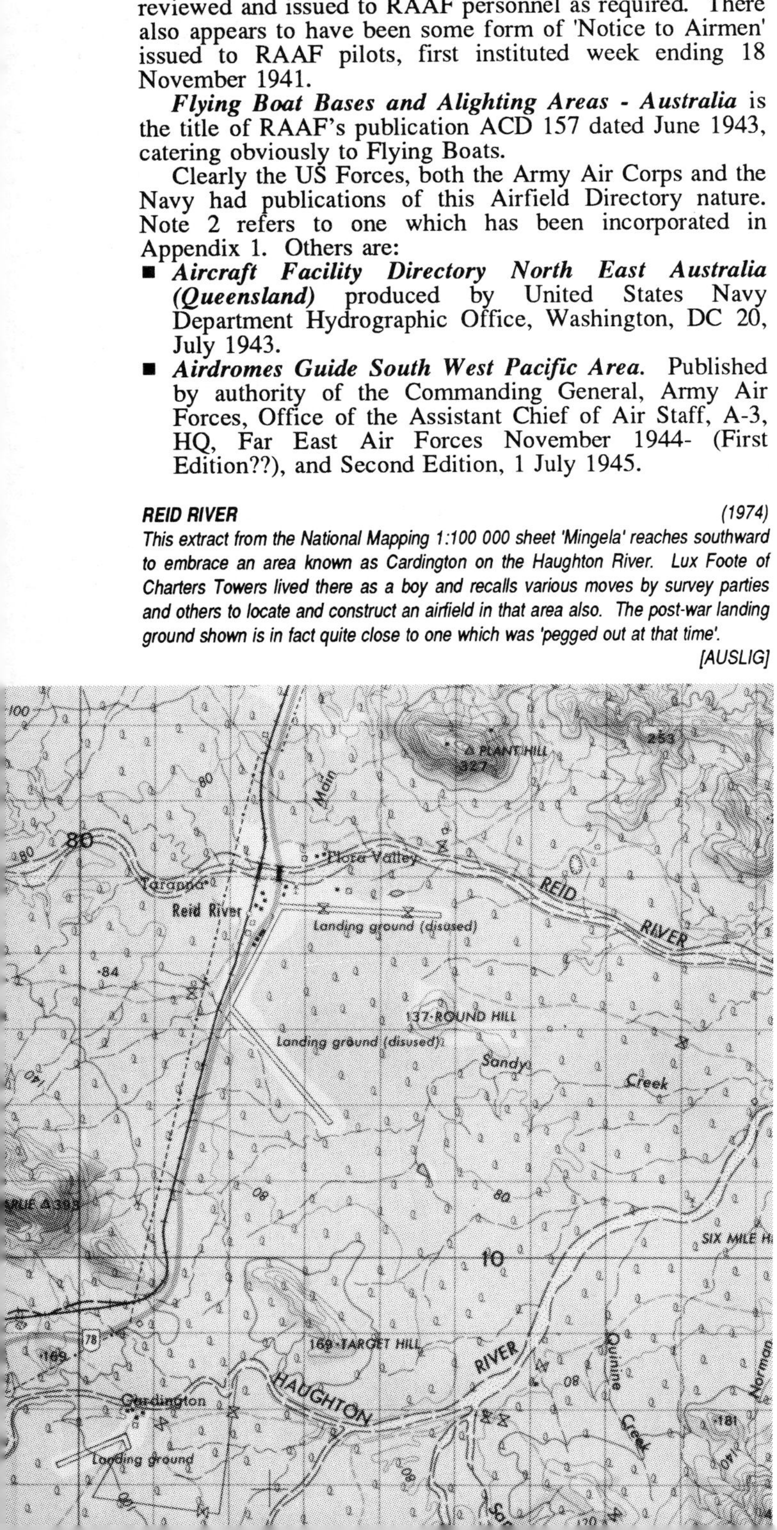

- ***Army Air Forces Installations Directory*** - **extracts** show origin as HQ Army Air Forces Washington DC and are dated 15 March 1946.

It is not clear what distribution such US documents had nor what was their perceived need for accuracy of the airfield information contained. I believe it is interesting to compare a RAAF and US listing as in Appendix 1.

In describing how the airfields are shown, if at all, in comparative mapping, I have attempted to display, firstly, the **actual words** and/or symbols used on the maps.

During the 1930s, Japanese pearling and fishing interests were highly active in Australia's northern waters. It is not surprising that they developed marine use charts of those waters and the continental coastline. It is doubtful that Japanese interests would have had aerial photographic coverage of the northern shores, which begs the question how can it be that, as Staff Sergeant Alan Elletson of NAOU Intelligence is reported, in ***Curtin's Cowboys***, to have observed -

'...it was a funny thing, but the original maps we used were all printed in Japan, prior to the war. They were Army maps, printed on very fine silk and pasted on cardboard, and anything of any note was on these maps.'

With that in mind I was later interested to hear from an Australian Army veteran of his part in a Bougainville Campaign during which a Japanese map was captured. It was of New Guinea and Queensland showing some form of symbol indication of airfields throughout Queensland.

A check of such map holdings at the Australian War Memorial (AWM) Canberra, revealed one map fitting this description. Literal translations of all relevant airfield notations are shown, for comparison, in Appendix 1.

The date of this map is apparently 25 September 1943 and it appears that Japan had access to DCA aviation service publications at least up until their entry into World War 2.

NOTE 1:

RAAF Publication No 694 - printed by RAAF Printing and Publications Unit; copy in National Library of Australia, Canberra, believed to be photocopy only of document held in the RAAF Museum Pt Cook.

NOTE 2:

United States Army Air Forces Publication - 2nd issue, dated 3 January 1944. Copy held by the RAAF Historical Section, Canberra.

NOTE 3:

Information from the BYRNE GOODRICK COLLECTION, held by National Library of Australia, Canberra and from Clive A Lynch, author of Aviation Historical Society of Australia's Newsletter inclusion of 20 August 1989 titled ***Australian Numbered Landing Grounds 1921/1948.***

NOTE 4:

Information from Map Holdings of National Library of Australia, Canberra, Queensland State Archives Brisbane, and State Library of New South Wales, Sydney.

NOTE 5:

Information from National Library of Australia Canberra, Sunmap Centre Queensland, Department of Lands, Brisbane and State Library of Queensland, Brisbane. In a few cases, Universal Business Directory (UBD) map information is used.

NOTE 6:

Australian War Memorial Series ORI No 2151 (3RS) Japanese Maps - Sheets 1072 A7-12 at 1:500 000 and 1103 A1-12 at 1:3 000 000 scale. Translation courtesy of Ms Bronwyn Dewar, Language Specialist, Queensland State Education Department, Brisbane.

AIRFIELD	LOCATION LATITUDE LONGITUDE	SHOWN IN RAAF A'DROME GUIDE (Jan'46) NOTE 1	SHOWN IN (US) AIR PILOT MANUAL OF PACIFIC ISLANDS (Jan'44) NOTE 2	DCA LG (Landing Ground) NUMBER DATE ASSIGNED NOTE 3	SHOWN IN WW 2 PERIOD MAPPING NOTE 4	SHOWN IN CURRENT MAPPING NOTE 5	SHOWN IN WW2 PERIOD JAPANESE MAPPING DATED 25/9/43 NOTE 6
HORN ISLAND	10°35'S 142°18'E	listed as RAAF ELG	map and legend RAAF MAIN LANDING GROUND	No 693 issue 2 8/43	by strip outline on map **Thursday Is** (C54-11-1) 1:63 360 dated 4/43 Grid 548353	shown as **Thursday Is Airport** on map **Thursday Is** (7376) 1:100 000 dated '78 Grid 412294	landing strip Army **Horn Is**
SOMERSET	10°44'S 142°36'E	not shown	not shown	No 448 revised 4/41	by legend ELG on map **Cape York(special)** (C54-12-2) 1:63 360 dated 1/43 Grid 919155	not shown see map **Cape York** (7476) 1:100 000 dated '76 Grid 740117	not shown
HIGGINS (JACKY JACKY) (RED ISLAND POINT)	10°55'S 142°30'E	diagram dated 10/45	map and legend CIVIL GROUND RAAF USED	not shown	**Runway** strip outline see map **Endeavour Strait** (C54-11-3) 1:63 360 dated 2/43 Grid 735920	As **Jacky Jacky Airstrip** on map **Thursday Is** (7376) 1:100 000 dated '78 Grid 585897	not shown
ELLIOT	11°15'S 142°23'E	not shown	not shown	not shown	not shown see map **Jardine R** (C54-15-5) 1:63 360 dated 7/43 Grid 630610	not shown see map **Vrilya Pt** (7375) 1:100 000 dated '77 Grid 510560	not shown
WEIPA MISSION	12°38'S 141°55'E	listed as Civil B Class	not shown	No 775 issue 1 4/44	by legend landing ground see map **Weipa** (D54-3) 1:250 000 dated '63 Grid 506381	not shown (now **Napranum Aboriginal Community**) see map **Weipa**(7272) 1:100 000 dated '80 Grid 965973	not shown
IRON RANGE	12°48'S 143°19'E	listed as RAAF ELG	diagram and photos	No 881 date unknown	**Gordon Strip, Claudie Strip, New Claudie Strip,** see map **Lloyd Bay** (D54-4-30) 1:63 360 dated '44 Grid 755665	**Lockhart River Airfield** see map **Cape Weymouth** (7572) 1:100 000 dated '78 Grid 510845	landing strip not public **Weymouth**
WENLOCK	13°06'S 142°59'E	not shown	not shown	No 597 issue 1 4/41	by legend landing ground **overgrown** see map **Coen** (D54-8) 1:250 000 dated '61 Grid 634329	as **landing ground not useable** see map **Wenlock** (7471) 1:100 000 dated '78 Grid 135498	landing strip not public **Wenlock**
AURUKUN MISSION	13°19'S 141°45'E	listed Civil B Class	not shown	No 776 issue 1 4/44	by legend landing ground see map **Aurukun** (D54-7) 1:250 000 dated '65 Grid 486301	as **landing ground** see map **Aurukun** (7271) 1:100 000 dated '80 Grid 778233	not shown
COEN (RAAF)	13°46'S 143°07'E	listed RAAF ELG	not shown	No 694 issue 1 3/43	by legend airport **Coen Aerodrome** see map **Coen** (D54-8) 1:250 000 dated '61 Grid 650250	as **Coen a'drome** see map **Coen** (7570) 1:100 000 dated '80 Grid 284775	landing strip not public

AIRFIELD	LOCATION LATITUDE LONGITUDE	SHOWN IN RAAF A'DROME GUIDE (Jan'46) NOTE 1	SHOWN IN (US) AIR PILOT MANUAL OF PACIFIC ISLANDS (Jan'44) NOTE 2	DCA LG (Landing Ground) NUMBER DATE ASSIGNED NOTE 3	SHOWN IN WW 2 PERIOD MAPPING NOTE 4	SHOWN IN CURRENT MAPPING NOTE 5	SHOWN IN WW2 PERIOD JAPANESE MAPPING DATED 25/9/43 NOTE 6
COEN (TOWN SITE)	13°57'S 143°12'E	not shown	not shown	No 598 issue 2 4/41	not shown see map **Coen** (D54-8) 1:250 000 dated '61 Grid 660228	not shown see map **Coen** (7570) 1:100 000 dated '80 Grid 377570	landing strip not public **Coen**
COOKTOWN (ENDEAVOUR RIVER MISSION)	15°26'S 145°12'E	diagram 10/45	diagram and photo	not shown	**Airfield** by strip outline on map **Cooktown** (D55-13-20) 1:63 360 dated '43 Grid 045475	**Cooktown airport** strip outline on map **Cooktown** (7967) 1:100 000 dated '64 Grid 043475	not shown
COOKTOWN (CIVIL DROME)	15°28'S 145°14'E	listed Civil B Class	map and legend RAAF MAIN LANDING GROUND and SEAPLANE ANCHORAGE	No 596 issue 3 7/43	**airfield** by strip outline on map **Cooktown** (D55-13-26) 1:63 360 dated '43 Grid 092436	**Landing Ground** strip outline on map **Cooktown** (7967) 1:100 000 dated '64 Grid 090445	landing strip public **Cooktown**
MITCHELL RIVER MISSION	15°29'S 141°45'E	listed Civil B Class	not shown	No 516 issue 3 12/42	by legend a'drome B class see map **Rutland Plains** (D54-15) 1:253 440 dated 3/44 Grid 488042	**Landing Ground** strip outline (**Kowanyama**) see map **Mitchell River** (7267) 1:100 000 dated '83 Grid 806876	landing strip not public **Mitchell River**
LAURA	15°33'S 144°26'E	not shown	not shown	not shown	not shown see map **Cooktown** (D55-13) 1:253 440 dated 1/43 Grid 218036	not shown - work believed to have commenced on map **Laura** (7766) 1:100 000 dated '82 Grid 258805	not shown
RUTLAND PLAINS	15°36'S 141°50'E	listed Civil B Class	not shown	No 518 revised 4/41	by legend a'drome B Class see map **Rutland Plains** (D54-15) 1:253 440 dated 3/44 Grid 499024	**Landing Ground** strip outline see map **Rutland Plains** (7266) 1:100 000 dated '79 Grid 905699	landing strip not public **Rutland Plains**
KOOLATAH	15°54'S 142°22'E	listed Civil C Class	not shown	No 521 issue 3 12/42	by legend ELG see map **Rutland Plains** (D54-15) 1:253 440 dated 3/44 Grid 569994	not shown - see map **Koolatah** (7366) 1:100 000 dated '89 Grid 543425	landing strip not public **Koolatah**
BLOOMFIELD STRIP (AYTON)	15°56'S 145°21'E	not shown	not shown	not shown	not shown see map **Ayton** (D55-13) 1:253 440 dated 1/43 Grid 322996	**Landing Ground** see map **Helenvale** (7966) 1:100 000 dated '82 Grid 210442	not shown
DUNBAR (ALL WEATHER)	16°00'S 142°22'E	listed Civil C Class	not shown	No 652 issue 1 3/41	by legend a'drome A Class see map **Galbraith** (E54-3) 1:253 440 dated 3/44 Grid 561978	**Landing Ground** strip outline see map **Dunbar** (7365) 1:100 000 dated '79 Grid 490289	landing strip not public **Dunbar**
DUNBAR (DRY WEATHER)	16°03'S 142°24'E	listed Civil A Class	not shown	No 808 issue 1 10/46 (effective) 3/45	not shown see map **Galbraith** (E54-3) 1:253 440 dated 3/44 Grid 563975	**Landing Ground** strip outline see map **Dunbar** (7365) 1:100 000 dated '79 Grid 490252	not shown

AIRFIELD	LOCATION LATITUDE LONGITUDE	SHOWN IN RAAF A'DROME GUIDE (Jan'46) NOTE 1	SHOWN IN (US) AIR PILOT MANUAL OF PACIFIC ISLANDS (Jan'44) NOTE 2	DCA LG (Landing Ground) NUMBER DATE ASSIGNED NOTE 3	SHOWN IN WW 2 PERIOD MAPPING NOTE 4	SHOWN IN CURRENT MAPPING NOTE 5	SHOWN IN WW2 PERIOD JAPANESE MAPPING DATED 25/9/43 NOTE 6
INKERMAN	16°16'S 141°23'E	listed Civil C Class	not shown	No 517 revised 2/41	by legend ELG see map **Galbraith** (E54-3) 1:253 440 dated 3/44 Grid 444950	not shown see map **Dinah Is** (7165) 1:100 000 dated '76 Grid 420020	landing strip not public **Inkerman**
GALBRAITH	16°24'S 141°32'E	listed Civil B Class	not shown	No 738 issue 1 3/42	a'drome A Class see map **Galbraith** (E54-3) 1:253 440 dated 3/44 Grid 463932	**Landing Ground (disused)** strip outline see map **Galbraith** (7265) 1:100 000 dated '76 Grid 573863	?not shown
WROTHAM PARK	16°40'S 144°00'E	not shown	not shown	not shown	by legend landing ground see map **Walsh** (E54-4) 1:250 000 dated '62 Grid 165896	**Landing Ground** strip outline see map **Walsh** (7644) 1:100 000 dated '78 Grid 185523	not shown
DENHAM ISLAND	16°40'S 139°14'E	listed Civil C Class	not shown	No 468 issue 4 12/42	by legend ELG see map **Mornington** (E54-1) 1:253 440 dated 9/44 Grid 186897	**Landing Ground (disused)** strip outline see map **Mornington** (6764) 1:100 000 dated '77 Grid 052554	landing strip not public **Denham**
MORNINGTON ISLAND	16°40'S 139°13'E	listed Civil B Class	not shown	No 814 issue 1 7/46 effective 2/44	**Landing Ground** strip outline see map **Mornington** (E54-1) 1:253 440 dated 9/44 Grid 185899	**Mornington Island Aerodrome** strip outline see map **Mornington** (6764) 1:100 000 dated '77 Grid 043568	not shown
CAIRNS	16°53'S 145°45'E	diagram dated 10/45	map and legend RAAF MAIN LANDING GROUND and SEAPLANE ANCHORAGE	No 423 issue 4 4/42	**Runway** strip outline see map **Cairns** (E55-2-52) 1:630 360 dated '46 Grid 717734	**Cairns Airport** strip outlines see map **Cairns** (8064) 1:100 000 dated '59 Grid 673327	landing strip Army **Cairns**
VAN ROOK STATION	16°58'S 141°56'E	listed Civil B Class	not shown	No 772 issue ? dated assigned not known	by legend ELG see map **Galbraith** (E54-3) 1:253 440 dated 3/44 Grid 510863	not shown see map **Van Rook** (7264) 1:100 000 dated '76 Grid 004235	not shown
DELTA DOWNS	16°59'S 141°18'E	listed Civil B Class	not shown	No 570 revised 3/41	by legend a'drome B Class see map **Galbraith** (E54-3) 1:253 440 dated 3/44 Grid 436861	**Landing Ground** strip outline see map **Macaroni** (7164) 1:100 000 dated '76 Grid 340211	landing strip not public **Delta Downs**
LOTUS VALE	17°01'S 141°21'E	listed Civil C Class	not shown	No 584 3/42	by legend ELG see map **Normanton** (E54-7) 1:253 440 dated 11/43 Grid 444853	**Landing Ground** strip outline see map **Double Lagoon** (7163) 1:100 000 dated '76 Grid 400149	landing strip not public **Lotus Vale**
MAREEBA (Hoevet Field)	17°03'S 145°25'E	listed RAAF airfield	by diagram and photo	not shown	**Runway(s)** strip outline see map **Dimbulah** (E55-5-56) 1:63 360 dated '43 Grid 322530	**Landing Ground** strip outline see map **Atherton** (7963) 1:100 000 dated '71 Grid 320119	not shown

AIRFIELD	LOCATION LATITUDE LONGITUDE	SHOWN IN RAAF A'DROME GUIDE (Jan'46) NOTE 1	SHOWN IN (US) AIR PILOT MANUAL OF PACIFIC ISLANDS (Jan'44) NOTE 2	DCA LG (Landing Ground) NUMBER DATE ASSIGNED NOTE 3	SHOWN IN WW 2 PERIOD MAPPING NOTE 4	SHOWN IN CURRENT MAPPING NOTE 5	SHOWN IN WW2 PERIOD JAPANESE MAPPING DATED 25/9/43 NOTE 6
MUNGANA	17°14'S 144°27'E (interpolated from US Map)	not shown	by map and legend CIVIL GROUND RAAF USED	not shown	not shown see map **Atherton** (E55-5) 1:253 440 dated '44 Grid 220 829	not shown see map **Mungana** (7763) 1:100 000 dated '80 Grid 280940	not shown
ATHERTON	17°15'S 145°30'E	not shown	not shown	No 314 revised 4/41	not shown see map **Bartle Frere** (E55-6-62) 1:63 360 dated '43 Grid 431290	not shown see map **Bartle Frere** (8063) 1:100 000 dated '69 Grid 416912	landing strip not public **Atherton**
MIRANDA DOWNS	17°22'S 141°51'E	listed Civil B Class	not shown	No 585 issue 3 12/42	by legend ELG see map **Normanton** (E54-7) 1:253 440 dated 11/43 Grid 501815	not shown see map **Stirling** (7263) 1:100 000 dated '76 Grid 920803	landing strip not public **Miranda Downs**
KARUMBA	17°28'S 140°50'E	listed RAAF ELG	not shown	not shown	not shown see map **Burketown** (E54-6) 1:253 440 dated 9/44 Grid 381804	**Landing Ground** strip outline see map **Karumba** (7063) 1:100 000 dated '79 Grid 820700	landing strip not public on map 1072
KARUMBA FLYING BOAT BASE	17°29'S 140°50'E	by reference FBB 1½ miles south of RAAF ELG	not shown	water a'drome No S5 issue 1 10/47	by legend seaplane base see map **Burketown** (E54-6) 1:253 440 dated 9/44 Grid 580803	not shown see map **Karumba** (7063) 1:100 000 dated '79 Grid 826670	landing strip Water Navy **Karumba**
INNISFAIL	17°34'S 146°01'E	not listed	not shown	No 565 revised 3/41	by legend ELG see map **Innisfail** (E55-6-69) 1:63 360 dated 10/43 Grid 017917	**Innisfail Airport** strip outline see map **Innisfail** (8162) 1:100 000 dated '69 Grid 580950	landing strip not public **Innisfail**
ABINGDON	17°38'S 143°10'E	listed Civil B Class	not shown	No 716 3/42	**Landing Ground** see map **Red River** (E54-8) 1:253 440 dated 5/43 Grid 650784	**Landing Ground** strip outline see map **Abingdon Downs** (7562) 1:100 000 dated '76 Grid 315518	not shown
MOUNT GARNET	17°41'S 145°06'E	not listed	map and legend CIVIL GROUND RAAF USED	No 609 date assigned not known	**Landing Strip** on road see map **Mt Garnet** (E55-5-67) 1:63 360 dated 2/44 Grid 960774	not shown see map **Ravenshoe** (7962) 1:100 000 dated '80 Grid 085445	not shown
NORMANTON	17°42'S 141°06'E	listed Civil B Class	map and legend RAAF MAIN LANDING GROUND	No 148 revised 4/41	by legend a'drome B Class see map **Normanton** (E54-7) 1:253 440 dated 11/43 Grid 408777	**Normanton Airport** strip outline see map **Normanton** (7162) 1:100 000 dated '79 Grid 074444	landing strip public **Normanton**
BURKETOWN	17°45'S 139°33'E	listed Civil B Class	not shown (outside scope of map)	No 501 revised 3/41	by legend a'drome B Class see map **Burketown** (E54-6) 1:253 440 dated 9/44 Grid 230768	**Landing Ground** strip outline see map **Burketown** (6862) 1:100 000 dated '76 Grid 443363	landing strip not public **Burketown**
DAGWORTH	17°52'S 143°42'E	listed Civil C Class	not shown	No 751 issue 2 12/42	**Landing Ground** see map **Red River** (E54-8)1:253 440 dated 5/43 Grid 135754	**Landing Ground** strip outline see map **Galloway** (7662) 1:100 000 dated '76 Grid 864222	not shown

AIRFIELD	LOCATION LATITUDE LONGITUDE	SHOWN IN RAAF A'DROME GUIDE (Jan'46) NOTE 1	SHOWN IN (US) AIR PILOT MANUAL OF PACIFIC ISLANDS (Jan'44) NOTE 2	DCA LG (Landing Ground) NUMBER DATE ASSIGNED NOTE 3	SHOWN IN WW 2 PERIOD MAPPING NOTE 4	SHOWN IN CURRENT MAPPING NOTE 5	SHOWN IN WW2 PERIOD JAPANESE MAPPING DATED 25/9/43 NOTE 6
DOOMADGEE MISSION	17°56'S 138°49'E	not shown	not shown (outside scope of map)	No 907 issue 1 8/47	by legend ELG see map **Westmoreland** (E54-5) 1:253 440 dated 9/44 Grid 147744	**Airfield** strip outline see map **Doomagee** (6662) 1:100 000 dated '74 Grid 686151	not shown
DUNK ISLAND	17°56.5'S 146°08.5'E	not shown	not shown	No 601 issue 1 3/41	by legend ELG see map **Clump Point** (E55-6-75) 1:63 360 dated 11/43 Grid 169457	**Landing Ground** strip outline see map **Innisfail** (8162) 1:100 000 dated '69 Grid 092162	landing strip not public **Dunk Island**
INVERLEIGH (CIVIL)	18°00'S 140°34'E	not shown	not shown (outside scope of map)	No 522 revised 3/41	by legend ELG see map **Donors Hill** (E54-10) 1:253 440 dated 12/42 Grid 350739	not shown see map **Magowra** (7062) 1:100 000 dated '76 Grid 539105	landing strip not public **Inverleigh**
INVERLEIGH (RAAF)	18°01'S 140°32'E	listed Civil A1 Class	not shown (outside scope of map)	No 823 issue 1 5/46	not shown see map **Donors Hill** (E54-10) 1:253 440 dated 12/42 Grid 345738	**Landing Ground** strip outline see map **Milgarra** (7061) 1:100 000 dated '80 Grid 510090	uncertain
MILGARRA	18°14'S* 140°55'E * correct latitude probably should have been 18°10.5'S	listed Civil C Class	not shown	No 355 revised 3/41	by symbol ELG see map **Donors Hill** (E54-10) 1:253 440 dated 12/42 Grid 389715	not shown see map **Milgarra** (7061) 1:100 000 dated '80 Grid 883854	landing strip public **Milgarra**
CROYDON	18°14'S 142°15'E	listed Civil B Class	not shown	No 359 revised 12/41	not shown see map **Croydon** (E54-11) 1:253 440 dated 1/43 Grid 544712	**Landing Ground** strip outline see map **Croydon** (7361) 1:100 000 dated '76 Grid 326850	landing strip not public **Croydon**
FORREST HOME STATION	18°14'S 143°03'E	listed Civil B Class	not shown	No 763 issue 1 12/42	by legend ELG see map **Georgetown** (E54-12) 1:253 440 dated '57 Grid 635709	**Landing Ground** strip outline see map **Forrest Home** (7561) 1:100 000 dated '73 Grid 160821	not shown
GEORGETOWN	18°14'S 143°28.5'E	not shown	not shown	No 614 revised 11/40 (deleted 12/41)	not shown see map **Georgetown** (E54-12) 1:253 440 dated '57 Grid 687708	not shown see map **Forrest Home** (7561) 1:100 000 dated '73 Grid 620820	landing strip not public **Georgetown**
PRESTWOOD	18°15'S 143°09'E	not shown	by map and legend CIVIL GROUND RAAF USED	not shown	not shown see map **Georgetown** (E54-12) 1:253 440 dated '57 Grid 653706	not shown see map **Forrest Home** (7561)1:100 000 dated '74 Grid 310800	not shown
AUGUSTUS DOWNS	18°30'S 139°53'E	not shown	not shown (outside scope of map)	No 851 issue 1 5/47	by legend ELG see map **Donors Hill** (E54-10) 1:253 440 dated 12/42 Grid 273675	**Landing Ground** strip outline see map **Augustus Downs** (6860) 1:100 000 dated '81 Grid 819530	not shown

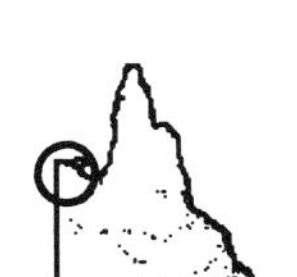

AIRFIELD	LOCATION LATITUDE LONGITUDE	SHOWN IN RAAF A'DROME GUIDE (Jan'46) NOTE 1	SHOWN IN (US) AIR PILOT MANUAL OF PACIFIC ISLANDS (Jan'44) NOTE 2	DCA LG (Landing Ground) NUMBER DATE ASSIGNED NOTE 3	SHOWN IN WW 2 PERIOD MAPPING NOTE 4	SHOWN IN CURRENT MAPPING NOTE 5	SHOWN IN WW2 PERIOD JAPANESE MAPPING DATED 25/9/43 NOTE 6
WONDOOLA	18°34'S 140°54'E	listed Civil C Class	not shown	No 362 revised 4/41	by legend ELG see map **Donors Hill** (E54-10) 1:253 440 dated 12/42 Grid 390670	**Landing Ground** strip outline see map **Donors Hill** (7060) 1:100 000 dated '77 Grid 887463	landing strip public **Wondoola**
FORSAYTH	18°35'S 143°34'E	listed Civil C Class	not shown	No 752 issue 1 6/42	**Forsayth A'drome** strip outline see map **Georgetown** (E54-12) 1:253 440 dated '57 Grid 120666	**Landing Ground** strip outline see map **Forsayth** (7600) 1:100 000 dated '73 Grid 716433	not shown
INGHAM	18°40'S 146°09'E	listed Civil B Class	not shown	No 611 revised 1/41	**ELG** see map **Ingham** (E55-10-93) 1:63 360 dated 3/44 Grid 173583	**Ingham a'drome** strip outline see map **Ingham** (8160) 1:100 000 dated '77 Grid 098358	landing strip not public **Ingham**
CLARAVILLE	18°40.5'S 141°42.5'E	not shown	not shown	No 579 revised 2/41	by legend ELG see map **Croydon** (E54-11) 1:253 440 dated 1/43 Grid 481657	not shown see map **Claraville** (7260) 1:100 000 dated '77 Grid 740350	landing strip not public **Claraville**
DONORS HILL	18°42'S 140°32'E	listed Civil C Class	not shown	No 577 issue 1 7/46	by legend a'drome B Class see map **Donors Hill** (E54-10) 1:253 440 dated 12/42 Grid 349650	**Landing Ground** strip outline see map **Donors Hill** (7060) 1:100 000 dated '77 Grid 515287	landing strip not public **Donors Hill**
PALM ISLES	18°42'S 146°35'E	not shown	map and legend SEA PLANE ANCHORAGE	not shown	not shown see map **Ingham** (E55-10) 1:253 440 dated 12/43 Grid 468655	not shown see map **Palm Is** (8260) 1:100 000 dated '80 Grid 555280	not shown
CARPENTARIA DOWNS (EINASLEIGH)	18°48'S 144°23'E	listed RAAF ELG	by map and legend CIVIL GROUND RAAF USED	No 883 date assigned not known	by legend a'drome A Class see map **Einasleigh** (E55-9) 1:253 440 dated 7/44 Grid 215639	**Landing Ground** strip outline see map **Einasleigh** (7760) 1:100 000 dated '73 Grid 250190	not shown
LYNDHURST	19°12'S 144°22'E	listed Civil C Class	not shown	No 732 issue 1 1/42	not shown see map **Clarke River** (E55-13) 1:253 440 dated nil! Grid 025056	**Landing Ground** strip outline see map **Lyndhurst** (7759) 1:100 000 dated '71 Grid 235750	not shown
MT ST JOHN	19°14'S 146°44'E	not shown	not shown	No 524 '38 (deleted 7/39)	not shown see map **Townsville (special)** (7/114) 1:63 360 dated '42 Grid 864870	not shown see map **Townsville** (8259) 1:100 000 dated '75 Grid 720714	not shown
GARBUTT	19°15'S 146°46'E	by diagram dated 10/45	by diagram and photo, and by map and legend SEAPLANE ANCHORAGE	No 537 issue 2 3/43	not shown see map **Townsville (special)** (7/114) 1:63 360 dated '42 Grid 880865	**Garbutt Airport** strip outline see maps **Magnetic** (8259-I) and **Alice River** (8259-III) 1:50 000 dated '73 Grid 750710	landing strip Army **Townsville**

AIRFIELD	LOCATION LATITUDE LONGITUDE	SHOWN IN RAAF A'DROME GUIDE (Jan'46) NOTE 1	SHOWN IN (US) AIR PILOT MANUAL OF PACIFIC ISLANDS (Jan'44) NOTE 2	DCA LG (Landing Ground) NUMBER DATE ASSIGNED NOTE 3	SHOWN IN WW 2 PERIOD MAPPING NOTE 4	SHOWN IN CURRENT MAPPING NOTE 5	SHOWN IN WW2 PERIOD JAPANESE MAPPING DATED 25/9/43 NOTE 6
STOCKROUTE (adjacent USAAF 4th Air Depot) (DALRYMPLE RD) (MT LOUISA)	19°16′S 146°46′E	not shown	by map and legend CIVIL GROUND RAAF USED	not shown	not shown see map **Townsville (special)** (7/114) 1:63 360 dated '42 Grid 877835	not shown see map **Alice River** (8259-III) 1:50 000 dated '73 Grid 750681	not shown
BOHLE RIVER (BLACK RIVER)	19°17′S 146°42′E	listed RAAF ELG	by map and legend CIVIL GROUND RAAF USED	not shown	as road see map **Townsville (special)** (7/114) 1:63 360 dated '42 Grid 802830	**Landing Ground** (abandoned) strip outline see map **Alice River** (8259-III) 1:50 000 dated '73 Grid 682675	not shown
ROSS RIVER	19°19′S 146°47′E	listed RAAF Airfield	by map and legend CIVIL GROUND RAAF USED	No 242 '36? (deleted 5/41)	not shown see map **Townsville (special)** (7/114) 1:63 360 dated '42 Grid 910890	**Landing Ground** (disused) strip outline see map **Alice River** (8259-III) 1:50 000 dated '73 Grid 781641	not shown
AITKENVALE (2nd WEIR) (THE WEIR STRIP)	19°19′S 146°45′E	listed RAAF ELG	by map and legend CIVIL GROUND RAAF USED	not shown	not shown see map **Townsville (special)** (7/114) 1:63 360 dated '42 Grid 847794	not shown see map **Alice River** (8259-III) 1:50 000 dated '73 Grid 724794	not shown
UPPER ROSS (KELSO FIELD)	19°25′S 146°44′E	not shown	not shown	not shown	not shown see map **Townsville (special)** (7/114) 1:63 360 dated '42 Grid 700660	not shown (part covered by Ross River Dam) see map **Townsville** (8259) 1:100 000 dated '64 Grid 720520	not shown
ANTIL PLAINS (MT STUART STRIP)	19°27′S 146°50′E	listed RAAF ELG	by map and legend CIVIL GROUND RAAF USED	not shown	not shown **Townsville (special)** (7/114) 1:63 360 dated '42 Grid 960645	**Landing Grounds** strips outlined see map **Townsville** (8259-II) 1:50 000 dated '76 Grid 827490	not shown
CANOBIE	19°29′S 140°56′E	listed Civil B Class	not shown	No 283 revised 4/41	by legend a'drome B Class see map **Dobbyn** (E54-14) 1:253 440 dated 2/43 Grid 393560	**Landing Ground** strip outline see map **Canobie** (7059) 1:100 000 dated '77 Grid 922460	landing strip public **Canobie**
GIRU	19°30′S 147°08′E	listed Civil ELG	map and legend CIVIL GROUND RAAF USED	not shown	white strips (cleared) in stippled green (scattered timber) see map **Mt Elliot (special)** 1:63 360 dated 7/42 Grid 242521	not shown see map **Ayr** (8358) 1:100 000 dated '72 Grid 240510	not shown
AYR	19°33.5′S 147°23′E	not shown	not shown	No 417 issue 3 1/41 - (deferred 7/41)	not shown see map **Ayr** (E55-15) 1:253 440 dated '43 Grid 561550	not shown see map **Ayr** (8358) 1:100 000 dated '72 Grid 420390	landing strip not public **Ayr**
WOODSTOCK	19°37′S 146°50′E	listed RAAF ELG	map and legend CIVIL GROUND RAAF USED	not shown	as roads, strip outlines see map **Mt Elliot (special)** (E55-14-123) dated 7/42 Grid 970420	**Landing Ground (disused)** strip outline see map **Mingela** (8258) 1:100 000 dated '74 Grid 830300	not shown

AIRFIELD	LOCATION LATITUDE LONGITUDE	SHOWN IN RAAF A'DROME GUIDE (Jan'46) NOTE 1	SHOWN IN (US) AIR PILOT MANUAL OF PACIFIC ISLANDS (Jan'44) NOTE 2	DCA LG (Landing Ground) NUMBER DATE ASSIGNED NOTE 3	SHOWN IN WW 2 PERIOD MAPPING NOTE 4	SHOWN IN CURRENT MAPPING NOTE 5	SHOWN IN WW2 PERIOD JAPANESE MAPPING DATED 25/9/43 NOTE 6
REID RIVER	19°46'S 146°50'E	listed RAAF ELG	by map and legend CIVIL GROUND RAAF USED	not shown	not shown see map **Mingela** (7/132) (E55-14-132) 1:63 360 dated '43 Grid 967240	**Landing Ground (disused)** strip outline see map **Mingela** (8258) 1:100 000 dated '74 Grid 830135	not shown
FANNING	19°55'S 146°32'E	listed RAAF ELG	by map and legend CIVIL GROUND RAAF USED	not shown	by legend ELG see map **Mingela** (7/132) (E55-14-132) 1:63 360 dated '43 Grid 614085	**Landing Ground (abandoned)** strip outline see map **Mingela** (8258) 1:100 000 dated '74 Grid 510995	not shown
CAMOOWEAL	19°55'S 138°08'E	by diagram dated 10/45	not shown (outside scope of map)	No 124 revised 2/41	by legend a'drome A Class see map **Camooweal** (E54-13) 1:253 440 dated '43 Grid 644505	**Camooweal a'drome** strip outline see map **Camooweal** (6558) 1:100 000 dated '75 Grid 987953	landing strip public **Camooweal**
BREDDAN (BREDDAN ISLAND) (LITTLE SANDY)	19°56'S 146°12'E	listed RAAF Airfield	by map and legend CIVIL GROUND RAAF USED	not shown	not shown see map **Dalrymple** (7/131) (E55-14-131) 1:63 360 dated 10/42 Grid 273023	**Breddan Landing Ground** strip outline see map **Dotswood** (8158) 1:100 000 dated '74 Grid 205942	not shown
MACROSSAN	20°02'S 146°29'E	listed RAAF ELG	by map and legend CIVIL GROUND RAAF USED	not shown	not shown see map **Charters Towers** (F55-2-141) 1:63 360 dated 9/42 Grid 560940	**Landing Ground (abandoned)** see map **Charters Towers** (8157) 1:100 000 dated '75 Grid 460860	not shown
BOWEN	20°03'S 148°13'E	by diagram dated 10/45	by map and legend RAAF MAIN LANDING GROUND	No 515 issue 4 7/43	a'drome and legend see map **Bowen** (F55-3-145) 1:63 360 dated 8/42 Grid 530925	**Airfield** strip outline see map **Bowen** (8557)1:100 000 dated '69 Grid 270859	landing strip Army **Bowen**
BOWEN FLYING BOAT MAINTENANCE BASE	20°03'S 148°13'E (3 km east of airfield)	not shown	by map and legend SEAPLANE ANCHORAGE	water a'drome No S8 issue 2 5/48	not shown see map **Bowen** (F55-3-145) 1:63 360 dated 8/42 Grid 569923	not shown see map **Bowen** (8557) 1:100 000 dated '69 Grid 302857	not shown
CHARTERS TOWERS (CORINDA)	20°03'S 146°17'E	listed RAAF ELG	by map and legend CIVIL GROUND RAAF USED	No 557 8/38 (deleted 2/40)	not shown see map **Charters Towers** (F55-2-141) 1:63 360 dated 9/42 Grid 320918	**Charters Towers Aerodrome** strip outline see map **Charters Towers** (8157) 1:100 000 dated '75 Grid 235832	not shown
SOUTHERN CROSS	20°07'S 146°06'E	not shown	not shown	not shown	not shown see map **Charters Towers** (F55-2-141) 1:63 360 dated 9/42 Grid 092840	not shown see map **Charters Towers** (8157) 1:100 000 dated '75 Grid 023781	not shown
POWLATHANGA	20°08'S 145°58'E	not shown	not shown	not shown	by legend ELG see map **Charters Towers** (F55-2) 1:253 440 dated 4/44 Grid 397480	not shown see map **Homestead** (8057) 1:100 000 dated '75 Grid 928720	not shown

AIRFIELD	LOCATION LATITUDE LONGITUDE	SHOWN IN RAAF A'DROME GUIDE (Jan'46) NOTE 1	SHOWN IN (US) AIR PILOT MANUAL OF PACIFIC ISLANDS (Jan'44) NOTE 2	DCA LG (Landing Ground) NUMBER DATE ASSIGNED NOTE 3	SHOWN IN WW 2 PERIOD MAPPING NOTE 4	SHOWN IN CURRENT MAPPING NOTE 5	SHOWN IN WW2 PERIOD JAPANESE MAPPING DATED 25/9/43 NOTE 6
TARANGIE	20°09'S 145°50'E	not shown	not shown	not shown	by legend ELG map **Charters Towers** (F55-2) 1:253 440 dated 4/44 Grid 381478	not shown see map **Homestead** (8057) 1:100 000 dated '75 Grid 793708	not shown
CLONAGH	20°10'S 140°41'E	listed Civil B Class	not shown	No 282 revised 7/41	by legend a'drome B Class **Cloncurry** (F54-2) 1:253 440 dated 4/43 Grid 364479	not shown see map **Clonagh** (7057) 1:100 000 dated '71 Grid 662707	landing strip not public **Clonagh**
BALFES CREEK	20°12'S 45°55'E	listed RAAF ELG	by map and legend CIVIL GROUND RAAF USED	not shown	by legend a'drome B Class see map **Charters Towers** (F55-2) 1:253 440 dated 4/44 Grid 391471	not shown **Homestead** (8057) 1:100 000 dated '75 Grid 870650	not shown
JESSIEVALE	20°15'S 40°27'E	not shown	not shown (outside scope of map)	No 389 issue 3 3/41	by legend a'drome B Class see map **Cloncurry** (F54-2) 1:253 440 dated 4/43 Grid 337467	not shown see map **Quamby** (6957) 1:100 000 dated '71 Grid 425619	landing strip not public **Jessievale**
STRATHMORE	20°30'S 47°38'E	listed Civil C Class	not shown	No 538 revised 4/41	by legend ELG see map **Bowen** (F55-3) 1:253 440 dated '43 Grid 585434	**Airfield** strip outline see map **Collinsville** (8456) 1:100 000 dated '71 Grid 648324	not shown
PROSERPINE	20°31'S 48°37'E	listed Civil B Class	not shown	No 699 issue 1 9/40	not shown see map **Midge Pt** (F55-4-3) 1:253 440 dated 3/44 Grid 285325	not shown see map **Calen** (8656) 1:100 000 dated '82 Grid 680314	landing strip not public **Proserpine**
CLONCURRY	20°40'S 40°30'E	by diagram dated 10/45	not shown (outside scope of map)	No 29?	by legend a'drome B Class see map **Cloncurry** (F54-2) 1:253 440 dated 4/43 Grid 345416	**Airfield** strip outline see map **Cloncurry** (7056) 1:100 000 dated '72 Grid 487145	landing strip public **Cloncurry**
JULIA CREEK	20°40'S 41°43'E	listed Civil A1 Class	by map and legend CIVIL GROUND RAAF USED	No 757 issue 1 10/43 NB: No 477 issue 2 3/40 is a B Class a'drome directly South of township	by legend a'drome B Class see map **Julia Creek** (F54-3) 1:253 440 dated '43 Grid 484415	**A'drome** strip outline see map **Julia Creek** (7256) 1:100 000 dated '77 Grid 756140	not shown
NONDA	20°40'S 42°29'E✱ ✱ WW2 strip was probably 2km west of present strip	not shown	not shown	not shown	by legend ELG see map **Julia Creek** (F54-3) 1:253 440 dated '43 Grid 568414	✱ **Landing Ground** strip outline see map **Nelia** (7356) 1:100 000 dated '77 Grid 550130	not shown
MT ISA	20°43'S 39°30'E	listed Civil B Class	not shown (outside scope of map)	No 121 revised 7/41	by legend a'drome B Class see map **Mt Isa** (F54-1) 1:253 440 dated '43 Grid 230404	**Landing Ground** strip outline see map **Mt Isa** (6756) 1:100 000 dated '71 Grid 439044	landing strip public **Mt Isa**

AIRFIELD	LOCATION LATITUDE LONGITUDE	SHOWN IN RAAF A'DROME GUIDE (Jan'46) NOTE 1	SHOWN IN (US) AIR PILOT MANUAL OF PACIFIC ISLANDS (Jan'44) NOTE 2	DCA LG (Landing Ground) NUMBER DATE ASSIGNED NOTE 3	SHOWN IN WW 2 PERIOD MAPPING NOTE 4	SHOWN IN CURRENT MAPPING NOTE 5	SHOWN IN WW2 PERIOD JAPANESE MAPPING DATED 25/9/43 NOTE 6
RICHMOND	20°43'S 143°08'E	listed Civil C Class	not shown	No 500 revised 3/41	by legend ELG see map **Richmond** (F54-4)1:253 440 dated '43 Grid 644405	**Rifle Range** strip outline see map **Richmond** (7556) 1:100 000 dated '77 Grid 225044	landing strip not public **Richmond**
TORRENS CREEK	20°46'S 145°00'E	listed RAAF ELG	by map and legend CIVIL GROUND RAAF USED	No 950 '47	by legend **RAAF (Class 2)** see map **Hughenden** (F55-1) 1:253 440 dated '43 Grid 287399	**Landing ground(s)/ (abandoned)** strip outline see map **Pentland**(7956) 1:100 000 dated '77 Grid 925000	not shown
PRAIRIE (STRATHGLASS) (EIGHT MILE)	20°48'S 144°38'E	not shown	not shown	not shown	by legend ELG see map **Hughenden** (F55-1) 1:253 440 dated '43 Grid 244401	white (cleared) strip in pale green (medium forest) see map **Prairie** (7856) 1:100 000 dated '77 Grid 525014	not shown
HUGHENDEN	20°50'S 144°12'E	not shown	not shown	No 424 revised 2/41	by legend ELG see map **Hughenden** (F55-1) 1:253440 dated '43 Grid 196394	**Hughenden Airfield** strip outline see map **Hughenden** (7756) 1:100 000 dated '77 Grid 113957	landing strip not public **Hughenden**
WARREAH (KAROON) (COTONVALE)	20°58'S 144°47'E	not shown	by map and legend CIVIL GROUND RAAF USED	not shown	by legend ELG **Karoon** see map **Hughenden** (F55-1) 1:253 440 dated '43 Grid 263379	white(cleared) strip in pale green (medium forest) see map **Prairie** (7856) 1:100 000 dated '77 Grid 696828	not shown
NATAL DOWNS	21°04.5'S 146°10'E	not shown	not shown	No 646 issue 1 1/41	by legend ELG see map **Buchanan** (F55-6) 1:253 440 dated 1/44 Grid 418363	by legend a'drome see map **Buchanan** (F55-6) 1:250 000 dated '89 Grid DS1267	landing strip not public **Natal Downs**
MACKAY	21°10'S 149°11'E	diagram dated 10/45	by map and legend RAAF MAIN LANDING GROUND AND SEAPLANE ANCHORAGE	No 260 issue 5 4/45	not shown see map **Mackay** (F55-8-7) 1:63 360 dated 6/42 Grid 939532	**Airport** strip outline see map **Mackay** (8755) 1:100 000 dated '71 Grid 265572	landing strip Army **Mackay**
McKINLAY	21°17'S 141°17'E	listed Civil ELG	not shown	No 28 revised 3/41	by legend ELG see map **McKinlay** (F54-7) 1:253 440 dated 2/43 Grid 434342	by legend a'drome see map **McKinlay** (F54-7) 1:250 000 dated '86 Grid WB3146	landing strip not public **McKinlay**
BURKE RIVER	21°19'S 139°50'E	listed Civil ELG	not shown (outside scope of map)	No 118 issue 3 3/42 **N.B.**:No 117 **Duchess** is located 8 km to SE	**Burke River ELG** see map **Duchess** (F54-6) 1:253 440 dated 2/43 Grid 268339	by legend a'drome **(abandoned)** see map **Duchess** (F54-6) 1:250 000 dated 7/87 Grid UB7844	landing strip not public shown in position of Burke River but named **Duchess**
MT COOLON	21°24'S 147°18'E	listed Civil C Class	not shown	No 415 revised 4/41	by legend ELG see map **Mt Coolon** (F55-7) 1:253 440 dated 6/44 Grid 548328	**Landing Ground** strip outline see map **Mt Coolon** (8355) 1:100 000 dated '72 Grid 344349	landing strip not public **Mt Coolon**

AIRFIELD	LOCATION LATITUDE LONGITUDE	SHOWN IN RAAF A'DROME GUIDE (Jan'46) NOTE 1	SHOWN IN (US) AIR PILOT MANUAL OF PACIFIC ISLANDS (Jan'44) NOTE 2	DCA LG (Landing Ground) NUMBER DATE ASSIGNED NOTE 3	SHOWN IN WW 2 PERIOD MAPPING NOTE 4	SHOWN IN CURRENT MAPPING NOTE 5	SHOWN IN WW2 PERIOD JAPANESE MAPPING DATED 25/9/43 NOTE 6
VELLUM DOWNS	21°30′S 143°34′E	not shown	not shown	No 425 4/41	by legend ELG see map **Manuka** (F54-8) 1:253 440 dated 6/44 Grid 123313	not shown see map **Manuka** (F54-8) 1:250 000 dated '80 Grid YB6419	landing ground not public **Vellum Downs**
WINTON	22°22′S 143°02′E	listed Civil A Class	by legend RAAF MAIN LANDING GROUND	No 499 revised 10/41	by legend ELG see map **Winton** (F54-12) 1:253 440 dated 1/44 Grid 629208	not shown see map **Winton** (F54-12) 1:250 000 dated '83 Grid YA0924	landing ground not public **Winton**
WINTON (WINTON EAST)	22°22′S 143°04.5′E	listed RAAF ELG	not shown	No 839 issue 1 11/46	by legend a'drome A Class see map **Winton** (F54-12) 1:253 440 dated 1/44 Grid 633209	by legend a'drome see map **Winton** (F54-12) 1:250 000 dated '83 Grid YA1425	not shown
GLENPRAIRIE	22°37′S 149°57′E	listed Civil	not shown	No 539 revised 3/41	by legend ELG see map **Styx** (F55-12-25) 1:63 360 dated 11/43 Grid 820798	**Landing Ground** strip outline see map **Marlborough** (8852) 1:100 000 dated '72 Grid 030965	not shown
CLERMONT	22°50′S 147°37′E	listed Civil C Class	by map and legend CIVIL GROUND RAAF USED	No 587 issue 2 3/41	by legend ELG see map **Clermont** (F55-11) 1:253 440 dated 2/44 Grid 582153	**Racecourse** see map **Clermont** (8452) 1:100 000 dated '72 Grid 630753	landing strip public **Clermont**
BOULIA	22°55′S 139°54′E	listed Civil B Class	not shown (outside scope of map)	No 761 issue 1 9/42	by legend landing ground see map **Boulia** (F54-10) 1:253 440 dated '59 Grid 276143	by legend a'drome see map **Boulia** (F54-10) 1:250 000 dated '70 Grid UV8766	not shown
CAPELLA	23°04′S 148°02′E	listed Civil C Class	by map and legend RAAF MAIN LANDING GROUND	No 387 revised 7/41	by legend ELG see map **Emerald** (F55-15) 1:253 440 dated 12/44 Grid 628123	**Landing Ground** strip outline see map **Capella** (8551) 1:100 000 dated '74 Grid 057480	landing strip not public **Capella**
MT HEDLOW	23°13.5′S 150°34′E	not shown	not shown	not shown	**emergency airfield** strip outline see map **Yeppoon** (F56-13-37) 1:63 360 dated '46 Grid 511058	not shown see map **Rockhampton** (9051) 1:100 000 dated '71 Grid 506294	not shown
ROCKHAMPTON (CONNOR PARK)	23°22′S 150°31′E	diagram dated 10/45	diagram only	No 241 issue 5 5/48	**Airfield** strip outline see map **Ridgelands** (F56-13-41) 1:63 360 dated 1/46 Grid 413873	**Airport** strip outline see map **Ridgelands** (8951) 1:100 000 dated '72 Grid 417124	landing strip Army **Rockhampton**
ROCKHAMPTON (seaplane)	23°25′S 150°35′E	not shown	by map and legend SEAPLANE ANCHORAGE	water a'drome No S 11 issue 1 10/47	not shown see map **Rockhampton** (F56-13-42) 1:63 360 dated 6/44 Grid 513859	not shown see map **Rockhampton** (9051) 1:100 000 dated '81 Grid 510110	not shown

AIRFIELD	LOCATION LATITUDE LONGITUDE	SHOWN IN RAAF A'DROME GUIDE (Jan'46) NOTE 1	SHOWN IN (US) AIR PILOT MANUAL OF PACIFIC ISLANDS (Jan'44) NOTE 2	DCA LG (Landing Ground) NUMBER DATE ASSIGNED NOTE 3	SHOWN IN WW 2 PERIOD MAPPING NOTE 4	SHOWN IN CURRENT MAPPING NOTE 5	SHOWN IN WW2 PERIOD JAPANESE MAPPING DATED 25/9/43 NOTE 6
LONGREACH	23°25′S 144°15′E	diagram dated 10/45	by map and legend RAAF MAIN LANDING GROUND	No 26 revised 7/41	by legend a'drome A Class see map **Longreach** (F55-13) 1:253 440 dated 6/44 Grid 207079	by legend a'drome see map **Longreach** (F55-13) 1:250 000 dated '79 Grid BQ2205	landing strip public **Longreach**
BARCALDINE	23°29′S 145°16′E and 23°34′S 145°18.5′E	listed Civil B Class	not shown	No 831 issue 1 2/46 issued for old site in error! New site to SE licence applied for 31/5/45	not shown see map **Longreach** (F55-13) 1:253 440 dated 6/44 Grid 323065	by legend a'drome see map **Longreach** (F55-13) 1:250 000 dated '79 Grid CP2793	not shown
EMERALD	23°31′S 148°10′E	listed Civil C Class	not shown	No 270 revised 1/41	**ELG** strip outline see map **Emerald** (F55-15-285) 1:63 360 dated '50 Grid 412691	not shown see map **Emerald** (8550) 1:100 000 dated '74 Grid 319067	landing strip not public **Emerald**
WESTWOOD	23°36.5′S 150°09′E	not shown	not shown	No 352 revised 10/40	not shown see map **Westwood** (F56-13-46) 1:63 360 dated '42 Grid 055590	not shown see map **Mt Morgan** (8950) 1:100 000 dated '71 Grid 090860	not shown
DUARINGA	23°44′S 149°41′E	listed Civil ELG	not shown	No 384 2/41	by legend ELG see map **Duaringa** (F55-16) 1:253 440 dated 12/43 Grid 251045	not shown see map **Duaringa** (8850) 1:100 000 dated '88 Grid 721735	not shown
GLADSTONE (seaplane)	23°50′S 151°15′E	not shown	by map and legend SEAPLANE ANCHORAGE	water a'drome No S 3 issue 1 10/47	by legend seaplane mooring area see map **Gladstone** (F56-13-54) 1:63 360 dated 12/42 Grid 281322	not shown see map **Gladstone** (9150) 1:100 000 dated '81 Grid 210630	not shown **Note**: landing strip not public is shown centred on Boyne Island
SPRINGSURE	24°08′S 148°06′E	listed Civil C Class	not shown	No 286 revised 4/41	by legend ELG see map **Springsure** (G55-3) 1:253 440 dated 10/44 Grid 633994	**Landing Ground** strip outline see map **Springsure** (8549) 1:100 000 dated '77 Grid 103308	landing strip not public. **Springsure**
CONNEMARA	24°14′S 142°17′E	listed Civil B Class	not shown	No 591 revised 10/41	not shown see map **Connemara** (G54-3) 1:250 000 dated '61 Grid 513983	not shown see map **Connemara** (G54-3) 1:250 000 dated '81 Grid XU3119	not shown
BLACKALL	24°26′S 145°25′E	listed RAAF ELG	by map and legend RAAF MAIN LANDING GROUND	No 852 ?	by legend a'drome A Class see map **Blackall** (G55-1) 1:253 440 dated 6/44 Grid 336961	by legend a'drome see map **Blackall** (G55-1) 1:250 000 dated '79 Grid CN4197	✶ symbol probably relates to civil ground
	24°26′S 145°28′E	listed Civil B Class	not shown	No 458 revised 10/40	by legend ELG see map **Blackall** (G55-1) dated 6/44 Grid 339959	not shown	landing strip public **Blackall** ✶

AIRFIELD	LOCATION LATITUDE LONGITUDE	SHOWN IN RAAF A'DROME GUIDE (Jan'46) NOTE 1	SHOWN IN (US) AIR PILOT MANUAL OF PACIFIC ISLANDS (Jan'44) NOTE 2	DCA LG (Landing Ground) NUMBER DATE ASSIGNED NOTE 3	SHOWN IN WW 2 PERIOD MAPPING NOTE 4	SHOWN IN CURRENT MAPPING NOTE 5	SHOWN IN WW2 PERIOD JAPANESE MAPPING DATED 25/9/43 NOTE 6
THANGOOL	24°30'S 150°34'E	listed Civil B Class	not shown	No 622 issue 1 4/41	by legend a'drome B Class see map **Monto** (G56-1) 1:253 440 dated '42 Grid 353952	**Airfield** strip outline see map **Biloela** (9049) 1:100 000 dated '72 Grid 544887	not shown
NORTHAMPTON DOWNS	24°36'S 145°46'E	not shown	not shown	No 308 ?	by legend a'drome B Class see map **Tambo** (G55-2) 1:253 440 dated 6/45 Grid 374941	not shown see map **Tambo** (G55-2) 1:250 000 dated '79 Grid CN7481	landing strip public **Northampton Downs**
MOORLANDS	24°44'S 152°11'E	not shown	not shown	not shown	by legend ELG see map **Baffle** (G56-2-77) 1:63 360 dated 12/42 Grid 315168	not shown see map **Bundaberg** (9348) 1:100 000 dated '75 Grid 176635	not shown
MEADOWVALE	24°47.5'S 152°17'E	not shown	by map and legend CIVIL GROUND RAAF USED	not shown	by legend a'drome B Class **Bundaberg** (G56-2-85) 1:63 360 dated 11/42 Grid 424148	not shown see map **Bundaberg** (9348) 1:100 000 dated '75 Grid 275576	not shown
JUNDAH	24°51'S 143°04'E	listed Civil B Class	not shown	No 815 issue 1 2/45	by legend landing ground see map **Jundah** (G54-4) 1:250 000 dated '62 Grid 628909	by legend a'drome see map **Jundah** (G54-4) 1:250 000 dated '81 Grid YT0951	not shown
TAMBO	24°53'S 146°16'E	listed Civil ELG	not shown	No 478 revised 4/41	by legend ELG see map **Tambo** (G55-2) 1:253 440 dated 6/45 Grid 429906	not shown see map **Tambo** (G55-2) 1:250 000 dated '79 Grid DN2549	not shown
THEODORE	24°53'S 150°06'E	listed Civil B Class	not shown	No 520 revised 4/41	by legend a'drome B Class see map **Monto** (G56-1) 1:253 440 dated '42 Grid 299904	not shown see map **Theodore** (8948) 1:100 000 dated '69 Grid 065440	not shown
YARAKA	24°53'S 144°04'E	not shown	by map and legend CIVIL GROUND RAAF USED	not shown	by legend ELG see map **Blackall** (G55-1) 1:253 440 dated 6/44 Grid 187902	not shown see map **Blackall** (G44-1) 1:250 000 dated '79 Grid BN0343	not shown
MONTO	24°54'S 151°06'E	listed Civil A Class	not shown	No 463 revised 3/41	by legend a'drome A Class see map **Monto** (G56-1) 1:253 440 dated '42 Grid 411904	**Airfield** strip outline see map **Monto** (9148) 1:100 000 dated '78 Grid 082452	not shown
BUNDABERG	24°55'S 152°20'E	diagram dated 10/45	by map and legend CIVIL GROUND RAAF USED and SEAPLANE ANCHORAGE	No 269 issue 3 3/43	by legend a'drome A1 Class see map **Bundaberg** (G56-2-85) 1:63 360 dated 1/43 Grid 459020	**Bundaberg Airport** strip outline see map **Bundaberg** (9348) 1:100 000 dated '75 Grid 313455	landing strip public **Bundaberg**
MOYEN STATION	25°05'S 142°54'E	listed Civil C Class	not shown	No 586 revised 3/41	not shown see map **Windorah** (G54-8) 1:250 000 dated '62 Grid 609880	by legend a'drome see map **Windorah** (G54-8) 1:250 000 dated '91 Grid XT9125	not shown

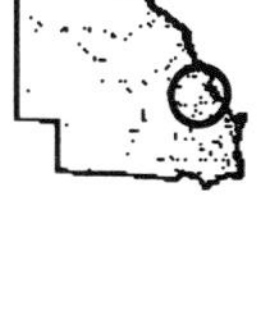

AIRFIELD	LOCATION LATITUDE LONGITUDE	SHOWN IN RAAF A'DROME GUIDE (Jan'46) NOTE 1	SHOWN IN (US) AIR PILOT MANUAL OF PACIFIC ISLANDS (Jan'44) NOTE 2	DCA LG (Landing Ground) NUMBER DATE ASSIGNED NOTE 3	SHOWN IN WW 2 PERIOD MAPPING NOTE 4	SHOWN IN CURRENT MAPPING NOTE 5	SHOWN IN WW2 PERIOD JAPANESE MAPPING DATED 25/9/43 NOTE 6
HIGHFIELDS	25°12'S 146°00'E	listed Civil C Class	not shown	No 418 revised 7/41	by legend ELG see map **Augathella** (G55-6) 1:253 440 dated 9/44 Grid 400866	not shown see map **Augathella** (G55-6) 1:250 000 dated '79 Grid CN9912	landing strip not public
CHILDERS	25°15'S 152°20'E	not shown	not shown	No 277 1/41 (withdrawn 8/41)	not shown see map **Dallarnil** (G56-6-94/103) 1:63 360 dated '42 Grid 464581	not shown see map **Childers** (9347) 1:100 000 dated '75 Grid 320057	landing strip not public **Childers**
MARYBOROUGH * (Seaplane anchorage also shown for Great Sandy Strait)	25°30'S 152°43'E	diagram dated 10/45	by map and legend RAAF MAIN LANDING GROUND *	No 249 issue 4 3/43	by legend a'drome B Class see map **Maryborough** (G56-6-113) 1:63 360 dated 12/42 Grid 884267	**Maryborough Airport** strip outline see map **Maryborough** (9446) 1:100 000 dated '78 Grid 711780	landing strip public **Maryborough**
MUNDUBBERA	25°35'S 151°18'E	listed Civil B Class	not shown	No 377 revised 4/41	not shown see map **Mundubbera** (G56-5-110) 1:63 360 dated 1/43 Grid 327200	not shown see map **Mundubbera** (9146) 1:100 000 dated '69 Grid 290690	not shown
LAGUNA STATION	25°35'S 146°18'E	listed Civil C Class	not shown	No 633 issue 1 3/41	by legend ELG see map **Augathella** (G55-6) 1:253 440 dated 9/44 Grid 432821	not shown see map **Augathella** (G55-6) 1:250 000 dated '79 Grid DM2973	not shown
THYLUNGRA	26°05'S 143°28'E	listed Civil B Class	not shown	No 779 issue 1 2/46	by legend landing ground see map **Eromanga** (G54-12) 1:250 000 dated '74 Grid 669755	by legend a'drome see map **Eromanga** (G54-12) 1:250 000 dated '87 Grid YS4613	not shown
GYMPIE	26°12'S 152°38"E	not shown	not shown	No 49 issue 1 1938 abandoned 11/39	by legend ELG see map **Gympie** (G56-10-132) 1:63 360 dated 1/43 Grid 800400	not shown see map **Gympie** (9445) 1:100 000 dated '82 Grid 647017	not shown
WONDAI	26°18'S 151°52'E	listed Civil B Class	not shown	No 484 issue 3 6/42	by legend ELG see map **Wondai** (G56-10-140) 1:63 360 dated 7/43 Grid 953321	not shown see map **Murgon** (9245) 1:100 000 dated '79 Grid 873896	landing strip not public
BONNIEVILLA	26°19'S 146°12'E	listed Civil ELG	not shown	No 151 issue 5 6/42	by legend ELG see map **Charleville** (G55-10) 1:253 440 dated 4/44 Grid 421731	not shown see map **Charleville** (G55-10) 1:250 000 dated '80 Grid DL2089	not shown
HOME CREEK	26°23'S 151°47'E	not shown	not shown	not shown	by legend ELG see map **Wondai** (G56-10-140) 1:63 360 dated 7/43 Grid 843227	not shown see map **Murgon** (9245) 1:100 000 dated '79 Grid 774809	not shown
MORVEN	26°24'S 147°07'E	listed Civil ELG	not shown	No 136 revised 1/41	by legend ELG see map **Mitchell** (G55-11) 1:253 440 dated 6/44 Grid 521720	not shown see map **Mitchell** (G55-11) 1:250 000 dated 11/86 Grid EL1180	landing strip not public **Morven**

AIRFIELD	LOCATION LATITUDE LONGITUDE	SHOWN IN RAAF A'DROME GUIDE (Jan'46) NOTE 1	SHOWN IN (US) AIR PILOT MANUAL OF PACIFIC ISLANDS (Jan'44) NOTE 2	DCA LG (Landing Ground) NUMBER DATE ASSIGNED NOTE 3	SHOWN IN WW 2 PERIOD MAPPING NOTE 4	SHOWN IN CURRENT MAPPING NOTE 5	SHOWN IN WW2 PERIOD JAPANESE MAPPING DATED 25/9/43 NOTE 6
CHARLEVILLE	26°25′S 146°16′E	diagram 10/45	by map and legend CIVIL GROUND RAAF USED	No 23 revised 6/41	by legend a'drome A Class see map **Charleville** (G55-10) 1:253 440 dated 4/44 Grid 428720	by legend a'drome see map **Charleville** (G55-10) 1:250 000 dated '80 Grid DL2678	landing strip public **Charleville**
LURNEA	26°25′S 146°41′E	listed Civil ELG	not shown	No 184 revised 1/41	by legend ELG see map **Charleville** (G55-10) 1:253 440 dated 4/44 Grid 475720	not shown see map **Charleville** (G55-10) 1:250 000 dated '80 Grid DL6878	not shown
MUNGALLALA	26°26′S 147°32′E	listed Civil ELG	not shown	No 181 revised 1/41	by legend ELG see map **Mitchell** (G55-11) 1:253 440 dated 6/44 Grid 586716	not shown see map **Mitchell** (G55-11) 1:250 000 dated 11/86 Grid EL5475	not shown
DULBYDILLA	26°26′S 147°22′E	listed Civil ELG	not shown	No 182 revised 3/41	by legend ELG see map **Mitchell** (G55-11) 1:253 440 dated 6/44 Grid 551716	not shown see map **Mitchell** (G55-11) 1:250 000 dated 11/86 Grid EL3776	not shown
MITCHELL	26°29′S 147°57′E	listed Civil C Class	not shown	No 748 issue 1 6/42	by legend ELG see map **Mitchell** (G55-11) 1:253 440 dated 6/44 Grid 613709	by legend a'drome see map **Mitchell** (G55-11) 1:250 000 dated 11/86 Grid EL9470	not shown
QUILPIE	26°34′S 144°19′E	listed Civil C Class	by map and legend RAAF MAIN LANDING GROUND (plotted 20 mins too far south)	No 741 3/42	by legend ELG see map **Quilpie** (D55-9) 1:253 440 dated 12/44 Grid 211694	not shown see map **Quilpie** (G55-9) 1:250 000 dated '80 Grid BL2753	not shown
ROMA	26°35′S 148°49′E	listed Civil B Class	not shown	No 180 revised 10/40	by legend a'drome B Class see map **Roma** (G55-12) 1:253 440 dated 6/44 Grid 164697	not shown see map **Roma** (G55-12) 1:250 000 dated 11/86 Grid FL8358	landing strip public **Roma**
KINGAROY	26°36′S 151°50′E	listed RAAF ELG	by map and legend RAAF MAIN LANDING GROUND	No 375 issue 3 7/43	not shown see map **Kumbia** (G56-10-150) 1:63 360 dated 1/43 Grid 916990	**Landing Ground** strip outline see map **Kingaroy** (9244) 1:100 000 dated '78 Grid 843590	not shown
MANNUEM (KUMBIA)	26°39′S 151°37′E	not shown	by map and legend CIVIL GROUND RAAF USED	not shown	not shown see map **Kumbia** (G56-10-150) 1:63 360 dated 1/43 Grid 670910	not shown see map **Kingaroy** (9244) 1:100 000 dated '79 Grid 615530	not shown
NANANGO	26°41.5′S 151°59.5′E	not shown	not shown	No 272 revised 10/40	by legend a'drome A Class see map **Kumbia** (G56-10-150) 1:63 360 dated 1/43 Grid 077856	**Landing Ground** strip outline see map **Kingaroy** (9244) 1:100 000 dated '78 Grid 993474	not shown

AIRFIELD	LOCATION LATITUDE LONGITUDE	SHOWN IN RAAF A'DROME GUIDE (Jan'46) NOTE 1	SHOWN IN (US) AIR PILOT MANUAL OF PACIFIC ISLANDS (Jan'44) NOTE 2	DCA LG (Landing Ground) NUMBER DATE ASSIGNED NOTE 3	SHOWN IN WW 2 PERIOD MAPPING NOTE 4	SHOWN IN CURRENT MAPPING NOTE 5	SHOWN IN WW2 PERIOD JAPANESE MAPPING DATED 25/9/43 NOTE 6
CHINCHILLA	26°44'S 150°37'E	not shown	by map and legend CIVIL GROUND RAAF USED	No 179 issue 2 '38	not shown see map **Chinchilla** (G56-9) 1:253 440 dated '42 Grid 357683	**Speedway** see map **Chinchilla** (9044) 1:100 000 dated '72 Grid 627419	landing strip not public **Chinchilla**
CONDAMINE	26°50'S 150°10'E	listed Civil ELG	by map and legend CIVIL GROUND RAAF USED	No 788 ?	not shown see map **Chinchilla** (G56-9) 1:253 440 dated '42 Grid 308666	**Landing Ground(s) /(abandoned)** see map **Miles** (8944) 1:100 000 dated '72 Grid 176308	not shown
TOOGOOLAWAH	27°06'S 152°27'E	listed Civil ELG	by map and legend CIVIL GROUND RAAF USED	No 791 ?	**under construction 9/43** strip outline see map **Cressbrook** (G56-14-171) 1:63 360 dated 9/44 Grid 583361	**Landing Ground** strip outline see map **Toogoolawah** (9343-11) 1:25 000 dated '82 Grid 463028	not shown
BRYMAROO (JONDARYAN)	27°13'S 151°38'E	listed RAAF ELG	by map and legend CIVIL GROUND RAAF USED	not shown	**Brymaroo a'drome** strip outline see map **Evergreen** (G56-14-170) 1:63 360 dated '46 Grid 655216	**Landing Ground (disused) Brymaroo** strip outline see map **Oakey** (9243) 1:100 000 dated '74 Grid 620880	not shown
WYANDRA	27°14'S 145°58'E	listed Civil ELG	not shown	No 109 revised 4/41	not shown see map **Wyandra** (G55-14) 1:250 000 dated '71 Grid 396620	not shown see map **Wyandra** (G55-14) 1:250 000 dated '79 Grid CK9887	not shown but landing strip not public symbol is shown for ROSEVALE 15km NW
PETRIE (STRATHPINE A-1)	27°17'S 153°00'E	listed RAAF ELG	by map and legend CIVIL GROUND RAAF USED	not shown	not shown see map **Brisbane** (G56-15-183) 1:63 360 dated '37 Grid 166125	not shown see map **Brisbane** (9543) 1:100 000 dated '72 Grid 997822	not shown
STRATHPINE A-2 (LAWNTON)	27°18'S 153°00'E	listed RAAF ELG	by map and legend CIVIL GROUND RAAF USED	not shown	not shown see map **Samford** (G56-14-182) 1:63 360 dated 12/30 Grid 157112	not shown see map **Caboolture** (9443) 1:100 000 dated '72 Grid 999810	not shown
STRATHPINE A-3	27°19.5'S 153°00'E	not shown	not shown	not shown	not shown see map **Samford** (G56-14-182) 1:63 360 dated 12/30 Grid 159077	not shown see map **Caboolture** (9443) 1:100 000 dated '72 Grid 983773	not shown
COOMINYA No 2 (A-6) (LOWOOD No 2)	27°24'S 152°29'E	not shown	by map and legend CIVIL GROUND RAAF USED	not shown	**emergency** strip outline see map **Crows Nest** (G56-14-181) 1:63 360 dated 5/45 Grid 600007	not shown see map **Esk** (9343) 1:100 000 dated '72 Grid 484706	not shown
COOMINYA No 1 (COOMINYA A-4) (COOMINYA)	27°24.5'S 152°31'E	not shown	by map and legend CIVIL GROUND RAAF USED	not shown	not shown see map **Samford** (G56-14-182) 1:63 360 dated 12/30 Grid 640985	not shown see map **Caboolture** (9443) 1:100 000 dated '72 Grid 513693	not shown

AIRFIELD	LOCATION LATITUDE LONGITUDE	SHOWN IN RAAF A'DROME GUIDE (Jan'46) NOTE 1	SHOWN IN (US) AIR PILOT MANUAL OF PACIFIC ISLANDS (Jan'44) NOTE 2	DCA LG (Landing Ground) NUMBER DATE ASSIGNED NOTE 3	SHOWN IN WW 2 PERIOD MAPPING NOTE 4	SHOWN IN CURRENT MAPPING NOTE 5	SHOWN IN WW2 PERIOD JAPANESE MAPPING DATED 25/9/43 NOTE 6
EAGLE FARM	27°25′S 153°06′E	listed USA Airfield	by map and legend CIVIL GROUND RAAF USED	not shown	not shown see map **Brisbane** (G56-15-183) 1:63 360 dated '37 Grid 255955	not shown see map **19 U.B.D.** 1:25 000 dated '91 Grid F8	not shown
HAMILTON REACH	27°27′S 153°05′E	not shown	not shown	water a'drome No S 21 issue 1 10/47	not shown see map **Brisbane** (G56-15-183) 1:63 360 dated '37 Grid 250925	not shown see map 19 U.B.D. 1:25 000 dated '91 Grid M7	not shown
OAKEY	27°25′S 151°45′E	listed RAAF ELG	by map and legend CIVIL GROUND RAAF USED	No 871 issue 1 5/47	strip outline see map **Jondaryan** (G56-14-180) 1:63 360 dated '43 Grid 795993	**Oakey a'drome** strip outline see map **Oakey** (9243) 1:100 000 dated '74 Grid 750682	not shown
LOWOOD	27°27′S 152°28′E	diagram dated 10/45	by map and legend CIVIL GROUND RAAF USED 'LOWOOD No 1'	not shown	**Lowood a'drome** strip outline see map **Crows Nest** (G56-14-181) 1:63 360 dated '45 Grid 610917	**Lowood Racing Circuit (disused)** see map **Esk** (9343) 1:100 000 dated '72 Grid 495620	not shown
WIVENHOE	27°27′S 152°36.5′E	not shown	not shown	not shown	not shown see map **Samford** (G56-14-182) 1:63 360 dated '58 Grid 732930	not shown see map **Caboolture** (9443) 1:100 000 dated '72 Grid 605633	not shown
TONGY	27°21′S 147°15′E	listed Civil B Class	not shown	No 498 revised 1/41	by legend landing ground see map **Homeboin** (G55-15) 1:250 000 dated '66 Grid 537607	not shown see map **Homeboin** (G55-15) 1:250 000 dated '85 Grid EK2675	not shown
CECIL PLAINS (TIPTON)	27°28′S 151°13′E	listed RAAF ELG	by map and legend CIVIL GROUND RAAF USED	No 888 issue 1 8/47	**Tipton a'drome** see map **Bowenville** (G56-13-179) 1:63 360 dated '52 Grid 226924	**Tipton a'field (abandoned)** strip outline see map **Dalby** (9143) 1:100 000 dated '73 Grid 230608	not shown
PRAIRIE STATION	27°31′S 144°21′E 24′	listed Civil C Class	not shown	No 547 revised 10/41	not shown see map **Toompine** (G55-13) 1:250 000 dated '69 Grid 225585	by legend a'drome see map **Toompine** (G55-13) 1:250 000 dated '86 Grid BK4254	not shown
TOOWOOMBA	27°32′S 151°55′E	listed Civil B Class	not shown but GATTON as CIVIL GROUND RAAF USED is! Interpolated Lat Long is 27°34′S 152°13′E	No 157 revised 10/40	strip outline see map **Toowoomba** (G56-14-190) 1:63 360 dated 2/44 Grid 985823	**Airport** strip outline see map **Toowoomba** (9242) 1:100 000 dated '74 Grid 930528	landing strip public **Toowoomba**
ARCHERFIELD	27°35′S 153°01′E	diagram dated 10/45	by map and legend CIVIL GROUND RAAF USED	No 280 issue 5 3/43	**Archerfield a'drome** see map **Beenleigh** (G56-15-193) 1:63 360 dated '47 Grid 170774	**Archerfield airport** strip outline see map **Beenleigh** (9542-IV) 1:50 000 dated '71 Grid 010500	landing strip public **Brisbane**

AIRFIELD	LOCATION LATITUDE LONGITUDE	SHOWN IN RAAF A'DROME GUIDE (Jan'46) NOTE 1	SHOWN IN (US) AIR PILOT MANUAL OF PACIFIC ISLANDS (Jan'44) NOTE 2	DCA LG (Landing Ground) NUMBER DATE ASSIGNED NOTE 3	SHOWN IN WW 2 PERIOD MAPPING NOTE 4	SHOWN IN CURRENT MAPPING NOTE 5	SHOWN IN WW2 PERIOD JAPANESE MAPPING DATED 25/9/43 NOTE 6
REDLAND BAY WATER A'DROME	27°36'S 153°17'E	not shown	by map and legend SEAPLANE ANCHORAGE BRISBANE SIMILAR LEGEND SHOWN AS MORETON BAY	?	not shown see map **Beenleigh** (G56-15-193) 1:63 360 dated '47 Grid 495735	**Brisbane Water Airport** see map **Redland Bay** (9542-1) 1:50 000 dated '67 Grid 313460	not shown
BREMER RIVER (FORKS A-7) (RIVERVIEW)	27°35'S 152°51'E	not shown	not shown	not shown	not shown see map **Ipswich** (G56-15-192) 1:63 360 dated 6/45 Grid 994776	not shown see map **Ipswich** (9442-1) 1:50 000 dated '71 Grid 845500	not shown
GAILES (GAILES A-9) (SERVICETON) (WACOL)	27°36'S 152°55'E	not shown	not shown	not shown	**ELG (overgrown)** strip outline see map **Ipswich** (G56-15-192) 1:63 360 dated '51 Grid 093727	not shown see map **Ipswich** (9442-1) 1:50 000 dated '71 Grid 936453	not shown
AMBERLEY	27°38'S 152°43'E	diagram dated 10/45	diagram and photo (MENTIONS JONDARYAN AS CLOSED IN WEATHER ALTERNATE)	not shown	not shown see map **Ipswich** (G56-15-192) 1:63 360 dated 6/45 Grid 850696	**Amberley airfield** strip outline see map **Ipswich** (9442-1) 1:50 000 dated '71 Grid 716424	not shown
WALLEN STATION	27°38'S 145°49'E	listed Civil C Class	not shown	No 807 issue 1 5/46	by legend landing ground see map **Wyandra** (G55-14) 1:250 000 dated '71 Grid 381571	not shown see map **Wyandra** (G55-14) 1:250 000 dated '79 Grid CK8344	not shown
KINGSTON A-11 (WOODRIDGE)	27°39'S 153°05'E	not shown	not shown	not shown	**ELG (overgrown)** strip outline see map **Beenleigh** (G56-15-193) 1:63 360 dated '47 Grid 264679	not shown see map **Beenleigh** (9542-IV) 1:50 000 dated '71 Grid 096415	not shown
LOGANLEA	27°40'S 153°09'E	not shown	by map and legend CIVIL GROUND RAAF USED	not shown	not shown see map **Beenleigh** (G56-15-193) 1:63 360 dated '47 Grid 316657	not shown see map **Beenleigh** (9542-IV) 1:50 000 dated '71 Grid 141398	not shown
ROSEWOOD	27°39.5'S 152°36.5'E	not shown	not shown	not shown	not shown see map **Ipswich** (G56-14-192) 1:63 360 dated 6/45 Grid 742683	not shown see map **Ipswich** (9442-1) 1:50 000 dated '71 Grid 619411	not shown
KINGSTON A-10	27°41'S 153°06'E	not shown	not shown	not shown	**ELG (overgrown)** strip outline see map **Beenleigh** (G56-15-193) 1:63 360 dated '47 Grid 272647	not shown see map **Beenleigh** (9542-IV) 1:50 000 dated '71 Grid 104383	not shown
KINGSTON A-12	27°42.5'S 153°05.5'E	not shown	not shown	not shown	**ELG (overgrown)** strip outline see map **Beenleigh** (G56-15-193) 1:63 360 dated '47 Grid 259610	not shown see map **Beenleigh** (9542-IV) 1:50 000 dated '71 Grid 090352	not shown

AIRFIELD	LOCATION LATITUDE LONGITUDE	SHOWN IN RAAF A'DROME GUIDE (Jan'46) NOTE 1	SHOWN IN (US) AIR PILOT MANUAL OF PACIFIC ISLANDS (Jan'44) NOTE 2	DCA LG (Landing Ground) NUMBER DATE ASSIGNED NOTE 3	SHOWN IN WW 2 PERIOD MAPPING NOTE 4	SHOWN IN CURRENT MAPPING NOTE 5	SHOWN IN WW2 PERIOD JAPANESE MAPPING DATED 25/9/43 NOTE 6
WATERFORD	27°41.5'S 153°08'E	not shown	not shown	not shown	**Emergency Airfield** strip outline see map **Waterford** (G56-15-193 (part)) 1:25 000 dated 5/45 Grid 298628	not shown see map **Beenleigh** (9542-IV) 1:50 000 dated '71 Grid 127370	not shown
WARRILL CREEK (A-8) (MUTDAPILLY) (TEN MILE CREEK) (WARWICK RD)	27°42'S 152°40'E	not shown	not shown	not shown	**ELG** strip outline see map **Ipswich** (G56-14-192) 1:63 360 dated 6/45 Grid 800634	**Rifle Range** part strip outline see map **Ipswich** (9442-1) 1:50 000 dated '71 Grid 672368	not shown
GOOLMAN (PURGA)	27°45'S 152°45'E	not shown	by map and legend CIVIL GROUND RAAF USED	not shown	**Emergency Landing Ground** see map **Ipswich** (G56-14-192) 1:63 360 dated 6/45 Grid 870571	**Landing Ground (abandoned)** strip outline see map **Ipswich** (9442-1) 1:50 000 dated '71 Grid 740314	not shown
KILLOWEN	27°55'S 145°44'E	listed Civil B Class	not shown	No 248 revised 2/41	not shown see map **Wyandra** (G55-14) 1:250 000 dated '71 Grid 371538	not shown see map **Wyandra** (G55-14) 1:250 000 dated '79 Grid CK7612	landing strip not public
LEYBURN	27°57'S 151°36'E	listed RAAF ELG	by map and legend CIVIL GROUND RAAF USED	not shown	**Airfield** strip outline see map **Clifton** (G56-14-200) 1:63 360 dated 3/44 Grid 650330	**Landing Ground (disused)** see map **Toowoomba** (9242) 1:100 000 dated '74 Grid 625075	not shown
THARGOMINDAH	27°59'S 143°46'E	listed Civil B Class	not shown - outside scope of map	No 594 revised 4/41	not shown see map **Thargomindah** (G54-16) 1:250 000 dated '64 Grid 158528	not shown see map **Thargomindah** (G54-16) 1:250 000 dated '86 Grid YQ7202	not shown
CUNNAMULLA (USA FIELD)	28°02'S 145°39'E	listed RAAF ELG	not shown	No 854 ?	by legend landing ground see map **Cunnamulla** (H55-2) 1:250 000 dated '76 Grid 359524	by legend a'drome see map **Cunnamulla** (H55-2) 1:250 000 dated '82 Grid CJ6499	landing strip not public
CUNNAMULLA (CIVIL)	28°04'S 145°42'E	Civil B Class	not shown	No 108 revised 10/41 NB: an ELG (No 733) first licensed 6/42) was located 3.5 km ESE of this ground?	not shown see map **Cunnamulla** (H55-2) 1:250 000 dated '76 Grid 367520	not shown see map **Cunnamulla** (H55-2) 1:250 000 dated '82 Grid CJ7296	not shown
ST GEORGE	28°03'S 148°36'E	listed Civil A Class	not shown	No 690 revised 4/41	by legend a'drome A Class see map **St George** (H55-4) 1:253 440 dated 5/45 Grid 141519	by legend Landing Ground see map **St George** (H55-4) 1:250 000 dated '79 Grid FJ5696	not shown
COOLANGATTA	28°09'S 153°30'E	listed Civil ELG	not shown	No 558 issue 1 1/41	not shown see map **Tweed Hds** (8/213A) 1:63 360 dated '35 Grid 693043	**Coolangatta Aerodrome** see map **Tweed Hds** (9641) 1:100 000 dated '83 Grid 496841	not shown

AIRFIELD	LOCATION LATITUDE LONGITUDE	SHOWN IN RAAF A'DROME GUIDE (Jan'46) NOTE 1	SHOWN IN (US) AIR PILOT MANUAL OF PACIFIC ISLANDS (Jan'44) NOTE 2	DCA LG (Landing Ground) NUMBER DATE ASSIGNED NOTE 3	SHOWN IN WW 2 PERIOD MAPPING NOTE 4	SHOWN IN CURRENT MAPPING NOTE 5	SHOWN IN WW2 PERIOD JAPANESE MAPPING DATED 25/9/43 NOTE 6
WARWICK	28°16'S 152°03'E	listed Civil C Class	not shown	No 479 revised 9/40	by legend ELG see map **Warwick** (H56-2) 1:253 440 dated '42 Grid 512495	not shown see map **Warwick** (9341) 1:100 000 dated '68 Grid 080730	not shown
INGLEWOOD	28°25'S 151°06'E	listed Civil C Class	not shown	No 749 issue 1 6/42	not shown see map **Goondiwindi** (H56-1) 1:253 440 dated '42 Grid 410477	**Showgrounds** see map **Inglewood** (9141) 1:100 000 dated '73 Grid 126564	not shown
CLOVER DOWNS	28°25'S 146°02'E	listed Civil C Class	not shown	No 707 issue 1 1/42	not shown see map **Cunnamulla** (H55-2) 1:250 000 dated '76 Grid 402477	not shown see map **Cunnamulla** (H55-2) 1:250 000 dated '82 Grid DJ0556	not shown
GOONDIWINDI	28°31'S 150°20'E	listed Civil B Class	not shown	No 621 revised 3/41	by legend a'drome B Class see map **Goondiwindi** (H56-1) 1:253 440 dated '42 Grid 328465	not shown now adjoins **Goondiwindi Airport** see map **Goondiwindi** (8940) 1:100 000 dated '75 Grid 382428	not shown

ARCHERFIELD *(1947-51)*

This map compilation is extracted from the Ipswich (51) and Beenleigh (47) topographic mapping. It is an opportunity to show the relationship to ARCHERFIELD of GAILES ELG (overgrown) in bottom left corner. It also is an opportunity to show just east of the Kerry Road hangar complex, the map indication of 'wireless tower and radar mast'. On grid references 184775 and 182774, dated 16/1/1943 and 19/9/1943 respectively, the RAAF Survey Negative Register lists photos of 'Archerfield AA Post(s)'. These photographs have not yet been found.

[Department of Defence]

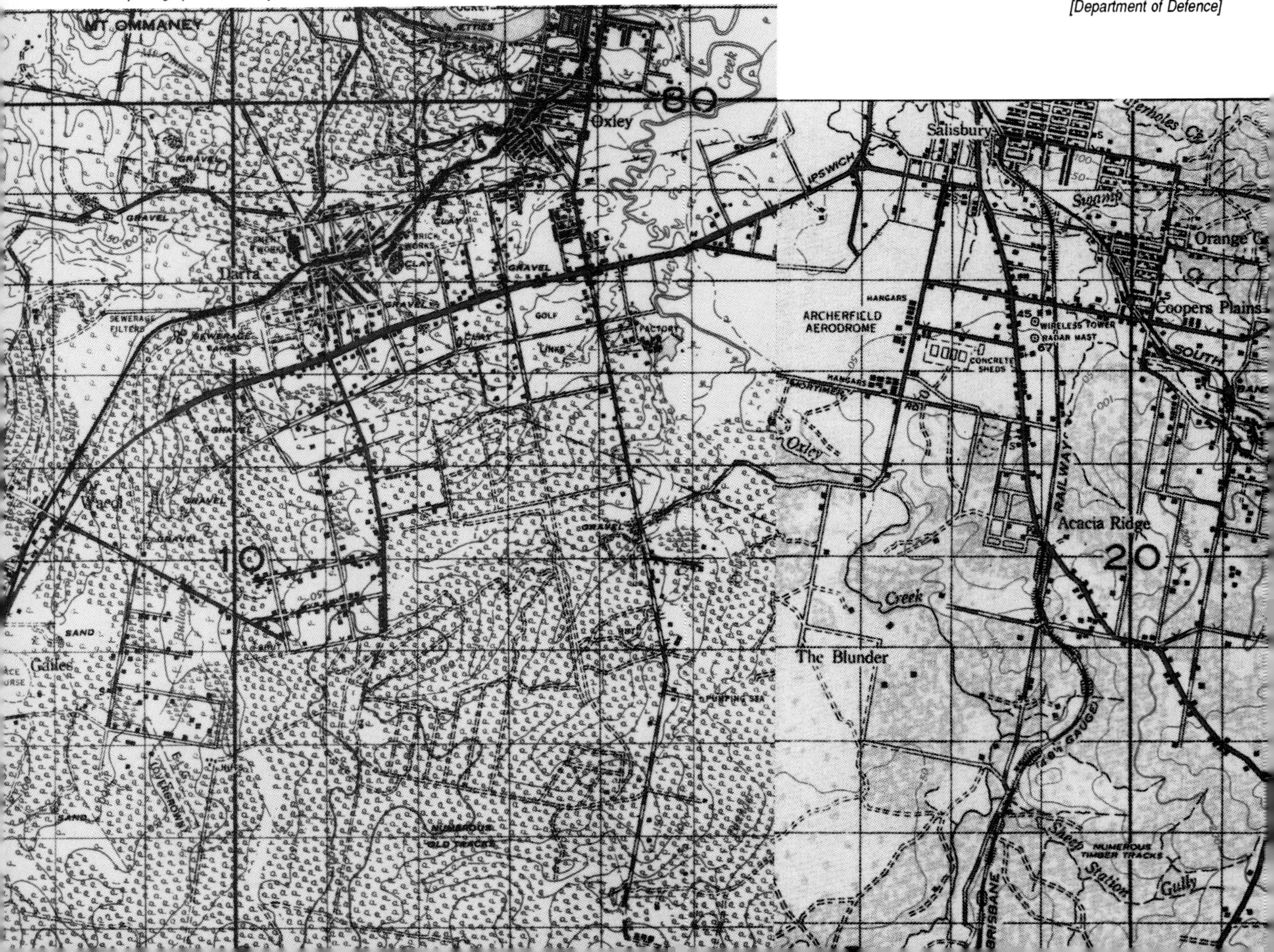

APPENDIX 2

GENERAL DESCRIPTION - FROM AIRFIELD DATA SHEETS

As a WW2 period system for the orderly recording of current detail applicable to each airfield, RAAF Directorate of Works and Buildings (DWB) maintained various Airfield Data Sheets (ADS).

The full scope of these ADS is not known to me. It appears that some of the sheets are:

Air Field Data Sheet referred to as 'number' **C**
Accommodation Data Sheet referred to as 'number' **G**
Schedule of Buildings referred to as 'number' **H**
Water Supply Data Sheet referred to as 'number' **J**
Mechanical and Electrical Services Data Sheet referred to as 'number' **K.**

Appendix 2 assembles the majority of 'C' sheets known to have applied. Clearly, space precludes the full display of information contained on front face/page of ADS 'C'. Only a précis of Runway detail is here attempted. The reverse face of 'C' lists Meteorological information, soil and subsoil, obstructions to flying, airfield suitability to aircraft - by type, and other general remarks.

Prominent well developed airfields required several of the sheets to fully record the 'current status'. New sites were assigned a number and a basic 'C' sheet at inception, and in cases where work did not proceed, this sheet probably remained on the register suitably inscribed.

TOOGOOLAWAH *(10/8/1943)*

This is the review date of information recorded on the RAAF Directorate of Works and Building Airfield Data Sheet, 225C. This is typical of the document 'C' which, in a full inspection review, would be accompanied by other sheets with other information (refer to Appendix 2 - General Description - From Airfield Data Sheets). *[RAAF Historical]*

ROYAL AUSTRALIAN AIR FORCE
DIRECTORATE OF WORKS AND BUILDINGS

When referring to this Document quote 225 C
Copy No. 1

AIR FIELD DATA SHEET

Name of Airfield TOOGOOLAWAH (QLD.)
Location 5 miles (by air) East of Toogoolawah Township and Railway Station, on western bank of Brisbane River Lat 27° 06' S Long 152° 28' E
See also Drawings Nos D. ... Brisbane/M.291. D.W.B. 42/43/1929. 42/43/2316
Parish Cressbrook County Cavendish Shire Esk
Military Map Crows Nest 1 miles Coordinates 583361 (Test Issue)
Civil Aviation No - R.A.A.F. No. -

General Description of Airfield to Include Dispersal Strips and Hideouts:

Surrounding Country Situated in slightly undulating country. Surrounding district is lightly timbered and farming land. Railway 5½ miles to West (Toogoolawah Rly. Stn.) Telephone line on road to South West of field.

Site 2 Hard surface Runways still under construction. After completion of grading out of flight strips, most of remainder of field would be usable for landings in dry weather.
No dispersal works carried out, although provided for in original scheme. Construction postponed or abandoned.
Drainage works incomplete, but pipes sufficient for drainage of 121° Runway are on site.
~~A.W.C. Construction camp buildings situated on main road on ... (See plan)~~.
No camouflage works.
It is proposed to complete gravelled consolidation of 121° Runway (about 5 days from date) and to complete drainage works on same. This Runway with good lightly grassed flight strips should then be in first class order. It is also proposed to complete - if possible - the gravelling of the NE arm of the 32° Runway, but no provision is being made for gravelling the SW arm. The surface of this arm, however, is excellent, and should be usable in all except heavy rainy weather. Grading out of flight strips on this Runway may not be completed unless further financial arrangements are made for the A.W.C. construction work to proceed.
ACCESS: Access by gravelled and consolidated all weather road to Toogoolawah Township 5½ miles.

STRIPS

Magnetic Bearing	Length Feet	Width Feet	Slopes Longtl	Surface	Max Poss Extension in Feet
121°	5000	125'	0.3%	Natural	1000-1200'
32°	3600	"	1.0%	"	Nil

RUNWAYS

Magnetic Bearing	Length Feet	Width Feet	Slopes Longtl.	Surface	Max Poss Extension in Feet
121°	5000'	150'	0.3%	Gravelled	1000-1200'
32°	3600'	150'	1.0%	partly gravelled. partly graded and rolled	Nil.
	Gravelling to 4" consolidated depth				

A.J. Lines
8 43

P.P.L.

MACKAY *(-/3/1945)*

Page 1 (of two page issue) of Department of Civil Aviation's Data Sheet for Landing Ground No 260. Note the southern and western areas described in this diagram as '6' tropical grass'. This is typical of the luxuriant growth in such an area renowned for its sugar cane. *[Civil Aviation Authority]*

(Confidential Document)

DEPARTMENT OF CIVIL AVIATION AUSTRALIA

DATA SHEET FOR LANDING GROUND NO.260
MACKAY, QLD.

ISSUE 5 - APRIL, 1945.
(2 pages)

GOVERNMENT AERODROME "A1" CLASS

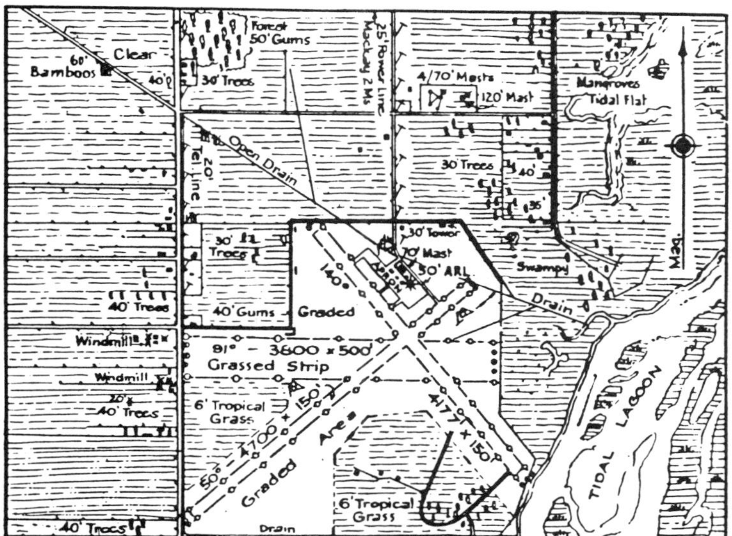

(For text see Page 2.)

W-645 W-435 P.3047

Generally, Appendix 2 does not include sites in this latter category.

It appears DWB's philosophy was to include on such file(s), data pertaining to all airfields considered to be useful for service aircraft. Hence some airfields owned by local authorities and private interests were relevant. Many of these were never able to operate heavy service aircraft during WW2, but, presumably, their retention on such register(s) meant their relative potential was quickly available for strategic review.

Appendix 2 has been compiled by including -

- virtually all airfields mentioned in RAAF Publication No 694, January 1946, *RAAF Aerodrome Guide, Australia and Adjacent Islands*,
- most of the sites which were hastily and/or partly developed and allowed to lapse prior to the January 1946 review, and
- a few airfields which are of special interest. These are either explained by brief note in their listing or elsewhere in the main text.

----- ' RUNWAYS - WHERE LISTED ON ADS ' -----

AIRFIELD	LOCATION LATITUDE LONGITUDE	RAAF DWB ADS REF No ... SHOWS INSPECTION /REV'N DATE... INSPECTING OFFICER...	MAGNETIC BEARING	LENGTH (feet)	WIDTH (feet)	SLOPES LONG'TL	SURFACE	MAXIMUM POSSIBLE EXTENSION IN FEET
HORN ISLAND	10°35'S 142°18'E	154C 27/11/44 F-O Torrance	80° 135°	3002 4256	150 150	slight slight	sealed sealed	Nil Nil
SOMERSET	10°44'S 142°36'E	214C 10/7/43 F-O N C Ray	---	---	---	---	---	---
HIGGINS (JACKY JACKY) (RED ISLAND POINT)	10°56'S 142°27'E	162C 22/11/44 F-O Torrance	133°	7079	150	up to 0.9% slight hump in centre	100' sealed with gravelled strip both sides	---
WEIPA MISSION	12°38'S 141°55'E	48C 3/7/43 F-O N Ray	109° 12°	3300 3250	200* 200* * strip only	level level	natural natural	nil nil
IRON RANGE	12°48'S 143°19'E	161C ?/?/44 ?	119° 120° (GORDON) 160° (OLD CLAUDIE) 160° (NEW CLAUDIE)	4000 6800 6800 9000	100 100 100 100	--- --- --- ---	natural surface (abandoned) sealed 6700' sealed - usable in emergency gravel	--- nil --- ---
WENLOCK	13°06'S 142°50'E	230C 9/8/43	---	---	---	---	---	---
AURUKUN MISSION	13°19'S 141°45'E	594C 4/7/43 F-O N Ray	136° 67°	3 000 under const'n	300 (strip only) ---	level ---	natural ---	nil ---

AIRFIELD	LOCATION LATITUDE LONGITUDE	RAAF DWB ADS REF No ... SHOWS INSPECTION /REV'N DATE... INSPECTING OFFICER...	MAGNETIC BEARING	LENGTH (feet)	WIDTH (feet)	SLOPES LONG'TL	SURFACE	MAXIMUM POSSIBLE EXTENSION IN FEET
COEN (RAAF)	3°46'S 143°07'E	?C 27/10/44 F-O Torrance	62° 108°	3 600 3 600	150 150	--- ---	gravel gravel	1 400 3 400
COOKTOWN (ENDEAVOUR RIVER MISSION)	15°26S 145°12'E	125C 17/2/45 F-O Torrance revised to 24/9/45 by 3 DWO	106°	7 080	100	less than 0.05%	5500' sealed 1580' at eastern end gravelled	3 000
COOKTOWN (CIVIL DROME)	15°28'S 145°14'E	124C 30/6/43 ?	37° 97° 127°	2 400 2 760 3 150	180 126 180	--- --- ---	gravel (1700') gravel (2100') gravel	--- --- ---
LAURA	15°33'S 44°26'E	811C 23/7/43 ?	180° 60°	8 500∗ 8 500∗	--- ---	--- ---	--- ---	∗ only minor work done before abandonment
RUTLAND PLAINS	15°36'S 41°50'E	212C 30/6/55 revised by CAD	82° 131° ∗ runs available	3 200∗ 3 000∗	--- ---	flat flat	natural red sandy clay	extension possible to 1½ miles all round
KOOLATAH	15°54'S 142°22'E	167C 4/8/43 ?	45° 159°	2 100∗ 1 740∗	--- ---	level	natural	∗ runs available
BLOOMFIELD STRIP (AYTON)	15°56'S 145°21'E	205C∗ 30/6/43						∗ possible site-proposed only (spelt ANTON on ADS)
DUNBAR (ALL WEATHER)	16°00'S 142°22'E	139C revised 30/6/44 CAD	N-S E-W	1 440∗ 2 160∗	--- ---	flat flat	natural	∗ runs available
DUNBAR (DRY WEATHER)	16°03'S 142°24'E	138C -/8/43 ?	NE-SW NW-SE	2 100∗ 1 800∗	--- ---	flat flat	natural	∗ runs available
INKERMAN	16°16'S 141°23'E	157C 25/9/43 F-O N C Ray	55° 128°	2 700∗ 2 350∗	--- ---	flat	natural clay pan	∗ runs available
GALBRAITH (GUM HOLES)	16°24'S 141°32'E	146C 25/9/43 F-O Ray	160°	2 340∗	300∗	less than 1%	natural	nil ∗ cleared strip
WROTHAM PARK	16°40'S 144°00'E	796C 15/7/43 ?	70° 149°	7 000 7 000	150 150	--- ---	graded partly constructed only	--- ---
DENHAM ISLAND	16°40'S 139°14'E	134C 30/6/44 revised CAD	48° 102° 138°	1 700 1 700 1 000	90 90 90	--- --- ---	sandy soil rough (as above) (as above)	--- --- ---
MORNINGTON ISLAND	16°40'S 139°13'E	944C 25/1/45 info' by CAD	NW-SE	3 300	180	even	hard red ironstone gravel	nil

AIRFIELD	LOCATION LATITUDE LONGITUDE	RAAF DWB ADS REF No ... SHOWS INSPECTION /REV'N DATE... INSPECTING OFFICER...	MAGNETIC BEARING	LENGTH (feet)	WIDTH (feet)	SLOPES LONG'TL	SURFACE	MAXIMUM POSSIBLE EXTENSION IN FEET
CAIRNS	16°53'S 145°45'E	107C revised 12/10/44 F-O Torrance	150° 116° 6°	5 160 3 700 3 650	150 150 150	less than 1% less than 1% less than 1%	sealed gravel W end sealed gravelling in progress	nil nil nil
VAN ROOK STATION	16°58'S 141°56'E	861C revised 30/6/44 CAD	15° 150°	2 130✱ 2 630✱	114 118	1% 1%	sandy loam sandy loam	450 to 4 500 ✱ strips only
DELTA DOWNS	16°59'S 141°18'E	133C 23/9/43 F-O N C Ray	78°	2 850	---	all over field	natural	to 5 500
LOTUS VALE	17°01'S 141°21'E	172C revised 30/6/44 CAD	95° 162°	2 100 2300	100 100	1% 1%	graded hard clay loam graded hard clay loam (grassed)	1-1½ mile E and W
MAREEBA (HOEVET FIELD)	17°03'S 145°25'E	180C 6/5/44 F-O Crouch	171° 104°	7 550 6 950	150 150	less than 1% less than 1%	sealed gravel sealed gravel	more than 10 000 2 000
MUNGANA	17°14'S 144°27'E (inter-polated from US Map)	195C✱ 30/6/43 ?	---	---	---	---	---	✱ blank sheet! lat-long given is Mungana Town
MIRANDA DOWNS	17°22'S 141°51'E	185C 23/9/43 F-O Ray	132° 80° 0°	2 790✱ 2 160✱ 1 560✱	--- --- ---	--- --- ---	clay pan ✱ runs available	up to 7 000 5 000 7 000
KARUMBA	17°28'S 140°50'E	164C 24/3/44 F-O E S Crouch	145° 20°	2 500✱ 2 500✱	300 300	level level	salt clay pan natural (as above)	✱ runs available
INNISFAIL	17°34'S 146°01'E	158C 4/8/43 ?	NE-SW NW-SE	1 650 2 550	--- ---	level level	graded surface	--- ---
ABINGDON	17°38'S 143°10'E	89C revised 30/6/44 CAD	60° 87° 133°	1 860 3 340 1 290	130✱ 130✱ 130✱	1% 2.5% 1%	red clay loam ✱ graded strips	nil nil nil
NORMANTON	17°42'S 141°06'E	198C -/10/43 F-O Ray	0° 142°	3 000 under const'n 2 500	150 150	less than 1% less than 1%	gravel gravel	1 000 3 500
BURKETOWN	17°45'S 139°33'E	106C 8/9/43 F-O N C Ray	36°	1 650	100	level	lightly gravelled	5 280
DAGWORTH	17°52'S 143°42'E	131C revised 30/6/44 CAD	47° 66°	2 330 1 800	93 ---	1½% 1½%	natural sandy loam	nil nil

AIRFIELD	LOCATION LATITUDE LONGITUDE	RAAF DWB ADS REF No ... SHOWS INSPECTION /REV'N DATE... INSPECTING OFFICER...	MAGNETIC BEARING	LENGTH (feet)	WIDTH (feet)	SLOPES LONG'TL	SURFACE	MAXIMUM POSSIBLE EXTENSION IN FEET
DOOMADGEE MISSION	7°56'S 138°49'E	790C 8/9/43 F-O N C Ray	118° 66°	5 000 3 500	350 350	level level	natural - grassed	7 000 5 000
INVERLEIGH (CIVIL)	8°00'S 140°34'E	842C 23/8/43 ?	51° 111° 153°	1 800✱ 1 200✱ 1 860✱	210 210 210	--- --- ---	graded graded graded	--- --- - ✱ strips only
INVERLEIGH (RAAF)	8°01'S 140°32'E	159C 11/9/43 F-O N C Ray	68° 138°	7 100 5 100	200 150	less than 1%	gravel (natural)	--- ---
MILGARRA	18°14'S✱ 140°55'E ✱ correct latitude probably should have been 18°10.5'S	184C -/10/43 F-O Ray	21° 136° 112°	2 100 1 740 1 950	75 75 75	level level level	natural natural natural	1 200 approx 2 000 approx 2 000
CROYDON	18°14'S 142°15'E	128C revised 30/6/44 CAD	NE-SW NW-SE	3 000✱ 2 190✱	--- ---	--- ---	natural - clay loam & gravel	--- ✱ runs available ---
FORREST HOME STATION	18°14'S 143°03'E	144C revised 30/6/44 CAD	32° 116°	2 100✱ 2 400✱	117 94	1% 1%	natural sandy clay ✱ runs available	450 900
GEORGETOWN	18°14'S 143°28.5'E	147C 30/6/43 ?	?	6 600	---	---	fair ?	nil
PRESTWOOD	18°15'S 143°09'E	202C 30/6/43 ?	108° 168°	7 200 7 200	150 150	--- ---	probably cleared only no reports for some time	--- ---
AUGUSTUS DOWNS	18°30'S 139°55'E	94C 7/9/43 F-O N C Ray	22°	6 000	100	level	gravel	10 000
WONDOOLA	18°34'S 140°54'E	235C revised 30/6/44 CAD	43° 108° 149°	1 600✱ 2 025✱ 1 700✱	90 85 80	1% 1% 1%	black soil black soil black soil	only to E 3 300 ✱ runs available
FORSAYTH	18°35'S 143°34'E	145C revised 30/6/44 CAD	89° 124°	2 000 1 920	164 164	falling both ways from strip inter-section	red clay loam	nil
INGHAM	18°40'S 146°09'E	155C revised 30/6/44 CAD	51°	2 580	100	nil	gravel	unlimited
CLARAVILLE	18°40.5'S 141°42.5'E	116C revised 30/6/44 CAD	171° 60°	1 650✱ 1 740✱	---	flat	clay pan	--- ✱ runs available
DONORS HILL	18°42'S 140°32'E	135C revised 30/6/44 CAD	170° 103°	2 200✱ 1 900✱	200 200	3% 1.5%	limestone gravel limestone gravel	nil ✱ runs available

AIRFIELD	LOCATION LATITUDE LONGITUDE	RAAF DWB ADS REF No ... SHOWS INSPECTION /REV'N DATE... INSPECTING OFFICER...	MAGNETIC BEARING	LENGTH (feet)	WIDTH (feet)	SLOPES LONG'TL	SURFACE	MAXIMUM POSSIBLE EXTENSION IN FEET
CARPENTARIA DOWNS (EINASLEIGH)	18°48'S 144°23'E	792C 11/10/43 F-O Ray	41° 108°	7 500 approx. 10 500∗	--- 400∗	--- less than 1%	lightly gravelled ∗ strip incomplete	--- ---
LYNDHURST	19°12'S 144°22'E	174C revised 30/6/44 CAD	6° 95°	2 475 2 150	245∗ 300∗	level level	sandy loam ∗ strips only sandy loam	nil without excessive work
MT ST JOHN	19°14'S 146°44'E	192C 30/6/43 ?	175°	4 500	150	---	graded grass	abandoned
GARBUTT	19°15'S 146°46'E	227C 10/8/44 F-Lt L G Watson	21° 67° 130°	7 088 7 020 6 995	150 150 150	--- --- ---	sealed sealed sealed	major e'works would be needed
STOCKROUTE (adjacent USAAF 4th Air Depot) (DALRYMPLE RD) (MT LOUISA)	19°16'S 146°46'E	218C 14/6/44 CAD	45°	3 600	100	---	sealed	NE end additional 400 gravel
BOHLE RIVER (BLACK RIVER)	19°17'S 146°42'E	98C 26/7/44 F-Lt L G Watson	26°	5 580	150	very even slight fall to south	sealed	nil
ROSS RIVER	19°19'S 146°47'E	210C revised 25/7/44 F-Lt Watson	76°	5 432	150	less than 1%	sealed gravel	1 568
AITKENVALE (2nd WEIR) (THE WEIR STRIP)	19°19'S 146°45'E	90C 29/2/44 F-O Crouch	49°	4 532	150	level	sealed	3 000
UPPER ROSS (KELSO FIELD)	19°25'S 146°44'E	797C 15/7/43 ?	16° 91° 142°	7 000∗ 7 000∗ 7 000∗	300 300 300	--- --- ---	cleared cleared cleared	∗ strips only
ANTIL PLAINS (MT STUART STRIP)	19°27'S 146°50'E	92C 28/7/44 L G Watson	35° 90°	4 992 5 014	100 100	depression about 1 800' from SW end - general fall to SW slight fall to east	sealed sealed	NE 900 SW 100 nil
CANOBIE	19°29'S 140°56'E	109C 2/9/43 F-O N C Ray	63° 156°	1 850∗ 2 250∗	50 50	level level	natural (gravelly) natural (gravelly)	5 000 5 000 ∗ runs available
GIRU	19°30'S 147°08'E	148C 22/9/44 L G Watson	41° 108° 168°	7 040 5 300 5 500	100 --- ---	--- --- ---	gravel (unconsolidated) cleared cleared	--- --- ---

AIRFIELD	LOCATION LATITUDE LONGITUDE	RAAF DWB ADS REF No ... SHOWS INSPECTION /REV'N DATE... INSPECTING OFFICER...	MAGNETIC BEARING	LENGTH (feet)	WIDTH (feet)	SLOPES LONG'TL	SURFACE	MAXIMUM POSSIBLE EXTENSION IN FEET
WOODSTOCK	19°37′S 146°50′E	236C 1/8/44 F-Lt Watson	70°	7 340	100	gentle fall from middle to W and E	sealed	2 500 to E
			104°	5 318	100	slight fall to W	sealed	3 000 to W
			175°	6 637	100	slight dip in middle	sealed	---
REID RIVER	19°46′S 146°50′E	✱ ADS not located. Information taken from drawing 42/43/1086[c]	24° ✱	5 500	100	flat	sealed	---
			87° ✱	7 350	100	---	sealed	---
			134° ✱	7 000	150	---	sealed	---
FANNING	19°55′S 146°32′E	142C -/4/44 F-O E S Crouch	95°	5 300	150	slight	sealed	nil
			71° ✱	5 300	---	---	71° cleared only-then abandoned	
CAMOOWEAL	19°55′S 138°08′E	108C revised 30/10/44 F-Lt Campbell	13°	3 100	132		sealed	7 000
			81°	3 260	100	less than 1%	sealed	7 000
			131°	4 090	132		sealed	7 000
BREDDAN (BREDDAN ISLAND) (LITTLE SANDY)	19°56′S 146°12′E	102C 22/6/45 F-Lt Torrance	60°	6 590	150	less than 1%	sealed	nil
			13°	6 500✱	150		graded-40′ gravel taxiway in centre	✱ strip only
MACROSSAN	20°02′S 146°29′E	176C revised 6/9/44 F-O Torrance	70°	6 100	150	slight fall to SW	sealed	1 500 NE
			56°	5 470✱	175		natural - gravelly	--- ✱ strip only
BOWEN	20°02′S 148°13′E	101C 15/9/44 F-O Torrance	45°	5 700✱	300	level	graded	nil
			120°	3 700	150	level	gravel-NW end sealed	nil
		✱ strips only	02°	2 600✱	150	level	graded	nil
CHARTERS TOWERS (CORINDA)	20°03′S 146°17′E	114C 17/4/44 F-O Crouch	009°	6 570	150	less than 1%	sealed	1 000
			56°	6 340	150		?	nil
SOUTHERN CROSS	20°07′S 146°06′E	215C 30/6/43	25°	7 000✱	300	---	graded ✱ strip only	---
POWLATHANGA	20°08′S 145°58′E	812C 23/7/43 ?	50°	800(?) approx. 7 000 shows cleared in photo 26/5/45	350	---	cleared then suspended	---
TARANGIE	20°09′S 145°50′E	816C ?	45°	7 000	---	---	cleared - abandoned	---
CLONAGH	20°10′S 140°41′E	118C 1/9/43 F-O N C Ray	4°	2 520✱	---	---	clay pan ✱ runs available	to 5 280
			75°	1 710✱	---	---		to 6 600

AIRFIELD	LOCATION LATITUDE LONGITUDE	RAAF DWB ADS REF No ... SHOWS INSPECTION /REV'N DATE... INSPECTING OFFICER...	MAGNETIC BEARING	LENGTH (feet)	WIDTH (feet)	SLOPES LONG'TL	SURFACE	MAXIMUM POSSIBLE EXTENSION IN FEET
BALFES CREEK	20°12'S 145°55'E	573C 12/8/44 F-Lt Watson	51°	6 000	150	---	gravel	* strips only - abandoned
			120°	7 000*	400	---	smoothed only	
			178°	7 000*	370	---	smoothed only	
STRATHMORE	20°30'S 147°38'E	219C 9/8/43 ?	NW-SE	2 400*	---	---	open black soil	* graded strips only
			WNW-ESE	2 340*	---	---		
PROSERPINE	20°31'S 148°37'E	203C 11/12/43 F-O N C Ray	ESE	3 060*	---	---	black soil	* runs available
			SSE	2 700*	---	---	black soil	
			ENE	2 400*	---	---	black soil	
CLONCURRY	20°40'S 140°30'E	119C 19/10/44 F-O Torrance	15°	5 090	150	fairly level	sealed	---
			64°	5 090	150	fairly level	sealed	---
			124°	7 000	150	fairly level	sealed	---
			164°	5 100	150	fairly level	sealed	---
JULIA CREEK	20°40'S 141°43'E	96C 24/8/43 F-O N Ray * strip only	11°	5 280*	132	level	natural (black soil)	possible essential gravel is at considerable distance from site
			55°	4 000*	132	level		
			145°	5 280*	132	level		
MT ISA	20°43'S 139°30'E	191C revised 30/10/44 F-Lt Campbell	15°	2 700*	100	less than 1%	all natural gravel	3 600
			68°	2 050*	100	less than 1%		2 500
			125°	4 000*	100	less than 1%	* graded strips only	4 300
			156°	3 600*	100	less than 1%		4 000
RICHMOND	20°43'S 143°08'E	207C 3/8/43 F-O N Ray	123°	3 500*	200	---	natural * strip only	nil
TORRENS CREEK	20°48'S 145°00'E	795C revised 26/10/43 F-O Torrance	76°	7 200	100	level	gravel ends*	* section (once with steel mat)
			13°	7 100	100	level	gravelled	
			141°	7 200	100	level	gravelled	
PRAIRIE (STRATHGLASS) (EIGHT MILE)	20°48'S 144°38'E	813C 3/8/43 F-O N Ray	140°	7 000*	300	less than 1%	natural * strip only	considerable extension possible
MACKAY	21°10'S 149°11'E	177C 14/9/44 F-O Torrance * strip only	140°	4 180	150	less than 1%	gravel 900' NW end sealed	2 000
			51°	4 500	150	less than 1%	similar, 900' SW end sealed	nil
			91° *	4 100*	533*	less than 1%	graded and grassed	nil
McKINLAY	21°17'S 141°17'E	182C 30/6/4? ?	?	3 strips up to 1380' each	?	level	open 'downs' country may be soft after heavy rain	?
BURKE RIVER	21°19'S 139°50'E	105C 27/8/43 F-O Ray * runs available	SE	2 640*	?	flat	natural surface	5 000
			NW?	1 320	?	flat	natural surface	2 640

AIRFIELD	LOCATION LATITUDE LONGITUDE	RAAF DWB ADS REF No ... SHOWS INSPECTION /REV'N DATE... INSPECTING OFFICER...	MAGNETIC BEARING	LENGTH (feet)	WIDTH (feet)	SLOPES LONG'TL	SURFACE	MAXIMUM POSSIBLE EXTENSION IN FEET
MT COOLON	21°24'S 147°18'E	190C -/8/43 ? ∗ runs available	NE-SW NW-SE	1 980∗ 1 650∗	--- ---	level	grassed black soil and clay on racecourse	--- ---
WINTON (WINTON EAST)	22°22'S 143°04.5'E	232C 2/11/44 F-O Crouch ∗ emergency only	25° ∗ 143°	? 6 000	--- 100	--- slight SE	graded-not maintained gravel	--- 2 000+
GLENPRAIRIE	22°37'S 149°57'E	149C 30/6/43 ?	N-S	3 750∗	---	---	generally good grass	--- ∗ strip only
CLERMONT	22°50'S 147°37'E	969C 6/3/45 ---	65° 110°	3 600∗ 3 800∗	200 200	flat flat	well graded surface in racecourse	∗ strips only
BOULIA	22°55'S 139°54'E	100C 21/7/43 ---	81° 121° 145°	1 953∗ 1 914∗ 2 907∗	--- --- ---	--- --- ---	∗ graded strips only	--- --- ---
CAPELLA	23°04'S 148°02'E	110C 7/12/43 F-O Ray	67° 113°	1 980 1 320	63 48	up to 1 in 40 up to 1 in 40	natural natural	1 500 600
ROCKHAMPTON (CONNOR PARK)	23°22'S 150°31'E	208C 18/9/43 P-O Lines	43° 148°	3 664 4 064	150 150	level 0.2%	metalled (sealed ends) metalled (NW end sealed)	--- ---
LONGREACH	23°25'S 144°15'E	171C 17/6/43 F-O L A Bannister	40° 87° 132°	8 000 7 000 2 250	100 150 100	0.75% 1.25% 0.5%	gravel gravel gravel	unlimited to NE 9 000 5 000 (need to move show ground building)
EMERALD	23°31'S 148°10'E	141C 28/10/43 P-O A J Lines	57° 136°	1 800 1 890	80 80	level level	grassed and lightly graded	nil nil
WESTWOOD	23°36.5'S 150°09'E	231C 10/8/43 ? ∗ runs available	NNE-SSW WNW-ESE	750∗ 2 250∗	--- ---	falls steeply to north	natural surface - fair	--- ---
DUARINGA	23°44'S 149°41'E	136C 27/10/43 P-O Lines	90° 180°	1 650∗ 2 000∗	330 330	2% 2%	graded and grassed ∗ strip only	nil 1 000 to S
SPRINGSURE	24°08'S 148°06'E	217C 18/10/43 P-O Lines ∗ strip only	28° 90° 128°	1 350∗ 1 650∗ 1 890∗	140 150 150	level falls away at ends level	natural gravel black soil natural gravel	nil nil to 500 SE on black soil
CONNEMARA	24°14'S 142°17'E	123C 21/10/43 P-O Lines	NE-SW NW-SE	3 420∗ 1 260∗	--- ---	flat flat	natural clay pan	1 000 --- ∗ runs available

AIRFIELD	LOCATION LATITUDE LONGITUDE	RAAF DWB ADS REF No ... SHOWS INSPECTION /REV'N DATE... INSPECTING OFFICER...	MAGNETIC BEARING	LENGTH (feet)	WIDTH (feet)	SLOPES LONG'TL	SURFACE	MAXIMUM POSSIBLE EXTENSION IN FEET
BLACKALL (USA Ferry)	24°26'S 145°25'E	844C 7/8/43 F-O Bannister	60° 120°	7 000 7 000	100 100	level level	graded and consol'd. gravel	7 500 7 000
BLACKALL (CIVIL)	24°26'S 145°28'E	794C 15/7/43 ?	NE-SW NW-SE	2 580✱ 2 880✱	--- ---	--- ---	graded graded	--- ✱ strip only
THANGOOL	24°30'S 150°34'E	221C 27/9/43 P-O Lines	139° 110° 72°	2 700✱ 2 400✱ 2 400✱	110 130 130	0.6% 0.6% 0.6%	graded light grass ✱ strip only	nil 1 000 SE nil
MEADOWVALE	24°47.5'S 152°17'E	183C 30/8/43 P-O A J Lines ✱ strip only	25° 100° 57°	2 600✱ 3 000✱ 2 700✱	660 660 660	0.4% 0.2% 0.2%	consol'd & grassed	nil nil nil
TAMBO	24°53'S 146°16'E	220C 21/10/43 P-O Lines	60°	1 920✱ ✱ strip only	300 min	level	natural clay pans	NE 1 200 SW 900
THEODORE	24°53'S 150°06'E	223C 14/10/43 P-O Lines	153° 25° 88°	2 400✱ 2 100✱ 1 700✱	100 110 indefinite	level level level	fair - graded and grassed ✱ strip only	1 000 but road crossing needed
MONTO	24°54'S 151°06'E	188C 28/9/43 P-O Lines	26° 63° 114°	2 400✱ 4 000✱ 2 600✱	100 100 100	0.6% level level	graded and grassed ✱ strip only	1 000 nil 1 000
BUNDABERG	24°55'S 152°20'E	104C 28/8/43 P-O A J Lines	140°	3 600	150	0.3%	gravelled	3 400
MOYEN STATION	25°05'S 142°54'E	193C 21/10/43 P-O Lines	all over field	2 200✱		level	grassed ✱ runs available	1 200 SE 1 200 SW
HIGHFIELDS	25°12'S 146°00'E	152C 21/10/43 P-O Lines	NW-SE N-S NE-SW	2 520✱ 2 040✱ 1 770✱	---	undulating	clay pan & black soil	450 to East ✱ runs available
MARYBOROUGH	25°30'S 152°43'E	181C 22/8/43 P-O Lines	35° 120° 165°	3 300 3 300 5 300	150 150 150	0.5% 0.3% 0.3%	sealed sealed sealed	500 SW nil 1 000 SE
MUNDUBBERA	25°35'S 151°18'E	194C 28/9/43 P-O Lines	NE-SW E-W	3 300✱ 2 400✱		2% to north	all over grassed field	--- ✱ runs available
LAGUNA STATION	25°35'S 146°18'E	168C 20/10/43 P-O Lines ✱ runs available	N-S NNW-SSE	1 980✱ 1 650✱		flat	all over grassed clay pan	300 S 600 NW
WONDAI	26°18'S 151°52'E	234C 17/8/43 P-O Lines ✱ runs available	NE-SW NW-SE E-W	3 564✱ 3 600✱ 2 190✱		undulat- ing 2% NW-SE	all over field well grassed surface	some to N

AIRFIELD	LOCATION LATITUDE LONGITUDE	RAAF DWB ADS REF No ... SHOWS INSPECTION /REV'N DATE... INSPECTING OFFICER...	MAGNETIC BEARING	LENGTH (feet)	WIDTH (feet)	SLOPES LONG'TL	SURFACE	MAXIMUM POSSIBLE EXTENSION IN FEET
BONNIEVILLA	26°19'S 146°12'E	99C 14/10/43 P-O Lines	0° 90°	1 590∗ 1 800∗	450	level	grassed ∗ runs available	7 000 5 000
HOME CREEK	26°23'S 151°47'E	153C 16/8/43 P-O Lines ∗ runs available	NE-SW NW-SE	4 200∗ 4 100∗		undulating	all over field - grassed	1 000 N and E
MORVEN	26°24'S 147°07'E	189C 14/10/43 P-O Lines ∗ runs available	N-S E-W	1 500∗ 1 500∗		level	all over field - grassed	5 000 7 000
CHARLEVILLE	26°25'S 146°16'E	112C 7/6/43 ?	86° 131° 176°	5 700 6 000 5 112	150 150 150	level level level	sealed sealed sealed	6 000 7 000 7 000
LURNEA	26°25'S 146°41'E	175C 18/10/43 P-O Lines ∗ runs available		1 980∗		level	all over natural surface - grassed	5 000 all but to north
MUNGALLALA	26°26'S 147°32'E	196C 18/10/43 P-O Lines ∗ strip only	0° 70°	1 575∗ 1 650∗	330 330	level level	natural - grassed	5 000 2 500
DULBYDILLA	26°26'S 147°22'E	137C 14/10/43 P-O Lines ∗ strip only	0° 90°	1 710∗ 1 710∗	330	level level	natural - grassed	2 500 5 000
MITCHELL	26°29'S 147°57'E	186C 5/8/43 ? ∗ runs available	N-S NW-SE E-W	1 620∗ 2 310∗ 1 770∗	--- --- ---	slight fall to south	all over field - smooth red loam	--- --- ---
QUILPIE	26°37'S 144°16'E	204C 21/10/43(?) P-O Lines ∗ strips only	76° 164°	2 070∗ 1 710∗	520 600	level	gravelly natural	1 200 E 1 200 W 500 S
ROMA	26°35'S 148°49'E	209C 5/11/43 P-O Lines ∗ strip only	0° 57° 135°	2 400 2 550 2 550∗	100 100 100	3% 1% level	gravel gravel graded	nil nil nil
KINGAROY	26°36'S 151°50'E	166C 5/8/43 P-O Lines ∗ runs available	NW-SE NE-SW N-S E-W	5 100∗ 4 800∗ 4 200∗ 4 800∗	--- --- --- ---	plus 2% to SE	all over field - well graded	--- --- --- ---
MANNUEM (KUMBIA)	26°39'S 151°37'E	179C -/8/43 P-O Lines ∗ runs available	NE-SW NW-SE	4 900∗ 5 100∗	--- ---	1½%	all over field - grass surface	1 000 N and E
NANANGO	26°41.5'S 151°59.5'E	197C 11/8/43 P-O A J Lines ∗ runs available	N-S NE-SW NW-SE	3 000∗ 3 300∗ 2 700∗	--- --- ---	3% fall to NW	all over field - well grassed	--- --- ---

AIRFIELD	LOCATION LATITUDE LONGITUDE	RAAF DWB ADS REF No ... SHOWS INSPECTION /REV'N DATE... INSPECTING OFFICER...	MAGNETIC BEARING	LENGTH (feet)	WIDTH (feet)	SLOPES LONG'TL	SURFACE	MAXIMUM POSSIBLE EXTENSION IN FEET
CONDAMINE	26°50'S 150°10'E	122C 30/6/43 ?	42° 135°	7 000 5 400	150 150	prac' level 1%	gravel - SW end sealed laterite NW end sealed	9 000 at least 5 600
TOOGOOLAWAH	27°06'S 152°27'E	225C 10/8/43 P-O A J Lines	121° 32°	5 000 3 600	150 150	0.3% 1.0%	gravelled partly gravelled and part graded and rolled	1 000- 1 200 nil
BRYMAROO (JONDARYAN)	27°13'S 151°38'E	163C 30/6/43 ?	54° 131°	5 200 7 000	150 150	1% 1.3%	sealed sealed	7 000 9 000 plus
WYANDRA	27°14'S 145°58'E	237C 14/10/43 P-O Lines	N-S NE-SW	1 650* 2 400*	--- ---	slight undulat-ing	natural surface	--- * runs available
PETRIE (STRATHPINE A-1)	27°17'S 153°00'E	766C 30/6/43 ?	122°	4 000*	400	0.4%	grassed * strip only	4 200 to SE only
STRATHPINE A-2 (LAWNTON)	27°18'S 153°00'E	766C 30/6/43 ?	6°	3 900	150	level	sealed	4 200 to N only
STRATHPINE A-3	27°19.5'S 153°00'E	815C 27/10/43 P-O A J Lines	55°	4 100	200	level	graded consol'd and grassed	1 000 to SW
COOMINYA No 2 (A-6) (LOWOOD No 2)	27°24'S 152°29'E	115C 7/5/43 F-O Bannister	105°	5 400	200	0.8%	graded and consol'd gravelly surface	7 000
COOMINYA No 1 (COOMINYA A-4) (COOMINYA)	27°24.5'S 152°31'E	126C 17/5/43 F-O L A Bannister * runs available	E-W N-S SE-NW	3 000* 2 500* 3 600*	--- --- ---	1% 1.6% 1.4%	all over field - grassed	3 300 3 000 4 000
EAGLE FARM	27°25'S 153°06'E	140C 8/11/43 P-O A J Lines	40° 74° 123°	6 000 7 735 5 000	150 150 150	max 1% max 1% max 1%	sealed unsealed gravel sealed	1 000 NE 2 000 NE 1 200 SE
OAKEY	27°25'S 151°45'E	798C 1/12/44 F-Lt A J Sneddon	45° 135°	5 100 6 000	150 150	prac' level	sealed sealed	at least 10 000 at least 10 000
LOWOOD	27°27'S 152°28'E	173C 15/5/43 F-O Bannister	128°	6 000	150	1.4%	sealed	---
WIVENHOE	27°27'S 152°36.5'E	233C 30/6/43 ?	---	---	---	undulating	all over field relief LG	---
TONGY	27°21'S 147°15'E	224C 20/10/43 P-O A J Lines * runs available	N-S E-W	2 700* 2 400*	--- ---	level	all over field graded surface	1 200 S 600 E 900 W
CECIL PLAINS (TIPTON)	27°28'S 151°13'E	111C 29/5/43 F-O Bannister	45° 135°	5 000 7 000	160 150	1% 0.75%	sealed sealed	7 000 8 500

AIRFIELD	LOCATION LATITUDE LONGITUDE	RAAF DWB ADS REF No ... SHOWS INSPECTION /REV'N DATE... INSPECTING OFFICER...	MAGNETIC BEARING	LENGTH (feet)	WIDTH (feet)	SLOPES LONG'TL	SURFACE	MAXIMUM POSSIBLE EXTENSION IN FEET
PRAIRIE STATION	27°31'S 144°24'E	201C 20/10/43 P-O A J Lines ∗ runs available	N-S	2 280∗	---	---	all over (rectangular) field lightly grassed	1 800 NE 600 SW
TOOWOOMBA	27°32'S 151°55'E	226C 26/7/43 P-O A J Lines ∗ runs available	E-W N-S	2 600∗ 2 600∗	--- ---	undulating 2½% falls to S and SW	all over field well grassed	nil
ARCHERFIELD	27°35'S 153°01'E	93C 22/7/43 P-O Lines ∗ runs available	NE-SW	6 000∗	---	2½% max to N and fall 2½% at SW	all over field - good turf	1 000 N 600 E
BREMER RIVER (FORKS A-7) (RIVERVIEW)	27°35'S 152°51'E	? 15/4/43 ? ∗ strip only	12°	3 300∗	200	---	graded and grassed	to be kept cleared as dummy drome
GAILES (GAILES A-9) (WACOL) (SERVICETON)	27°36'S 152°55'E	791C 6/8/43 P-O A J Lines	145°	5 200∗ ∗ strip only	250	up to 4%	natural soil	nil
AMBERLEY	27°38'S 152°43'E	91C 6/8/43 P-O A J Lines	45° 135°	5 000 7 900	150 150	0.5% 0.3%	sealed sealed	nil
KINGSTON A-11 (WOODRIDGE)	27°39'S 153°05'E	810C 22/7/43 ? ∗ strip only	53°	4 800∗	250	undulating slopes to 1:33 at NE end	graded and grassed	---
LOGANLEA	27°40'S 153°09'E	170C 30/5/43 ? ∗ strips only	40° 130°	2 800∗ 4 800∗	300 ---	0.8% 1.5%	grassed grassed	3 400 to SW need road development 5 800 to NW & SE - need road development
ROSEWOOD	27°39.5'S 152°36.5'E	814C 23/7/43 ?	65°	?	---	---	black soil 'strip probably graded and grassed'	site flood prone
KINGSTON A-10	27°41'S 153°06'E	--- ∗ strip only	122°	5 200∗	---	steep slopes	graded and grassed	N.B. to be kept graded as dummy 'drome
KINGSTON A-12	27°42.5'S 153°05.5'E	--- ∗ strip only	29°	4 500∗	---	---	graded and grassed	NB to be kept graded as dummy 'drome
WATERFORD	27°41.5'S 153°08'E	229C 30/6/43 ? ∗ strip only	70°	4 194∗	300	1%	cleared graded and rolled fairly well grassed	nil
WARRILL CREEK (A-8)	27°42'S 152°40'E	778C 5/8/43 P-O A J Lines	2°	4 000∗ ∗ strip only	300	almost level	gravelled	nil
GOOLMAN (PURGA)	27°45'S 152°45'E	150C 4/8/43 P-O A J Lines	21°	5 200	150	0.6%	gravelled 600' each end sealed	nil

AIRFIELD	LOCATION LATITUDE LONGITUDE	RAAF DWB ADS REF No ... SHOWS INSPECTION /REV'N DATE... INSPECTING OFFICER...	MAGNETIC BEARING	LENGTH (feet)	WIDTH (feet)	SLOPES LONG'TL	SURFACE	MAXIMUM POSSIBLE EXTENSION IN FEET
KILLOWEN	27°55'S 145°44'E	165C 18/10/43 P-O Lines	NW-SE NE-SW	3 300✶ 2 100✶	--- ---	level	natural surface ✶ runs available	5 000 3 000
LEYBURN	27°57'S 151°36'E	169C 30/6/43 ?	36° 138°	7 000 5 350	150 150	level	sealed sealed	9 000 plus 7 000 plus
THARGOMINDAH	27°59'S 143°46'E	222C 20/10/43 P-O Lines ✶ runs available	N-S E-W	2 250✶ 2 640✶	--- ---	level	all over field	300 N 600 S 1 200 E 900 W
CUNNAMULLA (USA Field)	28°02'S 145°39'E	762C 7/6/43 P-O Bannister	122°	6 000	100	level	gravelled	7 000
CUNNAMULLA (CIVIL)	28°04'S 145°42'E	129C 7/6/43 P-O Bannister ✶ strips only	60° 115° 132° 167°	2 200✶ 2 700✶ 2 100✶ 2 400✶	150 150 150 400	all level strips only	all over field graded and rolled patchy grass	3 000 4 000 4 000 5 000 plus
ST GEORGE	28°03'S 148°36'E	213C 25/11/43 P-O Lines ✶ runs available	---	3 600✶	---	---	all over field firm even grass	600 E 600 S
COOLANGATTA	28°09'S 153°30'E	729C 30/6/43 ?	175° 124° 75°	2 250 2 250 ---	100 100 ---	prac' level --- ---	gravel with patchy grass	need extensive swamp fill
WARWICK	28°16'S 152°03'E	228C 7/6/43 F-O Bannister ✶ runs available	N-S E-W	2 000✶ 1 800✶	--- ---	1% 2½%	all over field grassed	3 000 2 500
INGLEWOOD	28°25'S 151°06'E	156C 28/9/43 P-O Lines	---	1 900✶	---	level	all over field grassed	nil ✶ runs available
CLOVER DOWNS	28°25'S 146°02'E	120C 20/10/43 P-O Lines ✶ runs available	NNE-SSW NE-SW	2 880✶ 1 980✶	--- ---	level	all over field grassed natural surface	1 200 N 600 W
GOONDIWINDI	28°31'S 150°20'E	151C 28/9/43 P-O A J Lines ✶ runs available	---	2 100✶	---	level	all over field grassed	possible to West

APPENDIX 3

DRAWINGS

Usually, drawings in this context are hand worked documents by which instructions for new work are given, or by which a record of completed work is made. Usually they are to scale.

Whether they are legible in their surviving form, as found, after about 50 years is dependent on many factors. Principal among such factors is the draftsman's skill and application of care with the original or tracing. For a drawing to 'make it' through 50 years, not only must it endure doubtful stages of 'use', care and storage, it must also face the varying skills of various archivists!

Drawings complement other written records and photographs. While a photograph of an airfield may show indisputably the existence of a visibly distinguishable item such as a Runway or Hangar, a researcher may only ever learn the location of a 'camp area' through its notation on a drawing. The extension is also true - while a drawing may state or show an item, there is no guarantee that the item was 'proceeded with'. Hence W.A.E. - Work as Executed, drawings are of great value.

Most drawings I have found relevant to this study are referred to in files I've examined in the holdings of Australian Archives. Many I've located in the form of dyeline prints, pinned by the corner, folded and unfolded so many times as to make their preservation a nightmare to the Archive staff. When possible, I've obtained a photocopy by the cheapest 'section by section' means. It is unfortunate that such drawings do not lend themselves to reproduction in this book. My aim, in at least mentioning some of them in this Appendix, is to prompt an awareness in the reader. If you come to know of private holdings of drawings of this nature you believe may be worthy of preservation, I suggest you contact the staff of Australian Archives.

CHARTERS TOWERS *(29/7/1943)*

*Probably a 'code' designation for CHARTERS TOWERS airfield, **'A 23 Drawing compiled from aerial photographs taken 29 July 1943** and **produced by RAAF Intelligence Section NEA'**, shows the general layout of the field. Notice in the locality plan (bottom right) a name indication is placed against BREDDAN to the north and the temporary CEMETERY strip immediately to the west of the main CHARTERS TOWERS field. [North Queensland Air Force Museum (RAAF Official)]*

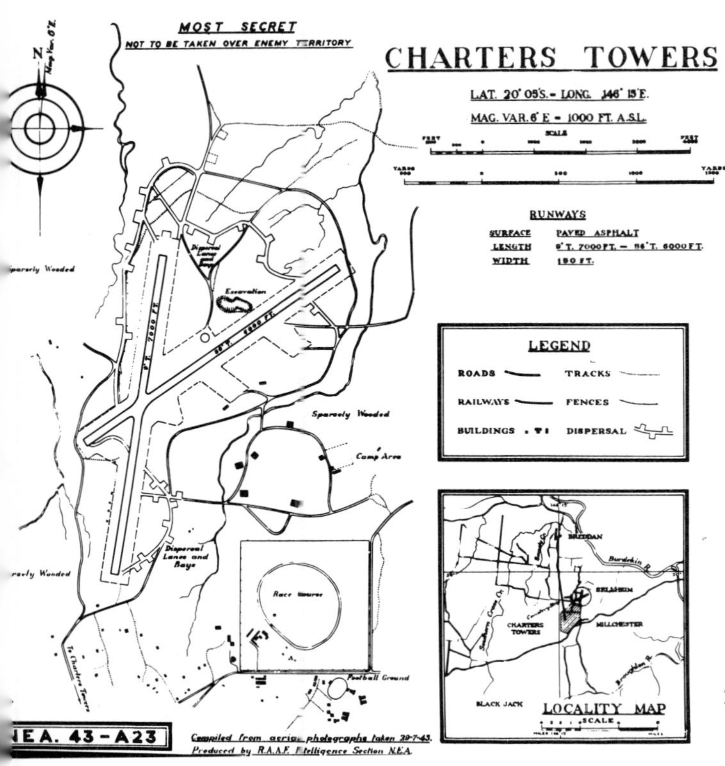

NOTE 1:

I have not sighted all of these drawings. Most of them are located at Australian Archives. Cataloguing systems governing creation and original filing of these drawings are not known to me. However, it seems obvious that 42/43 relates to financial year in which the work was planned. The following number is probably one assigned simply in order of work addressed. A following letter represents an alphabetically sequenced amendment.

DWO means Divisional Works Office, i.e. a de-centralisation of activity and staff of the RAAF's Directorate of Works. For example DWO 3 was located in Townsville and DWO 2 was located in Brisbane.

'BRIS-W' appears to relate to a system in use up to about the time of setting up the Allied Works Council (AWC) in February 1942.

U V W X Y* and *Z have all been encountered but their significance is not known to me.

CAD relates to Civil Aviation Department.

AT is possibly 'A' - AIR or AERODROME and 'T' almost certainly is TROPICAL, relating to drawings prepared by the North Queensland Division of the Dept of Interior's Works Branch. Prefixes **CT** and **ST** are also encountered in CRS J153. It appears 'C' relates to camouflage and 'S' may mean STRUCTURAL.

NEA-43-A13, typically is believed to relate to North East Area - 1943 - A (airstrip?). Number 13 apparently referred to Cairns. This series of drawings carries the wording,

> *'drawn from aerial photographs taken... 1943. Produced by Intelligence Section NEA... 1943.'*

The copies I've mentioned were noted in the care of the North Queensland Air Force Museum RAAF Townsville.

NOTE 2:

This original tracing is undated - its 'number' suggests it may have been created ≈mid 1945. This drawing extends southward only as far as Mackay. Location is only by legend placed near town name.

NOTE 3:

Copy held in John Oxley Library, Brisbane. 'Location' is by legend placed near town name.

ROSS RIVER *(-/8/1943)*

One of several drawings carefully produced by 'Intelligence Section NEA' in 1943. Precise use or distribution of this series of airfield drawings is not known. This airfield is shown as NEA 43-A20. Presumably '43' relates to the year of its compilation, but the significance of 'A20' is not clear. Known drawings in this series are mentioned in Appendix 3. [North Queensland Air Force Museum (RAAF Official)]

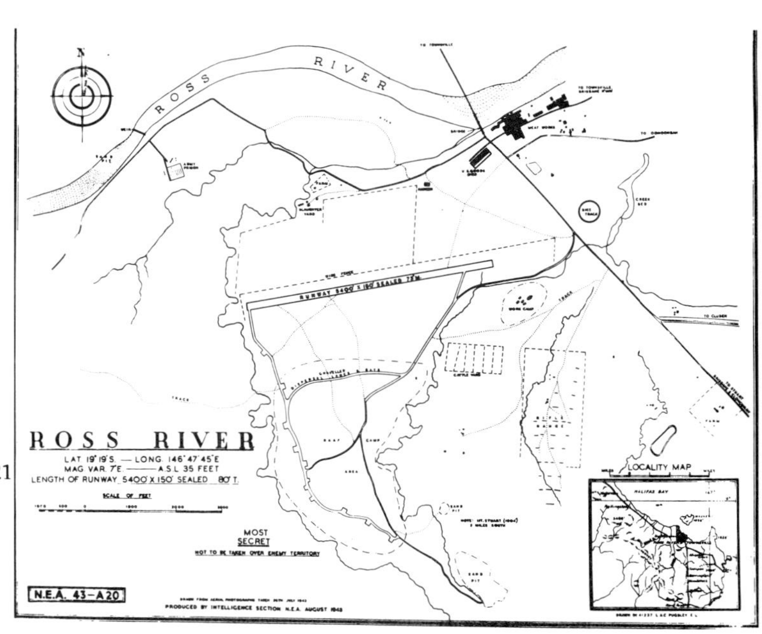

AIRFIELD	LOCATION LATITUDE LONGITUDE	RAAF AIRFIELD DATA SHEETS (ADS) REFER TO THESE DRAWINGS (TYPICALLY 42/43/2847) NOTE 1	SOME OTHER DRAWINGS BELIEVED RELEVANT NOTE 1	INDICATED IN AUSTRALIAN ARCHIVES TOWNSVILLE HOLDING CRS J153 REF.T1825 NOTE 2	INDICATED IN 'MAP' INCLUDED IN QUEENSLAND MAIN ROAD COMMISSION HISTORY 39-45 NOTE 3
HORN ISLAND	10°35'S 142°18'E	A.T. 44/45/129 42/43/1328 B	41/42/223A 41/42/619A 41/42/819A 43/44/208D 43/44/318 ST 616 45/46/712A	yes	yes (2 strips)
HIGGINS (JACKY JACKY) (RED ISLAND POINT)	10°55'S 142°30'E	A.T. 44/45/115 A.T. 44/45/141-142 42/43/2056C	RQM-1.7 (sheet 1 of 1)	yes (1 strip)	yes (2 strips)✶ ✶ believed to be in error!
WEIPA MISSION	12°38'S 141°55'E	42/43/2488	---	no	no
IRON RANGE	12°48'S 143°19'E	No.3 DWO A.T. 43/44/5 42/43/1628-30A 43/44/207 43/44/510	---	yes (2 strips)	yes (2 strips)
COEN (RAAF)	13°46'S 143°07'E	41/42/464E 42/43/2387C	X-276 X-332 41/42/464D	yes (2 strips)	yes (3 strips)
COOKTOWN (ENDEAVOUR RIVER MISSION)	15°26'S 145°12'E	42/43/1626E 44/45/232A 44/45/274 44/45/739	A.T. 43/44/157C	yes	yes (1 strip)
COOKTOWN (CIVIL DROME)	15°28'S 145°14'E	41/42/832 43/44/109	40/41/1489A X-274	no	yes (3 strips)
LAURA	15°33'S 144°26'E	---	---	yes	no - mentioned in MRC History
RUTLAND PLAINS	15°36'S 141°50'E	C.A.D. Z-1138	---	no	no
WROTHAM PARK	16°40'S 144°00'E	42/43/1221A (2 sheets)	---	yes (1 strip)	yes (2 strips)
DENHAM ISLAND	16°40'S 139°14'E	C.A.D. sketch A-117	---	no	no
MORNINGTON ISLAND	16°40'S 193°13'E	C.A.D. Z-560	---	no	no
CAIRNS	16°53'S 145°45'E	C.A.D. X-305 W-527-8 42/43/1540C 43/44/987	42/43/17B NEA-43-A13	yes	yes (3 strips)
VAN ROOK STATION	16°58'S 141°56'E	C.A.D. KSY-129	---	no	no
LOTUS VALE	17°01'S 141°21'E	C.A.D. KSZ-37	---	no	no
MAREEBA (HOEVET FIELD)	17°03'S 145°25'E	A.T. 42/43/358-9 A.T. 43/44/519 43/44/485A	42/43/1061-2 43/44/1955 NEA-43-A30	yes (2 strips)	yes (3 strips)✶ ✶ includes straight connecting taxiway
KARUMBA	17°28'S 140°50'E	A.T. 42/43/236B	---	no	no

AIRFIELD	LOCATION LATITUDE LONGITUDE	RAAF AIRFIELD DATA SHEETS (ADS) REFER TO THESE DRAWINGS (TYPICALLY 42/43/2847) NOTE 1	SOME OTHER DRAWINGS BELIEVED RELEVANT NOTE 1	INDICATED IN AUSTRALIAN ARCHIVES TOWNSVILLE HOLDING CRS J153 REF.T1825 NOTE 2	INDICATED IN 'MAP' INCLUDED IN QUEENSLAND MAIN ROAD COMMISSION HISTORY 39-45 NOTE 3
INNISFAIL	17°34′S 146°01′E	---	---	yes private/local authority L.G.	no
ABINGDON	17°38′S 143°10′E	C.A.D. KSX-49	---	yes private/local authority L.G.	no
NORMANTON	17°42′S 141°06′E	---	---	yes (2 strips)	yes (2 strips)
BURKETOWN	17°45′S 139°33′E	---	---	yes private/local authority L.G.	no
DAGWORTH	17°52′S 143°42′E	C.A.D. KSY-57	---	no	no
DUNK ISLAND	17°56.5′S 146°08.5′E	---	---	yes (1 strip)	yes (1 strip)
INVERLEIGH (RAAF)	18°01′S 140°32′E	42/43/2483B 43/44/195A	43/44/1954	yes (2 strips)	yes (2 strips)
MILGARRA	18°14′S✶ 140°55′E ✶ correct latitude probably should have been 18°10.5′S	---	---	no	no - mentioned in MRC History
CROYDON	18°14′S 142°15′E	C.A.D. KSY-38	---	yes private/local authority L.G.	no
FORREST HOME STATION	18°14′S 143°03′E	C.A.D.KSY-35	---	no	no
PRESTWOOD	18°15′S 143°09′E	42/43/2151 (2 sheets)	---	no	no
AUGUSTUS DOWNS	18°30′S 139°53′E	NEA-W-103 42/43/2257 42/43/582C	NEA-W-104	yes (1 strip)	yes (2 strips)✶ ✶ second one was probably the R/W constructed parallel and on western edge of first (temporary) strip
WONDOOLA	18°34′S 140°54′E	C.A.D. KSZ-73	---	no	no - mentioned in MRC History
FORSAYTH	18°35′S 143°34′E	C.A.D. Y-657	---	no	no
INGHAM	18°40′S 146°09′E	C.A.D. Z-1150	---	yes private/local authority L.G.	no
CLARAVILLE	18°40.5′S 141°42.5′E	C.A.D. Z-29	---	no	no
DONORS HILL	18°42′S 140°32′E	C.A.D. Z-765	---	no	no
CARPENTARIA DOWNS (EINASLEIGH)	18°48′S 144°23′E	42/43/2150A	---	yes (1 strip)	yes (2 strips)

AIRFIELD	LOCATION LATITUDE LONGITUDE	RAAF AIRFIELD DATA SHEETS (ADS) REFER TO THESE DRAWINGS (TYPICALLY 42/43/2847) NOTE 1	SOME OTHER DRAWINGS BELIEVED RELEVANT NOTE 1	INDICATED IN AUSTRALIAN ARCHIVES TOWNSVILLE HOLDING CRS J153 REF.T1825 NOTE 2	INDICATED IN 'MAP' INCLUDED IN QUEENSLAND MAIN ROAD COMMISSION HISTORY 39-45 NOTE 3
LYNDHURST	19°12'S 144°22'E	C.A.D. Y-524	---	no	no
MT ST JOHN	19°14'S 146°44'E	42/43/2148	---	no	yes (1 strip) misplaced near Mackay!
GARBUTT	19°15'S 146°46'E	42/43/1017E ∗43/44/1540F ∗ same drawing	A.T. 43/44/424∗ 43/44/388	yes	yes (3 strips)
STOCKROUTE (adjacent USAAF 4th Air Depot) (DALRYMPLE RD) (MT LOUISA)	19°16'S 146°42'E	42/43/611 42/43/1016B	A.T. 42/43/172 NEA-43-A3 43/44/723 46/47/361	yes (1 strip)	no - mentioned in MRC history
BOHLE RIVER (BLACK RIVER)	19°17'S 146°42'E	42/43/584C 43/44/1258B	A.T. 42/43/161	yes (1 strip)	yes (1 strip)
ROSS RIVER	19°19'S 146°47'E	42/43/597C 43/44/988	NEA-43-A20 (one of a series provided by Intelligence Section NEA)	yes (1 strip)	yes (1 strip)
AITKENVALE (2nd WEIR) (THE WEIR STRIP)	19°19'S 146°45'E	42/43/1014C 43/44/989	42/43/543	yes (1 strip)	yes (1 strip)
UPPER ROSS (KELSO FIELD)	19°25'S 146°44'E	42/43/1018	---	no	no - mentioned in MRC history
ANTIL PLAINS (MT STUART STRIP)	19°27'S 146°50'E	42/43/1015F 43/44/1538B	A.T. 42/43/227 NEA-43-A27	yes (2 strips)	yes (2 strips)
CANOBIE	19°29'S 140°56'E	---	---	no	no - mentioned in MRC history
GIRU	19°30'S 147°08'E	42/43/1627B 43/44/1924	---	yes (1 strip)	no
WOODSTOCK	19°37'S 146°50'E	42/43/1019D 43/44/1905	T153 A.T. 43/44/494 NEA-43-A14	yes (3 strips)	yes (1 strip)
REID RIVER	19°46'S 146°50'E	? ADS not located	NEA-43-A12 46/47/249	yes (3 strips)	yes (1 strip)
FANNING	19°55'S 146°32'E	42/43/591D 43/44/1260	A.T. 42/43/62	yes (1 strip)	yes (1 strip)
CAMOOWEAL	19°55'S 138°08'E	D.C.A. X-396	---	yes (3 strips)	yes (3 strips)
BREDDAN (BREDDAN ISLAND) (LITTLE SANDY)	19°56'S 146°12'E	42/43/586H 43/44/1259C	---	yes (2 strips)	yes (1 strip)
MACROSSAN	20°02'S 146°29'E	42/43/1085D 43/44/1261	41/42/2855 A.T. 42/43/21 NEA-43-A25	yes (1 strip)	yes (1 strip)
BOWEN	20°03'S 148°13'E	C.A.D. W-284 41/42/486A 42/43/1216C	40/41/1430A 42/43/759 42/43/293	yes (1 strip)	yes (2 strips)

AIRFIELD	LOCATION LATITUDE LONGITUDE	RAAF AIRFIELD DATA SHEETS (ADS) REFER TO THESE DRAWINGS (TYPICALLY 42/43/2847) NOTE 1	SOME OTHER DRAWINGS BELIEVED RELEVANT NOTE 1	INDICATED IN AUSTRALIAN ARCHIVES TOWNSVILLE HOLDING CRS J153 REF.T1825 NOTE 2	INDICATED IN 'MAP' INCLUDED IN QUEENSLAND MAIN ROAD COMMISSION HISTORY 39-45 NOTE 3
CHARTERS TOWERS (CORINDA)	20°03′S 146°17′E	42/43/588G 43/44/1309C L269/Base Section 2 USASOS-SWPA	42/43/27A NEA-43-A23 A.T. 44/45/167	yes (2 strips)	yes (3 strips) possibly includes temporary strip by cemetery
SOUTHERN CROSS	20°07′S 146°06′E	42/43/598B	---	yes (1 strip)	yes (1 strip)
POWLATHANGA	20°08′S 145°58′E	42/43/1220A	---	no	yes (1 strip)
TARANGIE	20°09′S 145°50′E	42/43/2876	---	no	no - mentioned in MRC History
BALFES CREEK	20°12′S 145°55′E	42/43/583C	---	yes (1 strip)	yes (2 strips)
PROSERPINE	20°31′S 148°37′E	---	---	yes (1 strip)	no
CLONCURRY	20°40′S 140°30′E	42/43/1541B 42/43/1765 A.T. 43/44/242A	C.A.D. W-472 C.A.D. X-303	yes (4 strips)	yes (2 strips)
JULIA CREEK	20°40′S 141°43′E	C.A.D. Y 667	---	yes private/local authority L.G.	no
NONDA	20°40′S 142°29′E✶ ✶ WW2 strip was probably 2km west of present strip	---	---	yes (1 strip)	no
MT ISA	20°43′S 139°30′E	---	C.A.D. Y-301	yes (2 strips)	yes (3 strips)
RICHMOND	20°43′S 143°08′E	---	---	yes private/local authority L.G.	no
TORRENS CREEK	20°48′S 145°00′E	42/43/1060B	---	yes (2 strips)	yes (3 strips)
PRAIRIE (STRATHGLASS) (EIGHT MILE)	20°48′S 144°38′E	42/43/596	---	yes (1 strip)	yes (1 strip)
HUGHENDEN	20°50′S 144°12′E	---	NEA-43-A28 (ex AA ACT)	yes private/local authority L.G.	no
WARREAH (KAROON) (COTONVALE)	20°58′S 144°47′E	---	---	yes (1 strip)	no - mentioned in MRC History
MACKAY	21°10′S 149°11′E	C.A.D. 344 W-524 41/42/556B 42/43/1330C	---	yes (1 strip)	yes (3 strips)
WINTON (WINTON EAST)	22°22′S 143°04.5′E	43/44/1215B A.T. 44/45/120	---	no	yes (2 strips)
CLERMONT	22°50′S 147°37′E	C.A.D. Z-1015	---	---	---

AIRFIELD	LOCATION LATITUDE LONGITUDE	RAAF AIRFIELD DATA SHEETS (ADS) REFER TO THESE DRAWINGS (TYPICALLY 42/43/2847) NOTE 1	SOME OTHER DRAWINGS BELIEVED RELEVANT NOTE 1	INDICATED IN AUSTRALIAN ARCHIVES TOWNSVILLE HOLDING CRS J153 REF.T1825 NOTE 2	INDICATED IN 'MAP' INCLUDED IN QUEENSLAND MAIN ROAD COMMISSION HISTORY 39-45 NOTE 3
MT HEDLOW	23°13.5'S 150°34'E	---	CR-CA-15	outside scope	no
ROCKHAMPTON (CONNOR PARK)	23°22'S 150°31'E	41/42/1853 42/43/1331A	C.A.D. Y-573 W-526 BRIS. Y-5	outside scope	yes (3 strips)
LONGREACH	23°25'S 144°15'E	C.A.D. W-583 42/43/1083A	---	outside scope	yes (2 strips
BARCALDINE	23°29'S 145°16'E	---	---	outside scope	yes (1 strip)
DUARINGA	23°44'S 149°41'E	42/43/2996	---	outside scope	no
BLACKALL	24°26'S 145°25'E	BRIS. Y-62	---	outside scope	yes (2 strips)
MOORLANDS	24°44'S 152°11'E	---	41/42/581	outside scope	no - mentioned in MRC History
MEADOWVALE	24°47.5'S 152°17'E	41/42/1770 42/43/3210A	---	---	no - mentioned in MRC History
BUNDABERG	24°55'S 152°20'E	BRIS. V6-A BRIS. W-35 BRIS. W-51A 42/43/2827(?)	40/41/550 H 40/41/551 42/43/2877	outside scope	yes (3 strips)
MARYBOROUGH	25°30'S 152°43'S	40/41/502I 42/43/1219C	BRIS. W-75	outside scope	yes (3 strips)
HOME CREEK	26°23'S 151°47'E	40/41/1474	---	outside scope	no
CHARLEVILLE	26°25'S 146°16'E	BRIS. U-2 42/43/1327A	41/42/1533	outside scope	yes (3 strips)
KINGAROY	26°36'S 151°50'E	BRIS. V-10 40/41/1319H 42/43/1218B	C.A.D. X-7	outside scope	yes (1 strip)
MANNUEM (KUMBIA)	26°39'S 151°37'E	40/41/1478A	---	outside scope	no
CONDAMINE	26°50'S 150°10'E	42/43/589B 42/43/1866	41/42/2743	outside scope	yes (2 strips)
TOOGOOLAWAH	27°06'S 152°27'E	BRIS. W-29A 42/43/1929A 42/43/2316	BRIS. W-2	outside scope	yes (2 strips)
BRYMAROO (JONDARYAN)	27°13'S 151°38'E	41/42/2744D 42/43/593A	---	outside scope	yes (2 strips)
PETRIE (STRATHPINE A-1)	27°17'S 153°00'E	BRIS. W-52	BRIS. W-49 42/43/3159 42/43/2648 43/44/379	outside scope	no
STRATHPINE A-2 (LAWNTON)	27°18'S 153°00'E	BRIS. W-52	BRIS. W-49 42/43/3159 43/44/379	outside scope	yes (1 strip)
STRATHPINE A-3	27°19.5'S 153°00'E	42/43/2628	43/44/379	outside scope	no

AIRFIELD	LOCATION LATITUDE LONGITUDE	RAAF AIRFIELD DATA SHEETS (ADS) REFER TO THESE DRAWINGS (TYPICALLY 42/43/2847) NOTE 1	SOME OTHER DRAWINGS BELIEVED RELEVANT NOTE 1	INDICATED IN AUSTRALIAN ARCHIVES TOWNSVILLE HOLDING CRS J153 REF.T1825 NOTE 2	INDICATED IN 'MAP' INCLUDED IN QUEENSLAND MAIN ROAD COMMISSION HISTORY 39-45 NOTE 3
COOMINYA No 2 (A-6) (LOWOOD No 2)	27°24'S 152°29'E	BRIS. W-184(18A?) 42/43/1916	42/43/2928 43/44/379	outside scope	yes (1 strip)
COOMINYA No 1 (COOMINYA A-4) (COOMINYA)	27°24.5'S 152°31'E	41/42/208A	43/44/379	outside scope	no
EAGLE FARM	27°25'S 153°06'E	42/43/1217B	43/44/379	outside scope	yes (3 strips)
OAKEY	27°25'S 151°45'E	42/43/993 42/43/2652A	42/43/1271^{L}	outside scope	yes (2 strips)
LOWOOD	27°27'S 152°28'E	42/43/1084C 42/43/1933A	BRIS.W-14F 43/44/379 46/47/358	outside scope	yes (1 strip)
WIVENHOE	27°27'S 152°36.5'E	---	43/44/379	outside scope	no
CECIL PLAINS (TIPTON)	27°28'S 151°13'E	41/42/2679E 42/43/587A	---	outside scope	yes (2 strips)
TOOWOOMBA	27°32'S 151°55'E	41/42/3211	---	outside scope	yes (3 strips)
ARCHERFIELD	27°35'S 153°01'E	BRIS. W-9 BRIS. W-15D 42/43/581C 42/43/1827A	43/44/379	outside scope	yes (3 strips)
BREMER RIVER (FORKS A-7) (RIVERVIEW)	27°35'S 152°51'E	---	43/44/258 43/44/379	outside scope	no
GAILES (GAILES A-9) (SERVICETON) (WACOL)	27°36'S 152°55'E	D.W.O. INDEX No.A/1-6 42/43/2848A	43/44/379	outside scope	no
AMBERLEY	27°38'S 152°43'E	BRIS. W-4A BRIS. W-62 W563 W880 42/43/580D MB 45D (USA)	43/44/379 M 69.2/1	outside scope	yes (3 strips)
KINGSTON A-11 (WOODRIDGE)	27°39'S 153°05'E	42/43/2827	43/44/379	outside scope	no
LOGANLEA	27°40'S 153°09'E	42/43/1329C Dept of Int L & S - 183	43/44/379	outside scope	yes (1 strip)
KINGSTON A-10	27°41'S 153°06'E	---	43/44/292 43/44/379	outside scope	no
KINGSTON A-12	27°42.5'S 153°05.5'E	---	43/44/254 43/44/379	outside scope	no
WATERFORD	27°41.5'S 153°08'E	43/44/139 MRC 38719	43/44/379	outside scope	yes (1 strip)
WARRILL CREEK **(A-8)**	27°42'S 152°40'E	D.W.O. INDEX No B/1-7 42/43/2633A	43/44/379	outside scope	no

AIRFIELD	LOCATION LATITUDE LONGITUDE	RAAF AIRFIELD DATA SHEETS (ADS) REFER TO THESE DRAWINGS (TYPICALLY 42/43/2847) NOTE 1	SOME OTHER DRAWINGS BELIEVED RELEVANT NOTE 1	INDICATED IN AUSTRALIAN ARCHIVES TOWNSVILLE HOLDING CRS J153 REF.T1825 NOTE 2	INDICATED IN 'MAP' INCLUDED IN QUEENSLAND MAIN ROAD COMMISSION HISTORY 39-45 NOTE 3
GOOLMAN (PURGA)	27°45'S 152°45'E	BRIS. W-37A C.A.D. C2 42/43/5920	43/44/379	outside scope	yes (1 strip)
LEYBURN	27°57'S 151°36'E	42/43/594B 44/45/623	41/42/3286C BRIS.W-135	outside scope	yes (2 strips)
CUNNAMULLA	28°02'S 145°39'E	BRIS. W-72 43/44/1022	---	outside scope	yes (1 strip)

BREDDAN *(26/9/1944)*
*This is the amendment date on Drawing 43-44/1259 C, titled **RAAF LANDING GROUND BREDDAN QUEENSLAND**. Only part of the drawing is shown here and its reproduction will in no way do it maximum credit. Near the runway intersection, two groups of buildings straddle the N-S taxiway. They are the Aircraft Repair Section to the west and the General Engineering Section to the east, with **Blacksmith and Welding, Machine Shop, Electro-plating, Instrument repair** and an **Armament** section. The (Aero) Engine Repair Section lies to the north. The drawing is quite detailed showing, for example, building names such as in the No 2 Camp and the large WAAAF Camp areas quite close to Runway 60°, the **'Dance Hall'** and another nearby, the **'C of E (Church of England) Padre's Hut'**. The extreme right part of the original drawing (not included in this extract) shows the No 1 Torpedo Maintenance Unit's Camp and a prominent loop road providing access to nine torpedo igloos. [AA A705/1 Item 7/1/1213]*

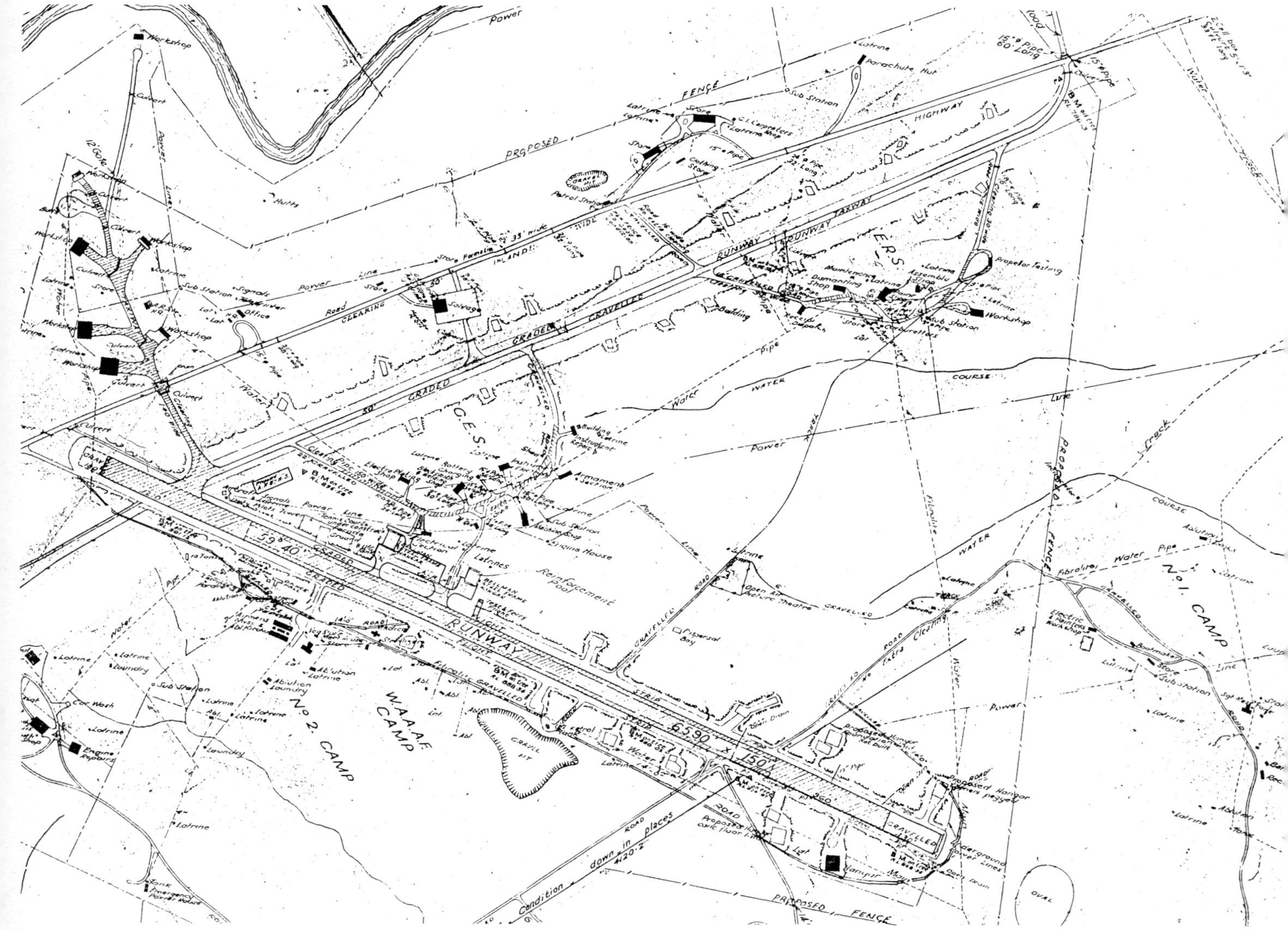

APPENDIX 4

AERIAL PHOTOGRAPHY - WW2 PERIOD

APPENDIX 4 represents a consolidation of known WW2 period aerial photography relevant to each airfield. The LISTING of particular photography does not of itself mean that prints, let alone negatives, exist to this day.

Written records are powerful sources of information. While a document may make reference to some aspect of an airfield's development or activity, the researcher will still require some degree of confirmation. Something as fundamental as the actual location of an airstrip has sometimes only been obtainable through searching photo-libraries.

The National Library of Australia (NLA) in Canberra, holds an extensive collection of RAAF and USAAF aerial photography of the Australian continent. Usually prints can

STOCKROUTE *(22/7/1943)*

The airstrip itself on bearing of 45° is running left to right in this photograph leaving Duckworth Street as the prominent taxiway running northward, i.e. diagonally upwards to the right. The many igloo structures are quite clearly discernible along with dozens of aircraft. The angle of the sun allows some interesting detail to be read by way of the shadows cast, e.g. the dome shape of the igloo structures themselves is evidenced by the shape of the shadow cast. Similarly, twin tailed aircraft can more easily be discerned by allowing the eye to 'read' the shadows.

[CPE (RAAF Official)

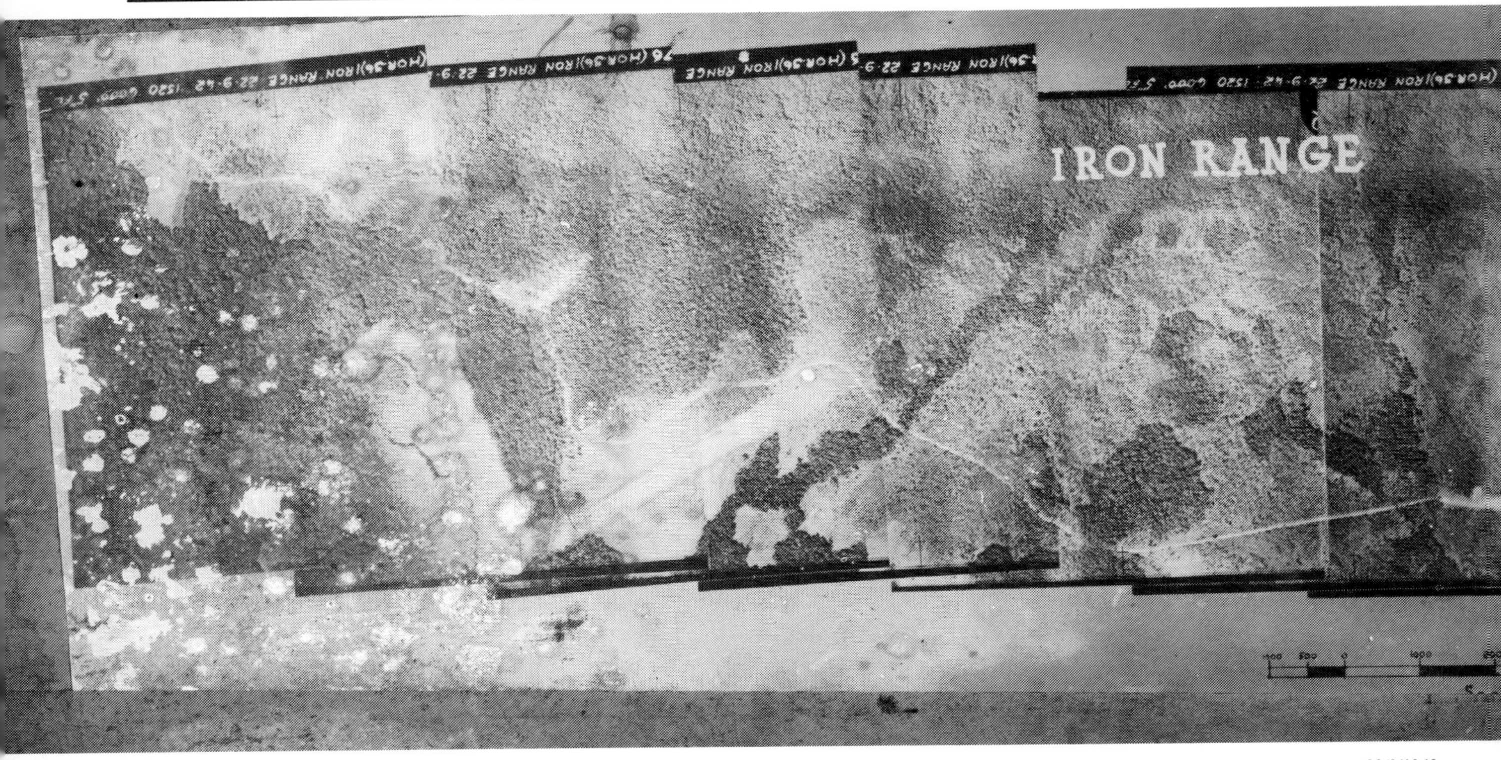

IRON RANGE *22/9/1942*

*Full edge detail reads **K63-76** (print numbers) **HOR 36** (HORN Is based photo mission '36') **IRON RANGE 22-9-42 1520** (hrs?) **6 000'** (altitude) **5" FL** (focal length of camera). This photo came unexpectedly from a RAAF veteran who wasn't 'there' during the war. Here reproduced half size, the original mosaic was 'souvenired' from the wall by Jack Cloake when his 26 OBU departed in 1946. Jack operated wireless equipment in the hut, located just south of Gordon strip and halfway along it. Being too long for his kit bag, Jack halved the mosaic with a pen knife. Nibbled by cockroaches and with its 'north point' partly obscured with spilled glue, this may be the sole surviving print of a 'run' flown just ten days after the US 22 BG moved in. At least seven B-26s are parked in the inserts along the 'C' dispersal linking GORDON with CLAUDIE. The northern one of these three strips is that which the 46 Engineers graded first in the clearing where, very likely, North Queensland Airways landed fortnightly in 1938.*

[RAAF Official courtesy J Cloake]

only be viewed at NLA. Copies may be ordered through Australian Surveying and Land Information Group (AUSLIG), Canberra. As a last resort I have purchased prints via NLA's Reprography Service.

It must be stressed firstly that the NLA's collection is not complete, and secondly that AUSLIG are in many cases not able to provide copies because they either cannot retrieve or simply do not hold the negative. In the latter case AUSLIG wisely advises that such photography **may** be obtained by contacting the RAAF.

The RAAF, through its Central Photographic Establishment (CPE), Laverton, in 1989 advised me in writing that after a full search, photographic coverage was then available for only 10 co-ordinate locations out of approximately 170 I'd asked about. Eight of these were dated 1952 and the other two were 1949. RAAF advised -

'...a fire during the 1950's has destroyed much of the early photography that still appears in records other than RAAF.'

Australian Archives (AA) ACT Regional Repository, holds what amounts, in general, to a duplicate collection of continental aerial photography recently handed over from the Australian Army Map Depot in Victoria. Under Accession No. CRS A8514 this, together with A1579 already in archive custody, represents a back-up to the NLA's collection. Again, prints only can be viewed, and AUSLIG remains the only copy source.

In Queensland the State Government's Department of Lands operates an aerial photo library. Its collection extends to prints of some RAAF and USAAF WW2 period photography, with copies obtainable only through AUSLIG.

Australian Archives' extensive document holdings contain sundry photo prints of relevance to this study. These are usually only obtainable through their Reprography Service.

The Australian War Memorial (AWM), in its Guide to Collections pertaining to aerial photographs advises -

'...as for 1939-45 holdings, the prints remain in packets as received by War Memorial in the late 1940's, and their sorting and indexing remains to be undertaken.'

The RAAF advises that it passed all negatives of its wartime activities, including a large quantity originating with the then Department of Information, to the AWM more than twenty years ago.

The AWM's position in the matter is such that while it has selected representative photographic material for lodgement in its core collection to which the public has access, it does not envisage any review of the remaining bulk, certainly that concerning continental Australia, for at least two to three years. Furthermore it rigidly denies access to 'non core' material by private researchers.

Cataloguing and retrieval systems remain essentially those used during the 1940s. Critical to retrieval is the MAP NUMBER, in particular that made standard pre WW2. It is typified by for example, G55-8-192 where 'G' is a latitude band, actually 24°S to 28°S and '55' is a longitude band 144°E to 150°E; '8' is one of 16 even rectangles each 1.5° of longitude wide by 1° of latitude high. A further regular numbered dissection results generally in 12 rectangles each 0.5° longitude wide by 0.25° of latitude high. Thus the full description, e.g. F55-3-145 identifies the 1:63 360 scale (1 inch = 1 mile) topographic map named Bowen, and ideally, all aerial photography of that rectangle, would be catalogued there. Similarly, where demographic and other considerations have limited WW2 period mapping, to the 1:253 440 (1 inch = 4 mile series) the higher altitude aerial photography related to the production of such maps is catalogued under simply, e.g. F54-2 Cloncurry.

Aerial photography generally has been flown for map production. Specific Flight Diagrams or 'Key Diagrams' are drawn to record and identify the numerous individual

frames/photos.

The RAAF has not used this earlier system of cataloguing for over 25 years, its present system being keyed directly to 'lat-longs'.

During the WW2 period there was ample opportunity for confusion to creep into such recording systems. High volumes of film identifications coupled with the pressing need to train personnel, to supervise and maintain their effectiveness, led inevitably to errors of marking and often to film being simply lost in the system. It may then be that various film negatives thought to be lost, will someday surface and prove to have simply been 'lost' in the system. Like a lot of older records or 'historical' material, such film has limited current day commercial value. Certainly, not many historical researchers can afford the high hourly fee claimed by organisations such as AUSLIG and the RAAF, to search for such possibly lost photographic records.

It is entirely possible that aerial photography of parts of the country has, over some years, been withheld and restricted from public access for reasons of security in the National interest. Whether lost by fire or lost in the filing system, purposefully withheld or just not flown, I can report a total blank in my search for 1942-43 vertical photography over D54-4-30 Lloyd Bay which contains **IRON RANGE** Airfield - except for a couple of *Pin Point* verticals surviving in Australian Archives, probably dated late 1945.

The 1:63 360 topo (Military) map margin records compilation; *from plane table survey by 2 Aust Field Survey Coy, October 1943*, with no mention of any contributing aerial photography. NLA have Trimetrogon (Trimet) runs covering most of the Queensland coast but not the section from Princess Charlotte Bay to Hunter Pt i.e. 14°S to 11°30′S. **IRON RANGE** is about 12°48′S.

An analysis of Key Diagrams (KDs) recording specific runs over WW2 period airfields and other areas of strategic importance, shows various gaps in the orderly identification of vertical photos during the obvious progress of a RAAF Survey Flight Aircraft assigned to the systematic photographic stock-take on airfields during 1943.

Logically, KD numbers were assigned consecutively, and while it is always possible that 'gaps' will exist through a simple decision such as abandonment, it is interesting to analyse such known KDs.

Where some prints of such airfield KD assignments have been located, the edge details, particularly on frames at the start and end of runs, often include reference to the aircraft being used. For example, during the period 21/7/43 to 11/8/43 aircraft A34-82, a DeHavilland 84 Dragon attached to RAAF Survey Flight, carried at least two cameras. Apparently these were designated 'A' for **Vertical** and 'B' for **Oblique** or **Pictorial.** A typical complete aerial photo record of inspection at an airfield included, say, obliques from four points of the compass, at 2 500' and a number of vertical runs from say 5 000'. The accumulated Verticals were given consecutive numbers and likewise the Pictorials.

By such deduction, it appears **IRON RANGE** airfield was visited on or about 7 August 1943. Prints P47 and P48 survive in NLA's collection, and for a time I felt sure the gap verticals between V666 (of **COEN**) and V677 (of **HIGGINS**), when found, would be **IRON RANGE** verticals associated with P47 and P48.

During a visit to AUSLIG, Canberra, in 1990, I was shown two very tattered hard covered folders entitled Survey Negative Register - Part 1 (Trimet and Oblique) and Part 2 (Vertical). In search of possible missing 1943 Verticals of **IRON RANGE**, I reviewed these folders more closely. Sure enough, while I've still not located the actual photos, this Register shows missing Verticals, V667-76 relate to **WENLOCK**. I had been suspicious because there was no gap in KDs - **COEN** was photographed on 7 August 1943 and that is the date of KD852. **HIGGINS FIELD** was photographed also on 7 August 1943, the result being assigned KD853.

KD 840 had some weeks before been photographed/assigned to the Map Area D54-4-26 Cape Weymouth, immediately north of **IRON RANGE** Airfield in D54-4-30 Lloyd Bay. I've not been able to locate such 1943 block photography over Lloyd Bay, though it seems inconceivable that it should not have been flown. Compilation dates of 11/43 and 10/43 appear on the 1" to 1 mile Military Maps for Cape Weymouth and Lloyd Bay respectively.

Aerial photography conducted for mapping usually involved systematic, roughly parallel E-W runs over the Map Area, tied at the ends and sometimes the centre with N-S runs, generally termed the East, West and Centre

Keys. It is a fact that there are two E Keys plotted on the Cape Weymouth Flight Diagram. Closer inspection of the extreme E Key photos shows it is actually an E Key run down the corresponding coastline of the Map Area to the south, i.e. Lloyd Bay.

To assist navigation in flying the E-W runs, it appears these N-S Keys were flown first. While this extreme E Key was definitely plotted in error on the Flight Diagram it seems unlikely it would have been flown in error. It does cover the coastline immediately north and south of wartime Lockhart River Mission with its small airstrip clearly evident in photos.

My inquiries to RAAF Historical Section in search of some sort of daily or mission oriented log of RAAF Survey Flight's aircraft's activities, resulted in advice that such records were not retained.

The review of the Survey Negative Register, held by AUSLIG, proved most valuable. In several cases, where isolated prints with the most meagre edge detail were found in NLA's holdings, reference to this Register has filled in vital date and other information. Probably, a thorough scrutiny of various registers which originated with CPE during the war, held either by AUSLIG or still with the RAAF, would clarify this apparent lack of 1940's aerial photography over Map Area Lloyd Bay.

REMEMBER - APPENDIX 4 represents a consolidation of known WW2 period aerial photography relevant to each airfield. The LISTING of particular photography does not of itself mean that prints, let alone negatives, exist to this day.

NOTE 1:

The repository where I have sighted such photo prints is indicated; e.g. **where NLA** appears, it means that prints have been sighted there, but it does not automatically mean that AUSLIG can supply prints. Where I have followed through, requested and received prints from AUSLIG, or have, without reference to NLA, obtained prints from AUSLIG, then AUSLIG appears.

Where CPE appears, it means I received such prints from CPE in the past. With continuing re-organisation of the RAAF's photo archives, these may now be available elsewhere, i.e. AUSLIG, or not now available for some reason known only to RAAF. CPE can to my knowledge, only be accessed by the public, through RAAF, Canberra.

Where PC appears it means private collection. As an example, original prints of P773 and 778, **TOOGOOLAWAH** obliques dated 17 July 1942, were held by a surviving RAAF veteran. To my knowledge, NLA, AUSLIG and the RAAF don't have them. AWM may well have them locked up in holdings inaccessible to the public.

Where --- appears it means the Survey Negative Register, held by AUSLIG, includes these photos but to date I've not located or seen them. Inclusion in APPENDIX 4 is aimed at providing a reference by which anyone finding such isolated photos may gauge their significance rather than discarding or destroying them!

NOTE 2:

This is generally the NAME of the particular map, e.g. the 1:63 360 (1" = 1 mile) rectangle of Barrington or if referring to the 1:253 440 (1" = 4 mile) rectangle including Barrington, the Area Name could be Charters Towers.

It can also be the project name, e.g. **HIGGINS FIELD**, on the photo margin or on the Flight or Key Diagram.

NOTE 3:

- **KD (Key Diagram)** Number. These are specific Flight Diagrams of aerial photo coverage of airfields or other strategic installation or area. More than one such mission may have been conducted over the same airfield over a period of time. Hence the KDs differ by number and date, and particular run detail.
- **MAP Number** indicates the photography was carried out during the period 1941 to 1946, expressly for the urgent production of mapping, tolerating up to 1° of roll, pitch and yaw. The number immediately following, is a particular mission, flight or sortie number, which may embrace several *Runs*. AUSLIG appear to have relevant registers of such information.
- **SVY Number** indicates photography carried out from 1946 to 1953, to a tighter or higher standard of control, allowing not more than 0.5° roll pitch and yaw. Termed SURVEY photography, it was utilised principally for the more accurate production of mapping. Again, each mission or sortie was assigned a number. AUSLIG appear to have relevant registers.
- **87 Squadron Number.** In 1948 RAAF Survey Fl was renamed 87 (Survey) Fl and in 1949 renamed again to become 87 (PR) Squadron. Hence, photography in late 1940's sometimes bears reference to 87 Squadron with individual mission numbers applying. Again AUSLIG appear to have relevant registers of these missions.
- **CA Number** relates to **Commonwealth of Australia** aerial mapping programme commenced around 1950 with particular States being referred to as B-C-D-E-F-G ie starting with Queensland (B) and progressing clockwise around the continent. Areas spanning state borders have appended the further letter e.g. CA B/G = Queensland/Northern Territory border area.
- **TRIMET** relates to a US system involving 3 cameras each simultaneously recording overlapping shots. A central camera was aimed vertically down, another aimed obliquely to the left and the third aimed obliquely to the right. The resulting photos were referred to as say Runs 1-293 V, 1-293 L and 1-293 R respectively. Interestingly, some P-38 Lockheed Lightning twin engine aircraft had 'L' and 'R' prominently white painted on the inner nacelle surfaces. Thus an oblique negative, showing 'L' on the nacelle, identified it as a left hand oblique (LOB). This served to avoid printing in reverse.

NOTE 4:

Some WW2 period photography carries edge marking such as CAN 8161-1. It is believed this related to storage of negative rolls in metal canisters.

The post war Commonwealth Aerial Mapping programme identification, e.g. CAB 2833 appears to be treated by AUSLIG as a *Film No.*

Clearly, the date of the photograph is often paramount to its proper storage and retrieval. In some cases the date on the KD differs from that on the photo margin, although other run and print numbers agree.

NOTE 5:

Typically a 4 digit print number such as **5008** indicates, by the initial 5, that it is a vertical photo. Other initial digits of 1 to 7 were once used in a system to identify various standard positions of camera mountings. The other digits were used to identify the individual frame in each run, e.g. frame **8** as above.

Various other systems of Print numbering have applied. A simple on-going number reaching up to 5 digits was used for mapping photography during WW2. Early in that period, for individual or 'project' sorties to photograph an airfield, it appears vertical and oblique shots were distinguished by the prefix VX and P for Vertical Exposure(?) and Pictorial(?) respectively.

Aerial and on ground photography of Queensland Airfields WW2 is known to still exist in USA. The USAF pre-1954 collection is currently on extended loan to the Smithsonian Institution's NASM in Washington DC. The US National Archives in Washington holds another significant collection originating with the 'Signal Corps'.

It is possible that aerial reconnaissance photography by Japanese aircraft during WW2 is available but, to date, my efforts to establish this have been unrewarded.

AIRFIELD NOTE 1	LOCATION LATITUDE LONGITUDE	AREA NAME NOTE 2	MAP/AREA NUMBER NOTE 3	FILM No. DATE NOTE 4	ALTITUDE (feet)	RUN	PRINT NUMBER NOTE 5
HORN ISLAND NLA	10°35′S 142°18′E	Horn Is Aerodrome (oblique)	C54-11-1	22/5/40	---	---	**FDP** 28,30-31
NLA		Horn Is A'drome (oblique)	Stn HQ Moresby	5/7/41 (?)	---	---	**FDP** 70(?)
NLA & CPE		Horn Is A'fld (36 Radar) (oblique)	7 Sqd HIG.60	8161-3 CPE 173 19/4/44	5 000	---	**P** 115-7 (0060)
AUSLIG		Horn Is	KD 660 Map 2780 NEA RAAF	6/11/42	3 000	1 2	81-84 91
---		Horn Is Defence Area	KD 770 RAAF 7 Sqd	-/1/43	(?)	3 W-E	26
---		Thursday Is	KD 880 RAAF 7 Sqd	-/1/43	(?)	8/7 N-S	5
NLA		Horn Is Strips	KD 854 Map 2819 RAAF HQ NEA	8124-17 7/8/43	3 200	1 2	**V** 698-704 **V** 705-714
AA		Horn Is Queensland (oblique)	CRS A705/1 7/1/1145	Prior to -/4/46	3 000 approx	---	---
AA		Horn Is Pin Point	CRS A705/1 7/1/1145	Prior to -/4/46	10 000 approx	---	---
SOMERSET ---	10°44′S 142°36′E	Cape York	KD 881 RAAF 7 Sqd C54-12-2	-/1/43	(?)	18	37(?) 38
HIGGINS (JACKY JACKY) (RED ISLAND POINT) NLA	10°55′S 142°30′E	Higgins Field Strip (oblique)	NEA HQ A34-82B C54-11-3	8115-9 7/8/43	3 000	---	**P** 51
AA		Higgins Field (oblique)	CRS A705/1 171/93/742	Prior to -/4/46	(?)	SWPA-ABSEC 6-28	27-2
---		Endeavour Str	KD 882 RAAF 7 Sqd	-/1/43	(?)	12 13	27 28,29
NLA		Higgins Field	KD 853 RAAF HQ NEA A34-82A	Can No. 8124-16 7/8/43	3 000	1 2	**V** 677-687 **V** 688-697
---		Camp Area gun positions DID, Engineers Dump, Ordnance Camp Area, Hospital	7 Sqd HIG.62	8161-9 29/4/44	2 000	---	**V** 3-50
AA		Higgins Field	CRS A705/1 171/93/742	Prior to -/4/46	(?)	SWPA-ABSEC 6-28	28-124
ELLIOT ---	11°15′S 142°23′E	Jardine R.	KD 884 RAAF 7 Sqd	-/1/43	(?)	11	14-16
NLA		Elliot Airstrip	87 Sqd 74 C54-15-7	22/7/50	11 500	---	5048

AIRFIELD NOTE 1	LOCATION LATITUDE LONGITUDE	AREA NAME NOTE 2	MAP/AREA NUMBER NOTE 3	FILM No. DATE NOTE 4	ALTITUDE (feet)	RUN	PRINT NUMBER NOTE 5
WEIPA MISSION CPE	12°38'S 141°55'E	Weipa Mission Strips (oblique)	NEA HQ A34-82B D54-3-23	8115-11 CPE 452 7/8/43	5 100	---	**P** 54-55
---		Weipa Mission Strip	NEA HQ A34-82A	8124-19 8/8/43	5 800	---	**V** 722-6
IRON RANGE NLA	12°48'S 143°19'E	Iron Range Strips (oblique)	NEA HQ A34-82B D54-4-30	8115-7 7/8/43	2 500	---	**P** 47,48
AA		Iron Range (oblique)	CRS A705/1 171/94/167	Prior to -/12/46	(?)	---	2796
PC		Iron Range	HOR 56	22/9/42	6 000	1	**K** 63-76 (mosaic)
AUSLIG		Cape Weymouth	KD 840 Map 1205 NEA HQ RAAF Survey Fl	15/7/43	16 000	11	**VD** 78328 shows first airstrip only
AA		Iron Range Pin Point (2 photos)	CRS A705/1 171/94/167	Prior to -/12/46	(?)	---	2796
WENLOCK ---	13°06'S 142°59'E	Wenlock Strip	NEA HQ A34-82A D54-8-33	8124-15 7/8/43	3 300	---	**V** 667-76
AURUKUN MISSION ---	13°19'S 141°45'E	Aurukun (oblique)	NEA HQ A34-82B	8115-12 8/8/43	6 000 and 7 300	---	**P** 56-57
---		Aurukun Mission Strips	NEA HQ	8124-20 8/8/43	7 300	---	**V** 727-35
---		Aurukun Trimet	Map 2387 D54-7-35	25/9/44	17 700	502	2 **V**, 2 **LOB**
NLA		Aurukun Trimet	Map 2389	27/9/44	18 300	504	2 **V**, 2 **ROB**
---		Aurukun Trimet	Map 2372	17/7/45	20 000	520	75 **V**, 75 **ROB**
COEN (RAAF) ---	13°46'S 143°07'E	Coen (oblique)	NEA HQ A34-82B	8115-5 6/8/43	4 200	---	**P** 44-45
CPE		Coen Strips	KD 852 RAAF NEA HQ D54-8-42	Can No. CPE 082 8124-14 7/8/43	5 400	1 2 3	**V** 651-655 **V** 656-660 **V** 661-666
COOKTOWN (ENDEAVOUR RIVER MISSION) NLA	15°26'S 145°12'E	North Cairns - Princess Charlotte Bay - Trimet	D55-13-26	24/8/42	25 000	132	1-928 **L**
---		Cooktown Mission (oblique)	NEA HQ	8115-5 6/8/43	2 700	---	**P** 43
AUSLIG		Cooktown	KD 839 Map 1013 RAAF Survey Fl	8/7/43	16 000	8	**VD** 77402,3
---		Cooktown Mission	KD 851 RAAF NEA HQ	Can No. 8124-13 6/8/43	2 500	1 2 3	**V** 621-629 **V** 630-638 **V** 639-650
COOKTOWN (CIVIL DROME) NLA	15°28'S 145°14'E	Cooktown and Vicinity Trimet	D55-13-26	16/5/42	5 000	133	116-120 **V,L & R**
---		Cooktown (oblique)	NEA HQ	8115-3 6/8/43	2 900	---	**P** 42
---		Cooktown Strips	NEA HQ	8124-12 6/8/43	2 700	---	**V** 615-20

AIRFIELD NOTE 1	LOCATION LATITUDE LONGITUDE	AREA NAME NOTE 2	MAP/AREA NUMBER NOTE 3	FILM No. DATE NOTE 4	ALTITUDE (feet)	RUN	PRINT NUMBER NOTE 5
COOKTOWN (cont...) (CIVIL DROME) AUSLIG	15°28′S 145°14′E	Cooktown	KD 839 Map 1013 RAAF Survey Fl	8/7/43	16 000	8	**VD** 77406
MITCHELL RIVER MISSION NLA	15°29′S 141°45′E	Carpentaria Gulf	Rutland Plains Map 3661 18 Photo Sqd 13 Air Force D54-15-67	887 2/8/43	15 000	15	**P** 1495 **V** **P** 1494 **LOB**
---		Mitchell River Strip (oblique)	NEA HQ	8115-13 9/8/43	3 500 & 4 000	---	**P** 58-59
---		Mitchell River Strip	NEA HQ	8127-2 8/8/43	3 800	---	**V** 724-47
LAURA ---	15°33′S 144°26′E	Laura	Q165 Adastra 1951 D55-13-28	31/7/51	12 750	2	118
RUTLAND PLAINS ---	15°36′S 141°50′E	Rutland Plains Strip (oblique)	NEA HQ	8115-14 9/8/43	4 800	---	**P** 60
---		Rutland Plains Strip	NEA HQ	8127-3 9/8/43	4 800	---	**V** 748-54
NLA		Carpentaria Gulf Trimet Rutland Plains	Map 3660 D54-15-72	887 43/44	15 000	14	**P** 1565 **R**
KOOLATAH NLA	15°54′S 142°22′E	Carpentaria Gulf Trimet Rutland Plains	Map 3659 D54-15-78	887 1943	15 000	13	**P** 1624 **R**
BLOOMFIELD STRIP (AYTON) NLA	15°56′S 145°21′E	North Cairns to Princess Charlotte Bay Trimet	D55-13-34	24/8/42	25 000	1	1-917 **V** 1-917 **L**
NLA		Bloomfield Strip (oblique)	NEA HQ A34-82B	8115-2 6/8/43	4 000 & 3 600	---	**P** 40,41
NLA		Bloomfield Strip	NEA HQ A34-82A	8124-11 6/8/43	4 000	1	**V** 606-614
NLA		Ayton	KD 1264 D55-13-34	10/5/44	17 000	East Key (25-34)	**M** 30
NLA		Ayton	KD 1264	14/6/45	17 000	4	**VD** 2663
DUNBAR (ALL WEATHER)NLA **DUNBAR** (DRY WEATHER)	16°00′S 142°22′E 16°03′S 142°24′E	Carpentaria Gulf Trimet	Rutland Plains Map 3656 18 Photo Sqd 13 Air Force D54-15	1943	15 000	12	1646 **V** 1646 **L**
INKERMAN NLA	16°16′S 141°23′E	Carpentaria Gulf	Galbraith Trimet Map 3651	-/6/43 or (?)2/8/43	15 000	10	**P** 1780 **R**
---		Inkerman Strip (oblique)	NEA HQ	8115-15 9/8/43	5 200	---	**P** 61
GALBRAITH NLA (GUM HOLES)	16°24′S 141°32′E	Carpentaria Gulf	Galbraith Trimet Map 3651	-/6/43 or (?) 2/8/43	15 000	10	**P** 1776 **L**
---		Galbraith (oblique)	NEA HQ	8115-16 9/8/43	4 700	---	**P** 62
---		Galbraith Strip	NEA HQ	8127-4 9/8/43	4 900	---	**V** 755-9

AIRFIELD NOTE 1	LOCATION LATITUDE LONGITUDE	AREA NAME NOTE 2	MAP/AREA NUMBER NOTE 3	FILM No. DATE NOTE 4	ALTITUDE (feet)	RUN	PRINT NUMBER NOTE 5
WROTHAM PARK ---	16°40′S 144°00′E	Wrotham Park (oblique)	NEA HQ	8114-18 5/8/43	4 300	---	**P** 33-34
AUSLIG		Wrotham Park	KD 849 Map 3557 RAAF NEA HQ E55-1-44	Can No. 8124-7 5/8/43	4 400	1 2	**V** 493-501 **V** 502-516
DENHAM ISLAND NLA	16°40′S 139°14′E	Carpentaria Gulf	Map 3698 18 Photo Sqd 13 Air Force	887 29/7/43	15 000	33 L	**P** 897 **L** (shows Mornington Is site prior to airstrip construction)
CAIRNS NLA	16°53′S 145°45′E	Cairns (oblique)	5 Sqd E55-2-52	8097-9 27/8/38	2 000 approx	---	**FPA** 33-35
AA		Cairns Aerodrome (oblique)	NEA 100 Sqn	618-1(?) 14/5/42	2 000 approx	C 1905 Box 16	**PA** 36 8 prints
NLA		30 miles N of Cairns to Townsville Trimet	---	16/5/42	22 000	1-163 to 264 (V,L & R)	1-183 **V** 1-184 **V**
NLA		Test Cairns and vicinity Trimet	---	16/5/42	10 000	92-113 V	1-112 **V** 1-109 **V**
PC		Airport at Cairns	---	Can No.142 25/5/42	20 000	---	1-99, 100
NLA		Cairns	KD 988 8 Photo Sqd USAAF	25/7/42	20 000	E [C5]	4 (shows approx 7 Catalinas moored in Trinity Inlet)
NLA		Cairns 'Drome (oblique)	NEA HQ A34-82B	8114-17 5/8/43	4 600	---	**P** 31-32
NLA		Cairns Strips	KD 848 RAAF NEA HQ E55-2-52	Can No. 8124-6 5/8/43	4 700	1 2	**V** 471-480 **V** 481-492
AA		Cairns Q Pin Point	CRS A705\1 7/1/963	prior to -/2/45	---	---	---
VAN ROOK STATION NLA	16°58′S 141°56′E	Carpentaria Gulf NB Flight Diagram not on AUSLIG Fiche of 5/5/90	18 Photo Sqd 13 Air Force E54-3-99	887 31/7/43	15 000	6	approx **P** 1250 **V** **P** 1250 **L**
DELTA DOWNS NLA	16°59′S 141°18′E	Carpentaria Gulf NB Flight Diagram not on AUSLIG Fiche of 5/5/90	18 Photo Sqd 13 Air Force E54-3-98	887 30/7/43	15 000	6	**P** 1286 **L**
---		Delta Downs Strip (oblique)	NEA HQ	8115-18 9/8/43	5 000	---	**P** 64
---		Delta Downs Strip	NEA HQ	8127-6 9/8/43	5 000	---	**V** 765-9
LOTUS VALE NLA	17°01′S 141°21′E	Carpentaria Gulf NB Flight Diagram not on AUSLIG Fiche of 5/5/90	18 Photo Sqd 13 Air Force E54-7-106	887 30/7/43	15 000	6	**V** 1289
---		Lotus Vale (oblique)	NEA HQ	8115-19 9/8/43	5 000	---	**P** 65
---		Lotus Vale Strip	NEA HQ	8127-7 9/8/43	5 100	---	**V** 770-2

AIRFIELD NOTE 1	LOCATION LATITUDE LONGITUDE	AREA NAME NOTE 2	MAP/AREA NUMBER NOTE 3	FILM No. DATE NOTE 4	ALTITUDE (feet)	RUN	PRINT NUMBER NOTE 5
MAREEBA (HOEVET FIELD) PC	17°03'S 145°25'E	Mareeba Airdrome	E55-5-56	25/7/42	20 000	---	1-717 **V**
NLA		Mareeba 'Drome (oblique)	NEA HQ A34-82B	8114-19 5/8/43	5 200 5 000	--- ---	**P** 35, **P** 36
NLA		Dimbulah	KD 813 Map 1068 1191 RAAF Survey Fl	11/7/43	16 000	1 2	76705-7 77160-4
NLA		Mareeba Strips	KD 850 NEA HQ A34-82A	8124-849 5/8/43	5 400	1 2 3 4 5	**V** 517-31 **V** 532-45 **V** 546-64 **V** 565-79 **V** 580-96
AUSLIG		Biboohra	KD 812 SVY 894	3/11/43 or 23/11/43?	16 000	10	**VD** 77874
ATHERTON NLA	17°15'S 145°30'E	Atherton Strip (oblique)	NEA HQ A34-82B E55-5-61	8114-20 5/8/43	3 800	---	**P** 37,38
MIRANDA DOWNS NLA	17°22'S 141°51'E	Carpentaria Gulf **N.B.** Flight Diagram for Mornington Trimet not on Auslig Fiche 5/5/90	18 Photo Sqd 13 Air Force E54-7-116	887 30/7/43	15 000	4	approx **V** 993 & 993 **L**
KARUMBA NLA	17°28'S 140°50'E	Carpentaria Gulf Trimet Plot 5	Map 3639? 18 Photo Sqd 13 Air Force E54-6-114	30/7/43	15 000	3	**P** 1086 **R**
---		Karumba (oblique)	NEA HQ A34-82B	8115-20 9/8/43	5 600	---	**P** 66 **P** 67
---		Karumba	NEA HQ	8127-8 9/8/43	5 600	---	**V** 773-81
KARUMBA FLYING BOAT BASE NLA	17°29'S 140°50'E	Meatworks Karumba	43 Sqd A24-44 E54-6-114	8139-15 1/12/43	200	---	**P** 1 **P** 2
INNISFAIL NLA	17°34'S 146°01'E	Test Cairns & vicinity	E55-6-69	Can No. 132 16/5/42	10 000	---	1-101 **L**
NLA		30 miles north of Cairns to Townsville Trimet	---	Can No. 132 16/5/42	22 000	---	1-205 **V**
AUSLIG		Innisfail	KD 819 Map 1177 RAAF Survey Fl	20/6/43	16 000	2	75797
---		Innisfail (oblique)	NEA HQ	8114-16 4/8/43	2 000	---	**P** 29-30
---		Innisfail Strip	NEA HQ	8124-5 4/8/43	2 000	---	**V** 465-70
ABINGDON AUSLIG	17°38'S 143°10'E	Red River Trimet	Map 2169 RAAF Survey Fl (?) E54-8-129	1/8/45	20 000	6	148 **LOB**
MOUNT GARNET ---	17°41'S 145°06'E	Mt Garnet	KD 815 RAAF Survey Fl E55-5-67	-/6/43	(?)	8	75212

AIRFIELD NOTE 1	LOCATION LATITUDE LONGITUDE	AREA NAME NOTE 2	MAP/AREA NUMBER NOTE 3	FILM No. DATE NOTE 4	ALTITUDE (feet)	RUN	PRINT NUMBER NOTE 5
NORMANTON NLA	17°42′S 141°06′E	Carpentaria Gulf Trimet	Map 3633 18 Photo Sqd 13 Air Force	887 3/7/43	15 000	2 (last half)	**V** 373
---		Normanton (oblique)	NEA HQ E54-7-125	8116-1 10/8/43	3 500	---	**P** 68
---		Normanton Strips	NEA HQ	8127-9 10/8/43	3 300	---	**V** 782-7
BURKETOWN NLA	17°45′S 139°33′E	Carpentaria Gulf Trimet Plot 5	Map 3632 18 Photo Sqd 13 Air Force E54-6-122	3/7/43	15 000	2 (first half)	**V** 306
DAGWORTH NLA	17°52′S 143°42′E	Red River Trimet	Map 2169 RAAF Survey Fl(?) E54-8-70	1/8/45	20 000	6	12 **ROB**
DOOMADGEE MISSION NLA	17°56′S 138°49′E	Westmoreland Trimet	Map 3630 18 Photo Sqd 13 Air Force E54-5	887 -/7/43	15 000	1 (last half)	**P** 182 **L**
DUNK ISLAND AUSLIG	17°56.5′S 146°08.5′E	Clump Point	KD 842 Map 450 RAAF Survey Fl	2/8/43	11 635	12	**VD** 81277
---		Dunk Island (oblique)	NEA HQ	8114-15 4/8/43	2 000	---	**P** 27-28
---		Dunk Island Strip	NEA HQ	8124-4 4/8/43	2 000	---	**V** 462-4
INVERLEIGH (RAAF) NLA	18°01′S 140°32′E	Carpentaria Gulf Burketown Trimet	Map 3630 18 Photo Sqd 13 Air Force E54-6-134	887 2/7/43	15 000	1 (first half)	**P** 93 **L**
NLA		Inverleigh (oblique)	NEA HQ A34-82B	8116-5 10/8/43	5 800	---	**P** 72
NLA		Inverleigh 'Drome	KD 855 NEA HQ RAAF A34-82A	Can No. 8127-13 10/8/43	6 500	1 2	**V** 801-809 **V** 810-820
MILGARRA NLA	18°14′S∗ 140°55′E ∗ correct latitude probably should have been 18°10.5′S	Milgarra	NEA HQ A34-82B	8116-4 10/8/43	5 800	---	**P** 71
NLA		Milgarra Strip	NEA HQ A34-82A E54-10-144	8127-12 10/8/43	5 600	1	**V** 796-800
PRESTWOOD ---	18°15′S 143°09′E	Georgetown	SVY 1146 E54-12-159	8/5/51	25 000	4 5	5157 5202
AUGUSTUS DOWNS PC	18°30′S 139°53′E	Augustus Downs (oblique)	E54-10-162	1/2/43	---	---	52
NLA		Augustus Downs (oblique)	NEA HQ A34-82B	8116-6 10/8/43	5 000	---	**P** 73
NLA		Augustus Downs	NEA HQ A34-82A	8127-14 10/8/43	4 900	1	**V** 821-26
WONDOOLA NLA	18°34′S 140°54′E	Wondoola (oblique)	NEA HQ A34-82B	8116-3 10/8/43	5 500	---	**P** 70
NLA		Wondoola Strip	NEA HQ E54-70-164	8127-11 10/8/43	5 600	---	**V** 793-5

AIRFIELD NOTE 1	LOCATION LATITUDE LONGITUDE	AREA NAME NOTE 2	MAP/AREA NUMBER NOTE 3	FILM No. DATE NOTE 4	ALTITUDE (feet)	RUN	PRINT NUMBER NOTE 5
INGHAM NLA	18°40′S 146°09′E	30 miles N of Cairns-Townsville Trimet	E55-10-93	Can No. 132 16/5/42	22 000	---	1-238 **R**
---		Ingham (oblique)	NEA HQ	8114-14 4/8/43	2 000	---	**P** 25-26
---		Ingham Strip	KD 847 NEA HQ RAAF	8124-3 4/8/43	2 000	1 2	**V** 452-6 **V** 457-61
AUSLIG		Ingham	KD 875 Map 1387	3/8/43	16 000	8	**VD** 81473
CLARAVILLE NLA	18°40.5′S 141°42.5′E	Claraville (oblique)	E54-11-166 NEA HQ A34-82B	8116-2 10/8/43	5 300	---	**P** 69
NLA		Claraville Strip	NEA HQ A34-82A	8127-10 10/8/43	5 300	1	**V** 788-92
DONORS HILL NLA	18°42′S 140°32′E	Donors Hill	E54-10-164 NEA HQ A34-82A	8127-15 10/8/43	4 500	1	**V** 827-9
CARPENTARIA DOWNS AUSLIG (EINASLEIGH)	18°48′S 144°23′E	Einasleigh	SVY-1376 E55-9-95	17/6/52	25 000	12	5110
MT ST JOHN AUSLIG	19°14′S 146°44′E	Townsville Trimet (30 miles north of Cairns to Townsville)	Map 3826 E55-14-106	Can No. 132 16/5/42	22 000	---	1-262 **V**
NLA		Mt St John Strip (oblique)	NEA HQ A34-82B	8114-8 22/7/43	4 000	---	**P** 14,17
AUSLIG		Townsville	Map 1381	8465-1 -/7/41	10 000	1	**VD** 33840
NLA		Mt St John Strip	NEA HQ A34-82A	8112-9 23/7/43	5 000	1	**V** 297-303
GARBUTT AUSLIG	19°15′S 146°46′E	Townsville Trimet (30 miles north of Cairns-Townsville)	Map 3826 E55-14-106/114	Can No. 132 16/5/42	22 000	---	1-262 **V**
NLA		Garbutt (oblique)	(?)	?31/3/42	3 000 approx.	---	**PA** 533,535, 536,537,539, 541,542
AUSLIG		Bundaberg-Townsville Trimet	Map 3829	Can No. 123 27/7/42	20 000	---	1-723 **V**
PC		Garbutt Strip	---	12/8/42	8 000	---	1-1295 **V**
NLA		Garbutt Strips (oblique)	NEA HQ A34-82B	8114-10 22/7/43	4 000	---	**P** 18,19
AUSLIG		Townsville	KD 561 Map 1387 RAAF Survey Fl	8465-1 -/7/41	10 000	1	**VD** 33837
NLA		Garbutt Strip	KD 825 NEA HQ A34-82A	Can No. 8112-8 23/7/43	5 000	1 2	**V** 284-288 **V** 292-296
AUSLIG		Garbutt Aerodrome	KD 1026 Map 3329 NEA HQ RAAF	Can No. 8161-1 24/4/44	6 000	1 2 3 4	**V** 1129-34 **V** 1137-42 **V** 1143-49 **V** 1151-57
---		Mt St John H. A/A	NEA HQ	8164-8 20/5/44	2 000	---	**V** 1174-6

AIRFIELD NOTE 1	LOCATION LATITUDE LONGITUDE	AREA NAME NOTE 2	MAP/AREA NUMBER NOTE 3	FILM No. DATE NOTE 4	ALTITUDE (feet)	RUN	PRINT NUMBER NOTE 5
STOCKROUTE (adjacent USAAF 4th Air Depot) AUSLIG	19°16'S 146°46'E	Townsville Trimet (30 miles north of Cairns-Townsville)	Map 3826 E55-14-106/114	Can No. 132 16/5/42	22 000	---	1-262 **V**
(DALRYMPLE RD) (MT LOUISA) NLA		Stockroute (oblique)	(?)	?31/3/42	2 000 approx	---	**PA** 543-548
AUSLIG		Bundaberg-Townsville Trimet	Map 3829	Can No. 123 27/7/42	20 000	---	1-723 **V**
PC		Stockroute	---	15/9/42	5 000	---	1-1463 **V**
---		Stockroute (oblique)	NEA HQ	8114-9 22/7/43	4 000	---	**P** 15-16
AUSLIG		Townsville	KD 561 Map 1387 RAAF Survey Fl	8465-1 -/7/41	10 000	1	**VD** 33837
CPE		Stockroute	NEA HQ A34-82A	8112-7 CPE 367 22/7/43	5 000	---	**V** 276-81
NLA		Garbutt Strip	KD 825 NEA HQ A34-82A	Can No. 8112-8 23/7/43	5 000	1 2	**V** 282-283 **V** 289-291
AUSLIG		Garbutt Aerodrome	KD 1026 Map 3329 NEA HQ RAAF	Can No. 8161-1 24/4/44	6 000	1 2 3 4	**V** 1120-34 **V** 1137-42 **V** 1143-49 **V** 1151-57
BOHLE RIVER (BLACK RIVER) NLA	19°17'S 146°42'E	Panoramic of all strips in Townsville Area	E55-14-114	Can No. 134 16/5/42	22 000	---	? **ROB**
NLA		Townsville Trimet 30 miles north of Cairns to Townsville	Map 3826	Can No. 132 16/5/42	22 000	---	1-262 **V**
NLA		Bohle River Strip (oblique)	NEA HQ A34-82B	8114-6 21/7/43	4 000	---	**P** 10,13
NLA		Bohle River Strip	KD 828 Map 2638 NEA HQ RAAF A34-82A	8112-10	5 000	1 22/7/43 2 26/7/43	**V** 304-9 **V** 310-15
ROSS RIVER NLA	19°19'S 146°47'E	30 miles north of Cairns to Townsville	Map 3826 E55-14-114	Can No. 132 16/5/42	22 000	---	1-264 **V**
AUSLIG		Townsville to Bundaberg Trimet	Map 3829	Can No. 123 27/7/42	20 000	---	1-725 **R**
NLA & CPE		Ross Ri Strip (oblique)	NEA HQ A34-82B	8114-7 21/7/43 & 8114-11 CPE 344 23/7/43	4 000	---	**P** 12 **P** 20 (**CPE** 1125)
NLA		Kissing Pt Townsville	KD 343 Map 1692 No 5 Sqd RAAF	5/7/38	8 000	2	3005
AUSLIG		Townsville	KD 561 Map 1338 RAAF Survey Fl	8468-1 1941	10 000	5	**VD** 34029
NLA		Ross Riv Strip	KD 829 NEA HQ A34-82A	8112-11 22/7/43	5 000	1 2 3 4	**V** 316-23 **V** 324-29 **V** 330-36 **V** 337-42

AIRFIELD NOTE 1	LOCATION LATITUDE LONGITUDE	AREA NAME NOTE 2	MAP/AREA NUMBER NOTE 3	FILM No. DATE NOTE 4	ALTITUDE (feet)	RUN	PRINT NUMBER NOTE 5
AITKENVALE (2nd WEIR) (THE WEIR STRIP) NLA	19°19'S 146°45'E	30 miles north of Cairns to Townsville Trimet	Map 3826 E55-14-106/114	Can No. 132 16/5/42	22 000	---	1-264 **V**
AUSLIG		Bundaberg to Townsville Trimet	Map 3829	Can No. 123 27/7/42	20 000	---	1-723 **V** 1-723 **R** 724 **R**
NLA		2nd Weir Strip (oblique)	NEA HQ A34-82B	8114-5 21/7/43	4 000	---	**P** 9,11
NLA		2nd Weir	KD 830 NEA HQ A34-82A	8112-6 22/7/43	5 000	1 2	**V** 256-66 **V** 267-75
UPPER ROSS (KELSO FIELD) NLA	19°25'S 146°44'E	30 miles north of Cairns to Townsville Trimet	Map 3826 E55-14-114	Can No. 132 16/5/42	22 000	---	1-262 **R**
AUSLIG		Bundaberg to T'ville Trimet	Map 3829	Can No. 123 27/7/42	20 000	---	1-723 **R**
NLA		Panoramic of all strips in Townsville Area	---	Can No. 133 16/5/42	22 000	---	1-281 **V**
AUSLIG		Townsville	Map 1384	-/6/42	10 000	12 13	73328 73393
NLA		Upper Ross Ri Strips (oblique)	NEA HQ A34-82B	8114-4 21/7/43	4 000	---	**P** 7-8
ANTIL PLAINS (MT STUART STRIP) NLA	19°27'S 146°50'E	Woodstock & Antel Plains Runway (oblique)	E55-14-114	31/3/42	1 500 approx	---	**PA** 550-569
NLA		Panoramic of all strips in Townsville Area	---	Can No. 133 16/5/42	22 000	---	1-267 **V**
PC		Antil Plains	---	1504-1 12/8/42	8 000	---	1-1291 **V**
NLA		Antil Plains (oblique)	NEA HQ A34-82B	8114-1 21/7/43	2 000	---	**P** 1-2
NLA		Stewart Ck Antil Plains T'ville	KD 343 Map 1692 No 5 Sqd RAAF	5/7/38	8 000	1 2	2973-82 2986-96
NLA		Antil Plains	KD 826 NEA HQ A34-82A	8112-12 21/7/43	5 000	1 21/7/43 2 21/7/43 3 26/7/43	**V** 343-52 **V** 353-62 **V** 363-71
---		Antil Plains (155mm guns)	NEA HQ	8164-7 20/5/44	1 000	---	**V** 1173
GIRU AUSLIG	19°30'S 147°08'E	Townsville-Bundaberg Trimet	Map 3829 E55-15-124	Can No. 123 27/7/42	20 000	---	1-733 **R**
AUSLIG		Ayr-Upstart	Map 654	16/5/45	17 000	1	1307
WOODSTOCK NLA	19°37'S 146°50'E	Woodstock & Antel Plains Runway (oblique)	E55-14-123	31/3/42	1 500 approx	---	**PA** 570-581
NLA		Panoramic of all strips in Townsville area	---	Can No. 132 16/5/42	22 000	---	---
PC		Woodstock	---	12/8/42	8 000	---	1-1293 **V**
PC		---	---	15/9/42	10 000	---	1-1439 **V**

AIRFIELD NOTE 1	LOCATION LATITUDE LONGITUDE	AREA NAME NOTE 2	MAP/AREA NUMBER NOTE 3	FILM No. DATE NOTE 4	ALTITUDE (feet)	RUN	PRINT NUMBER NOTE 5
WOODSTOCK (cont...) NLA	19°37'S 146°50'E	Woodstock Strip (oblique)	NEA HQ A34-82B	8114-2 21/7/43	3 000	---	**P** 3-4
NLA		Woodstock Strips	KD 824 NEA HQ A34-82A	8112-4 21/7/43	5 000	1 2 3	**V** 203-10 **V** 211-20 **V** 221-30
AUSLIG		Manton	KD 1288A Map 389 RAAF Survey Fl	16/6/45	17 000	3	VD 4650
REID RIVER PC	19°46'S 146°50'E	---	E55-14-132	15/9/42	10 000	---	1-1438 **V**
NLA & CPE		Reid Ri Strips (oblique)	NEA HQ A34-82B	8114-3 CPE 336 21/7/43	3 000	---	**P** 5-6 (**CPE** 1100)
AUSLIG		Mingela	KD 1261 Map 1552 RAAF Survey Fl	15/6/45	17 000	1	**VD** 3673
FANNING AUSLIG	19°55'S 146°32'E	Mingela	KD 1261 Map 1551 RAAF Survey Fl	15/6/45	17 000	4	**VD** 3540
AUSLIG		Dalrymple	KD 1258 Map 477 RAAF Survey Fl E55-14-131	9-10/5/45 (?)	17 000	4	**VD** 1426
CAMOOWEAL ---	19°55'S 138°08'E	Camooweal Drome (oblique)	NEA HQ E54-13-256	8117-11 19/9/43	6 700	---	**P** 117-8
---		Camooweal	NEA HQ	8128-13 19/9/43	6 500	---	**V** 988-91
AUSLIG		Camooweal	KD 1397A SVY-173	13-19/9/47	25 000	15	5471
BREDDAN (BREDDAN ISLAND)- (LITTLE SANDY)	19°56'S 146°12'E	Breddan Strips (oblique)	NEA HQ	8116-18 29/7/43	5 000	---	**P** 90-2
---		Breddan Strips	KD 837 NEA HQ E55-14-131	8127-20 11/8/43	7 500	1 2	**V** 859-65 **V** 866-76
AUSLIG		Dalrymple	KD 1258 Map 478 RAAF Survey Fl	9/5/45	17 000	5A	**VD** 1867
MACROSSAN ---	20°02'S 146°29'E	Macrossan (oblique)	NEA HQ F55-2-141/142	8114-13 29/7/43	6 000	---	**P** 23-24
---		Macrossan Strip	KD 834 Map 2685(?) NEA HQ A34-82A	8113-2 29/7/43	5 000	1 2	**V** 382-9 **V** 390-7
AUSLIG		Charters Towers	KD 1259 Map 220 RAAF Survey Fl	13/6/45	17 000	1A	**VD** 1876
AUSLIG		Ravenswood	KD 1308 RAAF Survey Fl	13/6/45	17 000	1	1992
BOWEN NLA & CPE	20°03'S 148°13'E	Bowen Aerodrome (oblique)	SHQ Archerfield F55-3-145	8072-4 CPE 037 26/6/40	3 000 approx	---	**FPZ** 24(18) 25(35)
AUSLIG		Townsville to Bundaberg Trimet	---	Can No. 123 27/7/42	20 000	---	1-762 **V**
---		Bowen	KD 861 9 Sqd RAAF	12/8/43	(?)	2 3	570-4 576-8

AIRFIELD NOTE 1	LOCATION LATITUDE LONGITUDE	AREA NAME NOTE 2	MAP/AREA NUMBER NOTE 3	FILM No. DATE NOTE 4	ALTITUDE (feet)	RUN	PRINT NUMBER NOTE 5
BOWEN (cont...) NLA	20°03′S 148°13′E	Bowen Aerodrome	KD 974 9 Sqd RAAF	8133-4 18/6/43	2 000	1 2 3 4	**V** 7-14 **V** 15-20 **V** 21-28 **V** 29-35
NLA		Bowen Drome (oblique)	NEA HQ A34-83B	8116-19 10/9/43	3 000	---	P 93,94
NLA		Bowen Drome	KD 887 NEA HQ A34-83A	8127-21 10/9/43	6 000	1 2	**V** 878-83 **V** 884-88
NLA		Bowen Drome (oblique)	Bow 72 9 Sqd	8168-10 20/7/44	2 000	---	**P** 132
NLA		Drome Bldgs	Bow 72 9 Sqd	8168-8 20/7/44	1 500	---	**V** 1020
AUSLIG		Bowen	KD 1269 Map 116 RAAF Survey Fl	24/6/45	17 000	1	**VD** 4047
BOWEN FLYING BOAT MAINTENANCE BASE AUSLIG	20°03′S 148°13′E (3 km east of airfield)	Townsville to Bundaberg Trimet	F55-3-145	Can No.123 27/7/42	20 000	---	1-762 **L**
AUSLIG		Bowen	KD 1269 Map 116 RAAF Survey Fl	24/6/45	17 000	1	4046
CHARTERS TOWERS (CORINDA) PC	20°03′S 146°17′E	Charters Towers	F55-2-141	12/8/42	---	---	**GA**-422
NLA & CPE		Charters Towers Strips (oblique)	NEA HQ A34-84B	8114-12 CPE 078 29/7/43	5 000	---	**P** 21,22
NLA		Charters Towers Strips	KD 832 NEA HQ A34-84A	Can No. 8113-1 29/7/43	6 000	1 2	**V** 372-6 **V** 377-81
AUSLIG		Charters Towers	KD 1259 Map 219	24/5/45	17 000	2	**VD** 1649
SOUTHERN CROSS NLA	20°07′S 146°06′E	Southern Cross (oblique)	NEA HQ A34-83B (?) F55-2-141	8116-17 29/7/43	5 000	---	**P** 88-9
NLA		Southern Cross	KD 835(?) NEA HQ A34-84A	8124-2 29/7/43	6 000	1	**V** 445-51
AUSLIG		Charters Towers	KD 1259 Map 219	24/5/45	17 000	3	**VD** 1677
POWLATHANGA ---	20°08′S 145°58′E	Powlathanga Strip (oblique)	NEA HQ F55-2-140	8116-16 29/7/43	5 000	---	**P** 86-7
---		Powlathanga Strip	KD 836 NEA HQ	8124-1 29/7/43	(?)	1 2	**V** 428-37 **V** 438-44
AUSLIG		Barrington	KD 1266A Map 1220 RAAF Survey Fl	26/5/45	17 000	4	**VD** 4259
TARANGIE NLA	20°09′S 145°50′E	Tarange Strip (oblique)	NEA HQ A34-84B	8116-15 29/7/43	5 000	---	**P** 84,85
NLA		Tarangie Strip	NEA HQ A34-84A	8113-3 29/7/43	6 000	1	**V** 422-7
AUSLIG		Barrington	KD 1266A Map 1220 RAAF Survey Fl F55-2-140	26/5/45	17 000	4	**VD** 4265

AIRFIELD NOTE 1	LOCATION LATITUDE LONGITUDE	AREA NAME NOTE 2	MAP/AREA NUMBER NOTE 3	FILM No. DATE NOTE 4	ALTITUDE (feet)	RUN	PRINT NUMBER NOTE 5
CLONAGH NLA	20°10'S 140°41'E	Clonagh Emer 'Drome (oblique)	NEA HQ A34-83B F54-2-224	8117-8 16/9/43	6 500	---	**P** 110-1
---		Clonagh Emerg' Strip	NEA HQ	8128-11 16/9/43	6 500	---	**V** 973-9
BALFES CREEK NLA	20°12'S 145°55'E	Balfes Creek (oblique)	NEA HQ A34-84B	8116-14 29/7/43	5 000	---	**P** 81,82,83
AUSLIG		Barrington	KD 1226A Map 1220 Raaf Survey Fl F55-2-140	26/5/45	17 000	5	**VD** 4306
JESSIEVALE NLA	20°15'S 140°27'E	Jessievale Emer' 'Drome (oblique)	NEA HQ A34-83B F54-2-223	8117-9 16/9/43	4 500 & 6 500	---	**P** 112-4
STRATHMORE NLA	20°30'S 147°38'E	Strathmore Emerg 'Drome (oblique)	NEA HQ A34-83B F55-3-154/164	8117-2 10/9/43	5 500 & 4 000	---	**P** 99-100
NLA		Strathmore Emerg 'Drome	NEA HQ A34-83A	8128-3 12/9/43	6 000	1	**V** 902-6
PROSERPINE ---	20°31'S 148°37'E	Proserpine	KD 859 9 Sqd RAAF	1/8/43	(?)	10	238
NLA		Proserpine Emerg 'Drome (oblique)	NEA HQ A34-83B	8117-1 11/9/43	3 000	---	**P** 97,98
NLA		Proserpine Emerg 'Drome	NEA HQ A34-83A	12/9/43	6 000	1	**V** 897-900
---		Proserpine 'Drome	KD 905 Map 1887 9 Sqd RAAF	11/11/43	(?)	7.5	**V** 999
AUSLIG		Proserpine	KD 1274 Map 120 RAAF Survey Fl F55-4-2/3	29/6/45	17 000	6	4873
CLONCURRY NLA	20°40'S 140°30'E	Cloncurry 'Drome (oblique)	SHQ Laverton F54-2-224	8143-3 22/11/41	---	---	**FPA** 20-21
PC		Cloncurry 'Drome (oblique)	---	22/9/42	---	---	**PA** 17/5
NLA		Cloncurry 'Drome	KD 856 NEA HQ A34-82A	8127-16 10/8/43	4 500 5 800	1 2	**V** 830-36 **V** 837-45
NLA		Cloncurry 'Drome (oblique)	NEA HQ A34-82B	8116-7 10/8/43	5 600	---	**P** 74
MT ISA NLA	20°43'S 139°30'E	Mt Isa 'Drome (oblique)	NEA HQ A34-83B	8117-10 19/9/43	6 600 6 400	--- ---	**P 115 NNW** **P 116 S**
NLA		Mt Isa 'Drome	NEA HQ A34-83A F54-1-241	8128-12 19/9/43	6 500	1	**V** 980-7
TORRENS CREEK PC	20°46'S 145°00'E	Torrens Creek	F55-1-168	15/9/42	10 000	---	1-1423 V and 1-1424 V
NLA		Torrens Creek (oblique)	NEA HQ A34-82B	8117-7 15/9/43	6 500	---	**P** 108-9
NLA		Torrens Creek 'Drome (oblique)	NEA HQ A34-83B	8116-12 12/8/43	7 400	---	**P 79 W**

AIRFIELD NOTE 1	LOCATION LATITUDE LONGITUDE	AREA NAME NOTE 2	MAP/AREA NUMBER NOTE 3	FILM No. DATE NOTE 4	ALTITUDE (feet)	RUN	PRINT NUMBER NOTE 5
TORRENS CREEK (cont...) NLA	20°46'S 145°00'E	Torrens Creek 'Drome	KD 879 NEA HQ A34-83A	8128-8 15/9/43	6 500 (?) --- --- 6 000	1 2 3 4	**V** 933-39 **V** 940-47 **V** 948-55 **V** 956-64
AA		Torrens Creek (oblique)	CRS A705/1 171/93/532	(?) prior to - /4/46	1 500 approx	SWPA ABSEC 6-19	27-94 (1 print! to SW)
AA		Torrens Creek Pin Point	CRS A705/1 171/93/532	(?) prior to - /4/46	(?)	SWPA ABSEC 6-19	26-58 (1 print)
PRAIRIE (STRATHGLASS) --- (EIGHT MILE)	20°48'S 144°38'E	Prairie Strip (oblique)	NEA HQ	8116-11 11/8/43	5 200	---	**P** 78
---		Prairie Strip (oblique)	NEA HQ	8117-5 15/9/43	5 800	---	**P** 105-6
---		Prairie Emergency Drome	NEA HQ	8128-6 15/9/43	5 800	---	**V** 921-6
AUSLIG		Hughenden	SVY-1277 F55-1-168	1/10/51	25 000	12A	5103
HUGHENDEN NLA	20°50'S 144°12'E	Hughenden Strip (oblique)	NEA HQ A34-82B (?)	8116-10 11/8/43	5 300	---	**P** 77
NLA		Hughenden Strip	NEA HQ A34-82A F55-1-167	8127-19 11/8/43	7 600	1	**V** 856-8
NLA		Hughenden Strip	NEA HQ A34-83A	8128-9 15/9/43	6 000	1	**V** 965-8
WARREAH (KAROON) NLA (COTONVALE)	20°58'S 144°47'E	Koroon Strip (oblique)	NEA HQ A34-83B	8117-6 15/9/43	6 200	---	**P** 107 **NW**
NLA		Koroon Emer 'Drome	NEA HQ A34-83A F55-1-168	8128-7 15/9/43	5 900	1	**V** 927-32
MACKAY AUSLIG	21°10'S 149°11'E	Townsville to Bundaberg Trimet	F55-8-7	Can No.123 27/7/42	20 000	---	1-799 **L**
AUSLIG		Townsville to Bundaberg Trimet	---	27/7/42	20 000	---	1-800 **V**
---		Mackay Drome (oblique)	NEA HQ	8116-20 10/9/43	3 000	---	**P** 95-6
---		Mackay Drome	KD 877 NEA HQ	8128-1 11/9/43	6 000	1 2	**V** 889-93 **V** 894-96
AUSLIG		Mackay	KD1407c SVY 201	6-7/10/47	16 000	5	5452
WINTON ---	22°22'S 143°02'E	Winton Drome (oblique)	NEA HQ	8117-13 20/9/43	3 000 to 5 000	---	**P** 121-4
WINTON EAST AUSLIG	22°22'S 143°04.5'E	Winton	SVY 1202 F54-12-319	8/7/51	25 000	6	5029
MT HEDLOW Lands Dpt Qld	23°13.5'S 150°34'E	Yeppoon	F56-13-37	Q1165 22/8/61	12 000	7	61-63
ROCKHAMPTON (CONNOR PARK) NLA	23°22'S 150°31'E	Connor Park Aerodrome	KD 417 Map 1895 23 Sqd	8075-14 22/8/39	---	1	**FVN** 743-8
---		AOB Rockhampton (oblique)	SHQ Archerfield	8074-11 7/11/41	---	---	**FPZ** 535-8

AIRFIELD NOTE 1	LOCATION LATITUDE LONGITUDE	AREA NAME NOTE 2	MAP/AREA NUMBER NOTE 3	FILM No. DATE NOTE 4	ALTITUDE (feet)	RUN	PRINT NUMBER NOTE 5
ROCKHAMPTON (CONNOR PARK) (cont...) ---	23°22'S 150°31'E	Ridgelands	KD 604A Map 1154 RAAF Survey Fl	25/1/42	10 000	11	48264
AUSLIG		Townsville to Bundaberg Trimet	F56-13-41	Can No. 123 27/7/42	20 000	---	1-863 **R**
NLA		Rockhampton 'Drome (oblique)	NEA HQ A34-83B	8117-21 27/10/43	5 000 4 200	--- ---	**P** 134 **N** **P** 135 **W**
NLA		Rockhampton 'Drome	KD 915 NEA HQ A34-83A	8139-9 27/10/43	8 000	1 2	**V** 1042-47 **V** 1048-53
LONGREACH NLA	23°25'S 144°15'E	Longreach (oblique)	SHQ Archerfield F55-13-267	8073-27 18/8/41	800 approx	---	**FPZ** 427-430
NLA		Longreach	SHQ Archerfield	8080-11 18/8/41	2 500 approx	---	**FVN** 1124 '29' Vertical over airfield
---		Longreach Drome (oblique)	NEA HQ	8117-14 23/9/43	5 700	---	**P** 125-6
NLA		Longreach 'Drome	KD 878 NEA HQ A34-83A	8128-17 23/9/43	5 700	1 2 3	**V** 1010-19 **V** 1020-27 **V** 1028-35
BARCALDINE AUSLIG	23°34'S 145°18.5'E	Longreach	SVY 1224 F55-13-279	11/8/51	25 000	9	5097
WESTWOOD NLA	23°36.5'S 150°09'E	Westwood (oblique)	NEA HQ F56-13-46	8117-17 28/9/43	3 700	---	**P** 129
BLACKALL (WEST) and --- (CIVIL)	24°26'S 145°25'E 24°26'S 145°28'E	Blackall Civil (oblique)	SVY 1217 SHQ Archerfield	8073-29 18/8/41	---	---	**FPZ** 435-438
---		Blackall	SHQ Archerfield	8080-11 18/8/41	---	---	**FVN** 1126
AUSLIG		Blackall	G55-1-309	31/7/52	25 000	7	5050
MOORLANDS ---	24°44'S 152°11'E	Moorlands Strip	23 Sqd	8016-8 9/7/43	11 267	---	**VX** 24287-9
AUSLIG		Baffle	KD 1440 SVY 483	17/11/48	15 000	6	5042
MEADOWVALE ---	24°47.5'S 152°17'E	Meadowvale Strip	23 Sqd	8016-8 9/7/43	11267	---	**VX** 24290-2
AUSLIG		Bundaberg	SVY 483 G56-2-85	17/11/48	15 000	2	5088
MONTO ---	24°54'S 151°06'E	Monto Aerodrome (oblique)	SHQ Archerfield	8074-18 10/11/41	---	---	**FPZ** 564-7
---		Monto Aerodrome	SHQ Archerfield	8080-14 10/11/41	---	---	**FVN** 1152
AUSLIG		Cannindah (Monto)	KD-1422 SVY 476 G56-1-83	17/11/48	15 000	4	5063
BUNDABERG ---	24°55'S 152°20'E	Bundaberg Aerodrome	SHQ Archerfield	8076-10 14/8/40	---	---	**FVN** 655-7
RAAF Historical		Bundaberg Aerodrome (oblique)	SHQ Archerfield	8074-10 10/11/41	---	---	**FPZ** 531-4
RAAF Historical		Bundaberg RAAF 'Drome (oblique)	SHQ Archerfield	8074-29 12/11/41	---	---	**FPZ** 592-6

AIRFIELD NOTE 1	LOCATION LATITUDE LONGITUDE	AREA NAME NOTE 2	MAP/AREA NUMBER NOTE 3	FILM No. DATE NOTE 4	ALTITUDE (feet)	RUN	PRINT NUMBER NOTE 5
BUNDABERG (cont...) ---	24°55'S 152°20'E	Bundaberg Aerodrome	SHQ Archerfield	8080-12 10/11/41	---	---	**FVN** 1145
---		Bundaberg 'Drome (oblique)	23 Sqd	8012-15 19/7/42	1 000	---	PX 2200,1,2 & 4
---		Bundaberg Drome	23 Sqd	8009-3 19/7/42	10 000	---	**VX** 2165-71 **VX** 2183-90
AA		Bundaberg 'Drome	23 Sqd CRS A705 171/53/2	8015-19 8/6/43	10 000	---	**VX** 03441-3
AWM		RAAF Base Bundaberg (oblique)	8 SFTS	8188-5 6/12/44	2 000	---	**P** 279/64/14, 17,18 [AWM] **P** 68-71 [RAAF]
AUSLIG		Bundaberg	SVY 483 G56-2-85	17/11/48	15 000	4	5141
MARYBOROUGH AUSLIG	25°30'S 152°43'S	Maryborough	KD 495 Map 1226 RAAF Survey Fl	14/7/40	13 000	2	9944
---		Maryborough Aerodrome	SHQ Archerfield	8076-11 14/8/40	---	---	**FVN** 658-61
---		Maryborough Aerodrome (oblique)	SHQ Amberley	8122-4 15/5/41	---	---	**PA** 35-43
---		Maryborough Aerodrome (oblique)	SHQ Archerfield	8074-9 10/11/41	---	---	**FPZ** 527-30
---		Maryborough Aerodrome	SHQ Archerfield	8080-12 10/11/41	---	---	**FVN** 1145
---		Maryborough Aerodrome	5 Sqd	8091-2 15/3/43	7 600	---	**V** 4604-6
AWM		RAAF Base Maryborough (oblique)	8 SFTS G56-6-113	8188-2 6/12/44	2 000	---	**P** 279/64/12,15 [AWM] **P** 56-9 [RAAF]
GYMPIE CPE	26°12'S 152°48'E	Gympie A'fld	G56-10-132	CPE 162 28/3/40	1 500 approx	---	0478
CPE		Gympie Aerodrome (oblique)	SHQ Archerfield	8074-13 10/11/41	---	---	**FPZ** 544-7
CPE		Gympie Aerodrome	SHQ Archerfield	8080-12 10/11/41	---	---	**FVN** 1148
BONNIEVILLA NLA	26°19'S 146°12'E	Bonnievilla (obliques)	SHQ Archerfield G55-10-391	8073-9 18/8/41	300 approx	---	**FPZ** 353-357
NLA		Bonnievilla	SHQ Archerfield	8080-8 18/8/41	---	---	**FVN** 1106(34)
MORVEN AUSLIG	26°24'S 147°07'E	Morven (oblique)	SHQ Archerfield G55-11-393	8073-18 18/8/41	---	---	**FPZ** 390-393
---		Morven	SHQ Archerfield	8080-12 18/8/41	---	---	**FVN** 1115
---		Mitchell	SVY 564 G55-11-393	21/10/48	25 000	6	5220
CHARLEVILLE NLA	26°25'S 146°16'E	Charleville (oblique)	SHQ Archerfield G55-10-391	8073-6 18/8/41	---	---	**FPZ** 337-341

AIRFIELD NOTE 1	LOCATION LATITUDE LONGITUDE	AREA NAME NOTE 2	MAP/AREA NUMBER NOTE 3	FILM No. DATE NOTE 4	ALTITUDE (feet)	RUN	PRINT NUMBER NOTE 5
CHARLEVILLE (cont...) NLA	26°25'S 146°16'E	Charleville Racecourse (oblique) (Civil 'drome in b'ground)	SHQ Archerfield	8073-7 18/8/41	300 approx	---	**FPZ** 342-347
NLA		Charleville 'Drome (obliques)	5 Sqd (?)	22/2/43	1 000 (?)	---	**P** 978 **S** 979 **E** 980 **NE** 981 **N**
NLA		Charleville Aerodrome (obliques)	23 Sqd (?)	8015-1 26/6/43	2 000	---	**PX** 244 **NW** 243 **SW** 245 **E** 246 **S**
NLA		Charleville	SHQ Archerfield	8080-8 18/8/41	---	---	**FVN** 1103 (29)
NLA		Charleville Racecourse (Civil drome on edge)	SHQ Archerfield	8080-8 18/8/41	---	---	**FVN** 1104 (30)
---		Charleville Aerodrome	5 Sqd	8090-14 23/2/43	8 000	---	**V** 3680-3
CPE		Charleville Afld	23 Sqd	8016-2 (077) 26/6/43	13 000	---	**VX** 24237,9 (0023)
LURNEA NLA	26°25'S 146°41'E	Lurnea (oblique)	SHQ Archerfield G55-10-392	8073-20 18/8/41	(?)	---	**FPZ** 398-401
NLA		Lurnea	SHQ Archerfield	8080-10 18/8/41	---	---	**FVN** 1117 (89 ?)
MUNGALLALA ---	26°26'S 147°32'E	Mungallala	SHQ Archerfield G55-11-394	8073-17 18/8/41	---	---	**FPZ** 386-389
---		Mungallala	SHQ Archerfield	8080-10 18/8/41	---	---	**FVN** 1114
AUSLIG		Mitchell	SVY 455	22/10/45	25 000	7	5101
DULBYDILLA ---	26°26'S 147°22'E	Dulbydilla	SHQ Archerfield G55-11-393	8073-12 18/8/41	---	---	FPZ 366-369
---		Dulbydilla	SHQ Archerfield	8080-9 18/8/41	---	---	**FVN** 1109
AUSLIG		Mitchell	SVY 455	22/10/45	25 000	7	5105
MITCHELL AUSLIG	26°29'S 147°57'E	Mitchell	SVY 455 G55-11-394	22/10/45	25 000	8	5067
QUILPIE AUSLIG	26°34'S 144°19'E	Quilpie	SVY 1214 G55-9-397	30/7/52	25 000	9	5082
ROMA ---	26°35'S 148°49'E	Roma	SHQ Archerfield G55-12-144	8073-23 18/8/41	---	---	**FPZ** 410-413
---		Roma	SHQ Archerfield	8080-11 18/8/41	---	---	**FVN** 1120
AUSLIG		Roma	SVY 468	9/11/48	15 000	2	5148
KINGAROY ---	26°36'S 151°50'E	Kingaroy Aerodrome (oblique)	SHQ Amberley G56-10-150	8097-15 23/3/41	---	---	**PB** 38-43
---		Kingaroy Aerodrome (oblique)	SHQ Archerfield	8074-24 12/11/41	---	---	**FPZ** 588-91

AIRFIELD NOTE 1	LOCATION LATITUDE LONGITUDE	AREA NAME NOTE 2	MAP/AREA NUMBER NOTE 3	FILM No. DATE NOTE 4	ALTITUDE (feet)	RUN	PRINT NUMBER NOTE 5
KINGAROY (cont...) ---	26°36′S 151°50′E	Kingaroy Drome (oblique)	5 Sqd	8039-13 28/7/42	4 500	---	**PC** 821-4
---		Kingaroy Drome	5 Sqd	8034-16 28/7/42	6 100	---	**V** 3055-9
---		Kingaroy Drome	23 Sqd	8010-19 9/3/43	11 267	---	**VX** 03175-7
CPE		Kingaroy Afld	---	CPE 188 23/3/43	(?)	---	0023
AWM		RAAF Base Kingaroy (oblique)	8 SFTS	8188-4 6/12/44	2 000	---	**P** 279/64/13, 19-21 [AWM] **P** 64-7 [RAAF]
AUSLIG		Kumbia	KD 1345 Map 202 RAAF Survey Fl	18/7/45	17 000	2	11694
MANNUEM (KUMBIA) AUSLIG	26°39′S 151°37′E	Kumbia	KD 1345 Map 204 RAAF Survey Fl G56-10-150	18/7/45	17 000	4 5	14750 14758
NANANGO ---	26°41.5′S 151°59.5′E	Nanango Aerodrome (oblique)	SHQ Amberley G56-10-150	8122-1 3/4/41	---	---	**PB** 65-6
---		Nanango Aerodrome (oblique)	SHQ Amberley	8122-1 30/4/41	---	---	**PA** 5-6
---		Nanango Aerodrome (oblique)	SHQ Archerfield	8074-22 10/11/41	---	---	**FPZ** 580-3
---		Nanango Aerodrome	SHQ Archerfield	8080-14 10/11/41	---	---	**FVN** 1156
AUSLIG		Kumbia	KD 1345 Map 204 RAAF Survey Fl	30/9/48	17 000	5	14774
CONDAMINE ---	26°50′S 150°10′E	Condamine (oblique)	23 Sqd G56-9-157	8013-9 18/1/43	4 000	---	**PX** 44-7
---		Condamine Drome	23 Sqd	8010-3 18/1/43	13 000	---	**VX** 02993-5
AUSLIG		Condamine	Map 688	27/9/45	17 000	2	13833
TOOGOOLAWAH PC	27°06′S 152°27′E	Toogoolawah (oblique)	5 Sqd RAAF G56-14-171	8039-11 17/7/42	300	---	**P** 773 **SE** **P** 778 **NW**
---		Toogoolawah Drome	5 Sqd	8034-13 19/7/42	10 000 5 000	--- ---	**V** 2792-5 **V** 2796-804
CPE		Toogoolawah Afld	---	8035-3 CPE 426 16/8/42	8 000	---	**V** 3117-26
---		Toogoolawah Drome	5 Sqd	8090-13 9/2/43	8 000	---	**V** 3672-5
RAAF Historical		Toogoolawah Aerodrome	23 Sqd	8010-20 23/3/43	10 000	---	**VX** 03183-5
AUSLIG		Cressbrook	KD 1248 Map 1538	7/7/45	17 000	3	11578
BRYMAROO (JONDARYAN) NLA	27°13′S 151°38′E	Brymaroo Drome (obliques)	23 Sqd RAAF G56-14-170	8013-8 16/1/43	2 000	---	**PX** 40 **W** **PX** 41 **S** **PX** 42 **E** **PX** 43 **N**

AIRFIELD NOTE 1	LOCATION LATITUDE LONGITUDE	AREA NAME NOTE 2	MAP/AREA NUMBER NOTE 3	FILM No. DATE NOTE 4	ALTITUDE (feet)	RUN	PRINT NUMBER NOTE 5
BRYMAROO (cont...) (JONDARYAN) ---	27°13′S 151°38′E	Brymaroo Drome	23 Sqd	8010-1 16/1/43	10 000	---	**VX** 03062-6
AA(NSW)		Dispersal Hideouts- Jondaryan (obliques)	23 Sqd Dept of Home Security- Camouflage	19/6/43	800 approx	---	**PX** 212-32
AA(NSW)		Dispersal Hideouts Jondaryan	23 Sqd Dept of Home Security- Camouflage	8016-3 19/6/43	4 267	---	**VX** O3499-513
AUSLIG		Evergreen	KD 1247 Map 216	30/6/45	17 000	6	**VD** 11491
PETRIE CPE (STRATHPINE A-1)	27°17′S 153°00′E	Petrie Afld	23 Sqd G56-15-183	8010-10 CPE 303	10 000	11/3/43 14/3/43	**VX** 03099-101 **VX** 23784-7 (1500)
---		Brisbane	KD 1036 RAAF Survey Fl	16/5/44	17 000	2	19
STRATHPINE A-2 NLA	27°18′S 153°00′E	Proposed 'Drome Site Strathpine	SHQ Archerfield G56-15-183	8080-5 18/9/40	70 000	---	**FVN** 688-699
NLA		Strathpine Proposed LG	SHQ Archerfield	8077-7 8/1/41	5 200	---	**FVN** 854-62
---		Strathpine A2	South Petrie Landing Strip (oblique)	23 Sqd	8012-7 17/7/42	1 000	**PX** 84-7
---		Strathpine A2	SE of Petrie Strip (oblique)	23 Sqd	8012-9 17/7/42	1 000	**PX** 92-5
---		Brisbane	KD 1036 RAAF Survey Fl	16/5/44	17 000	3	44
AWM		RAAF Base Strathpine	8 SFTS	6/12/44	2 500 approx	---	**P** 279/64/16 NE [AWM]
AUSLIG		Samford	KD 1526 SVY 572	28/11/49	16 000	2	5130
STRATHPINE A-3 ---	27°19.5′S 153°00′E	West Bald Hills Strip (oblique)	23 Sqd G56-14-182	8012-8 17/7/42	1 000	---	**PX** 88-91
AUSLIG		Samford	KD 1526 SVY 572	28/6/49	16 000	3	5095
COOMINYA No 2 (A-6) --- (LOWOOD No 2)	27°24′S 152°29′E	Coominya Landing Strip	23 Sqd G56-14-181	8009-14 17/7/42	10 000	---	**VX** 22924-6
---		Coominya Strip & Drome (oblique)	23 Sqd	8012-18 20/7/42	1 000	---	**PX** 2292-8
AA(NSW)		Coominya Strip	23 Sqd Lowood	8010-7 26/2/43	8 000 approx	C1905 T1 Box 15	**VXO** 3164-5
AUSLIG		Crows Nest	Map 2155	25/4/44	5 000	L	107 **LOB**
AUSLIG		Crows Nest	Map 2141	26/4/44	10 000	B	114 **LOB**
COOMINYA No 1 (COOMINYA A-4) (COOMINYA) AUSLIG	27°24.5′S 152°31′E	Crows Nest Exp	Map 1541	26/4/44	17 000	2	5

AIRFIELD NOTE 1	LOCATION LATITUDE LONGITUDE	AREA NAME NOTE 2	MAP/AREA NUMBER NOTE 3	FILM No. DATE NOTE 4	ALTITUDE (feet)	RUN	PRINT NUMBER NOTE 5
EAGLE FARM PC	27°25′S 153°06′E	Eagle Farm (obliques)	SHQ Archerfield G56-14-182 & -15-183	8073-30 2/9/41	400 approx	---	**FPZ** 453-456 **E,W,N** etc
---		Eagle Farm	SHQ Archerfield	8080-12 2/9/41	6 600	---	**FVN** 1127
NLA		Brisbane	KD 1039	20/12/41	6 600	7	147
---		Eagle Farm 'Drome (oblique)	23 Sqd	8012-16 20/7/42	1 000	---	**PA** 2284-7
PC		Eagle Farm 'Drome	23 Sqd	8009-8 20/7/42	10 000	---	**VX** 2247-53 **VX** 2256-63
---		Eagle Farm 'Drome	23 Sqd	8009-17 21/8/42	8 000	---	**VX** 23106-8
---		Eagle Farm 'Drome	23 Sqd	8010-12 14/3/43	10 000	---	**VX** 03107-9
NLA		Eagle Farm Aerodrome (obliques)	23 Sqd	17/3/43	2 000	---	**PX** 98,99
---		Brisbane	KD 1036 B RAAF Survey Fl	16/5/44	17 000	5 6	129 14
NLA		Eagle Farm Aerodrome (obliques)	32 Sqd (?)	11262-4 30/6/45	2 000 approx	---	**P** 124 **SE**
NLA		Brisbane Metro	KD 1517 SVY 520	13/12/48	10 500	4	5076-7
OAKEY ---	27°25′S 151°45′E	Oakey	KD 810 23 Sqd RAAF G56-14-180	24/5/43	(?)	1 2 3	03324-8 03329-33 03334-38
AUSLIG		Jondaryan	Map 723	2/10/45	17 000	5	14847
LOWOOD ---	27°27′S 152°28′E	Tarampa Proposed 'Drome Site	SHQ Archerfield G56-14-181	8078-8 3/6/41	6 600	---	**FVN** 1011-20
CPE		Mt Tarampa Afld (oblique)	23 Sqd	8002-1 CPE 255 17/7/42	1 000	---	**PX** 22465,8,70, 71
---		Lowood (oblique)	23 Sqd	8013-11 11/2/43	2 000	---	**PX** 72-5
CPE		Lowood Afld	23 Sqd	8010-5 CPE 203 11/2/43	10 000	---	**VX** 03061-5 (1445)
---		Lowood 'Drome	23 Sqd	8016-11 10/7/43	10267	---	**VX** 24316-8
AUSLIG		Crows Nest Trimet	Map 2155 Map 2141 Map 2141	25/4/44 25/4/44 26/4/44	5 000 10 000 10 000	Run D} Run V} Run E Run D	98 **ROB** 96-7 Vert 55 Vert 103 **ROB**
CECIL PLAINS (TIPTON) NLA	27°28′S 151°13′E	Tipton	5 Sqd G56-13-179	8034-7 18/7/42	11 000	---	**V** 2766
NLA		Tipton Dispersal Point (oblique)	5 Sqd	8039-17 1/9/42	2 000	---	**P** 885-8
---		Tipton Dispersal	5 Sqd	8036-8 1/9/42	5 000	---	**V** 3211-7

AIRFIELD NOTE 1	LOCATION LATITUDE LONGITUDE	AREA NAME NOTE 2	MAP/AREA NUMBER NOTE 3	FILM No. DATE NOTE 4	ALTITUDE (feet)	RUN	PRINT NUMBER NOTE 5
CECIL PLAINS --- (TIPTON) (cont...)	27°28′S 151°13′E	Tipton Strip	5 Sqd	8036-8 1/9/42	5 000	---	**V** 3218-22
---		Cecil Plains Landing Strip (oblique)	23 Sqd	8013-5 28/10/42	1 500	---	**PX** 02583-6
NLA		Aerodrome Cecil Plains (oblique)	23 Sqd	8013-2 23/10/42	1 000	---	**PX** 02762-5
NLA		Cecil Plains Aerodrome (obliques) (vertical)	23 Sqd	8013-10 19/1/43	2 000	---	**PX** 48 **W** **PX** 49 **E** **PX** 50 **S** **PX** 51 **N**
CPE		Bowenville (Cecil Plains 'Drome)	23 Sqd	8010-2 CPE 074 19/1/43	9 600	---	**VX** 03086-8
AUSLIG		Bowenville	KD 1343A Map 684 RAAF Survey Fl	3/10/45	17 000	6	15432
TOOWOOMBA ---	27°32′S 151°55′E	Toowoomba Aerodrome (oblique)	SHQ Amberley	8097-14 3/3/41	---	---	**PB** 22-3
---		Toowoomba Aerodrome (oblique)	SHQ Amberley	8122-1 3/5/41	---	---	**PA** 1,7-11
---		Toowoomba (oblique)	SHQ Archerfield	8073-24 18/8/41	---	---	**FPZ** 414-8
---		Toowoomba	SHQ Archerfield	8080-11 18/8/41	---	---	**FVN** 1121
---		Toowoomba A Flight Dispersal (oblique)	5 Sqd	8039-6 7/6/42	800	---	**PC** 708-11
---		Toowoomba RAAF Camp (oblique)	5 Sqd	8039-7 7/6/42	800 & 600	---	**PC** 712-16
---		Toowoomba 'Drome (oblique)	5 Sqd	8032-4 21/7/42	10 000	---	**PC** 798
---		Toowoomba 'Drome (oblique)	5 Sqd	8039-15 16/8/42	2 000	---	**PC** 847,853, 855-6
---		Toowoomba Camp Site (oblique)	5 Sqd	8039-16 21/8/42	3 000	---	**PC** 859-63
NLA		Toowoomba Trimet	KD 1116 RAAF Survey Fl G56-14-190	25/4/44 25/4/44	17 000 11 000	A A	51 **LOB** 34 **LOB**
AUSLIG		Toowoomba	Map 227	5/10/45	17 000	2	16536
ARCHERFIELD NLA	27°35′S 153°01′E	Archerfield RAAF Station (oblique)	SHQ Amberley G56-15-193	8097-13 2/3/41	---	---	**PB** 14-16
---		Archerfield (oblique)	SHQ Archerfield	8075-1 7/10/41	---	---	**FPZ** 487-489
---		Archerfield RAAF Station (oblique)	SHQ Archerfield	8074-7 3/11/41	---	---	**FPZ** 611-614

AIRFIELD NOTE 1	LOCATION LATITUDE LONGITUDE	AREA NAME NOTE 2	MAP/AREA NUMBER NOTE 3	FILM No. DATE NOTE 4	ALTITUDE (feet)	RUN	PRINT NUMBER NOTE 5
ARCHERFIELD (cont...) NLA	27°35′S 153°01′E	Archerfield (Decoy Aircraft) (oblique)	SHQ Archerfield	8074-30 27/11/41	---	---	**FPZ** 506-509
---		Archerfield (oblique)	SHQ Archerfield	8075-10 12/3/42	---	---	**FPZ** 717-720
---		Archerfield 'Drome (oblique)	23 Sqd	8012-2 17/7/42	2 000	---	**PX** 2223-2226
NLA		Archerfield 'Drome (oblique)	23 Sqd	8010-15 21/3/43	2 000	---	**PX** 110 **N**
NLA		Archerfield	SHQ Archerfield	8080-3 28/3/40	(?)	---	**FVZ** 479
---		Royal Australian Air Force Station Archerfield	KD 469 23 Sqd	8075-20 26/4/40	2 000	1 2 3	**FVN** 486-493 **FVN** 494-500 **FVN** 501-508
NLA		Archerfield	SHQ Archerfield	8078-1 31/1/41	6 600	---	**FVN** 863-8
---		Archerfield Aerodrome	SHQ Archerfield	8080-7 4/3/41	8 000	---	**FVN** 909
NLA		Archerfield	SHQ Archerfield	8078-13 8/9/41	4 000	---	**FVN** 1128-33
NLA		Archerfield	SHQ Archerfield	8080-12 7/10/41	8 000	---	**FVN** 1140
NLA		Brisbane Area	KD 1036A (1039?) SHQ Archerfield	8079-9 20/12/41	6 600	1	**FVN** 1206 (3)
NLA		Archerfield	SHQ Archerfield	8080-15 12/3/42	8 000	---	**FVN** 1385-7
NLA		Archerfield A/D	SHQ Archerfield	8079-12 23/3/42	3 000	---	**FVN** 1391-1452
---		Archerfield 'Drome	23 Sqd	8009-1 17/7/42	10 000	---	**VX** 2206-22
---		Archerfield 'Drome	23 Sqd	8009-16 21/8/42	8 000	---	**VX** 23104-5
NLA		Archerfield Aerodrome	23 Sqd	8010-15 21/3/43	14 000	---	**VX** 23801-7
AUSLIG		Beenleigh	KD 1049 Map 496 RAAF Survey Fl	9228-1 27/4/44	17 000	2	**VD** 402
BREMER RIVER (FORKS A-7) --- (RIVERVIEW)	27°35′S 152°51′E	Bremer River Strip	23 Sqd	8011-5 15/4/43	10 267	---	**VX** 23886-8
AUSLIG		Ipswich	Map 1543	17/5/44	17 000	4A	179
GAILES (GAILES A-9) --- (SERVICETON) (WACOL)	27°36′S 152°55′E	Gailes 'Drome (oblique)	23 Sqd	8012-3 17/7/42	2 000	---	**PX** 2227-30
---		Gailes	23 Sqd	8009-6 17/7/42	10 000	---	**VX** 2205
---		Gailes Strip	23 Sqd	8011-4 15/4/43	10 267	---	**VX** 23883-5
AUSLIG		Beenleigh	SVY 496	27/4/44	17 000	3	355
AUSLIG		Ipswich	Map 1542	17/5/44	17 000	5	49

AIRFIELD NOTE 1	LOCATION LATITUDE LONGITUDE	AREA NAME NOTE 2	MAP/AREA NUMBER NOTE 3	FILM No. DATE NOTE 4	ALTITUDE (feet)	RUN	PRINT NUMBER NOTE 5
AMBERLEY NLA	27°38'S 152°43'E	Amberley 'Drome	G56-14-192	8080-2 24/1/40	10 000	---	**FV** 451
NLA		Amberley	---	8080-3 28/3/40	---	---	**FVZ** 476
NLA		Royal Australian Air Force Station Amberley	KD 468 Map 2113,4 23 Sqd SHQ Archerfield	8076-1 8076-2 1/5/40	--- ---	1 2 3 4 5 6	**FVN** 512-22 **FVN** 523-32 **FVN** 533-41 **FVN** 542-50 **FVN** 551-58 **FVN** 559-67
NLA		Amberley	SHQ Archerfield	8077-2 4/12/40	6 600	---	**FVN** 745-50
NLA		Amberley RAAF Station (oblique)	SHQ Amberley	8097-13 2/3/41	---	---	**PB** 1, 18-20
NLA		Amberley Hangars & Work Shops (oblique)	SHQ Amberley	8097-14 3/3/41	---	---	**PB** 33
NLA		Amberley (oblique)	SHQ Amberley	8097-14 3/3/41	---	---	**PB** 36
NLA		Amberley	---	8080-7 4/3/41	8 000	---	**FVN** 908
NLA		Amberley (oblique)	SHQ Amberley	8097-14 21/3/41	---	---	**PB** 37
NLA		Inst Bldg Area	HQ Amberley	8097-12 18/6/41	---	---	**VA** 2
NLA		Amberley 'Drome	SHQ Archerfield	8079-1 8/9/41	6 600	---	**FVN** 1134-9
NLA		RAAF Amberley	SHQ Amberley	8097-12 14/9/41	---	---	**VA** 13
NLA		Amberley Drome (obliques)	23 Sqd	8012-6 17/7/42	1 000	---	**PX** 02199 **S** **PX** O2202 **N**
---		Amberley Landing Strip	23 Sqd	8009-7 17/7/42	10 000	---	**VX** 2191-3
---		Amberley 'Drome	23 Sqd	8008-15 23/7/42	10 000	---	**VX** 2377-86
NLA		Amberley Drome	23 Sqd	8009-15 21/8/42	8 000	---	**VX** 23100-1
NLA		Operations Amberley (obliques)	23 Sqd G56-14-192	8013-12 26/2/43	1 500	---	**PX** 79 **NE** **PX** 80 **SW** **PX** 81 **SW**
NLA		Ops Amberley	23 Sqd	8010-6 26/2/43	6 000	---	**VX** 03155-6
NLA		Amberley Aerodrome	23 Sqd	8010-13 14/3/43	10 000	---	**VX** 03116-8
NLA		Amberley Drome (obliques)	23 Sqd	8014-2 21/3/43	3 000	---	**PX** 114 **N** **PX** 115 **E** **PX** 116 **S** **PX** 117 **W**
---		Amberley 'Drome	23 Sqd	8016-7 2/7/43	10 267	---	**VX** 24258
NLA		Amberley (oblique)	23 Sqd	8015-15 26/7/43	3 000	---	**P** 279
C-M		Amberley (oblique)	(?)	-/1/46?	200	---	24

AIRFIELD NOTE 1	LOCATION LATITUDE LONGITUDE	AREA NAME NOTE 2	MAP/AREA NUMBER NOTE 3	FILM No. DATE NOTE 4	ALTITUDE (feet)	RUN	PRINT NUMBER NOTE 5
AMBERLEY cont.... AUSLIG	27°38′S 152°43′E	Ipswich-Samford	SVY-430	29/9/48	---	4	5060
KINGSTON A-11 (WOODRIDGE) NLA	27°39′S 153°05′E	Kingston NW Landing Strip (obliques)	23 Sqd G56-15-193	8012-5 17/7/42	2 000	---	**PX** 2420 **NE** **PX** 2421 **NW** **PX** 2422 **SW** **PX** 2423 **SE**
NLA		Kingston Landing Strip	23 Sqd	17/7/42	1 000	---	**VX** 2096-2100
NLA		Kingston Strip (obliques)	23 Sqd	18/7/43	3 000	---	**P** 276 **N** **P** 277 **S**
NLA		Kingston Strip	23 Sqd	8016-12 18/7/43	9 267	---	**V** 03887-9
AUSLIG		Beenleigh	KD 1049 SVY 496 RAAF Survey Fl	27/4/44	17 000	5	314
LOGANLEA NLA	27°40′S 153°09′E	Kingston E Landing Strip (obliques)	23 Sqd G56-15-193	8012-15 17/7/42	2 000	---	**PX** 2432 **NE** 2433 **W** 2434 **SW** 2435 **S**
NLA		Kingston Landing Strip	23 Sqd	17/7/42	1 000(?) 10 000?	---	2 **VX** 3.2 **SE**?
AUSLIG		Beenleigh	KD 1049 SVY 496 RAAF Survey Fl	27/4/44	17 000	5	312
ROSEWOOD ---	27°39.5′S 152°36.5′E	Rosewood Emergency LG (oblique)	SHQ Amberley G56-14-192	8097-14 3/3/41	---	---	**PB** 34
NLA		Amberley Landing Strip (obliques)	23 Sqd	8012-6 17/7/42	1 000	---	**PX** O2203 **S** **PX** 02204 **NE** **PX** O2206 **SW**
NLA		Amberley Landing Strip	23 Sqd F24/3	8009-7 17/7/42	10 000	---	**VX** O2191
KINGSTON A-10 NLA	27°41′S 153°06′E	Kingston SW Landing Strip (obliques)	23 Sqd G56-15-193	8012-5 17/7/42	2 000	---	**PX** 2424 **N** 2425 **W** 2426 **SW** 2427 **SE** 2428 **E**
NLA		Kingston Landing Strip	23 Sqd	---	1 000 ?	---	**VX SW**?
AUSLIG		Beenleigh	KD 1049 SVY 496 RAAF Survey Fl	27/4/44	17 000	5	314
KINGSTON A-12 NLA	27°42.5′S 153°05.5′E	Kingston SSW Landing Strip (obliques)	23 Sqd G56-15-193	8012-5 17/7/42	2 000	---	**PX** 2429 **N** 30 **E** 31 **S**
AUSLIG		Beenleigh	KD 1049 Map 496	27/4/44	17 000	6	423
WATERFORD NLA	27°41.5′S 153°08′E	Kingston S Landing Strip (obliques)	23 Sqd G56-15-193	8012-5 17/7/42	2 000	---	**PX** 2436 **SE** 7 **NE** 8 **NW** 9 **SW**
NLA		Kingston Landing Strip	23 Sqd	17/7/42	1 000	---	**VX S**
AUSLIG		Beenleigh	KD 1049 Map 496 RAAF Survey Fl	27/4/44	17 000	6	423

AIRFIELD NOTE 1	LOCATION LATITUDE LONGITUDE	AREA NAME NOTE 2	MAP/AREA NUMBER NOTE 3	FILM No. DATE NOTE 4	ALTITUDE (feet)	RUN	PRINT NUMBER NOTE 5
WARRILL CREEK (A-8) NLA	27°42'S 152°40'E	Amberley Landing Strip (obliques)	23 Sqd G56-14-192	8012-6 17/7/42	1 000	---	**PX** 02197 **S** **PX** 02198 **N**
NLA		Amberley Landing Strip	23 Sqd	17/7/42	10 000	---	**VX** 02192
---		Warrill Creek Strip	23 Sqd	8011-2 15/4/43	10 267	---	**VX** 23877-9
GOOLMAN (PURGA) NLA	27°45'S 152°45'E	Amberley Landing Strip (oblique)	23 Sqd G56-14-192	8012-6 17/7/42	1 000	---	**PX** 02194 **N**
NLA		Amberley Landing Strip	23 Sqd	17/7/42	10 000	---	**VX** 02193
---		Goolman Strip	23 Sqd	8011-3 15/4/43	10 267	---	**VX** 23880-2
LEYBURN ---	27°57'S 151°36'E	Leyburn 'Drome (oblique)	23 Sqd	8013-3 23/10/42	1 000	---	**PX** 02766-9
---		Leyburn Aerodrome	23 Sqd	8008-12 23/10/42	10 000	---	**VX** 02772-4
---		Leyburn Strip	23 Sqd	8010-4 1/2/43	10 000	---	**VX** 03106-8
CPE		Leyburn Afld (oblique)	G56-14-200	CPE 197 5/3/43	---	---	4267
AUSLIG		Clifton	Map 686 RAAF Survey Fl	4/10/45	17 000	5	15693
CUNNAMULLA AUSLIG	28°02'S 145°39'E	Cunnamulla	SVY 1409 H55-2-460	24/7/52	25 000	1	5157
CUNNAMULLA (CIVIL) ---	28°04'S 145°42'E	Cunnamulla	SHQ Archerfield	8073-19 18/8/41	---	---	**FPZ** 394-397
---		Cunnamulla	SHQ Archerfield	8080-10 18/8/41	---	---	**FVN** 116
AUSLIG		Cunnamulla	SVY 1409 H55-2-460	24/7/52	25 000	1	5158
COOLANGATTA CPE	28°09'S 153°30'E	Coolangatta Afld	SHQ Archerfield	8076-12 CPE 092 2/9/40	7 000	---	**FVN** 662
---		Coolangatta Aerodrome (oblique)	SHQ Archerfield	8097-15 23/3/41	---	---	**PB** 48-50
---		Coolangatta Aerodrome (oblique)	6 Sqd H56-3-213A	8019-1 29/6/41	---	---	**FPA** 408 **SE** 409 **E**
AUSLIG		Tweed Hds	KD 1137 Map 599	2/4/46	16 000	1	20
GOONDIWINDI ---	28°31'S 150°20'E	Goondiwindi Aerodrome (oblique)	SHQ Amberley H56-1-228	8122-3 3/5/41	---	---	**PA** 21-6
---		Goondiwindi Aerodrome (oblique)	SHQ Archerfield	8073-3 18/8/41	---	---	**FPZ** 324-327
---		Goondiwindi Aerodrome	SHQ Archerfield	8080-8 18/8/41	---	---	**FVN** 1100
AUSLIG		Goondiwindi	SVY 448	16/10/48	15 000	1	5008

APPENDIX 5

UNITS IN OCCUPATION

A primary source used in compiling this table was -
BRL No 493 (3rd Edition)
AWM Library, Class No 222.13
RAAF - Units by Areas & Groups (1/9/39-1/9/45)

On the subject of copying such records, the following experience remains etched in my memory of this project. After an early session trying to locate relevant AWM 'core collection' photographs via the computer catalogue, I was guided by an attendant to a Finding Aid - the above document!

I recognised its value to the compilation of this Appendix 5. For that immediate photo search it meant I could enter as a keyword, the Unit, say '24 Squadron' in search of photos relating to STOCKROUTE airstrip, where the photo caption may not even include the strip name. Even if the strip name was in the catalogued photo caption, there seemed to be no guarantee it would be 'Keyworded'.

Time beat me so I lodged a photocopy order, paid the fee and postage for dispatch in due course. After a couple of months with no response, I wrote enquiring progress. Some days later, I was advised by phone, that the order was held up because the document was not cleared!! Three months after lodging my order, the item was opened for public access - i.e. 'cleared' - and I received my copies.

The first version of this Table was then compared with a similar listing kindly furnished by RAAF Historical Canberra, dated 20 May 1991. I had already begun to add detail of some US Units as drawn from the two important volumes edited by Maurer Maurer for the USAF Historical Division, Air University, Department of the (US) Air Force, in the 1960s. Cross referencing began to clarify some entries. The US practice of organising various Squadrons into Groups means one has to be careful to differentiate between for example, the 19th Bombardment Sqn (BS) and the 19th Bombardment Group (BG).

From various 'Station lists' or 'Status reports' in microfilmed records from the USAF HRC at Maxwell AL, some more detailed information is available. In fact, such a wealth of spot detail appears, that it becomes necessary to judiciously discard reference to some 'smaller units' and/or those with short periods at or near these airfields. So too for smaller RAAF formations.

Bill Bentson gave the benefit of his mark-up of RAAF Historical's list and more recently I was fortunate to learn of unpublished listings by Dennis Whiley of Adelaide, SA and Ian Jenkins (et al) of the AWM. These have all been helpful in refinement of this Appendix.

Lastly, direct information from veterans in terms of 'our unit spent "time" at a particular airfield' has been included with care. Such care has not been so necessary where the documentary evidence of an individual's log books or diaries is available. However, where it's only fragile memory supporting a statement, it's needed.

Total accuracy is not claimed - maybe it can never be.

EAGLE FARM *(27/11/1943)*

This is the date appearing on these two drawings prepared by George V Rimmington, one of the first RAAF serviceman to be associated with the US 5th Air Force's Technical Air Intelligence Unit at EAGLE FARM. Mr Rimmington's role was as a draftsman, collecting, measuring and constructing drawings both detail and silhouette for Allied Forces' needs, particularly in the realm of aircraft recognition. The aircraft pictured is by title shown as a Type 1F 'Oscar', similar to an aircraft which Mr Col Pay of Scone NSW, held for several years with a view to restoration to flying condition. It has since passed to a New Zealand collection. *[Rimmington Family]*

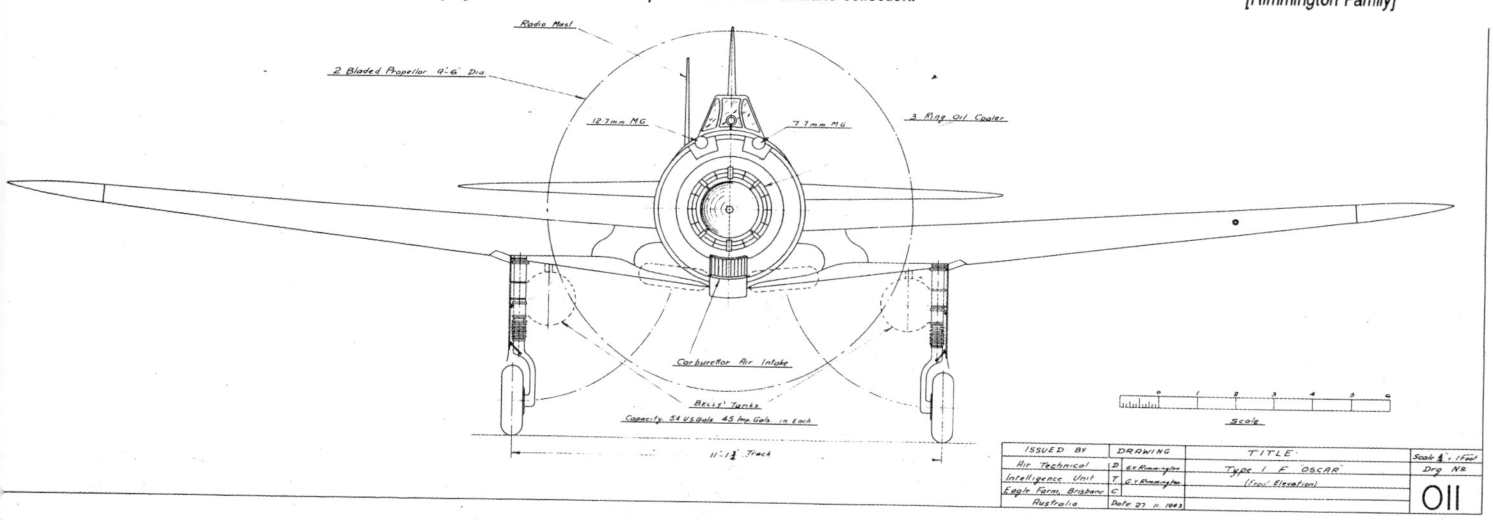

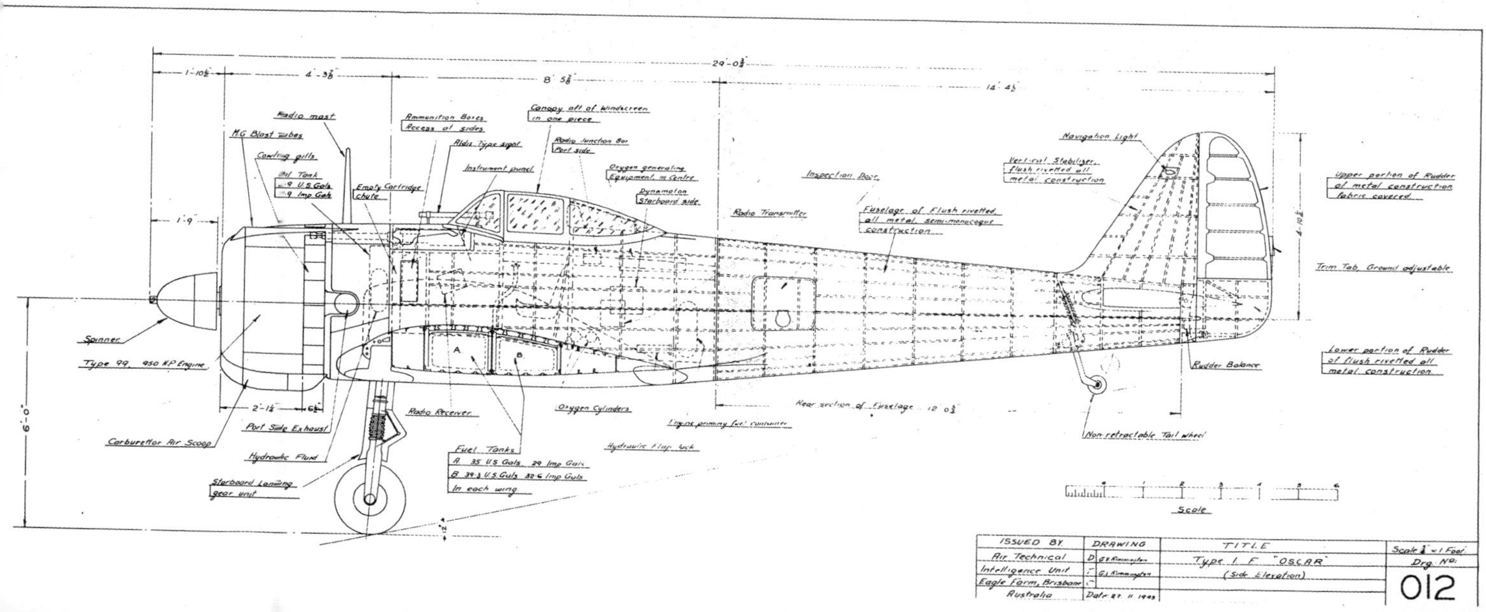

AIRFIELD	UNIT		PERIOD
HORN ISLAND	**RAAF**		
	6 Sqn		21/8/42-6/10/42
	7 (GR/B) Sqn (Det)		28/1/43-12/2/43, 20/7/43-22/9/43
	23 Sqn		31/3/44-29/6/44
	32 Sqn		26/4/42-18/11/42
	75 Sqn		1/10/42-13/12/42
	84 (I/F) Sqn		4/4/43-22/5/44
	28 OBU		15/5/42-29/12/44
	4 Works Maintenance Unit (Det)		5/9/42-14/2/43
	9 Mobile Fighter Sector HQ		9/10/42-3/11/42
	112 Mobile Fighter Sector HQ		28/6/43-6/4/44
	2 Air Ambulance Unit (Det)		1/3/44-30/4/44
	36 Radar Stn		27/8/43-9/10/45
	USAAF		
	49 FG	7 FS	14/3/42-19/4/42
	38 BG(M)(Det's)	71 BS	9/9/42-29/10/42
		405 BS	9/9/42-29/10/42
HIGGINS (JACKY JACKY) (RED ISLAND POINT)	**RAAF**		
	7 (GR/B) Sqn		26/3/44-26/10/44
	33 OBU		13/10/43-31/7/46
	1 Repair & Salvage Unit		16/6/43-18/6/44
	5 RSU		-/5/44--/3/45
	52 Radar Stn (Muttee Head)		29/3/43-29/9/45
IRON RANGE	**RAAF**		
	26 OBU		7/6/42-9/12/44
	43 Radar Stn (Portland Roads)		8/10/42-29/9/45
	USAAF		
	22 BG	19 BS	15/9/42-4/2/43
		33 BS	29/9/42-4/2/43
	43 BG(H)	64 BS	12/10/42-c8/11/42
		65 BS	13/10/42-7/11/42
		403 BS	15/9/42-22/10/42
	90 BG(H)	319 BS	13/11/42-2/2/43
		320 BS	13/11/42-21/1/43
		321 BS	19/11/42-10/2/43
		400 BS	13/11/42-c22/3/43
	35 Air Base G	350 Service S(Prov)	29/9/42-23/10/42
		28 Service S	23/10/42-19/1/43
COEN (RAAF)	**RAAF**		
	OB		-/6/42-17/2/45
	USAAF		
	3 BG(Det)		c-/4/42--/5/42
COOKTOWN (ENDEAVOUR RIVER MISSION)	**RAAF**		
	12 (DB) Sqn		-/7/43--/9/43
	13 (GR/B) Sqn		10/6/44-25/8/44
	27 OBU		c-/12/42-26/4/46
COOKTOWN (CIVIL DROME)	**RAAF**		
	27 OBU		24/4/42-c-/12/42
	44/56 Radar Stn		15/11/42-19/12/45
MITCHELL RIVER MISSION	**RAAF**		
	320 Radar Stn		9/6/43-5/9/44
MORNINGTON ISLAND	**RAAF**		
	313 Radar Stn		25/3/43-3/9/43
CAIRNS	**RAAF**		
	10 Comm'n Unit		11/9/44-10/12/44
	11 (GR/FB) Sqn		11/11/42-3/7/44
	20 (GR/FB) Sqn		11/11/42-13/9/44
	41 (T/Sea) Sqn		1/7/44-27/9/45
	75 Sqn		13/12/42-8/2/43
	113 Air-Sea Rescue Flt		10/1/45-10/3/45
	114 Air-Sea Rescue Flt		18/7/45-27/9/46
	1 Aircraft Performance Unit (Det)		21/8/44-
	25 OBU		25/5/42-7/5/46
	39 OBU		12/10/44-16/11/44
	17 Air Obs Post Flt		1/10/44-9/12/45

AIRFIELD	UNIT		PERIOD
CAIRNS (cont...)	9 Mobile Fighter Sector HQ		3/11/42-11/2/43
	12 Mobile Fighter Sector HQ		24/3/43-27/6/43
	3 Mobile Torpedo Unit		1/3/44-9/4/44
	USAAF		
	374 T/C G	33 T'p Carrier S(Det)	c1/11/42-10/12/42
MAREEBA (HOEVET FIELD)	**RAAF**		
	5 (A/Co-op) Sqn		11/6/43-17/11/44
	5 Comm'n Unit (Det)		1/9/44-21/5/45
	100 Sqn		22/5/42-2/7/42
	8 SFTS (Det)		29/11/44-25/3/45
	24 OBU		25/6/42-1/4/46
	16 Air Obs Post Flt (Det)		23/2/45-19/4/45
	Mobile Paratroop Maint'ce Unit		29/5/44-30/10/45
	USAAF		
	8 FG	35 FS	24/2/43-10/5/43
		36 FS	22/2/43-22/5/43
		80 FS	6/2/43-21/3/43
	19 BG(H)	28 BS	24/7/42-c18/11/42
		30 BS	24/7/42-c10/11/42
		93 BS	23/7/42-c25/10/42
	43 BG(H)	63 BS	20/8/42-c22/1/43
		64 BS	c8/11/42-c19/1/43
		65 BS	7/11/42-c19/1/43
		403 BS	c21/1/43-c11/5/43
	36 Service G	49 Service S	c8/3/43-?
	46 Service G		1/10/42-6/3/43
KARUMBA FLYING BOAT BASE	**RAAF**		
	43 (GR/FB) Sqn		24/8/43-9/4/44
	34 OBU		16/12/43-20/5/46
DUNK ISLAND	**RAAF**		
	27 Radar Stn		5/10/42-19/12/45
INVERLEIGH (RAAF)	**RAAF**		
	29 OBU		5/7/43-11/12/44
AUGUSTUS DOWNS	**RAAF**		
	29 OBU		-/11/42--/7/43
GARBUTT	**RAAF**		
	5 Comm'n Flight		30/11/42-24/5/46
	22 Sqn		22/9/42-19/10/42
	24 Sqn		14/10/40-1/12/41,26/2/42-30/7/42
	35 (T/Land) Sqn		31/3/45-10/6/46
	36 (T/Land) Sqn		20/2/44-19/8/46
	40 (T/Sea) Sqn		31/3/44-22/7/44
	41 (T/Sea) Sqn		21/8/42-15/5/44
	75 Sqn		4/3/42-21/3/42, 11/5/42-17/5/42
	77 Sqn		26/1/43-13/2/43
	82 Sqn		30/4/44-25/9/44
	2 Air Ambulance Unit		3/4/43-10/1/44
	23 OBU		3/7/42-15/12/44
	33 OBU		28/4/43-6/6/46
	44 OBU		23/12/42-4/3/43
	13 Aircraft Repair Depot-(Det)		11/43-12/43
	5 RSU		26/5/44-11/3/45
	12 Mobile Fighter Sector HQ		11/3/43-24/3/43
	43 Radar Stn		24/8/42-8/10/42
	USAAF		
	8 FG	35 FS	27/7/42-18/9/42
		36 FS	4/4/42-26/4/42, 30/6/42-18/9/42
	35 FG	39 FS	26/7/42-18/10/42
		40 FS	-/4/42-2/6/42, 30/7/42-c25/11/42
	19 BG(H)	435 BS ('Kangaroo S')	-/2/42--/11/42
	374 T/C G	6 Troop Carrier S	2/10/43-c26/8/44
	374 T/C G	22 Troop Carrier S	11/10/42-24/1/43, 4/10/43-29/8/44
		33 Troop Carrier S	5/10/43--/4/44
	317 T/C G	39 Troop Carrier S	22/1/42-22/2/42
		40 Troop Carrier S	23/1/43-4/10/43

AIRFIELD	UNIT		PERIOD
GARBUTT (cont...)	**USAAF**		
	317 T/C G	41 Troop Carrier S	23/1/43-5/10/43
		46 Troop Carrier S	23/1/43-1/10/43
	6 Recon G	8 Photo S-'A' Flt	8/5/42--/8/42
STOCKROUTE (adjacent USAAF 4th Air Depot) (DALRYMPLE RD) (MT LOUISA)	**RAAF**		
	24 Sqn		24/3/42-18/6/42
	33 Sqn		-/7/42-c-/12/42
	36 Sqn		c-/12/42-c20/2/44
	USAAF		
	22 BG(M) using dispersals only		
		19 BS	28/3/42-c4/7/42
	4 Air Depot G		2/10/42-31/12/44
	2 ADG		14/4/43-c/30/6/45
	5 ADG		18/6/43-c31/12/44
	45 Service G		8/3/43-?
BOHLE RIVER (BLACK RIVER)	**RAAF**		
	8 (T/B) Sqn		5/6/43-15/9/43
	30 Sqn		17/8/42-14/9/42
	86 (I/F) Sqn		13/5/44-26/5/44,10/1/45-20/12/45
	100 Sqn		c22/9/42-c22/11/42
	44 OBU		-/2/43--/4/43
	22 RSU		11/12/43-10/2/44
	1 Mobile Fighter Sector HQ		11/12/43-10/2/44
	USAAF		
	35 FG	40 FS	20/4/42
ROSS RIVER	**RAAF**		
	7 Sqn		6/11/42-26/3/44
	24 Sqn		18/6/42-c-/7/42
	84 (I/F) Sqn		17/11/44-29/1/46
	4 RSU		14/5/44-12/3/45
AITKENVALE (2nd WEIR) (THE WEIR STRIP)	**RAAF**		
	76 Sqn		c9/4/42-c15/7/42
	80 Sqn		10/9/43-21/2/44
	86 (I/F) Sqn		6/5/43-5/7/43
	33 OBU (Det)		15/5/43-2/10/43
	USAAF		
	35 FG	39 FS	20/4/42
ANTIL PLAINS (MT STUART STRIP)	**USAAF**		
	8 FG	36 FS	6/5/42
	35 FG	40 FS	8/5/42-28/5/42
	22 BG(M)	33 BS	7/4/42-20/7/42
WOODSTOCK	**USAAF**		
	8 FG	35 FS	8/5/42, 29/6/42-27/7/42
	22 BG(M)	19 BS	4/7/42-15/9/42, 4/2/43-c11/7/43
		33 BS	20/7/42-29/9/42, 4/2/43-15/10/43
	35 FG	39 FS	20/4/42-2/6/42
	345 BG(M)	498 BS	16/5/43-c20/6/43
REID RIVER	**USAAF**		
	22 BG	2 BS	9/4/42-9/10/43
		408 BS	12/4/42-15/10/43
	345 BG(M)	499 BS	16/5/43-c20/6/43
		500 BS	16/5/43-c20/6/43
FANNING	**RAAF**		
	1 Signals Task Unit		-/3/44--/7/44
BREDDAN (BREDDAN ISLAND) (LITTLE SANDY)	**RAAF**		
	1 Torpedo Maintenance Unit		18/11/42-26/6/44
	5 Central Recovery Depot		15/6/44-29/6/45
	13 Aircraft Repair Depot (ARD) (Det)		-/3/43-19/8/43
	13 ARD		19/8/43-20/11/47
	USAAF		
	38 BG(M)	71 BS	12/8/42-1/10/42
		405 BS	7/8/42-c30/9/42
MACROSSAN	**RAAF**		
	84 (I/F) Sqn		22/6/44-17/11/44
	86 (I/F) Sqn		25/6/44-10/1/45
	12 RSU		-/4/43--/9/43
	13 ARD (Det)		-/9/43--/1/44
	8 Stores Depot		1/6/44-1/3/49
BOWEN	**RAAF**		
	9 (Fleet Co-op) Sqn		13/1/43-3/12/44
	22 OBU		1/8/42-10/10/43
	1 Aircraft Performance Unit (Det)		3/44-8/44
	114 Air-Sea Rescue Flight (Det)		6/8/45-31/8/45
	Chemical Research Unit		15/8/44-24/12/45
	42/55 Radar Stn		5/3/43-24/4/46
BOWEN FLYING BOAT MAINTENANCE BASE	**RAAF**		
	11 (GR/FB) Sqn		30/7/42--/11/42
	20 (GR/FB) Sqn		7/5/42-12/11/42
	43 (GR/FB) Sqn		1/5/43-24/8/43
	1 Flying Boat Maintenance Unit		10/10/43-14/3/47
CHARTERS TOWERS (CORINDA)	**RAAF**		
	10 RSU		25/11/42-21/1/43
	12 RSU		15/5/42--/10/42
	USAAF		
	3 BG	8 BS	c17/3/42-31/3/42, c9/5/42-28/1/43
		13 BS	10/3/42-c10/1/43
		89 BS	8/3/42-c1/9/42
		90 BS	8/3/42-28/1/43
	38 BG(M)	822 BS	c1/8/43-8/10/43
		823 BS	c1/8/43-8/10/43
	345 BG(M)	501 BS	16/5/43-c20/6/43
	475 FG	431 FS	14/5/43-c1/7/43
		432 FS	14/5/43-14/6/43
		433 FS	14/5/43-17/6/43
	35 Service G	47 Service S	c10/3/42-c10/3/44
		370 Service S	17/12/42-c7/2/44
		350 Service S (Prov)	6/9/42-24/9/42
	V (Fifth) Fighter Com'd Replacement Ctr		6/4/43-c-/5/44
	V (Fifth) Bomber Com'd Replacement Ctr		6/4/43-c-/5/44
CLONCURRY	**RAAF**		
	29 OBU		22/6/42-5/7/43
	30 OBU		12/12/42-19/9/46
	107 Radar Stn		1/6/42-11/9/42
	108 Radar Stn		1/6/42-11/9/42
	USAAF		
	19 BG(H)	28 BS	c28/3/42-c5/5/42
		30 BS	c27/3/42-c13/5/42
		93 BS	29/3/42-18/5/42
TORRENS CREEK	**USAAF**		
	43 BG(H)	63 BS	3/8/42-20/8/42
		65 BS	15/8/42-13/10/42
		403 BS	27/8/42-c17/10/42
MT HEDLOW	**US ARMY**		
	41 Div (Artillery Spotting)		-/3/43
ROCKHAMPTON (CONNOR PARK)	**RAAF**		
	21 OBU		2/5/42-20/2/46
LONGREACH	**USAAF**		
	19 BG(H)	28 BS	c5/5/42-24/7/42
		30 BS	c13/5/42-c24/7/42
		93 BS	18/5/42-23/7/42
BUNDABERG	**RAAF**		
	32 Sqn 'B' Flight		1/5/43-4/9/44
	66 (Reserve) Sqn		20/5/43-6/1/44
	71 (Reserve) Sqn - 'B' Flight		1/5/42-26/1/43
	88 OBU		14/6/45-26/4/46
	12 EFTS		16/10/41-12/1/42

AIRFIELD	UNIT	PERIOD
BUNDABERG (cont...)	**RAAF**	
	8 SFTS	14/12/41-25/7/45
	NEIAF	
	Personnel & Equipment Pool	25/6/45-26/4/46
MARYBOROUGH	**RAAF**	
	3 Recruit Depot	3/4/42-16/7/43
	6 Recruit Depot	15/11/44-19/6/45
	3 Wireless Air Gunners School	18/9/41-15/12-44
	1 Radar School (RDF School)	11/11/44-23/11/45
CHARLEVILLE	**RAAF**	
	15 OBU	6/3/43-14/12/44
	USAAF	
	43 BG(H) 63 BS	15/6/42-3/8/42
	45 Air Base G 8 Materiel S	9/3/42-28/7/42
KINGAROY	**RAAF**	
	1 (Attack) Sqn	15/1/45-22/5/45
	2 Air Ambulance Unit	26/2/43-6/9/44
	4 Sqn	6/8/42-1/11/42
	5 (A/Co-Op) Sqn	2/2/43-11/6/43
	13 Comm'n Unit	17/7/45-22/10/45
	75 Sqn	11/5/42-3/7/42
	92 (Attack) Sqn	25/5/45-4/1/46
	93 (Attack) Sqn	22/1/45-23/5/45
	84 OBU	10/1/45-9/5/45
	90 OBU	23/7/45-28/11/45
	3 ITS	24/7/42-4/9/45
TOOGOOLAWAH	**RAAF**	
	5 (A/Co-op) Sqn	7/11/42-2/2/43
BRYMAROO (JONDARYAN)	**RAAF**	
	99 (HB) Sqn	31/3/45-13/9/45
PETRIE (STRATHPINE A-1)	**USAAF**	
	8 FG 80 FS	10/5/42-20/7/42
	US ARMY	
	1st Cavalry Div (Artillery Spotting Unit)	c1943
	RAF	
	549 (I/F) Sqn	15/12/43-24/5/44
STRATHPINE A-2	**RAAF**	
	12 (DB) Sqn	22/8/44-14/12/44
	23 Sqn (Det)	c8/2/43
	83 (I/F) Sqn	26/2/43-19/1/44
	114 Mobile Fighter Control Unit	29/1/45-3/4/45
	HB Replacement Training Unit	12/9/44-21/2/45
	13 Signals Unit	7/12/43-30/1/44
	RAF	
	548 (I/F) Sqn	15/12/43-28/5/44
STRATHPINE A-3	**US ARMY**	
	1st Cavalry Div (Artillery Spotting Unit)	c1943
COOMINYA No 2 (A-6) (LOWOOD No 2)	**RAAF**	
	11 RSU	28/12/43-17/1/44
EAGLE FARM	**USAAF**	
	38 BG(M) HQ & HQ Sqn	c8/5/42
	81 ADG HQ & HQ Sqn	28/5/42-21/4/44
	81 Depot Repair S	-/8/42--/4/44
	81 Depot Supply S	-/8/43--/4/44
	44 Depot Repair S	14/6/43- ?
	455 Service S	19/10/43- ?
OAKEY	**RAAF**	
	6 Aircraft Depot	18/10/43-1/2/46

AIRFIELD	UNIT	PERIOD
LOWOOD	**RAAF**	
	21 (DB) Sqn	30/11/43-19/1/44
	23 (DB) Sqn	6/6/42-19/1/44
	24 (HB) Sqn	13/4/44-9/6/44
	32 (GR/B) Sqn	10/5/44-30/11/45
	71 (Reserve) Sqn	21/6/43-28/12/43
	'A' Flight	1/5/42-8/6/44
	75 Sqn	3/7/42-26/7/42
	Survey Flight	26/3/43-25/1/46
	14 OBU	10/11/42-17/2/47
	47 OBU	9/12/43-17/1/44
	10 RSU	1/9/42-23/11/42
	12 EFTS	12/1/42-18/4/42
	USAAF	
	8 FG 36 FS	13/3/42-4/4/42
	80 FS	28/3/42-28/5/42
CECIL PLAINS	**RAAF**	
	12 (HB) Sqn	19/12/44-1/5/45
	102 (HB) Sqn	30/5/45-18/3/46
TOOWOOMBA	**RAAF**	
	5 (A/Co-op) Sqn	11/5/42-11/9/42
ARCHERFIELD	**RAAF**	
	18 Sqn (NEIAF/RAAF)	-/2/42-4/4/42
	22 Sqn	14/9/42-22/9/42
	23 Sqn	30/8/39-6/5/42
	38 (T/Land) Sqn	27/12/44-15/8/46
	76 Sqn	14/3/42-16/4/42
	2 Air Ambulance Unit	7/9/44-7/1/46
	4 Comm'n Flight	7/9/42-16/4/46
	3 FTS/2 EFTS	6/11/39-24/4/42
	USAAF	
	374 T/C G 21 Troop Carrier S	3/4/42-18/2/43; 28/9/43-26/8/44
	317 T/C G 39 Troop Carrier S	22/2/42-28/9/43
	45 Air Base G 8 Materiel Sqn	-/1/42--/4/42
	22 Air Base G 30 Materiel Sqn	8/5/42-28/5/42
	NEIAF	
	19 Sqn (1 NEI T'port Sqn-unofficial)	-/5/45-15/8/45
	RAF	
	194 Staging Post	1/8/45-7/2/46
	1315 Servicing Echelon	-/6/45--/8/45
	1315 Flight	2/5/45--/3/46
AMBERLEY	**RAAF**	
	23 Sqn	6/5/42-6/6/42
	24 Sqn	17/6/40-15/10/40
	85 OBU	8/3/45-25/4/45
	3 Aircraft Depot	16/3/42-S/A
	3 Central Recovery Depot	11/1/44-12/9/46
	10 RSU	26/6/42-1/9/42
	3 Recruit Depot	2/7/40-3/4/42
	6 Recruit Depot	24/8/42-1/12/42
	3 SFTS	21/10/40-20/4/42
	USAAF	
	3 BG	7/3/42-7/4/42
	35 Air Base G 2 Materiel S	6/5/42-
	22 Service Group	22/11/42-1945
	US Air T'port Command (Det)	-/1/43-1945
GOOLMAN	**RAAF**	
	11 RSU (Det)	31/5/43-22/10/43
LEYBURN	**RAAF**	
	21 (HB) Sqn	16/7/44-27/12/44
	23 (HB) Sqn	19/10/44-6/3/45
	99 (HB) Sqn	1/2/45-30/4/45
	200 Special Duties Flt	15/2/45-12/4/46

APPENDIX 6

ARCHIVE SOURCES

Some Australian Government Archive holdings in which I've found material, are listed here. Some are not because I've considered their contribution to be of insufficient significance. Others will not be listed simply because I've failed to identify and study them.

Archive searching is like life itself - after you've spent the time and come to show what you've achieved, disappointment often appears. Frequently the search could have, indeed should have, been tackled differently.

I doubt there is any easy way to find relevant historical material you may be seeking. Depending on its precise nature, and at the risk of displaying gross naivety, I'd suggest the following order of search - a dictionary and/or encyclopedia, a careful keyword and/or bibliographic search at suitable libraries, followed by the attendant reading of published work. Then and only then, should you tackle the raw and relatively disordered state of Archive searching. Sure, you can hope for a suitable 'finding aid booklet' and please, please, a helpful 'reader adviser', but it is time and patience with which you must go armed.

Just as amorous expectation can be quenched by a headache, so, often is the Archive search stalled by... 'that item's not cleared'! Ninety plus days later, I've found it's usually opened for public access, because my study period of approximately fifty years ago is well beyond the general thirty year proviso. If you reach this stage, just hope the item turns out to be relevant and helpful!

The holdings of the Australian War Memorial (AWM) are, in effect, an extension of the Australian Government Archives. WW2 prompted the beginnings of a national archives system in Australia with the appointment of a War Archives Committee in 1942. Around war's end, the AWM was designated as the official archival authority for all war records, including those of Commonwealth Government Departments concerned with Defence. This was altered in 1952 but the AWM continued as the official repository for the operational records of the services. For me, research at the AWM has been disadvantaged, primarily by not living in Canberra. My short visits have been directed to the AWM's vast photographic collection and to a lesser extent, the Operational Record Books (ORB's), or 'Form A50' - Unit History Sheets, of various RAAF Units. I was unfortunate in finding the former in a drawn-out state of reorganisation towards an optical disc storage and retrieval system. The latter was then plagued by various Squadron records being listed as 'OPEN - (Returned to AIR FORCE Office... as far back as 1976!)'. Now the RAAF Historical Section has microfilmed nearly all ORBs making copy reels available for purchase.

I have corresponded with the US Air Force Historical Research Centre at Maxwell AL. I enquired where such US records (duplicate microfilms etc) may currently be held in Australia, for public access. To date I've not learned of any such place. I made a modest outlay of some $A200 for the purchase of several microfilm reels. I was warned that the quality and legibility was regrettably poor in some of these reels, and the warning was not wrong! I shall lodge these reels in a suitable repository with public access.

The following holdings apply broadly and as such, don't generally appear in the specific airfield listings -

Australian Archives, (Qld) -

BP1/1 Minutes, Allied Works Council (Qld), 1942-45. While some early documents in this large holding are carbon copies, several sheets removed from the typewriter key, the majority are legible.

BP293/2 General Index Cards of the Works and Services Branch of the Dept of Interior Queensland. These are a mix of 'Property' and 'Subject' cards. In some cases they are the only record of the date and nature of work done, that I have found.

BP375 ITEM QL680, Land Acquisition in relation to Air Force wartime landing grounds (1944-45). This is relevant to many airstrip sites in that the holding contains original dye-line prints/drawings and basic detail. Most of these drawings feature a block of information appropriate to a stated 'serviceability date'. Unfortunately much of it is faint and difficult to read.

Australian Archives (NSW) -

CRS C1707 P1 BOX 7 Camouflage Reports DSCC Queensland, Minutes of Meetings etc., i.e. of the State Defence Camouflage Committee. This incomplete set touches on most relevant Queensland Airfields. It is of interest along with other Boxes in the same holding, and items in - **CRS C1904 T1** and **CRS C1905 T1.**

An item particular to say, IRON RANGE, with that name displayed in the title, has been assigned a **Bold** Ref number in the right hand column. Non-bold Ref numbers apply to items with general titles. My approach in the 'citation' of all items is somewhere between 'full' and 'abbreviated'. It should serve to introduce the reader interested in a particular Queensland airfield WW2 to this phase of research.

A word of advice. While an item I've mentioned may not contain the information you seek, something as simple as an ***allied file reference*** or a ***fifty year old handwritten margin note*** may lead to it.

COOMINYA *(29/4/1943)*
This sketch shows the nature of such appreciation drawings prepared by Department of Home Security Camouflage Officers. This particular drawing is of note in that it records the location of some thirteen US arch type fighter hideouts constructed during 1942 for/by(?) the US Forces. *[AA CRS A705 Item 7/1/1663]*

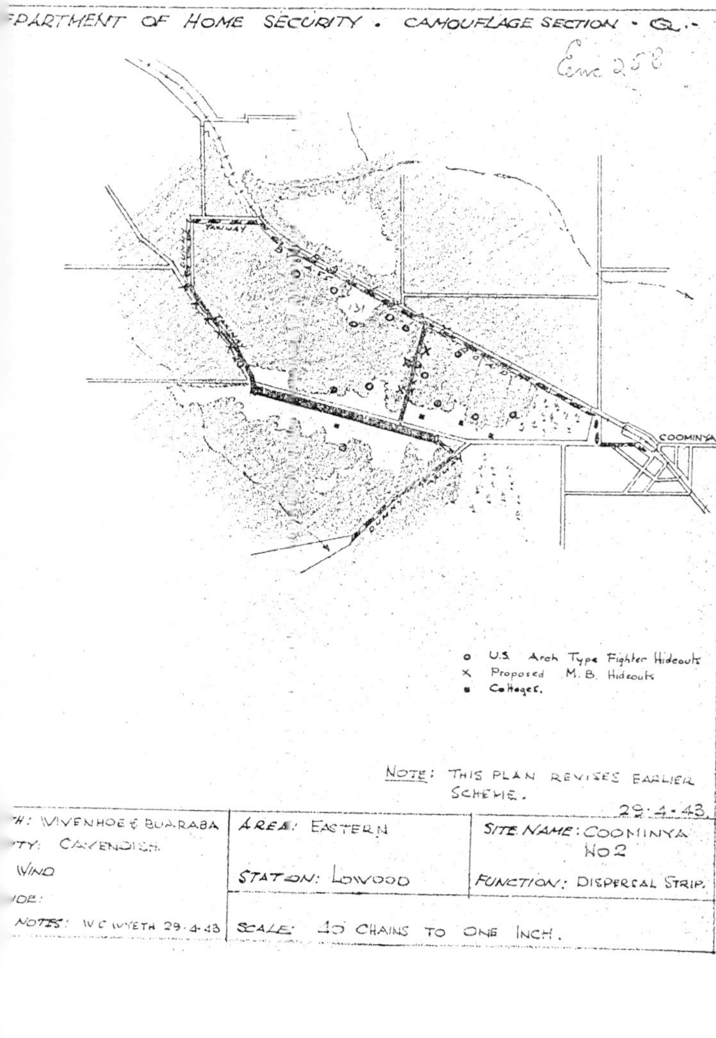

AIRFIELD	ARCHIVE	ACCESSION / ITEM NUMBER / TITLE / DATE RANGE	REF.
HORN ISLAND	AA (ACT) AA (Qld) AA (ACT)	A705 ITEM 7/1/774 : PARTS 1,2,3 - RAAF A'DROME HORN ISLAND (37-46) A705 ITEM 7/1/872 : PARTS 1,2 - RAAF AOB HORN ISLAND BUILDINGS & SERVICES (41-47) A705 ITEM 7/1/1145 : HORN ISLAND AOB SITE ACQUISITION (41-46) BP375 ITEM QL734 : PART 1,1A - HORN ISLAND A'DROME AQUISITION OF SITE (39-74) A705 ITEM 7/1/693 : RAAF AOB HUTMENTS - PROGRESS REPORTS (41-?) A705 ITEM 7/1/1258 : A'DROMES NORTH EAST AREA-GENERAL (42-45) A705 ITEM 7/1/1819 : RAAF A'DROMES CIVIL AVIATION REQUIREMENTS & TRANSFERS DCA (44-48) A705/1 ITEM 151/1/703 : PARTS 1,2 - SITE SELECTION - NEW A'DROMES TOWNSVILLE AREA (42-44) A705/1 ITEM 151/1/738 : SELECTION OF SITES - CAPE YORK PENINSULA (42-45) Plus Refs. **33, 39, 42, 51, 72, 75, 92, 127**	**1** **2** **3** **4** 171 172 173 174 175
HIGGINS (JACKY JACKY) (RED ISLAND POINT)	AA (ACT) AA (Qld) AA (AWM)	A705 ITEM 7/1/1768 : HIGGINS FIELD A'DROME (43-48) A705/1 ITEM 171/93/742 : HIGGINS FIELD OB HIRING OF SITE (43-47) BP375 ITEM QL656 : PARTS 1A,2 - HIGGINS FIELD (JACKY JACKY) RAAF A'DROME (43-70) AWM 54 ITEM 13/2/2 : LIST OF CIVIL A'DROMES,ELG's & F/B Bases AUST & NEW GUINEA (42) Plus Refs. **1, 2, 10, 33, 51, 72, 127,** 172, 173, 174, 175	**5** **6** **7** 176
IRON RANGE	AA (ACT) AA (Qld) ∗T'ville	A373/1 ITEM 6856 : IRON RANGE - GENERAL SECURITY (43-44) A705 ITEM 7/1/1484 : RAAF A'DROME IRON RANGE (42-46) A705 ITEM 171/94/167 : IRON RANGE OB ACQUISITION OF SITE (43-54) BP375 ITEM QL473 : IRON RANGE A'DROME - SITE SURVEY BY AUST' ARMY FOR US (43-59) BP243/1 ITEM KS178 : OP BASES NQ - CONSTRUCTION & IMPROVEMENT (42-48) ∗J153 : TRACINGS FROM DEPT INTERIOR OPERATIONS TOWNSVILLE (42-45+) Plus Refs. **1, 12, 25, 33, 47, 51, 72, 127,** 172, 173, 174, 175	**8** **9** **10** **11** 177 178
COEN (RAAF) **and COEN** (TOWN SITE)	AA (ACT) AA (Qld)	A705 ITEM 7/1/672 : RAAF COEN A'DROME WORKS (39-45) BP375/1 ITEM QL1288 : COEN A'DROME ACQUISITION (39-49) Plus Refs. **2, 39, 42, 74, 92,** 171, 172, 175	**12** **13**
COOKTOWN (ENDEAVOUR RIVER MISSION) **and** **COOKTOWN** (CIVIL DROME)	AA (ACT) AA (Qld)	A705 ITEM 7/1/1483 : RAAF COOKTOWN MISSION (42 - 46) A705/1 ITEM 7/1/1016 : COOKTOWN A'DROME (41-46) A705/1 ITEM 7/1/1020 : PARTS 1,2 - RAAF AOB COOKTOWN BUILDINGS & SERVICES (41-48) A705/1 ITEM 171/93/743 : COOKTOWN MISSION OB HIRING OF SITES (43-52) BP375 ITEM QL687 : COOKTOWN MISSION RAAF BASE (45-51) Plus Refs. **8, 33, 39, 42, 47, 51, 72, 74, 85,** 171, 172, 173, 174, 175, 185	**14** **15** **16** **17** **18**
LAURA	AA (ACT)	A705/1 ITEM 151/1/738 : SELECTION OF SITES - CAPE YORK PENINSULA (42-45)	175
WROTHAM PARK	AA (ACT)	A705 ITEM 7/1/1258 : A'DROMES NORTH EAST AREA - GENERAL (42-45) Plus Refs. 173, 174, 175	172
MORNINGTON ISLAND	AA (ACT)	A705 ITEM 7/1/1800 : MORNINGTON ISLAND LG (44-55)	**19**
CAIRNS	AA (ACT) AA (Qld) AA (ACT)	A705 ITEM 7/1/608 : PARTS 1 AND 2 - CAIRNS A'DROME WORKS (39-46) A705 ITEM 7/1/784 : PART 3 - RAAF STATION CAIRNS BUILDINGS & SERVICES (44-46) BP375/1 ITEM QL2033 : CAIRNS A'DROME (34-51) A1969/100/3 ITEM 24/47/AIR : A'DROME AND LANDING GROUND CAIRNS (42-43) Plus Refs. **16, 33, 39, 42, 51, 72, 74,** 171, 172, 173, 174, 175, 185	**20** **186** **21** **22**
MAREEBA (HOEVET FIELD)	AA (ACT) AA (Qld)	A705/1 ITEM 7/1/1480 : RAAF MAREEBA LG (42-46) A705/1 ITEM 171/89/78 : RAAF STATION MAREEBA BUILDINGS & SERVICES (43-45) A705/1 ITEM 175/6/109 : PARTS 1,2 - MAREEBA A'DROME HIRING OF SITE (42-52) BP375/1 ITEM QL877 : MAREEBA A'DROME (43-51) Plus Refs. **33, 47, 51,** 172, 173, 174, 175	**23** **24** **25** **26**
KARUMBA and KARUMBA FLYING BOAT BASE	AA (ACT)	A705 ITEM 171/94/173 : KARUMBA FLYING BOAT BASE ACQUISITION OF SITE (43-) A816 ITEM 14/301/252 : DEFENCE OF KARUMBA AND GROOTE EYLANDT F/B BASES (39-43) Plus Refs. **20, 72,** 172	**27** **28**
INNISFAIL	AA (ACT)	A705 ITEM 7/1/1424 : INNISFAIL A'DROME (42-52)	**29**
MOUNT GARNET	AA (ACT)	A705 ITEM 7/1/1541 : RAAF A'DROME MT GARNET LG RSU (42-43)	**30**
NORMANTON	AA (Qld)	BP375 ITEM QL43 : NORMANTON A'DROME (36-60)	**31**
BURKETOWN	AA (Qld)	BP375/1 ITEM QL1296 : BURKETOWN A'DROME - LEASE FROM STATE (37-50)	**32**
INVERLEIGH (CIVIL) **and** **INVERLEIGH** (RAAF)	AA (ACT) AA (Qld)	A705 ITEM 7/1/1653 : RAAF INVERLEIGH LG (42-46) BP243/1 ITEM K1528 : INVERLEIGH A'DROME - CONSTRUCT LG & CAMP (42-47) Plus Refs. **1, 36, 47,** 172, 173	**33** **34**
PRESTWOOD	AA (ACT)	A705 ITEM 7/1/1258 : A'DROMES NORTH EAST AREA - GENERAL (42-45) Plus Refs. 173, 175	172

AIRFIELD	ARCHIVE	ACCESSION / ITEM NUMBER / TITLE / DATE RANGE	REF.
AUGUSTUS DOWNS	AA (ACT)	A705 ITEM 7/1/1422 : RAAF AOB AUGUSTUS DOWNS A'DROME WORKS (42-45) A705/1 ITEM 171/93/1095 : AUGUSTUS DOWNS A'DROME HIRING OF SITE (43-47) AA1969/100/3 ITEM 1/202/AIR : REPORT ON OPS AUGUSTUS DOWNS (OB29),27-30/NOV 42 Plus Refs. **33, 34, 52,** 172, 173, 174	**35** **36** **37**
CARPENTARIA DOWNS (EINASLEIGH)	AA (ACT)	A705 ITEM 7/1/1258 : A'DROMES NORTH EAST AREA - GENERAL (42-45) A705 ITEM 7/1/1819 : RAAF A'DROMES CIVIL AVIATION REQUIREMENTS & TRANSFERS DCA (44-48) Plus Refs. **36,** 174, 175	172 173
GARBUTT	AA (ACT) AA (Qld)	A705 ITEM 171/19/43 : PART 1 - RAAF STATION TOWNSVILLE BUILDINGS & SERVICES (39-40) A705 ITEM 171/19/54 : PARTS 1,2,4 - RAAF STATION TOWNSVILLE BUILDINGS & SERVICES (40-45) A705/1 ITEM 171/19/299 : GARBUTT EXTENSION TO A'DROME (43-50) A1196/2 ITEM 42/501/172 : GARBUTT FIELD PROJECT (USA) BUILDINGS & SERVICES (42-45) BP243/1 ITEM D39 : RAAF TOWNSVILLE AWC CIV ENG (39-43) BP243/1 ITEM KS57 : AAC(?) TOWNSVILLE AIR PROJECT NO 2 GARBUTT (42-47) BP375 ITEM QL622 : GARBUTT SITE ACQUISITION FOR USAAC & RAAF (38-55) Plus Refs. **24, 50, 52, 57, 72, 79,** 172, 173, 174	**38** **39** **40** **41** **42** **43** **44**
STOCKROUTE (adjacent USAAF 4th Air Depot) (DALRYMPLE RD) (MT LOUISA)	AA (ACT) AA (Qld)	A705/1 ITEM 175/6/8 : PARTS 1,2 - TOWNSVILLE US HANGARS & WK'SHOPS HIRE SITE (42-49) BP243/1 ITEM USM312 : 4 AD, GARBUTT (43-47) Plus Refs. **24, 35, 47, 50, 52, 53, 57, 72, 79,** 172, 173, 174, 187	**45** **46**
BOHLE RIVER (BLACK RIVER)	AA (ACT) AA (Qld)	A705 ITEM 7/1/1523 : RAAF BOHLE RIVER LANDING GROUND (42-53) A705/1 ITEM 171/19/248 : BOHLE RIVER A'DROME HIRING OF SITE (43-51) BP375/1 ITEM QL1847 : BOHLE RIVER - RAAF REMOTE RECEIVING STATION (41-68) Plus Refs. **33, 50, 51, 57, 72,** 172, 173, 174	**47** **48** **49**
ROSS RIVER	AA (ACT)	A705/1 ITEM 7/1/1521 : RAAF ROSS RIVER LG (42-46) A705/1 ITEM 171/19/261 : ROSS RIVER A'DROME HIRING OF SITE (42-49) Plus Refs. **33, 47, 57, 72,** 172, 173, 174	**50** **51**
AITKENVALE (2nd WEIR) (THE WEIR STRIP)	AA (ACT)	A705/1 ITEM 7/1/1522 : RAAF A'DROME AITKENVALE (42-46) A705/1 ITEM 171/19/121 : WEIR DISPERSAL STRIP - HIRING OF SITE AT AITKENVALE (42-49) A705/1 ITEM 171/19/117 : AITKENVALE - HIRE GLEESON's HOUSE & PART OF POR 63 (42-47) Plus Refs. **33, 35, 47, 50, 51, 57, 72,** 172, 173, 174	**52** **53** **55**
UPPER ROSS (KELSO FIELD)	AA (ACT)	A705/1 ITEM 7/1/1521 : RAAF ROSS RIVER LG (42-46) Plus Refs. **51,** 172, 174	**50**
ANTIL PLAINS (MT STUART STRIP)	AA (ACT) AA (Qld)	A705 ITEM 7/1/1348 : RAAF ANTIL PLAINS A'DROME WORKS (42-45) A705/1 ITEM 171/93/517 : ANTIL PLAINS A'DROME HIRING OF SITE (42-53) A705/1 ITEM 175/6/143 : ANTIL PLAINS USAAC LG ACQUISITION OF SITE (43-51) BP243/1 ITEM KS95 : WESTERN DROMES (42-48) Plus Refs. **35, 47, 51, 72, 79,** 172, 173, 174	**57** **58** **59** 179
GIRU	AA (ACT)	A705/1 ITEM 171/93/1026 : GIRU A'DROME HIRING OF SITE (43-48) A705 ITEM 7/1/1585 : A'DROMES EAST AREA - GENERAL (42-46) Plus Refs. 172, 173, 174	**60** 180
WOODSTOCK	AA (ACT) AA (Qld)	A705/1 ITEM 171/19/262 : WOODSTOCK A'DROME HIRING OF SITE (43-52) A705/1 ITEM 171/94/126 : WOODSTOCK A'DROME ACQUISITION OF SITE BP375 ITEM QL651 : WOODSTOCK SITE FOR A'DROME (42-56) Plus Refs. **35, 47, 50, 51, 57, 72, 79,** 172, 173, 174, 179	**61** **62** **63**
REID RIVER	AA (ACT)	A705/1 ITEM 175/6/122 : REID RIVER - USAAC A'DROME - HIRING OF SITE (42-52) Plus Refs. **33, 47, 51, 57, 78, 79, 88,** 172, 173, 174, 179	**64**
FANNING	AA (ACT) AA (Qld)	A705 ITEM 7/1/1407 : RAAF FANNING LANDING GROUND (42-44) A705/1 ITEM 171/93/1001 : FANNING A'DROME HIRING OF SITE (43-47) BP243/1 ITEM K1058 : FANNING LG & DISPERSAL - WORKS & BUILDINGS (42-48) Plus Refs. **33, 52, 57, 71, 72,** 172, 173, 174, 179, 187	**65** **66** **67**
CAMOOWEAL	AA (Qld)	BP375 ITEM QL613 : CAMOOWEAL A'DROME-LAND ACQUISITION & EXTENSION TO RUNWAYS (27-51)	**68**
BREDDAN (BREDDAN ISLAND) (LITTLE SANDY)	AA (ACT) AA (Qld)	A705/1 ITEM 7/1/1213 : RAAF BREDDAN A'DROME WORKS (41-46) BP375 ITEM QL649 : BREDDAN A'DROME (43-53) Plus Refs. **47, 50, 66, 67, 71, 72, 77, 78, 80, 114,** 172, 173, 174, 187	**69** **70**
MACROSSAN	AA (ACT) AA (Qld)	A705 ITEM 7/1/1408 : RAAF MACROSSAN LANDING GROUND (10 RSU) (42-47) BP243/1 ITEM K1070/1 : RAAF STATION MACROSSAN (42-50) BP375 ITEM QL704 : MACROSSAN A'DROME (42-56) Plus Refs. **33, 57, 66, 67, 78,** 172, 173, 174, 179, 187	**71** **72** **73**

AIRFIELD	ARCHIVE	ACCESSION / ITEM NUMBER / TITLE / DATE RANGE	REF.
BOWEN and BOWEN FLYING BOAT MAINTENANCE BASE	AA (ACT) AA (Qld) ∗T'ville	A705 ITEM 7/1/580 : RAAF BOWEN A'DROME WORKS (39-45) A705 ITEM 7/1/788 : RAAF AOB BOWEN - BUILDINGS & SERVICES (40-44) BP375 ITEM QL735 : BOWEN A'DROME (40-52) BP243/1 ITEM K2645 : RAAF F/BB - ADDITIONAL BUILDINGS & SERVICES (44-48) ∗J153 : TRACINGS FROM DEPT INTERIOR OPERATIONS TOWNSVILLE (42-45+) Plus Refs. **16, 24, 39, 72, 84, 92,** 171, 172, 174, 175	**185** **74** **75** **76** 181
CHARTERS TOWERS (CORINDA)	AA (ACT) AA (Qld)	A705 ITEM 7/1/1111 : CHARTERS TOWERS DISTRICT - PROPOSED A'DROME SITE SELECTION (41) A705 ITEM 7/1/1210 : RAAF CHARTERS TOWERS A'DROME WORKS (41-46) A705 ITEM 7/1/1217 : CHARTERS TOWERS RAAF A'DROME - HIRING OF SITE (41-47) A705 ITEM 175/6/146 : PART 8 - CHARTERS TOWERS A'DROME (43-48) BP243/1 ITEM KS55 : CHARTERS TOWERS - R/WS, NEW A'DROME & CAMP (42-47) BP375 ITEM QL635 : CHARTERS TOWERS RAAF A'DROME SITE (41-48) Plus Refs. **25, 47, 50, 57, 66, 67, 69, 71, 88,** 172, 173, 174, 175, 181, 187	**77** **78** **79** **80** **81** **82**
SOUTHERN CROSS	AA (ACT)	A705/1 ITEM 7/1/1522 : RAAF A'DROME AITKENVALE (42-46) Plus Refs. 172, 173, 174, 179	**52**
BALFES CREEK	AA (ACT)	A705/1 ITEM 171/93/479 : BALFES CREEK A'DROME HIRING OF SITE (42-49) Plus Refs. **50, 71,** 172, 173, 174, 179	**83**
PROSERPINE	AA (ACT)	A705 ITEM 7/1/786 : PROSERPINE A'DROME (40-41)	**84**
CLONCURRY	AA (ACT) AA (Qld)	A705 ITEM 7/1/1208 : PART 1 - RAAF CLONCURRY A'DROME WORKS (41-60) BP375 ITEM QL614 : CLONCURRY A'DROME (32-53) Plus Refs. **33, 72, 102,** 173, 174, 187	**85** **86**
NONDA	AA (ACT)	A705 ITEM 7/1/1258 : A'DROMES NORTH EAST AREA - GENERAL (42-45)	172
TORRENS CREEK	AA (ACT)	A705/1 ITEM 171/93/532 : TORRENS CREEK A'DROME HIRING OF SITE (42-49) Plus Refs. **36, 47, 57, 71, 88,** 172, 173, 174, 180	**87**
PRAIRIE (STRATHGLASS) (EIGHT MILE)	AA (ACT)	A705 ITEM 7/1/1351 : RAAF A'DROME PRAIRIE (8 MILE) (42-44) Plus Refs. **50,** 172, 173, 174, 179	**88**
HUGHENDEN	AA (ACT) AA (Qld)	A705 ITEM 7/1/1782 : RAAF A'DROME HUGHENDEN (43) BP243/1 ITEM K1070 : HUGHENDEN A'DROME (42-48) Plus Refs. **72,** 174	**89** **187**
WARREAH (KAROON) (COTONVALE)	AA (ACT)	A705 ITEM 7/1/1350 : RAAF A'DROME KAROON (42-44) Plus Refs. **50, 88,** 172, 173, 174	**90**
MACKAY	AA (ACT)	A705 ITEM 7/1/1186 : RAAF MACKAY AOB A'DROME WORKS (41-45) A705 ITEM 7/1/785 : RAAF AOB MACKAY BUILDINGS & SERVICES (40-46) Plus Refs. **2, 16, 39, 72, 74,** 171, 172, 185	**91** **92**
WINTON (WINTON EAST)	AA (ACT) AA (Qld)	A705 ITEM 175/6/79 : WINTON A'DROME - HIRING OF SITE (42-48) BP375/1 ITEM QL432 : WINTON A'DROME - LONGREACH A'DROME (42-52) Plus Refs. **102,** 174	**93** **94**
MT HEDLOW	AA (Qld) AA (NSW)	BP198/1 : CORPS OF ENGINEERS US ARMY. CAMP AREAS ROCKHAMPTON DISTRICT (44) C1707 (BOX 7) : PERIODIC REPORTS DEPT HOME SECURITY CAMOUFLAGE SECT QLD (42-43)	**95** 182
ROCKHAMPTON (CONNOR PARK) **and ROCKHAMPTON** (seaplane)	AA (ACT) AA (Qld)	A705 ITEM 7/1/607 : RAAF ROCKHAMPTON A'DROME - BOB'S CREEK (39-42) A705 ITEM 7/1/1277 : RAAF A'DROME ROCKHAMPTON (42-46) BP243/1 ITEM KS54 : ROCKHAMPTON AAC - R/WS & BLDGS & SERVICES (RAAF A'DROME) (42-47) BP375 ITEM QL422 : ROCKHAMPTON A'DROME (39-45) Plus Refs. **2, 72, 91, 92, 100,** 171, 174, 175	**96** **97** **98** **99**
LONGREACH	AA (ACT)	A705 ITEM 7/1/1218 : RAAF LONGREACH A'DROME WORKS (42-51) Plus Refs. **102,** 172, 174	**100**
BARCALDINE	AA (ACT)	A705/1 ITEM 7/1/1907 : BARCALDINE A'DROME (45-50) Plus Refs. **94,** 174	**101**
BLACKALL	AA (ACT)	A705/1 ITEM 171/93/229 : BLACKALL A'DROME HIRING OF SITE (42-48) Plus Ref. **119,** 173	**102**
MOORLANDS	AA (ACT)	A705 ITEM 171/53/5 : RAAF A'DROME MOORLANDS NO 8 SFTS BUNDABERG	**103**
MEADOWVALE	AA (ACT)	A705 ITEM 171/53/31 : MEADOWVALE RLG ACQUISITION OF SITE	**104**
MONTO	AA (ACT)	A705 ITEM 7/1/750 : A'DROME MONTO (40-42)	**105**

AIRFIELD	ARCHIVE	ACCESSION / ITEM NUMBER / TITLE / DATE RANGE	REF.
BUNDABERG	AA (ACT) AA (Qld)	A705 ITEM 7/1/546 : RAAF NO 8 SFTS BUNDABERG A'DROME WORKS (39-45) A705 ITEM 171/53/1 : PARTS 1,2 - RAAF NO 8 SFTS BUNDABERG BUILDINGS & SERVICES (40-51) A705 ITEM 171/53/4 : BUNDABERG A'DROME HIRING OF SITE (41-46) A705 ITEM 171/53/38 : BUNDABERG A'DROME UTILISATION & GUARDING (44-49) BP243/1 ITEM K161 : RAAF STATION BUNDABERG (41-44) BP243/1 ITEM K3889 : RAAF BUNDABERG - PROPOSED R/W (45-48) BP375 ITEM QL529 : BUNDABERG A'DROME (41-49) Plus Refs. **114, 115, 117, 122**	**106** **107** **108** **109** **110** **111** **112**
MARYBOROUGH	AA (ACT) AA (Qld)	A705 ITEM 7/1/581 : RAAF A'DROME MARYBOROUGH (39-45) A705 ITEM 171/54/1 : PARTS 1,2 - RAAF NO 3 ANS - NO 3 WAGS M'BORO BLDGS & SERVICES (40-47). A705 ITEM 171/54/2 : PART 1 - MARYBOROUGH A'DROME ACQUISITION OF SITE (40-49) A705/1 ITEM 171/89/113 : ROYAL NAVY NAA MARYBOROUGH BUILDINGS & SERVICES (45) BP243/1 ITEM K144 : RAAF MARYBOROUGH BUILDINGS & SERVICES (41-50) Plus Refs. **106, 122**	**113** **114** **115** **116** **117**
CHARLEVILLE	AA (Qld)	BP243/1 ITEM KS56/3 : AAC EAGLE FARM - HANGAR ERECTION - ALSO AT CHARLEVILLE (42-45) BP243/1 ITEM KS59 : CHARLEVILLE USAAC PROJECT NO 2 (42-48) BP375/1 ITEM QL437 : CHARLEVILLE A'DROME (32-51) Plus Refs. **72, 106,** 173, 174	**118** **119** **120**
QUILPIE	AA (ACT)	A705/1 ITEM 171/93/229 : BLACKALL A'DROME HIRING OF SITE (42-48)	**102**
ROMA	AA (NSW)	C1707 (BOX 7) : PERIODIC REPORTS DEPT HOME SECURITY CAMOUFLAGE SECT QLD (42-43)	182
KINGAROY	AA (ACT) AA (Qld)	A705 ITEM 7/1/833 : RAAF NO 3 SFTS KINGAROY A'DROME WORKS (40-46) BP243/1 ITEM K338 : RAAF NO 3 SFTS KINGAROY (42-48) BP375 ITEM QL391 : KINGAROY A'DROME (41-47) Plus Refs. **72, 106, 115, 117,** 172	**121** **122** **123**
CHINCHILLA	AA (ACT) AA (Qld)	A705 ITEM 7/1/1518 : RAAF A'DROME CHINCHILLA (42) BP243/1 ITEM K1070 : HUGHENDEN A'DROME (42-48)	**188** **187**
CONDAMINE	AA (ACT) AA (Qld)	A705 ITEM 7/1/1338 : RAAF CONDAMINE A'DROME WORKS (42-46) A705/1 ITEM 171/93/5 : CONDAMINE A'DROME HIRING OF SITE (42-50) BP243/1 ITEM K1066 : RAAF A'DROME CONDAMINE - CONSTRUCTION (42-46) Plus Refs. **149, 168,** 172, 173, 179, 188	**124** **125** **126**
TOOGOOLAWAH	AA (ACT)	A705 ITEM 7/1/902 : TOOGOOLAWAH A'DROME WORKS (31-46) A705 ITEM 7/1/1664 : TOOGOOLAWAH - NAVIGATION SYSTEMS (42) A705/1 ITEM 171/93/327 : TOOGOOLAWAH A'DROME HIRING OF SITE (42-48) A705/1 ITEM 7/1/1912 : POST WAR POLICY - AIRFIELD REQUIREMENTS (45-50) Plus Refs. **106,** 172, 173	**127** **128** **129** 183
BRYMAROO (JONDARYAN)	AA (Qld) AA (NSW)	BP375/1 ITEM QL368 : BRYMAROO A'DROME - ACQUISITION OF SITE (42-50) C1905 T1 : Boxes 13, 17 - AIRCRAFT HIDEOUT PHOTOGRAPHS Plus Refs. **124, 126, 127, 149, 168,** 172, 173, 179, 180, 188	**130** **131**
PETRIE (STRATHPINE A-1) **and** **STRATHPINE A-2** (LAWNTON)**and** **STRATHPINE A-3**	AA (ACT) AA (Qld) AA (ACT)	A705 ITEM 7/1/858 : RAAF STRATHPINE A'DROME WORKS (40-53) A705/1 ITEM 171/94/177 : STRATHPINE DROME STRIPS A1,A2 ACQUISITION OF SITE (42-46) A705/1 ITEM 175/6/100 : PARTS 1,2 - STRATHPINE (STRIPS A1,A2) HIRE SITES (42-49) BP243/1 ITEM K1735 : RAAF STRATHPINE (43-46) A705/1 ITEM 171/81/1 : PART 1 - RAAF INSTALLATIONS BRISBANE BUILDINGS & SERVICES Plus Refs. **72, 106, 127,** 173, 180	**132** **133** **134** **135** 184
COOMINYA No 2 (LOWOOD No 2) (A-6) **and** **COOMINYA No 1** (COOMINYA A-4) (COOMINYA)	AA (ACT)	A705/1 ITEM 7/1/1663 : RAAF A'DROME COOMINYA NO 2 LG (42-50) Plus Refs. **106,** 173, 180, 183	**136**
EAGLE FARM	AA (ACT) AA (Qld)	A705 ITEM 7/1/1263 : RAAF A'DROME EAGLE FARM - MARGINAL AREAS (42-43) A705 ITEM 7/1/1398 : EAGLE FARM 1ST EXTENSION TO A'DROME (42-45) A705/1 ITEM 171/93/706 : EAGLE FARM A'DROME EXTENSION - HIRING OF SITE (40-51) A705/1 ITEM 175/6/6 : EAGLE FARM BUILDING AREA HIRING OF SITE (42-44) BP243/1 ITEM K353 : ARCHERFIELD RAAF, EXTENSION OF RLG EAGLE FARM (41-45) BP243/1 ITEM KS56/3 : AAC EAGLE FARM - HANGAR ERECTION - ALSO AT CHARLEVILLE (42-45) BP375 ITEM QL805 : EAGLE FARM A'DROME (42-43) Plus Refs. **41, 119, 127, 132,** 172, 173	**137** **138** **139** **140** **141** **142** **143**
OAKEY	AA (ACT) AA (Qld)	A705 ITEM 171/100/5 : PARTS 1,2 - OAKEY RAAF NO 6 AD ACQUISITION OF SITE (43-57) A705 ITEM 171/106/191 : PART 1 - DISPOSAL OF SURPLUS ASSETS-OAKEY BP243/1 ITEM K1065 : PARTS 1,2 - RAAF NO 6 AD OAKEY (42-48) Plus Refs. **72, 149,** 173, 188	**144** **145** **146**

AIRFIELD	ARCHIVE	ACCESSION / ITEM NUMBER / TITLE / DATE RANGE	REF.
LOWOOD	AA (Qld)	BP243/1 ITEM K2175 : LOWOOD - CONSTRUCT RUNWAY AND TAXIWAY (43-47) BP375 ITEM QL618 : LOWOOD A'DROME - RAAF NO 12 EFTS (41-51) Plus Refs. **20, 106, 115, 117, 121, 122, 127, 136, 155,** 173, 180, 183, 188	**147** **148**
CECIL PLAINS (TIPTON)	AA (ACT) AA (Qld)	A1196/1 ITEM 42/501/106 : RAAF CECIL PLAINS A'DROME WORKS (42-45) BP375/1 ITEM QL349 : CECIL PLAINS A'DROME ACQUISITION (42-51) Plus Refs. **124, 126, 127, 168,** 172, 173, 179, 180	**149** **150**
TOOWOOMBA	AA (ACT) AA (Qld)	A705 ITEM 7/1/1341 : TOOWOOMBA EXTENSION TO A'DROME (42-44) - HIRING RANKINS'S P'TY A705 ITEM 7/1/1395 : TOOWOOMBA A'DROME (42-44) BP375/1 ITEM QL875 : TOOWOOMBA A'DROME (27-66) Plus Refs. **106, 127**	**151** **152** **153**
ARCHERFIELD	AA (ACT) AA (Qld)	A705 ITEM 171/16/147 : RAAF STATION ARCHERFIELD - SQUADRON BOMB DUMP (41) BP243/1 ITEM K169 : ARCHERFIELD NO 2 EFTS EXTENSION (41-46) BP243/1 ITEM K353 : ARCHERFIELD RAAF, EXTENSION OF RLG EAGLE FARM (41-45) BP243/1 ITEM K923 : ARCHERFIELD RAAF - EXTENSION OF LANDING AREA N & W SIDES (42-46) BP375 ITEM QL278 : ARCHERFIELD A'DROME - LEASE ANA HANGAR NO 6 (44-53) BP375 ITEM QL718 : ARCHERFIELD A'DROME - AQ'N AIRFRAME REPAIR UNIT & OPS STORE (43-51) Plus Ref. **106, 107, 114, 121, 127, 132, 137, 139, 162, 165, 168,** 172, 175, 180	**154** **155** **156** **157** **158** **159**
BREMER RIVER (FORKS A-7) (RIVERVIEW)	AA (ACT) AA (Qld)	A705 ITEM 7/1/546 : RAAF NO 8 SFTS BUNDABERG A'DROME WORKS (39-45) A705 ITEM 7/1/1545 : RAAF A'DROME - ROSEWOOD LG (42) BP375/1 ITEM QL423 : AMBERLEY RAAF A'DROME EXTENSIONS (44-52)	**106** **167** **165**
AMBERLEY	AA (ACT) AA (Qld)	A705 ITEM 171/16/155 : RAAF STATION AMBERLEY SQD BOMB DUMP BUILDINGS & SERVICES (41) BP243/1 ITEM USM218 : AMBERLEY HANGARS, CAMP BUILDINGS & SERVICES (42) BP243/1 ITEM K1849 : AMBERLEY NO 3 AD (42-46) BP375 ITEM QL345 : AMBERLEY OPS BUILDING,ADDITIONAL LAND ACQUISITION (43-83) BP375 ITEM QL406 : PART 1,1A - AMBERLEY RAAF STATION (37-58) BP375/1 ITEM QL423 : AMBERLEY RAAF A'DROME EXTENSIONS (44-52) Plus Refs. **106, 109, 114, 119, 121, 122, 127, 132, 154, 167, 168,** 172, 173, 180	**160** **161** **162** **163** **164** **165**
KINGSTON A-11 (WOODRIDGE) **and** **KINGSTON A-10 and** **KINGSTON A-12**	AA (ACT)	A705 ITEM 7/1/781 : RAAF OP BASE KINGSTON ACQUISITION OF SITE (40-42) Plus Refs. **106, 132,** 180	**166**
LOGANLEA	AA (ACT)	A705 ITEM 7/1/546 : RAAF NO 8 SFTS BUNDABERG A'DROME WORKS (39-45) Plus Refs. **127,** 173, 180, 183	**106**
ROSEWOOD	AA (ACT)	A705 ITEM 7/1/1545 : RAAF A'DROME - ROSEWOOD LG (42) Plus Refs. **106, 132, 165**	**167**
WATERFORD	AA (ACT)	A705 ITEM 7/1/546 : RAAF NO 8 SFTS BUNDABERG A'DROME WORKS (39-45) Plus Refs. 173, 183	**106**
WARRILL CREEK (A-8) (MUTDAPILLY) (TEN MILE CK) (WARWICK RD)	AA (ACT) AA (Qld)	A705 ITEM 7/1/546 : RAAF NO 8 SFTS BUNDABERG A'DROME WORKS (39-45) A705 ITEM 7/1/1545 : RAAF A'DROME - ROSEWOOD LG (42) A705 ITEM 7/1/1819 : RAAF A'DROMES CIVIL AVIATION REQUIREMENTS & TRANSFERS DCA (44-48) BP375/1 ITEM QL423 : AMBERLEY RAAF A'DROME EXTENSIONS (44-52)	**106** **167** 173 **165**
GOOLMAN (PURGA)	AA (ACT)	A705 ITEM 7/1/546 : RAAF NO 8 SFTS BUNDABERG A'DROME WORKS (39-45) Plus Refs. **127, 132, 165, 167,** 173, 180	**106**
LEYBURN	AA (ACT)	A705 ITEM 7/1/1339 : RAAF LEYBURN A'DROME WORKS (41-46) A705/1 ITEM 171/89/37 : RAAF STATION LEYBURN BUILDINGS & SERVICES (42-45) Plus Refs. **124, 127, 149,** 172, 179, 180, 187, 188	**168** **169**
CUNNAMULLA (USA FIELD) **and** **CUNNAMULLA** (CIVIL)	AA (ACT)	A705/1 ITEM 171/93/227 : CUNNAMULLA LANDING GROUND HIRING OF SITE (42-49) A705 ITEM 7/1/1819 : RAAF A'DROMES CIVIL AVIATION REQUIREMENTS & TRANSFERS DCA (44-48) A705/1 ITEM 171/93/229 : BLACKALL A'DROME HIRING OF SITE (42-48)	**170** 173 **102**

BIBLIOGRAPHY

ALCORN J S ***The Jolly Rogers - History of the 90th Bomb Group during World War II.*** USA:Historical Aviation Album, 1981.

ATKINSON J P E ***From Aircraft Assistant to Chief Engineer and Airport Manager 1931-1956.*** Unpublished. (Obtained 1991).

AUSTRALIAN PARLIAMENT ***Report of Joint Parliamentary Committee on War Expenditure-Defence Construction in North Queensland and Northern Territory (27 November 1944).*** Parliamentary Papers 1943-44 and 1944-45, Vol II.

AVERY N L ***B-25 Mitchell. The Magnificent Medium.*** USA:Phalanx Publishing Co Ltd., 1992.

BALFE J ***...And Far From Home. Flying RAAF Transports in the Pacific War and After.*** Melbourne:Macmillan, 1985.

BETTIENS R G ***Brief History of Air Force Activities at Bundaberg.*** Unpublished holding in care of Bundaberg and District Historical Society Inc. 1978.

BRERETON L H ***The Brereton Diaries.*** New York:William Morrow, 1946.

BROOK W H ***Demon to Vampire : The Story of 21 Squadron.*** Victoria:Demonvamp Publications, 1986.

BROWNE Waveney ***A Man of Achievement.*** Brisbane:PEP Enterprises, 1974.

BUETTEL N E ***A History of Maryborough 1842-1976.*** 1976.

CARDELL R G ***Wings Around Us.*** Brisbane:Amphion Press with Author, 1991.

CASEY H J ***Engineers of the Southwest Pacific 1941-1945.*** Volume I-VIII. Washington DC:US Govt Printing Office, 1947.

CRAVEN W F and CATE J L ***The Army Air Forces in World War II.*** Chicago:The University of Chicago Press, 1955. Vols I,IV,VI,VII.

DOD K C ***The Technical Services. The Corps of Engineers : The War Against Japan.*** Washington DC:US Government Printing Office, 1966.

FARRIOR W D ***Individual Flight Record (AAF Form No 5).*** January 1942 to January 1946 (with exceptions). Unpublished.

FIELDS K W ***Kangaroo Squadron - Memories of a Pacific Bomber Pilot.*** An Interview with Lt Col J W Fields USAF (ret'd) 435 Squadron, 19 BG. ca 1990.

FYSH W Hudson ***Qantas Rising - The autobiography of the Flying Fysh.*** Sydney:Angus & Robertson Ltd.1965.
——————— ***Qantas at War.*** Sydney:Angus & Robertson Ltd. 1968.

EAGLE FARM *ca(-/8/1944)?*
Bob Mikesh followed-up his work on the (Allied) TAIU at EAGLE FARM, with his publication in 1993 of ***Japanese Aircraft Code Names and Designations****. He included photos of two RAAF members, Sqn Ldr Norm Clappison and F-Sgt George Rimmington. Bill Bentson, ex US Army, produced some wartime 'telephone directories' he'd kept. I gave the half dozen RAAF names listed to the Discharged Personnel Records section in Canberra and they came up trumps! I've since met and interviewed Harry Bolton and Noel Millar. Surviving family of Clarence Roxby also kindly responded. W Cdr (ret'd) Millar produced this photo of an RAAF pilot 'Jell' Cuming in the cockpit of 'Oscar' MkI preparing for a flight. [N Millar ex RAAF]*

EAGLE FARM *ca(-/8/1944)*
According to Noel Millar, this is left to right Sam Hepford and Frank McCoy, USAAF with an RAAF pilot, Derek R 'Jell' Cuming. The aircraft in the background is an RAAF Beaufighter. Jell Cuming was chief test pilot in the RAAF's 1 APU.
[N Millar ex RAAF]

GALLOP A ***The Bush Engineer - Early Road Construction and Development in Northern & Western Queensland.*** Cairns:A E & R D Gallop, 1979.

GENERAL HQ ARMY PACIFIC ***Anti Aircraft Artillery Activities in the Pacific War.*** 1946. (Section only - obtained from US Cairns Historical Society).

GILLETT R ***Australia's Military Aircraft.*** Australia:Aerospace Publications P/L, 1987.

GILLISON D ***Royal Australian Air Force 1939-1942.*** Canberra:AWM, 1962.

GRAHAM B and SMYTH F ***A Nation Grew Wings - The Graphic Story of the Australian-Built Beauforts of the Royal Australian Air Force in New Guinea.*** Melbourne:Winterset House, 1946.

GUNN J ***Challenging Horizons - QANTAS 1939-1954.*** St Lucia Qld:University of Queensland Press, 1987.

GWYNN-JONES T ***Heroic Australian Air Stories.*** Adelaide:Rigby Publishers Ltd, 1981.

HALL R ***The Black Diggers.*** North Sydney:Allen & Unwin Aust P/L, 1989.

HENRY J H ***A History of the 38th Bombardment Group (M) November 20, 1940 - April 23, 1946.*** USA:Military Affairs/Aerospace Historian Publishing, 1978.

HICKEY L J ***Warpath Across the Pacific.*** USA:International Research & Publishing Corp., 1984.

HOCH I ***Barcaldine 1846-1986.*** Barcaldine:Barcaldine Shire Council, 1986.

JOHNSON D H ***Torres Strait to Coral Sea - The Defence of North Queensland.*** Charters Towers:D H Johnson, 1992.

JOHNSON L B *Diary Pages.* 3 to 10 June 1942 inclusive. Lyndon Baines Johnson Library. Johnson City, TX.

JOHNSON 2nd Lt W H ***Squadron Diary - for 88 Recon Squadron (H)*** **(later 435 BS, 19 BG).** Unpublished:ex USAF Archives. Microfilm reel number unknown.

JONES E ***HMS Nabstock : The Royal Navy in Maryborough 1945.*** Maryborough:Produced under the auspices of the Naval Association of Australia 1991.

KIELLY W O J ***RAAF Sandgate At the Mouth of the Pine 50th Anniversary.*** Sandgate and District Historical Soc and Museum Inc. 1990. (3rd printing 26/1/1994).

LINDBERGH C A ***The Wartime Journals of Charles A Lindbergh.*** New York:Harcourt Brace Jovanovich Inc, 1970.

LONG B ***Operation Semut 1.*** Sydney:Author in cooperation with Transpareon Press, 1989.

MIKESH R C ***Made in Japan...Tested in America!*** USA:Published in two parts - Pt 1:*WINGS* Magazine, June 1982. Pt 2:*AIRPOWER* Magazine, July 1982.

__________ ***Japanese Aircraft Code Names and Designation.*** USA:Schiffer Publishing Ltd. Atglen PA, 1993.

MITCHELL J H ***On Wings We Conquer.*** USA:GEM Publishers, 1990.

MITCHELL Rick ***Airacobra Advantage : The Flying Cannon.*** USA:Pictorial Histories Publishing Company, 1992.

MOFFAT A ***The Longreach Story - A History of Longreach and Shire.*** Brisbane:Jacaranda Press for Longreach Shire Council, 1987.

MROZEK D, HIGHAM R and ALLEN-NEWELL J ***The Martin Marauder and the Franklin Allens - A War Time Love Story.*** Kansas:Sunflower University Press, 1980.

ANTIL PLAINS *(12/8/1942)*
A magnifying glass over a tattered surviving print showed vacant dispersals around the fringe clearing, consistent with the US 33 BS, 22 BG, having departed for WOODSTOCK. But, halfway down the western edge of the 35° strip (see top right), this would have to be 40-1491, the B-26 on which Lt R W 'Robbie' Robinson failed the nose wheel strut taxiing over suspect ground. Field repair was underway. This extract is enlarged 4.5 times (see page 82). *[RAAF Official courtesy I Jenkins]*

ODGERS G ***Air War Against Japan 1943-1945.*** Canberra:AWM, 1957.

PARNELL N M and LYNCH C A ***Australian Air Force Since 1911.*** Sydney:A H & A W Reed, 1976.

PENTLAND G ***RAAF Camouflage and Markings 1939-45.*** Volumes 1, 2. Melbourne:Kookaburra Technical Publication, Vol 1, 1980, Vol 2, 1989.

PRIEST Joan ***Virtue in Flying - A Biography of Pioneer Aviator Keith Virtue.*** Australia:Angus & Robertson, 1975.

QUEENSLAND MAIN ROADS COMMISSION ***The History of the Queensland Main Roads Commission during World War II 1939-1945.*** Brisbane:1949.

RAAF HISTORICAL SECTION ***History of No 7 Squadron.*** Unpublished, Defence Department 1990.

ROBERTSON G B J ***The War Diary of P-O Graham Brindley John Robertson 412717 RAAF 1914-1942.*** Unpublished. (Obtained 1993).

RORRISON J D ***Nor the Years Contemn - Air War on the Australian Front 1941-42.*** Brisbane:Palomar Publications, 1992.

SIMPSON J A ***47/479 Service Squadron - 1944 to 1984 Reunion and 1985 Reunion.*** Arizona, 1984, 1985.

TAYLOR C R ***I Sustain the Wings. A History of No. 11 Repair and Servicing Unit, RAAF 1942-1945.*** Victoria:Author on behalf of 11 RSU RAAF History Book Fund. 1992.

TRENCH H V ***Airfield Selection North Queensland 1942-1943.*** Unpublished. (Obtained 1989).

TUNNY J N ***Gateway to Victory.*** Brisbane:Tunny, 1991.
__________ ***Fight Back from the North.*** Brisbane:Tunny, 1992.

WIXTED E P ***The Americans Arrive at Archerfield.*** Unpublished biographical note 1992.

WOODWARD J ***Three Times Lucky.*** Brisbane:Boolarong Publications with Author, 1991.

81 ADG ***There Will Always be an 81st.*** Brisbane:Printed by W R Smith & Patterson Ltd. 1943.

81 ADG ***81st Air Depot Group 1943-44.*** Brisbane:Printed by W R Smith & Patterson Ltd. 1944.

AMBERLEY *(26/8/1994)*
No 'field repair' about this exercise! The northernmost Bellman hangar is now 'workshop' to the 'Boston Project'. Pictured from left, Airframe/Engine fitter (ATECH) Raymond J Maxwell and W-O Dennis Doggett OAM, both of the RAAF's 23 (City of Brisbane) Active Reserve Sqn confer on 'some small part' of this world class WW2 aircraft restoration project. Right is Martin Lening, a Manual Arts Teacher-aide at Runcorn State High School. Martin is a highly skilled metal machinist, one of a few civilian volunteers in this RAAF team. With 29 years in the Service, some of it at Wagga NSW, bringing the AWM's 'Zero' fighter restoration to its impeccable conclusion, Dennis Doggett is now 'hands on' guiding this project. Cooperation between the RAAF and the PNG Government is the key here - the USAAF A-20G in the background is headed back to Museum display in Papua New Guinea while the ex 22 Sqn A28-8 (left foreground) will join the RAAF Museum's impressive collection. Doggett's philosophy here is well established in Australia's Defence Force. Constant proficiency needs balanced training which costs money. Where 'going through the motions' can save a fisherman at sea or serve in disaster relief, great! In the absence of such prior call, 'training dollars' directed towards aviation heritage keeps grass-roots technical skills alive. *[Author]*

INDEX

BUILDINGS

SOME RAAF UNITS

SOME US UNITS

SUNDRY

Some ships mentioned:

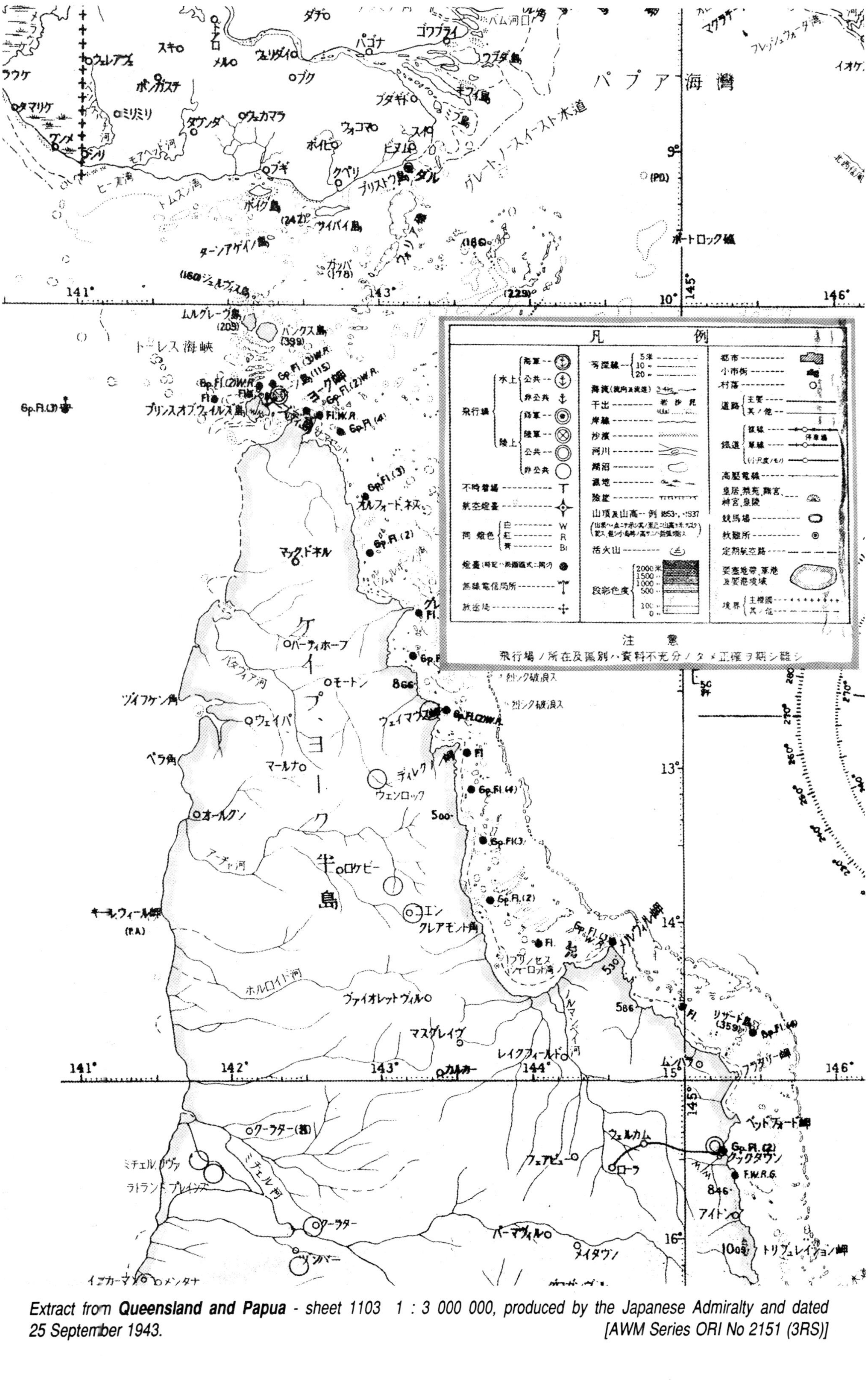

*Extract from **Queensland and Papua** - sheet 1103 1 : 3 000 000, produced by the Japanese Admiralty and dated 25 September 1943.* *[AWM Series ORI No 2151 (3RS)]*

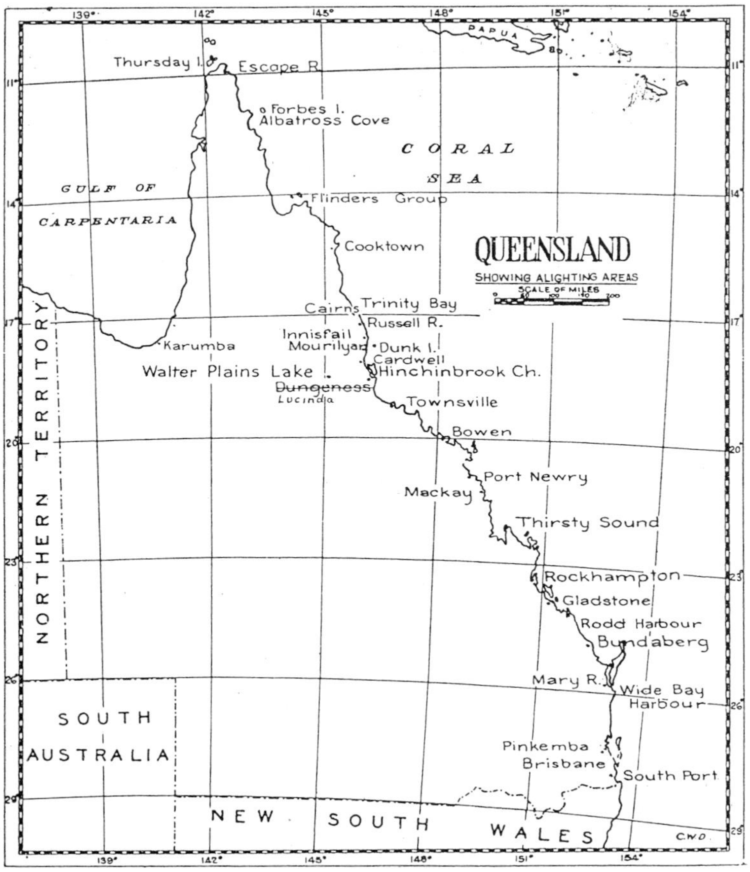

Extract from publication ACD 157 ***Flying Boat Bases and Alighting Areas - Australia****, produced by HQ, RAAF June 1943.* *[RAAF Museum, Pt Cook Vic]*

CAIRNS *(16/5/1942)*
Enlarged approximately seven times, the 'three white Flying Boats' moored in Trinity Inlet become more interesting. The top two appear identical and are probably Catalinas. The other is larger and probably a Short Empire boat. [NLA (RAAF Official)]

BOWEN FB MAINTENANCE BASE *ca (1945)*
The identity of the RAAF 9 Sqn Walrus is not clear in this photo taken amongst the clutter on the concrete apron. Can that be a 'dinkum Aussie dunny' in right foreground? The RAAF Museum Point Cook, is currently restoring a Walrus.
[J Cloake ex RAAF]

CAIRNS *(16/5/1942)*
This vertical is one of a Trimet test originating in Cairns, flown approximately one week after the Battle of the Coral Sea. None of the intense waterfront building programme to service naval and other forces had commenced, i.e. the building programme near Smiths Creek further up the inlet. *[NLA (RAAF Official)]*